Vol. 29. **The Analytical Chemistry of Sulfur and Its Compounds** (*in three parts*). By J. H. Karchmer

Vol. 30. **Ultramicro Elemental Analysis.** By Günther Tölg

Vol. 31. **Photometric Organic Analysis** (*in two parts*). By Eugene Sawicki

Vol. 32. **Determination of Organic Compounds: Methods and Procedures.** By Frederick T. Weiss

Vol. 33. **Masking and Demasking of Chemical Reactions.** By D. D. Perrin

Vol. 34. **Neutron Activation Analysis.** By D. De Soete, R. Gijbels, and J. Hoste

Vol. 35. **Laser Raman Spectroscopy.** By Marvin C. Tobin

Vol. 36. **Emission Spectrochemical Analysis.** By Morris Slavin

Vol. 37. **Analytical Chemistry of Phosphorus Compounds.** Edited by M. Halmann

Vol. 38. **Luminescence Spectrometry in Analytical Chemistry.** By J. D. Winefordner, S. G. Schulman and T. C. O'Haver

Vol. 39. **Activation Analysis with Neutron Generators.** By Sam S. Nargolwalla and Edwin P. Przybylowicz

Vol. 40. **Determination of Gaseous Elements in Metals.** Edited by Lynn L. Lewis, Laben M. Melnick, and Ben D. Holt

Vol. 41. **Analysis of Silicones.** Edited by A. Lee Smith

Vol. 42. **Foundations of Ultracentrifugal Analysis.** By H. Fujita

Vol. 43. **Chemical Infrared Fourier Transform Spectroscopy.** By Peter R. Griffiths

Vol. 44. **Microscale Manipulations in Chemistry.** By T. S. Ma and V. Horak

Vol. 45. **Thermometric Titrations.** By J. Barthel

Vol. 46. **Trace Analysis: Spectroscopic Methods for Elements.** Edited by J. D. Winefordner

Vol. 47. **Contamination Control in Trace Element Analysis.** By Morris Zief and James W. Mitchell

Vol. 48. **Analytical Applications of NMR.** By D. E. Leyden and R. H. Cox

Vol. 49. **Measurement of Dissolved Oxygen.** By Michael L. Hitchman

Vol. 50. **Analytical Laser Spectroscopy.** Edited by Nicolo Omenetto

Vol. 51. **Trace Element Analysis of Geological Materials.** By Roger D. Reeves and Robert R. Brooks

Vol. 52. **Chemical Analysis by Microwave Rotational Spectroscopy.** By Ravi Varma and Lawrence W. Hrubesh

Vol. 53. **Information Theory As Applied to Chemical Analysis.** By Karel Eckschlager and Vladimír Štěpánek

Vol. 54. **Applied Infrared Spectroscopy: Fundamentals, Techniques, and Analytical Problem-solving.** By A. Lee Smith

Vol. 55. **Archaeological Chemistry.** By Zvi Goffer

Vol. 56. **Immobilized Enzymes in Analytical and Clinical Chemistry.** By P. W. Carr and L. D. Bowers

Vol. 57. **Photoacoustics and Photoacoustic Spectroscopy.** By Allan Rosencwaig

Vol. 58. **Analysis of Pesticide Residues.** Edited by H. Anson Moye

Vol. 59. **Affinity Chromatography.** By William H. Scouten

Vol. 60. **Quality Control in Analytical Chemistry.** By G. Kateman and F. W. Pijpers

Vol. 61. **Direct Characterization of Fineparticles.** By Brian H. Kaye

Vol. 62. **Flow Injection Analysis.** By J. Ruzicka and E. H. Hansen

Vol. 63. **Applied Electron Spectroscopy for Chemical Analysis.** Edited by Hassan Windawi and Floyd Ho

Vol. 64. **Analytical Aspects of Environmental Chemistry.** Edited by David F. S. Natusch and Philip K. Hopke

(*continued on back*)

Determination of Molecular Weight

CHEMICAL ANALYSIS

A SERIES OF MONOGRAPHS ON
ANALYTICAL CHEMISTRY AND ITS APPLICATIONS

VOLUME 103

A WILEY-INTERSCIENCE PUBLICATION

JOHN WILEY & SONS

New York / Chichester / Brisbane / Toronto / Singapore

Determination of Molecular Weight

Edited by

ANTHONY R. COOPER

Chemistry Department
Lockheed Palo Alto Research Laboratories
Palo Alto, California

A WILEY-INTERSCIENCE PUBLICATION

JOHN WILEY & SONS

New York / Chichester / Brisbane / Toronto / Singapore

Library of Congress Cataloging in Publication Data:

Determination of molecular weight/edited by Anthony R. Cooper.
p. cm.—(Chemical analysis, ISSN 0069-2883; v. 103)

"A Wiley-Interscience publication."
Bibliography: p.
Includes index.
1. Molecular weights—Measurement. 2. Polymers and polymerization. 3. Distribution (Probability theory) I. Cooper, Anthony R. II. Series.

QD381.9.M64D47 1989
541.2′22—dc19

88-25881
ISBN 0-471-05893-9 CIP

Printed in the United States of America
10 9 8 7 6 5 4 3 2 1

To Audrey

CONTRIBUTORS

Karin D. Caldwell, Department of Bioengineering, University of Utah, Salt Lake City, Utah 84112

Benjamin Chu, Departments of Chemistry and of Materials Science and Engineering, State University of New York at Stony Brook, Stony Brook, New York 11794

Hans Coll, Research Laboratories, Eastman Kodak Company, Rochester, New York 14650

Anthony R. Cooper, Chemistry Department, Lockheed Palo Alto Research Laboratories, Palo Alto, California 94304

J. Calvin Giddings, Department of Chemistry, University of Utah, Salt Lake City, Utah 84112

Georg S. Greschner, An der Nonnenwiese 59, 6500 Mainz 1, Federal Republic of Germany

Archie E. Hamielec, McMaster Institute for Polymer Production Technology, McMaster University, Hamilton, Ontario, Canada L8S 4L7

Robert E. Harris, The BFGoodrich Research and Development Center, Brecksville, Ohio 44141

Kenji Kamide, Fundamental Research Laboratory of Fibers and Fiber Forming Polymers, Asahi Chemical Industry Company Ltd., Hacchonawatecho, Takatsuki, Osaka 569, Japan

Laya F. Kesner, Department of Chemistry, University of Utah, Salt Lake City, Utah 84112

Jack L. Koenig, Department of Macromolecular Science, Case Western Reserve University, Cleveland, Ohio 44106

Robert P. Lattimer, The BFGoodrich Research and Development Center, Brecksville, Ohio 44141

J. E. Mark, Department of Chemistry and the Polymer Research Center, The University of Cincinnati, Cincinnati, Ohio 45221

Shigenobu Matsuda, Fundamental Research Laboratory of Fibers and Fiber Forming Polymers, Asahi Chemical Industry Company Ltd., Hacchonawatecho, Takatsuki, Osaka 569, Japan

C. E. M. Morris, Materials Research Laboratory, Defence Science and Technology Organisation, Ascot Vale, Victoria 3032, Australia

Jean-Pierre Queslel, Manufacture Francaise des Pneumatiques Michelin Centre d'Essais et de Recherche de Ladoux, 63040 Clermont-Ferrand Cedex, France

Randal W. Richards, Department of Pure and Applied Chemistry, University of Strathclyde, Glasgow G1 1XL, Scotland

Bruce E. Richter, Lee Scientific Applications Laboratory, Salt Lake City, Utah 84123

Masatoshi Saito, Fundamental Research Laboratory of Fibers and Fiber Forming Polymers, Asahi Chemical Industry Company Ltd., Hacchonawatecho, Takatsuki, Osaka 569, Japan

Hans-Rolf Schulten, Fachhochschule Fresenius, Department of Trace Analysis, Dambachtal 20, 6200 Wiesbaden, Federal Republic of Germany

Leslie E. Smith, Polymers Division, National Bureau of Standards, Gaithersburg, Maryland 20899

Mark G. Styring, ICI Advanced Materials, Research and Technology, Wilton, Middlesbrough, Cleveland, TS6 8JE, England

Peter H. Verdier, Polymers Division, National Bureau of Standards, Gaithersburg, Maryland 20899

A. Marie Zaper, Department of Macromolecular Science, Case Western Reserve University, Cleveland, Ohio 44106

FOREWORD

The ever-expanding use of polymeric materials has increased the interest in techniques for determination of molecular weight and molecular weight distribution. These are the properties that are most important in the end-use application of the polymer. It is no longer sufficient simply to determine one of the average molecular weights, that is, number, viscosity, weight, or *Z*. Determining these molecular weights, however, remains extremely important as a method of calibrating and cross-checking on methods that determine the entire molecular weight distribution. These classical methods, although they usually are time consuming and require specialized equipment, remain extremely important.

The development of new characterization techniques has been necessitated by the need to determine molecular weight distribution rapidly and accurately. These techniques must necessarily be coupled with automated data analysis to produce the results as soon as practical and, indeed, on-line monitors are now becoming available.

An understanding of the relationship between physical and mechanical properties and their relationship to molecular weight distribution is a field that is receiving increased attention. Additionally, in order to understand and predict lifetimes of polymeric products, it is essential to characterize and standardize the raw materials used in their production. Furthermore, processing operations must be monitored in order to assure a reproducible product.

Thus the accurate and timely measurement of molecular weight distribution is necessary to control the quality of the product. This book, therefore, serves a very timely need.

JULIAN F. JOHNSON

Institute of Materials Science
University of Connecticut, Storrs
June, 1989

PREFACE

The characterization of polymeric materials continues to be an active area for research. This volume is concerned with established and new techniques for determination of molecular weight. In the early days of polymer science many techniques were developed which produced absolute values for a particular average molecular weight. More recently the determination of molecular weight distribution has become increasingly important. Most of the methods available are not absolute methods but must be calibrated with appropriate standards. The status of the older methods and the theory and applications of some newer methods are reviewed in this volume.

Chapter 1 covers the concepts of molecular weight averages and distribution functions for polydisperse polymer systems. Chapters 2, 3, and 4 summarize various techniques for determining number average molecular weight, collectively known as colligative property measurements. These include various physical and chemical ways to count the number of molecules and the well-established techniques of vapor pressure osmometry and membrane osmometry. Classical light scattering (Chapter 5) has been used for many years to determine weight average molecular weight, and more recently X-ray and neutron-scattering techniques (Chapter 6) have been similarly applied. Importantly, dynamic light scattering is now being applied to determine molecular weight distribution. The ultracentrifuge (Chapter 7), although it was one of the earliest instruments for determining the molecular weight of polymers, continues to be applied and improved. Ultracentrifugation is perhaps unique in the sense that it is capable of establishing molecular weight distribution by an absolute measurement. Viscometric determination of molecular weight is still widely used, and Chapter 8 discusses applications of the method and its limitations.

Fractionation techniques based on solution phase equilibrium have been effective methods for characterizing molecular weight distribution, and Chapter 9 presents theoretical analyses and optimum methods. Chromatographic techniques, especially gel permeation chromatography (Chapter 10), are widely used and the newer technique of phase distribution chromatography (Chapter 11), is particularly useful for characterizing narrow molecular weight distribution samples. Chapter 12 covers field-flow fractionation techniques, also known as single-phase chromatography. These methods are capable of separation over the whole molecular weight range and are particularly useful at extremely high molecular weight.

At the other end of the molecular weight range are oligomeric mixtures, whose full characterization remains a challenge. Chapter 13 reviews supercritical fluid chromatography, which was developed more than 20 years ago and is currently an area of renewed interest. Developments in the use of mass spectrometry to characterize molecular weight in the oligomer region have also occurred and are described in Chapter 14. Methods to characterize insoluble polymers have been actively pursued and are reviewed in terms of physical and chemical methods (Chapter 15), as well as mechanical and solvent-swelling techniques (Chapter 16). The final chapter surveys the various types of standard polymers which are available from commercial sources.

The book is intended for a wide audience of polymer scientists and technologists in many diverse fields of research and applications. It would also be useful in teaching institutions at the undergraduate and graduate levels.

The editor is indebted to the authors for their outstanding contributions, and any remaining errors or errors of omission are the responsibility of the editor. Comments from readers will always be appreciated.

ANTHONY R. COOPER

Palo Alto, California
April, 1989

CONTENTS

Determination of Molecular Weight

CHAPTER

1

MOLECULAR WEIGHT AVERAGES AND DISTRIBUTION FUNCTIONS

ANTHONY R. COOPER

Chemistry Department Lockheed Palo Alto Research Laboratories Palo Alto, California

MOLECULAR WEIGHT AVERAGES

Because of the statistical nature of the polymerization process, most polymeric materials are composed of mixtures of molecules having a range of molecular weights. A complete description of the molecular weight distribution of a polymer is important to understand its physical, rheological, and mechanical properties. Peebles (1) has summarized the dependence of molecular weight distribution on the polymerization kinetics and mechanism. If the polymer is a copolymer or terpolymer, then, in addition to molecular weight distribution, there is the possibility of a compositional distribution which must also be determined to fully characterize these materials and understand their properties.

Certain techniques for molecular weight determination are capable of yielding only one of the molecular weight averages of the distribution. These averages are defined in terms of the molecular weight M_i and the number of moles n_i, or the weight w_i, of the component molecules by equations (1)–(4).

Number average molecular weight

$$\bar{M}_n = \frac{\sum n_i M_i}{\sum n_i} = \frac{\sum w_i}{\sum w_i / M_i} \tag{1}$$

Weight average molecular weight

$$\bar{M}_w = \frac{\sum n_i M_i^2}{\sum n_i M_i} = \frac{\sum w_i M_i}{\sum w_i} \tag{2}$$

Z average molecular weight

$$\bar{M}_z = \frac{\sum n_i M_i^3}{\sum n_i M_i^2} = \frac{\sum w_i M_i^2}{\sum w_i M_i} \tag{3}$$

$Z + 1$ average molecular weight

$$\bar{M}_{z+1} = \frac{\sum n_i M_i^4}{\sum n_i M_i^3} = \frac{\sum w_i M_i^3}{\sum w_i M_i^2} \tag{4}$$

Viscosity average molecular weight

$$\bar{M}_v = \left[\frac{\sum n_i M_i^{1+a}}{\sum n_i M_i}\right]^{1/a} = \left[\frac{\sum w_i M_i^a}{\sum w_i}\right]^{1/a} \tag{5}$$

All these molecular weight averages except the viscosity average have unique values for a given polymer. The viscosity average molecular weight has a value which depends on the particular solvent and temperature conditions used for the measurement. This is an important practical molecular weight average derived from viscometry. To calculate this average from equation (5), one must know the exponent a of the Mark–Houwink relationship, equation (6), relating intrinsic viscosity $[\eta]$ to molecular weight M.

$$[\eta] = KM^a \tag{6}$$

The value of a lies between 0.5 and 1.0 for random coils, depending on the solvent and temperature employed. With these limits for a, it may be seen that $\bar{M}_v$ is always larger than $\bar{M}_n$ but can be equal to $\bar{M}_w$ when the upper limit of a is reached for random coils. For rigid rod molecules, the value of a is expected to be ~ 1.8.

If molecular weight is considered to be a continuous variable, the molecular weight distribution may be described by a set of moments μ_r, given by the integrals

$$\mu_r = \int_0^\infty M^r f(M)\, dM \tag{7}$$

with $r = 0, 1, 2, 3, \ldots$; $f(M)$ the number density distribution, and $f(M)dM$ the number of moles of molecules with molecular weight between M and $(M + dM)$.

The average molecular weights are defined as follows:

$$\bar{M}_n = \mu_1/\mu_0 \tag{8}$$

$$\bar{M}_w = \mu_2/\mu_1 \tag{9}$$

$$\bar{M}_z = \mu_3/\mu_2 \tag{10}$$

$$\bar{M}_{z+1} = \mu_4/\mu_3 \tag{11}$$

Min (2) has described a method for determining molecular weight averages from moments.

MOLECULAR WEIGHT DISTRIBUTION FUNCTIONS

Various mathematical functions have been employed to describe the distribution of molecular weights. Some of the more common functions are shown in Table 1, written in terms of the mole fraction X.

Table 1. Molecular Weight Distribution Functions

Name	Function	Comments
Gaussian	$X(M) = \dfrac{1}{\sigma_n(2\pi)^{1/2}} \exp\left[-\dfrac{(M - M_m)^2}{2\sigma_n^2}\right]$	M_m, median value equal to $\bar{M}_n$
Log-normal	$X(M) = \dfrac{1}{M\sigma_n(2\pi)^{1/2}} \exp\left[-\dfrac{(\ln M - \ln M_m)^2}{2\sigma_n^2}\right]$	M_m, geometric mean
Poisson	$X(M) = v^{M-1} \dfrac{\exp(-v)}{\Gamma(M)}$	$v = \bar{M}_n - 1$; $\Gamma(M)$ is the gamma function
Flory–Schulz	$X(M) = \dfrac{\beta^{k+1} M^{k-1} \bar{M}_n \exp(-\beta M)}{\Gamma(k+1)}$	k = degree of coupling; $\beta = k/\bar{M}_n$; $\Gamma(k+1)$ is the gamma function
Most Probable	$X(M) = P^{M-1}(1 - P)$	P = fraction of functional groups reacted

Width of Molecular Weight Distributions

The width of the molecular weight distribution (MWD) may be calculated from the molecular weight averages. This allows the width of the molecular weight distribution to be determined without determining the molecular weight distribution as long as the distribution function is known.

The width of the Gaussian distribution function may be expressed in terms of the standard deviation of the mole fraction MWD function σ_n or the mass fraction MWD function σ_w as.

$$\sigma_n = (\bar{M}_w \bar{M}_n - \bar{M}_n)^{0.5} \tag{12}$$

$$\sigma_w = (\bar{M}_z \bar{M}_w - \bar{M}_w)^{0.5} \tag{13}$$

The standard deviation is an absolute measure of the width for the Gaussian function only. Widths of other molecular weight distribution functions must be calculated in each case from the distribution function itself.

RELATIONSHIPS BETWEEN MOLECULAR WEIGHT AVERAGES AND THE PARAMETERS OF MOLECULAR WEIGHT DISTRIBUTION FUNCTIONS

Gaussian Distribution Function

Mole fraction distribution function:	Median value $= \bar{M}_n$
Mass fraction distribution function:	Median value $= \bar{M}_w$

Log-Normal Distribution Function

Mass fraction distribution function: Median value $= \bar{M}_m$

$$\bar{M}_n = \bar{M}_m \exp[(\sigma_w)^2/2] \tag{14}$$

$$\bar{M}_w = \bar{M}_m \exp[3(\sigma_w)^2/2] \tag{15}$$

$$\bar{M}_z = \bar{M}_m \exp[5(\sigma_w)^2/2] \tag{16}$$

which leads to

$$\exp(\sigma_w)^2 = \bar{M}_w/\bar{M}_n = \bar{M}_z/\bar{M}_w \tag{17}$$

thus the ratios of two adjacent averages are constant.

Poisson Distribution Function

The ratio of the weight average to number average molecular weight is given by

$$\bar{M}_w/\bar{M}_n = 1 + 1/\bar{M}_n - (1/\bar{M}_n)^2 \tag{18}$$

and thus depends only on $\bar{M}_n$.

Flory–Schulz Distribution Function

The molecular weight averages for the Flory–Schulz function are related by

$$\frac{\bar{M}_n}{k} = \frac{\bar{M}_w}{k+1} = \frac{\bar{M}_z}{k+2} \tag{19}$$

where k is the coupling constant, defined as the number of independently growing chains required to form one dead chain.

Most Probable Distribution Function

$$\bar{M}_n = \frac{1}{(1-p)}, \qquad \bar{M}_w = \frac{1+p}{1-p}, \qquad \text{and} \qquad \frac{\bar{M}_w}{\bar{M}_n} = 1 + p$$

The viscosity average is also related to the number average by the relation $\bar{M}_v = \bar{M}_n \Gamma(2+a)$, where a is the Mark–Houwink exponent. For the case of a theta solvent where $a = 0.5$, $\bar{M}_n : \bar{M}_v : \bar{M}_w = 1:1.33:2$ when $p \to 1$.

POLYDISPERSITY

Traditionally polydispersity Q has been defined as

$$Q = \bar{M}_w/\bar{M}_n = U + 1 \tag{20}$$

and U, the molecular inhomogeneity, has a numerical value one less than Q. The width of molecular weight distributions increases with increasing Q and U. Except for the case of the Gaussian distribution function, the standard deviation is only a relative measure of distribution width. Pyun (3, 4) has further analyzed this problem.

REFERENCES

1. L. H. Peebles, *Molecular Weight Distributions in Polymers*, Wiley-Interscience, New York, 1971.
2. K. W. Min, *J. Appl. Polym. Sci.*, **22**, 589 (1978).
3. C. W. Pyun, *J. Polym. Sci., Polym. Phys. Ed.*, **17**, 2111 (1979).
4. C. W. Pyun, *J. Polym. Sci., Polym. Phys. Ed.*, **24**, 229 (1986).

CHAPTER

2

DETERMINATION OF NUMBER AVERAGE MOLECULAR WEIGHT BY END-GROUP ANALYSIS, VAPOR PRESSURE MEASUREMENTS, AND CRYOSCOPY

ANTHONY R. COOPER

Chemistry Department Lockheed Palo Alto Research Laboratories Plot Alto, California

Number average molecular weights may be determined by absolute methods which determine the number of molecules present. Direct analysis of a specific chemical group in a polymer of known structure is one way to do this. Colligative properties are those which depend on the number of molecules present rather than their kind. These techniques count the number of molecules present and thus lead to methods for determination of number average molecular weight.

END GROUP ANALYSIS

When the concentration of an end group is measured by a suitable technique (1; 2; 3, Chap. 8; 4; 5) the value for $\bar{M}_n$ may be calculated from the known structure of the polymer. In condensation polymers one or both ends of the polymer have specific groups which may be titrated by suitable means. Polyesters (4), polyamides (4), and polyurethanes (6) have been subjected to this analysis. Polyethers have been analyzed by hydroxyl group titrations similar to those used for polyesters. This has also been carried out by reaction with an excess of phenyl isocyanate followed by reaction with excess di-*n*-butylamine, which was back-titrated with perchloric acid (7). Nonaqueous titrations (8) have been used to characterize polysulfones with $\bar{M}_n$ values up to 25,000. Spectrophotometric analyses have been widely used, notably ultraviolet–visible (5), infrared techniques (5), and more recently nuclear magnetic resonance spectrometry (9). Recent examples include the determination of phenolic end groups in polymers by UV spectroscopy (10) and potentiometric titration (8). Infrared spectroscopy has been used (11) to estimate carboxyl groups in the presence of carbonyl groups by reacting the former with SF_4 and the resulting thionyl halides were quantitatively measured. NMR spectroscopy (12) has been used to characterize $\bar{M}_n$ of hydroxyl end capped polystyrenes. The single proton on the hydroxyl group was reacted to yield a trimethylsilyl group containing 9 protons. The enhanced sensitivity allowed the determination of $\bar{M}_n$ values up to 80,000. A similar method has been reported for hydroxy-terminated polybutadienes (13), polyester polyurethanes (14), and epoxides (15).

In certain cases, extremely high sensitivity for determination of number average molecular weight may be achieved using radioactive labeling (16). Using radiolabeled bisulfite initiator (^{35}S), the number average molecular weight of Teflon samples was determined to be in the 389,000–8,900,000 range (17). This technique is particularly useful technique for insoluble polymers.

MEASUREMENT OF COLLIGATIVE PROPERTIES

Colligative properties are those that depend on the number of species present rather than their kind. From thermodynamic arguments it may be shown that for very dilute ideal solutions

$$\ln a_1 = -X_2 \tag{1}$$

where a_1 is the activity of the solvent in a dilute ideal solution and X_2 is the mole fraction of solute. From this relationship the molecular weight of the solute may be calculated if the weight fraction w_2 is known, namely.

$$X_2 = \frac{n_2}{n_1} = \frac{w_2}{M_2}\frac{M_1}{w_1} \tag{2}$$

This equation demonstrates that the activity, which may be measured by several methods, is proportional to the number of solute molecules. Thus, in the case of polydisperse solutes, the number average molecular weight is the average which is being determined. At higher concentrations and molecular weights, the solutions become nonideal and higher powers of the concentration of solute are introduced into the equations. Measurements of several concentrations of solute are required and appropriate extrapolation techniques must be used.

EXPERIMENTAL METHODS FOR DETERMINATION OF NUMBER AVERAGE MOLECULAR WEIGHT

Vapor Pressure Lowering

The partial vapor pressure p_1 of solvent 1 over a solution is lower than the vapor pressure over the pure solvent p_1^0. This is expressed by Raoult's law

$$p_1 = X_1 p_1^0 \tag{3}$$

where X_1 is the mole fraction of the solvent.

For a binary solution containing a mole fraction X_2 of solute then

$$X_2 = \frac{p_1^0 - p_1}{p_1^0} = \frac{\Delta p_1}{p_1^0} \tag{4}$$

and for a dilute solution

$$X_2 = \frac{n_2}{n_1} = \frac{w_2 M_1}{M_2 w_1} \tag{5}$$

Combining equations (5) and (4) yields

$$M_2 = w_2 \frac{M_1}{w_1} \frac{p_1^0}{\Delta p_1} \tag{6}$$

The unknown molecular weight of solute M_2 may be calculated from the lowering of the vapor pressure caused by the addition of w_2 grams of solute to form a binary solution. The technique has been reviewed (18, p. 113; 19; 20; 21, p. 275) and $\bar{M}_n$ values up to 1000 may be determined.

A variation of this technique, the isopiestic method (18, p. 114; 19, p. 4448; 21, p. 268), was devised to avoid the difficulty of accurately determining the small difference in vapor pressure caused by the addition of solute. In this method two limbs of a container joined by a common vapor space (22) are filled, one with a reference solution of known weight concentration w_s and molecular weight M_s and the other with the unknown solution of known weight concentration w_2 and unknown molar concentration. The apparatus is immersed in a constant temperature bath controlled to $\pm 0.001\,°C$; the absolute temperature is not important. Distillation of the solvent continues until the vapor pressures above the solutions in each limb are equal and thus the solutions contain equal mole fractions of solute. At equilibrium, which may require several weeks, the final volumes V_s and V_2 of the known and unknown solutions are read from the calibrations on each limb. Assuming ideal solution behavior, the unknown molecular weight is calculated from

$$\bar{M}_n = \frac{w_2}{V_2}\frac{M_s V_s}{w_s} \tag{7}$$

Values up to 20,000 have been determined for $\bar{M}_n$ by this method.

Ebulliometry

Ebulliometry (18, p. 113; 19, p. 4448; 21, p. 113) is another technique for determining the depression of the activity of the solvent by the solute. In this case, the elevation of the boiling point is determined. Several recent reviews (23; 24, p. 105; 25; 26) summarize the current state of the art. The boiling point elevation ΔT_b is measured with sensitive thermocouples or matched thermistors in a Wheatstone bridge. The molecular weight $\bar{M}_n$ is calculated from

$$\bar{M}_n = \frac{K_b C}{\Delta T_b} \tag{8}$$

where c is the concentration of solute in grams per 1000 g of solvent and

$$K_b = \frac{R T_b^2 M}{1000 \Delta H_e} \tag{9}$$

is the molal ebullioscopic constant; M is the molecular weight of the solvent, T_b is its boiling point, ΔH_e is the molar latent heat of vaporization, and R is the gas constant. The ebullioscopic constant may be evaluated directly from equation (9) and is thus an absolute method for molecular weight determination. However, the value for K_b is often determined by using a high purity solute of known molecular weight.

Currently a sensitivity of 1×10^{-5}°C is attainable using a 160 junction thermocouple (27) or 2.4×10^{-6}°C using thermistors (28). The upper limit of molecular weight which may be determined by this method also depends on the solvent and solution nonideality. Currently it is claimed that with careful measurements and good ebulliometer design $\bar{M}_n$ values up to 100,000 may be determined (29). The sample used in this work was isotactic polypropylene and the apparatus is shown in Figure 1. The application of this technique to aqueous solutions has been reported (30). An ebulliometer of simple design with a temperature sensitivity of 50×10^{-6}°C has recently been reported (31).

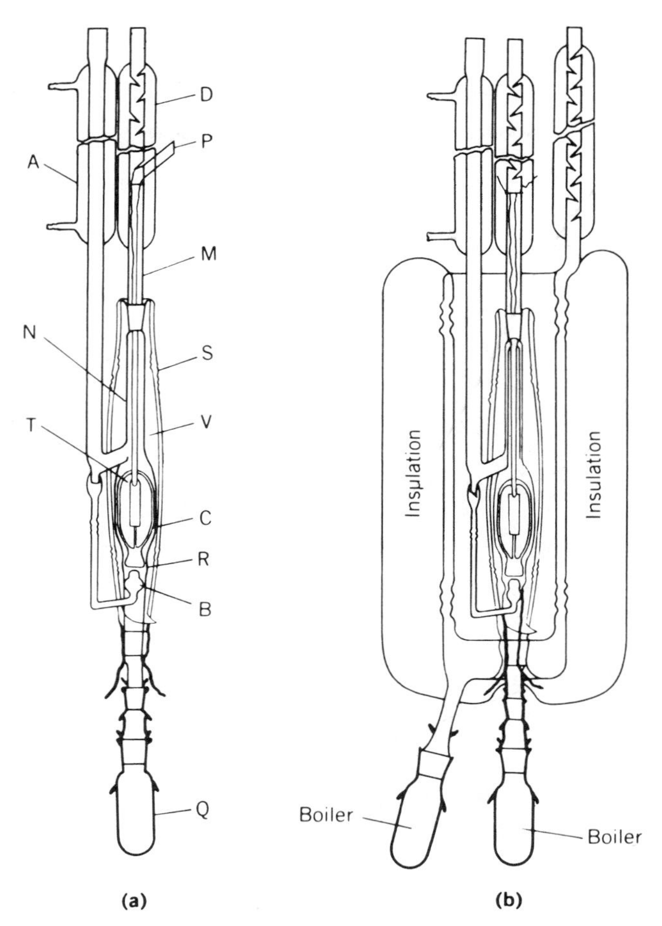

Figure 1. (*a*) Ebulliometer design for molecular weight determination up to $\bar{M}_n = 100{,}000$. (*b*) Complete ebulliometer with the adiabatic jackets and of the insulating apparatus (29). (Reproduced by permission of Hüthig Wepf Verlag, Basel.)

A rotating ebulliometer has also been described (32) which attempts to overcome the superheating problem. Errors in the methods of treating ebullioscopic data to obtain molecular weight have been discussed (33).

Cryscopy

The freezing point of a solution is depressed below that of the pure solvent by an amount which is proportional to the mole fraction of solute. The value for

$\bar{M}_n$ is obtained from

$$\bar{M}_n = \frac{K_f C}{\Delta T} \tag{10}$$

where C is the concentration of solute in grams per 1000 g of solvent, and

$$K_f = \frac{R T_f^2 M}{1000 \Delta H_f} \tag{11}$$

is the molal cryoscopic constant

where M is the molecular weight of the solvent,
T_f is its melting point,
ΔH_f is the molar latent heat of fusion of the solvent, and
R is the gas constant.

Again K_f may be calculated from equation (11) and cryoscopy is an absolute method, or K_f may be obtained by calibration with a compound of known molecular weight. Some materials such as camphor have very large cryoscopic constants and may be used to increase the sensitivity of the method. The application of the method to determine $\bar{M}_n$ for polymers has been reviewed (3, p. 225; 18, p. 125; 21, p. 275; 24, p. 124; 34). Recent applications to determine molecular weights of polyesters (35) and ethylene–vinyl acetate polymers (36) have been described.

A novel variation of this technique (37) involves depression of the first-order nematic–isotropic melting transition of *N*(*p*-ethoxybenzylidence)*p*-*n*-butylaniline. Polystyrene and polyethylene oxide are soluble in both phases, and $\bar{M}_n$ values up to 10^6 have been studied.

REFERENCES

1. N. C. Billingham, *Molar Mass Measurements in Polymer Science*, Halsted Press, New York, 1977, p. 234.
2. G. F. Price, in *Techniques of Polymer Characterization*, P. W. Allen, Ed., Butterworth, London, 1959, Chap. VII.
3. S. R. Rafikov, S. A. Pavlova, and I. I. Tverdokhlebora, in *Determination of Molecular Weights and Polydispersity of High Polymers*, Israel Programme for Scientific Translation, Jerusalem, 1964.
4. R. G. Garmon, in *Polymer Molecular Weights*, P. E. Slade, Jr. Ed., Dekker, New York, 1975, Chap. 3.

5. J. Urbanski, N. Czerwinski, K. Janicka, F. Majewska, and H. Zowall, *Handbook of Analysis of Synthetic Polymers and Plastics*, Wiley, New York, 1977.
6. D. J. David and H. B. Staley, *Analytical Chemistry of Polyurethanes*, Wiley, New York, 1959.
7. D. H. Reed, F. E. Critchfield, and D. K. Elder, *Anal. Chem.*, **35**, 571 (1963).
8. A. J. Wnuk, T. F. Davidson, and J. E. McGrath, *J. Appl. Polym. Sci., Appl. Polym. Symp.*, **34**, 89 (1978).
9. J. R. Ebdon, in *Analysis of Polymer Systems*, L. S. Bark and N. S. Allen, Eds., Applied Science Publishers, Barking, 1982, p. 21.
10. E. Shchori and J. E. McGrath, *J. Appl. Polym. Sci., Appl. Polym. Symp.*, **34**, 103 (1978).
11. J. F. Heacock, *J. Appl. Polym. Sci.*, **7**, 2319 (1963).
12. A. D. Edwards and M. J. R. Loadman, unpublished work at Malaysian Rubber Producers' Research Association, 1976.
13. G. Fages and Q. T. Pham, *Makromol. Chem.*, **180**, 2435 (1979).
14. F. W. Yeager and J. W. Becker, *Anal. Chem.*, **49**, 722 (1977).
15. W. B. Moniz and C. F. Poranski, Jr., in *Epoxy Resin Chemistry*, R. S. Bauer, Ed. (ACS Symposium Series No. 114), American Chemical Society, Washington, DC., 1979.
16. K. L. Berry and J. H. Peterson, *J. Am. Chem. Soc.*, **73**, 5195 (1951).
17. R. C. Dobau, A. C. Knight, J. H. Peterson, and C. A. Sperati, paper presented at the 130th Meeting, American Chemical Society, Atlantic City, September 1956.
18. D. F. Rushman, in *Techniques of Polymer Characterization*, P. W. Allen, Ed., Butterworths, London, 1959.
19. I. M. Kolthoff and P. J. Elving, Eds., *Treatise on Analytical Chemistry*, Part 1, *Theory and Practice*, Vol. 7, Wiley-Interscience, New York, 1965, p. 4446.
20. G. Radakoff, *Z. Chem.*, **1**, 135 (1961).
21. R. U. Bonnar, M. Dimbat, and F. H. Stross, *Number Average Molecular Weights*, Wiley-Interscience, New York, 1958.
22. R. L. Parette, *J. Polym. Sci.*, **15**, 450 (1955).
23. M. Ezrin, in *Characterization of Macromolecular Structure*, M. McIntyre, Ed. (NAS Publication No. 1573), National Academy of Sciences, Washington, DC, 1968, p. 3.
24. C. A. Glover, in *Polymer Molecular Weights*, Part I, P. E. Slade, Jr., Ed., Dekker, New York, 1975.
25. C. A. Glover, in *Polymer Molecular Weight Methods*, M. Ezrin, Ed. (ACS Advances in Chemistry Series No. 125), American Chemical Society, Washington DC, 1973, p. 1.
26. G. Davison, in *Analysis of Polymer Systems*, L. S. Bark and N. S. Allen, Eds., Applied Science Publishers, Barking, 1982, p. 209.
27. C. A. Glover, *Adv. Analyt. Chem. Instrum.*, **5**, 1966.
28. E. Zichy, SCI Monograph No. 17, 122 (1963).

29. P. Parrini and M. S. Vacanti, *Makromol. Chem.*, **175**, 935 (1974).
30. W. De Oliveira, "Differential Ebulliometry," Ph.D. thesis, Clarkson College of Technology, 1975.
31. W. De Oliveira and G. Francisco, *Chem. Biomed. Environ. Instrum.*, **10**, 189 (1980).
32. J.-T. Chen, H. Stobayashi, and F. Asmussen, *Colloid Polym. Sci.*, **259**, 1202 (1981).
33. J. Melsheimer and H. Sotobayashi, *Makromol. Chem.*, **179**, 2913 (1978).
34. E. J. Newitt and V. Kokle, *J. Polym. Sci., A2*, **4**, 705 (1966).
35. L. Carbonnel, R. Guieu, C. Ponge, and J. C. Rosso, *J. Chim. Phys. Physicochim. Biol.*, **70**, 1400 (1973).
36. A. Ya Ryasnyanskaya, S. L. Lyubimova, V. V. Kalashnikov, R. A. Terteryan, V. N. Mosastyrskii, and B. V. Gryaznov, *Neftepererab Neftekhim*, Moscow, 1973, p. 69.
37. B. Kronberg and D. Patterson, *Macromolecules*, **12**, 916 (1979).

CHAPTER

3

VAPOR PRESSURE OSMOMETRY

C. E. M. MORRIS

Materials Research Laboratory, Defence Science and Technology Organisation
Ascot Vale, Victoria, Australia

INTRODUCTION

Vapor pressure osmometry (VPO) is one of a number of molecular weight techniques which depend on the measurement of a colligative property of solutions; it therefore leads, for a polydisperse system, to number average molecular weight ($\bar{M}_n$) of the solute. The relative simplicity of the procedure has resulted in widespread use of this technique in polymer science and also in various biological applications. A review of the various approaches to colligative property measurements has been provided by Glover (1) together with an examination of the sources of experimental and computational error in these methods.

The basic principle of the VPO method is isothermal distillation arising from concentration differences between solvent and solution. The solvent transfer is detected and measured indirectly from the heat of condensation of the solvent in the more concentrated solution. The initial idea for vapor pressure osmometry was reported by the physiologist Hill (2) in 1930. Experimental modifications, notably the use of thermistors to detect temperature differences and a null detector (3, 4), greatly increased the utility of the technique. Its practical use in polymer science dates from the late 1960s.

INSTRUMENTS

The performance of a laboratory-constructed apparatus (5), and its effectiveness in the determination of $\bar{M}_n$ for a series of narrow molecular weight distribution polystyrene samples, was demonstrated (6) in 1969. The basic design of the measuring chamber of this equipment, used to determine $\bar{M}_n$ values exceeding 10^4 daltons, is shown in Figure 1. It consists of a temperature-controlled, insulated chamber containing a solvent-saturated

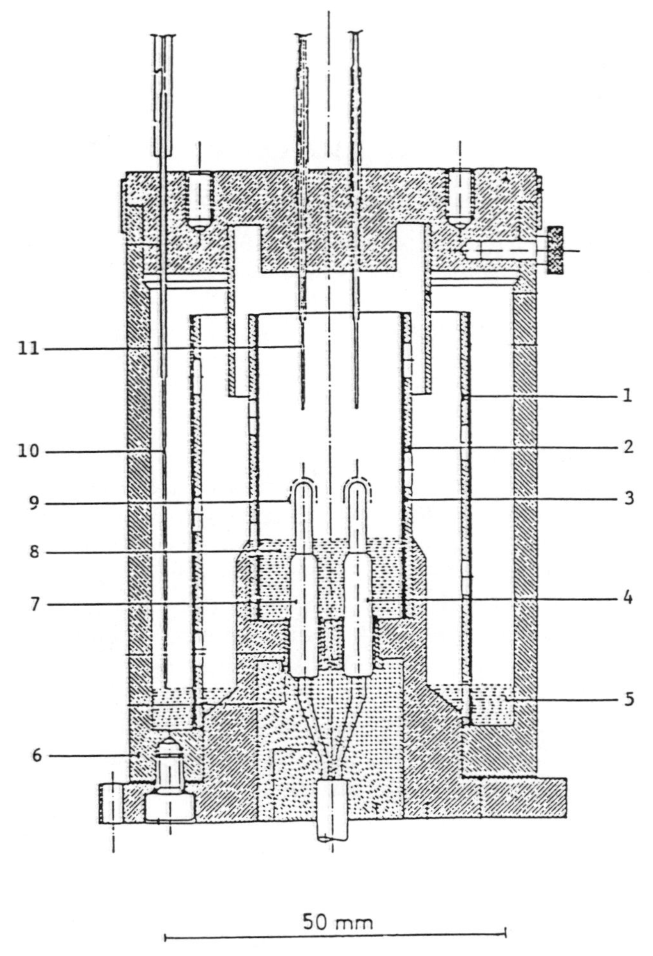

Figure 1. Schematic diagram of the measuring chamber of the apparatus designed by Dohner, Wachter, and Simon. (After Ref. 5.) 1, 2, Perforated copper cylinder; 3, filter paper; 4, specimen thermistor; 5, outer solution/solvent reservoir; 6, cell wall (copper); 7, reference thermistor; 8, inner solution/solvent reservoir; 9, platinum gauze; 10, extraction ducts for solvent; 11, inner ducts.

atmosphere. Two matched, inverted thermistors, suspended in the atmosphere, are connected to two arms of a Wheatstone bridge. Drops of solution or solvent can be placed on the thermistors through inlet ducts. In principle, with a solvent drop on one thermistor and a solution drop on the other, there is a net condensation of solvent on the solution drop because of the lower activity

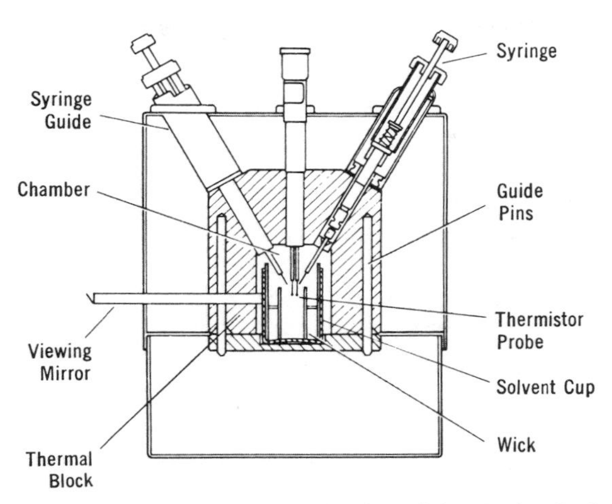

Figure 2. Schematic diagram of the measuring chamber of the Hewlett-Packard (Mechrolab) model 302B vapor pressure osmometer. (From Ref. 7.)

of the latter, thereby producing a temperature difference, which is determined via the Wheatstone bridge.

Various commercial instruments, based broadly on these principles, became available; the solvent chamber of the Hewlett-Packard (Mechrolab) instrument is shown schematically in Figure 2. Since the conditions within the solvent chamber in these instruments are never strictly adiabatic, true equilibrium is not achieved, but in spite of this a constant temperature difference is reached in quite a short time. The heat flow processes occurring on the thermistors have been the subject of various analyses (4, 8, 9). In practice, measurements are made with a series of solutions of known concentration of the sample of interest and of a calibration compound of known molecular weight (see below).

After some years of experience with the Hewlett-Packard instruments, certain problems, arising from the complexity of the physical situation within the solvent chamber, began to be realized. One of the earliest of these was a drop size effect (9–11), in which larger drops gave smaller readings of the bridge imbalance. This focused attention on the need to exercise extreme care in the placing of drops on the thermistors. This problem was reduced in later instruments produced by Hitachi Perkin-Elmer and by Knauer in which the thermistors were invented, thereby giving a more readily reproducible drop size. A further modification has been made in the more recent Wescan instruments, in which as a further aid to the achievement of controlled, reproducible drop sizes, the inverted thermistors are covered with a fine platinum mesh, as in the original apparatus of Dohner et al. (5).

Another problem has been with the calibration of the equipment. The temperature difference between the two thermistors, measured as a resistance difference ΔR in the earlier instruments and in terms of voltage ΔV in the later models, is assumed to be related to the weight concentration C of the solution (10) such that

$$\Delta R \text{ (or } \Delta V) = kC/\bar{M}_n + A_2 C^2 + \cdots \quad (1)$$

where k is a constant assumed to be independent of the solute and is referred to as the calibration constant of the particular instrument for a given solvent and temperature combination, and A_2 is the second virial coefficient. Commonly, a low molecular weight, pure compound (such as benzil) is used for this determination. A number of workers have demonstrated a dependence of the calibration constant on the molecular weight of the solute, thereby leading to values of $\bar{M}_n$ for polymers that were significantly in error (12–18). An example of the variation of the calibration constant with solute molecular weight is shown in Figure 3. Various procedural manipulations have been advocated to overcome this difficulty (12, 15, 16). It has also been shown, in some cases, that the molecular weight for a particular sample depends on the solvent in which the determination was performed (14).

It has been claimed that all these problems arise, at least in part, from factors connected with the instrumental design. One aspect which has been considered in some detail by various workers is thermistor self-heating effects (9, 19, 20). Other aspects are the degree of saturation of the vapor within the

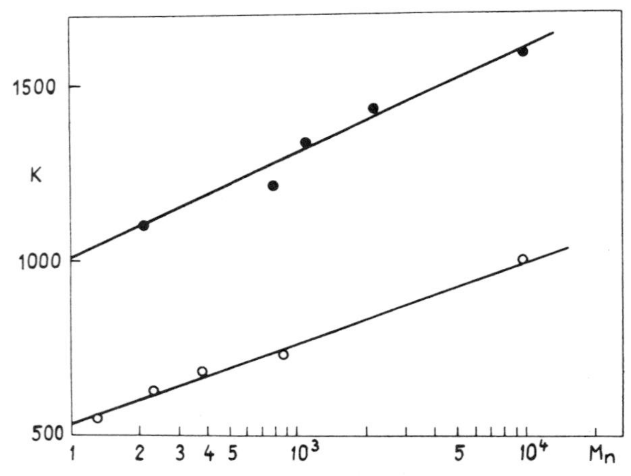

Figure 3. Dependence of the calibration constant on molecular weight: upper curve, determinations at 37°C in chloroform; lower curve, determinations at 45°C in benzene. (From Ref. 16.)

chamber (8) and diffusion effects within the drop on the thermistor (20). Certain modifications to the earlier instruments, advocated by Kamide et al. (19), evidently overcame these difficulties. Similarly, the newer Wescan instruments, based more closely on the design features of the apparatus of Dohner et al. (5), apparently do not suffer from these problems, although why this occurs is not altogether evident (20). In both these approaches satisfactory determination of $\bar{M}_n$ for polymers in excess of 10^5 daltons in favorable cases has been demonstrated (20, 21).

TREATMENT OF EXPERIMENTAL DATA

Originally, instrument manufacturers recommended that the data be assessed using the slope of the plot ΔR (or ΔV) against C or (better) from a plot of $(\Delta R/C)$ versus C extrapolated to infinite dilution. Departures from a linear relationship in either of these cases signified departures from ideal behavior and the necessity to incorporate higher order terms in the equation.

An alternative expression for $\bar{M}_n$ is

$$(\Delta V/C)_{C=0} = K\bar{M}_n^{-a} \tag{2}$$

where K and a are constants for a given solvent and temperature (17). The quantity $1 - a$ represents the deviation of the system from ideal behavior. Values for a as low as 0.7 have been reported (22, 23). It has been proposed that where data follow equation (2) and a differs from unity, the molecular weight average obtained by the VPO method exceeds the true number average molecular weight to a degree dependent on the polydispersity of the sample (23, 24). This proposal was based on extensive experimental and theoretical work on a Hewlett-Packard model 302B instrument, and it would perhaps be desirable to see this finding confirmed with data on a newer, improved instrument.

Some attention has been paid to the way in which $(\Delta V/C)$ versus C should be extrapolated to infinite dilution. Burge (20) demonstrated that for high molecular weight samples ($\sim 10^5$ daltons) it was advisable to plot $(\Delta V/C)^{1/2}$ versus C, obtain the best fit to a straight line, determine $(\Delta V/C)^{1/2}_{C=0}$, and then square this value. This procedure corrects for the contribution from the third virial coefficient, which should not be ignored with high molecular weight solutes.

It has also been pointed out the number average molecular weight for a polydisperse sample determined by VPO may be lower than that determined by, say, membrane osmometry, because in the latter case the smallest molecules may be able to pass through the membrane. The presence of traces

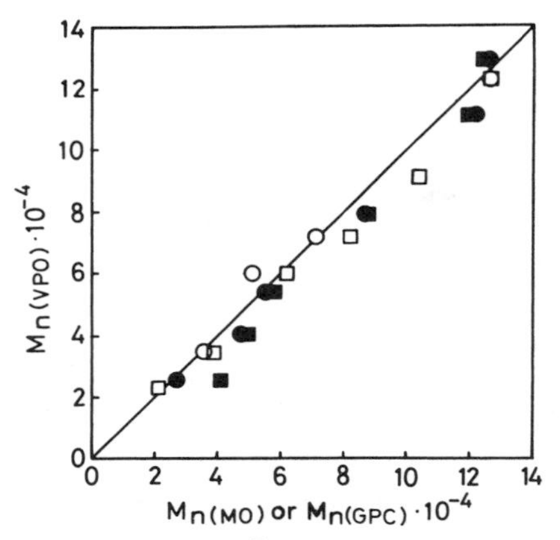

Figure 4. Correlation between the values of $\bar{M}_n$ determined by various methods for polystyrene (open symbols) and cellulose acetate (closed symbols). Circles, VPO versus MO data; squares, VPO versus GPC data. (From Ref. 21.)

of impurities can be especially significant in this regard (25). There is, possibly, a suggestion of this in the data of Figure 4. The polydispersity ($\bar{M}_w/\bar{M}_n$) of the samples used in that work (21) was in the 1.2–1.6 range.

APPLICATIONS

Vapor pressure osmometry has, for many years, been a widely used method for the determination of molecular weight in the 10^2–10^4 dalton range. Its popularity arises from its sensitivity and ease of use and from the wide range of temperatures and solyents that can be employed. VPO has been applied to the study of substances of many types, such as lipids, emulsifiers, and bitumens, as well as synthetic polymers. Because of the instrumental shortcomings outlined above, results of VPO determinations undertaken before the full extent of these pitfalls was recognized should be approached with considerable caution. These studies do, however, illustrate the utility of the technique, when correctly applied.

Molecular Weight of Polymers

Although the early apparatus of Dohner et al. was used for studies on standard polystyrene samples of number average molecular weights well in excess of 10^5 daltons (6), the manufacturers of the earlier commercial instruments proposed 10^2 to 2.5×10^4 daltons as the practical range using organic solvents. For more recent instruments, an upper limit of about 10^5 daltons is proposed.

Aqueous Systems

The sensitivity of VPO instruments with water as the solvent is substantially less than that with organic solvents. Thus the upper solute molecular weight which can be determined with acceptable accuracy is considerably lower in aqueous, compared with organic, systems in a similar solute concentration range. The manufacturers of the newer, Wescan instruments, for example, suggest an upper limit of 2×10^4 daltons for solutes in water, compared with 1×10^5 daltons in organic solvents. It has been claimed that the use of an alcohol–water mixture as the solvent to some degree alleviates this problem, at the expense of a significant increase in the equilibration time (26).

Virial Coefficients

The use of VPO measurements on polymers as a means of determining higher order virial coefficients has been long recognized (27), although the values so obtained on older VPO instruments may be in some doubt because of the

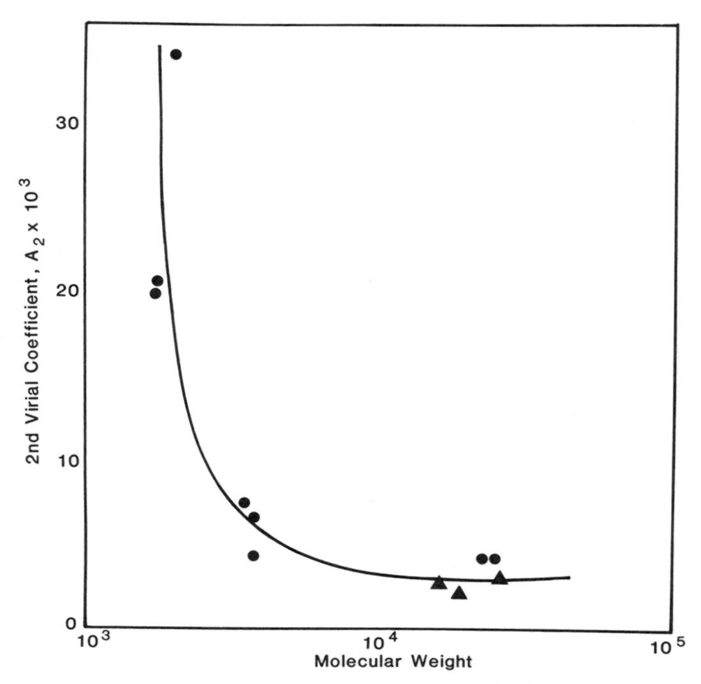

Figure 5. Variation of the second virial coefficient A_2 (mol·cm^3/g^2) with molecular weight for polyethylene: Circles, narrow, linear polyethylene; triangles, whole, branched PE. (After Ref. 25.)

problems already discussed. However, more recently, VPO data obtained on a Wescan model 232A instrument was used to demonstrate that the second virial coefficient for polyethylene increases very steeply with decreasing molecular weight below about 3×10^3 daltons (Fig. 5) (25).

Association

VPO measurements can be applied to the study of the extent of self-association by various solutes under different conditions. Emulsifiers and biological compounds, such as proteins and enzymes, have been examined in this context (28). Vapor pressure osmometry has been widely used in the characterization of asphaltenes, which are brown to black amorphous solids obtainable from bituminous materials such as crude oil or coal. Molecular weights are generally believed to be in the range 10^3 to 10^4 daltons, but their claimed propensity for association into micellelike aggregates in dilute solutions in nonpolar solvents has led to reported values of up to 3×10^5 daltons for some molecular weight methods (29, 30). These effects, coupled with instrumental and procedural problems of the types outlined above, have resulted in considerable confusion (31, 32). A new study, taking these problems into account, may now permit resolution of this confusion. In all these cases, because of the deviations from ideal solution behavior, care is required in the manipulation of the raw experimental data.

Osmolality

VPO instruments are also extensively used in the determination of the osmolality of a wide range of biological fluids. This, and related clinical applications (33), are considered to be outside the scope of this chapter.

REFERENCES

1. C. A. Glover, in *Polymer Molecular Weights*, Part I, P. E. Slade, Ed., Dekker, New York, 1975, p. 79.
2. A. V. Hill, *Proc. R. Soc., A*, **127**, 9 (1930).
3. A. P. Brady, H. Huff, and J. W. McBain, *J. Phys. Collid Chem.*, **55**, 304 (1951).
4. C. Tomlinson, C. Chylewski, and W. Simon, *Tetrahedron*, **19**, 949 (1963).
5. R. E. Dohner, A. H. Wachter, and W. Simon, *Helv. Chim. Acta*, **50**, 2193 (1967).
6. A. H. Wachter and W. Simon, *Anal. Chem.*, **41**, 90 (1969).
7. Hewlett-Packard, Avondale, PA. USA brochure, Vapor Pressure Osmometer, 1968.

8. K. Kamide, K. Sugamiya, and C. Nakayama, *Makromol. Chem.*, **133**, 101 (1970).
9. B. H. Bersted, *J. Appl. Polym. Sci.*, **17**, 1415 (1973).
10. A Adicoff and W. J. Murbach, *Anal. Chem.*, **39**, 302 (1967).
11. A. C. Meeks and I. J. Goldfarb, *Anal. Chem.*, **39**, 908 (1967).
12. J. Brzezinski, M. Glowala, and A. Kornas-Calka, *Eur. Polym. J.*, **9**, 1251 (1973).
13. B. H. Bersted, *J. Appl. Polym. Sci.*, **18**, 2399 (1974).
14. C. E. M. Morris, *J. Appl. Polym. Sci.*, **21**, 435 (1977).
15. B. H. Bersted, *J. Appl. Polym. Sci.*, **23**, 633 (1979).
16. I. Kucharikova, *J. Appl. Polym. Sci.*, **23**, 3041 (1979).
17. M. Marx-Figini and R. V. Figini, *Makromol. Chem.*, **181**, 2401 (1980).
18. L. Mrkvicakova and S. Pokorny, *J. Appl. Polym. Sci.*, **30**, 1211 (1985).
19. K. Kamide, T. Terakwa, and H. Uchiki, *Makromol. Chem.*, **177**, 1447 (1976).
20. D. E. Burge, *J. Appl. Polym. Sci.*, **24**, 293 (1979).
21. K. Kamide, T. Terakawa, and S. Matsuda, *Br. Polym. J.*, **15**, 91 (1983).
22. R. V. Figini, *Makromol. Chem.*, **181**, 2409 (1980).
23. R. V. Figini and M. Marx-Figini, *Makromol. Chem.*, **182**, 437 (1981).
24. M. Marx-Figini, M. Tagliabue, and R. V. Figini, *Makromol. Chem.*, **184**, 319 (1983).
25. F. M. Mirabella, *J. Appl. Polym. Sci.*, **25**, 1775 (1980).
26. J. Szabo-Lakatos and I. Lakatos, *Acta Chim (Budapest)*, **89**, 1 (1976).
27. K. Kamide, K. Sugamiya, and C. Nakayama, *Makromol. Chem.*, **132**, 75 (1970).
28. E. T. Adams, P. J. Wan, and E. F. Crawford, *Method Enzymol.*, **48**, 69 (1978).
29. S. E. Moschopedis, J. F. Fryer, and J. G. Speight, *Fuel*, **65**, 227 (1976).
30. I. Schweger, W. C. Lee, and T. F. Yen, *Anal. Chem.*, **49**, 2365 (1977).
31. K. E. Chung, L. L. Anderson, and W. H. Wiser, *Fuel*, **58**, 847 (1979).
32. I. Lang and P. Vavrecka, *Fuel*, **60**, 1176 (1981).
33. Wescan Instruments, Inc. Santa Clara CA. USA brochure, "VPO—Modern Technology for Clinical Research Applications," 1984.

CHAPTER

4

MEMBRANE OSMOMETRY

HANS COLL

Research Laboratories Estman Kodak Company Rochester, NY

INTRODUCTION

The term *osmosis* (Greek for the act of pushing or thrusting) refers to the spontaneous flow of solvent across a membrane into a solution, thereby diluting the latter. This flow can be counteracted by applying a hydrostatic pressure on the solution side of the membrane. The pressure which just suffices to stop the flow is called the osmotic pressure, which is conventionally designated by the letter π. The membrane must have the property of being semipermeable; that is, it must allow the flow of solvent while retaining the solute.

Osmotic experiments were conducted as early as 1748 by Abbé Nollet, but a quantitative formulation, on which rests the determination of molecular weight, was not given before 1875, when H. van't Hoff showed that π is proportional to the molar concentration n of the solute

$$\pi = nRT = cRT/M \tag{1}$$

where R is the gas constant, T the absolute temperature, c the mass concentration, and M the molecular weight of the solute. According to equation (1), π is a colligative property. As will be shown below, M is the number average molecular weight (usually designated as $\bar{M}_n$). Equation (1) is a limiting law which holds for ideal solutions. It serves as an approximation for real solutions only at extreme dilution.

In principle, the measurement of osmotic pressure requires relatively simple apparatus, and a variety of so-called manual osmometers were designed before 1965. In more recent years, high speed electronic osmometers have replaced the manual devices. Osmometry is best suited for the determination of the (number average) molecular weight of macromolecular solutes in the range of 10,000 to 500,000 daltons. (Throughout this chapter "osmometry" always refers to measurements made with the help of membranes.)

Osmometry has been used in many studies of proteins, but between 1940 and 1970 it found an even wider application in the characterization of synthetic polymers, often in conjunction with light scattering, which yielded the weight average molecular weight ($\bar{M}_w$). The ratio $\bar{M}_w/\bar{M}_n$ served as an index for the breadth of the molecular weight distribution. With the advent of size exclusion (gel permeation) chromatography, osmometry has seen a sharp decline. Even in its heyday, the reliability of osmometry was sometimes questioned. Difficulties were encountered when the macromolecular solute was not fully retained by the membrane, and the effect of incomplete solute retention on the osmotic pressure was often not appreciated. A better understanding of the errors introduced by solute permeation was given by the theoretical work of Staverman (1, 2).

Osmometry has certain advantages over other methods of molecular weight determination. It is an absolute method capable of very high precision, particularly suitable for solutes of high molecular weight. The chemical nature of the solute is immaterial as long as it is fully retained by the membrane. The extensive contribution of a small amount of impurities of low molecular weight (such as residual monomer, plasticizer, or antioxidants) in a polymer to the number average molecular weight can be eliminated by measuring the osmotic pressure after these materials have diffused through the membrane. This is a clear-cut advantage over vapor pressure osmometry. Similarly, the presence of supporting electrolyte in a solution of macroions does not interfere. Finally, besides being a method for molecular weight determination, osmometry provides a measure for solvent activity having thermodynamic implications over the entire concentration range of a macromolecular or colloidal solution.

An extensive literature on osmometry accumulated up to 1970, with only few contributions thereafter, as the interest in the subject waned with the rise of size exclusion chromatography. A general treatment of osmometry together with other measurements of colligative properties was published in 1958 by Bonnar, Dimbat, and Stross (3). Tombs and Peacocke (4), in a later book, give special attention to biological macromolecules. Biological applications have also been reviewed by Kupke (5) and by Kelly and Kupke (6). Other accounts, as chapters of books, have been written by Krigbaum and Roe (7), Helbig (8; in German), Ulrich (9), Vink (10), and Jeffrey (11), the last-mentioned author dealing mostly with the theory of protein interactions with reference to osmometry and to other methods. A lucid treatment of theory can be found in the textbook by Tanford (12). In 1964 Chiang (13) reviewed techniques of high polymer characterization, including osmometry, emphasizing measurements at elevated temperatures. An account of the state of the art of equilibrium osmometry up to 1967 and of dynamic measurements can be found in Reference 14. Coll (15) has reviewed the osmometry of membrane-permeating solutes.

MANUAL MEASUREMENTS

The osmometric experiment is best described by a *static* measurement with a capillary osmometer. The same principle applies to electronic osmometers, where minute displacements of a strain gage, for instance, correspond to changes of meniscus in a capillary.

Figure 1 schematically depicts a single-chamber (manual) capillary osmometer. The cell containing the solution is separated from the surrounding solvent by two membranes, the larger membrane area allowing for faster equilibration. A second capillary, with a bore identical to the reading capillary, dips into the surrounding solvent. The osmometer is carefully thermostated. If, after temperature equilibration, the liquid level is the same in both capillaries, solvent will pass through the membranes into the osmometer cell and the meniscus in the reading capillary will rise until the hydrostatic head is just sufficient to stop the further influx of solvent. The meniscus height (H) can be followed with a cathetometer. Equilibrium is reached when H remains constant over an extended time interval. The osmotic pressure (π) is then calculated from the meniscus levels in the capillaries

$$\pi = g(\rho H - \rho_0 H_0), \tag{2}$$

where g is the gravitational acceleration (980 cm/s^2), ρ is the density, and the subscript zero refers to the solvent. In equation (2), solvent and solution density may be different, H and H_0 are measured from a mark near the center of the (vertical) membranes. In practice, the density difference is usually small enough to be ignored, and the difference of the meniscus levels can be measured directly. Errors caused by differences of surface tension in the two

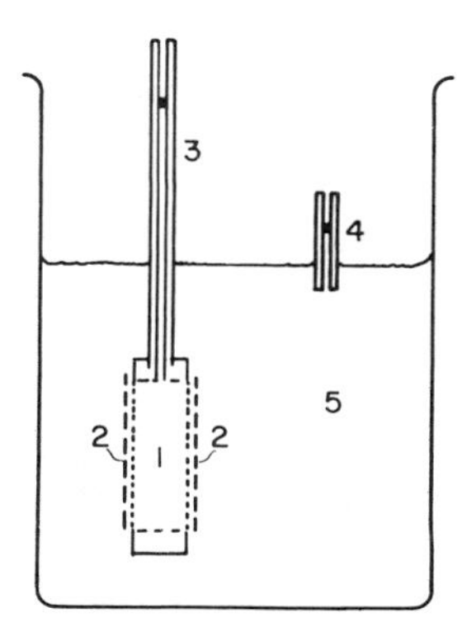

Figure 1. One-cell osmometer (schematic). 1, solution compartment; 2, membranes; 3, reading capillary; 4, reference capillary; 5, solvent.

capillaries must be negligible, and the influx of solvent during equilibration must not significantly change the solute concentration.

Equilibration in this static determination of osmotic pressure is slow, requiring hours, if not days, for one measurement. Membranes that allow for fast solvent permeation (while fully retaining the solute) and narrow bore capillaries will shorten the time of the experiment. The approach to equilibrium is described by the equation

$$a\,dh = (\pi - h)AP\,dt \tag{3}$$

where h is the difference of meniscus levels in the two capillaries, a is the cross-sectional area of the capillary, and A is the membrane area. Integration of equation (3) yields

$$\ln\left(\frac{h_2 - \pi}{h_1 - \pi}\right) = \frac{AP(t_2 - t_1)}{a} \tag{4}$$

where h_1 and h_2 are the hydrostatic pressures measured at times t_1 and t_2. The osmotic pressure π is expressed as a height, which must be multiplied by $g\rho$ to give the dimension of pressure.

We may consider as typical a membrane area of $10\,cm^2$ with a permeation constant of $10^{-8}\,cm^3/s$ per square centimeter and cm H_2O, and a capillary diameter of 0.05 cm ($a = 0.002\,cm^2$). Then, at an initial hydrostatic head of 5 cm below π, it takes approximately 30 hours to approach π to within 0.02 cm. Decreasing the capillary diameter to one-half reduces the time by one-fourth, and seemingly rather fast measurements can be made with sufficiently narrow capillaries. Apart from experimental difficulties associated with such capillaries, a limit is set by the flexibility of the membrane, the distortion (ballooning) of which can be expressed as a volume change ΔV, which for sufficiently small hydrostatic pressures h may be written as

$$\Delta V = K_1 \Delta h \tag{5}$$

where K_1 is a constant. By adding ΔV to the volume change ha in the reading capillary, one obtains an effective cross-sectional area, $a' = a + K_1$, to replace a in equation (3). By providing a rigid support for the membrane, ballooning can be kept to a minimum. A typical value for a well-supported flexible membrane with an area of $10\,cm^2$ may be $K_1 = 0.005\,cm^3/cm\ H_2O$ (equal to an average displacement of only $5\,\mu m/cm\ H_2O$).

Based on equations (3) and (4), *dynamic* methods of measurement have been developed for estimating the osmotic pressure during the approach to equilibrium rather than from the point of equilibrium itself. This saves time and is likely to give better (but nor correct) results when the solute molecules are not fully retained by the membrane. More about "leaking" membranes is said in a later section. Details of dynamic methods are not given here, since manual osmometers, for which these procedures were developed, have become less important in molecular weight determinations. For a full account of the methods, the reader is referred to the literature.

Dynamic methods are generally restricted to osmometers with completely rigid membranes or to osmometers for which the membrane is firmly supported on both sides.

The method of Bruss and Stross (16) exemplifies a dynamic procedure. Once the osmometer, filled with solution, has reached temperature equilibrium, the rate of meniscus change ($u = \Delta h/\Delta t$) is measured for different arbitrary values of meniscus height h, followed by an extrapolation to $u = 0$. This follows from equation (3)

$$h = \pi - uK' \tag{6}$$

where $K' = a/(AP)$. It is necessary that K' remain constant for the duration of one set of measurements.

This method is fast and very convenient if a cathetometer telescope with two parallel cross hairs is used, the distance Δh being defined by their spacing. The rate u is then immediately obtained by determining the time that the meniscus requires to pass between the cross hairs.

A quick adjustment of the meniscus height in the capillary is desirable. This is conveniently done with a push rod acting as a piston on the solution in the osmometer.

Other dynamic methods that have been widely applied in the past are the procedure of Philipp (17), the half-sum method of Fuoss and Mead (18), and the method of Elias (19), which was developed for work with porous glass osmometers (19, 20).

A comparison of several dynamic methods has been made by Bruss and Stross (21).

THEORY

Two-Component Solutions

Consider the static osmotic experiment described in the preceding section with the osmometer cell containing a solution of electrically neutral macromolecules. Solvent at atmospheric pressure P_0 is at equilibrium with the macromolecular solution on the other side of the semipermeable membrane.

Zero net flow of solvent across the membrane requires that an excess pressure π be applied to the solution. Stated differently, the chemical potential of the solvent μ_1'', having been lowered by the presence of the macromolecular solute, must be raised by an applied hydrostatic pressure to equal the chemical potential of the pure solvent μ_1':

$$\mu_1' = \mu_1'' + \int_{P_0}^{P_0+\pi} \left(\frac{\partial \mu_1}{\partial P}\right)_T dP = \mu_1'' + \bar{V}_1 \pi \tag{7}$$

The derivative under the integral sign is equal to the partial molar volume $\bar{V}_1$, which can always be treated as a constant over the modest pressure range considered here. The difference between μ_1'' and μ_1' is given by

$$\mu_1'' - \mu_1' = RT \ln a_1 \tag{8}$$

where R is the gas constant, T the absolute temperature, and a_1 the solvent activity in the solution, the pure solvent being in the reference state. From equations (7) and (8) we obtain

$$\pi = -(RT/\bar{V}_1) \ln a_1 \tag{9}$$

At extreme dilution the mole fraction of the solvent N_1 approaches unity and a_1 becomes equal to $N_1 = 1 - N_2$, N_2 being the mole fraction of the solute. Also, $\bar{V}_1$ can be replaced by the molar volume of the solvent, V_1. Expansion of the logarithm with neglect of all but the first term and change of the concentration units c to grams of solute per cubic centimeter yields

$$\pi = N_2 RT/V_1 = cRT/M \tag{10}$$

where M is the molecular weight of the solute. This is van't Hoff's limiting law (eq. 1).

If the solute consists of a mixture of i species of different molecular weights with a total concentration $c = \sum_i c_i$ (all solute molecules being retained by the membrane), the osmotic pressures due to the solute components are additive so that

$$\pi = RT \sum \frac{c_i}{M_i} = RTc \frac{\sum c_i/M_i}{\sum c_i} = RTc \frac{1}{\bar{M}_n} \tag{11}$$

$\bar{M}_n$ being the number average molecular weight defined by

$$\bar{M}_n = \frac{\sum n_i M_i}{\sum n_i} = \frac{\sum c_i}{\sum c_i/M_i} \tag{12}$$

Measurements of osmotic pressure must be made at concentrations where the limiting law generally does not hold. The deviations are then accounted for by a virial expansion, stated here in three equivalent forms,

$$\pi/c = RT/M + Bc + Cc^2 + \cdots \tag{13a}$$

$$\pi/c = (RT/M)(1 + \Gamma_2 c + \Gamma_3 c^2 + \cdots) \tag{13b}$$

$$\pi/c = RT(1/M + A_2 c + A_3 c^2 + \cdots) \tag{13c}$$

The dimensions of the virial coefficients, which are different in the three equations, are evident from a comparison with the first term.

By plotting π/c as a function of c, we can determine $\bar{M}_n$ from the intercept obtained by extrapolation to $c = 0$. The limiting slope of the plot is equal to the second virial coefficient B. Linearity of the plot over the entire range of concentrations indicates that the third and higher virial coefficients are negligibly small.

In the case of an ideal (or theta) solution of macromolecules, the second, and higher, virial coefficients of the osmotic pressure are negligibly small, but not strictly zero, because of the truncated expansion of the logarithm in equation (11).

In a real solution the virial coefficients depend on the excluded volume Q of the solute *molecules*, and the second virial coefficient is (12)

$$A_2 = N_A Q/(2M^2) \tag{14}$$

N_A being Avogadro's number. For a dilute system of uncharged solid spheres of density ρ, $Q = 8M/(\rho N_A)$. As an example, we may choose $M = 10^5$, $V_1 = 100$, and $c = 0.01$, and sphers of unit density. Then, $A_2 c/(1/M) = 0.04$. Osmometry can provide some structural information, since this ratio becomes considerably larger if the particles are highly elongated (12, 22, 23), or swollen with solvent, as is typical for a long chain polymer in a good solvent.

The classical expressions for the excluded volume of a random coil polymer were derived by Flory and Krigbaum (24). A more recent treatment has been given by Yamakawa (25). The rather lengthy mathematical expression of the Flory–Krigbaum theory, which can be found in Tanford's book (12), is not reproduced here. Of importance is the concept of the theta temperature and the thermodynamic solvent strength. The excluded volume, hence the second virial coefficient, of a random coil polymer is zero at the theta temperature. In a good solvent, where the coil is highly expanded, these quantities assume rather large values. Negative values can be observed in a limited temperature range below the theta point.

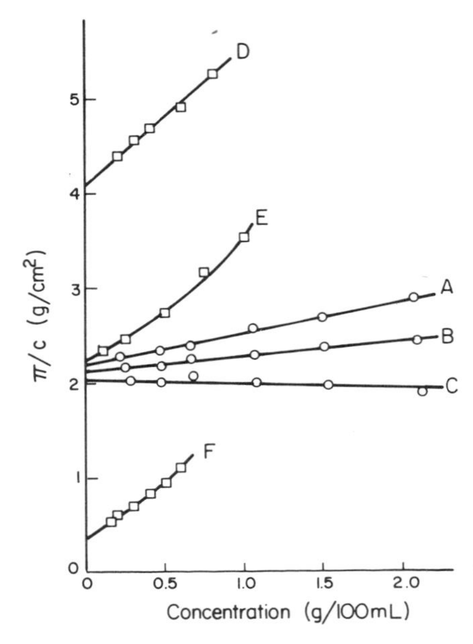

Figure 2. Reduced osmotic pressure as a function of concentration. Curves A, B, C: polystyrene (MW 125,000) in cyclohexane at 50, 40, and 30°C. Curves D, E, F: polystyrene (MW 61,500, 120,000, and 612,000) in toluene at 30°C. (From Refs. 26 and 27.)

An illustration of solvent power on the reduced osmotic pressure (26, 27) is given in Figure 2 for polystyrene fractions in toluene (a good solvent) and in cyclohexane (a theta solvent at 34°C). The pronounced nonideality of the toluence solutions is evident, as is the need to include the third virial coefficient in the extrapolation of π/c at the higher molecular weights. A slight negative virial coefficient is indicated for the cyclohexane solution at 30°C. The intercepts of the cyclohexane plots are not identical because of temperature differences. The second virial coefficient A_2 for polystyrene of $M = 10^5$ in toluence at room temperature is $0.00049\ \text{cm}^3 \cdot \text{mol/g}^2$ (26); so that for $c = 0.01$ the ratio $A_2c/(1/M) = 0.49$. Thus, the osmotic pressure is nearly 50% higher than in a theta solvent.

Contributions from the third virial coefficient A_3 are no longer negligible once A_2c has approached the magnitude of $1/M$. The third virial coefficient for solutions of chain molecules has been discussed by Stockmayer and Casassa (28), and their results have been taken as a justification for the approximation $\Gamma_3 = 0.25\Gamma_2^2$, or $A_3 = 0.25\ MA_2^2$. In the example given in the preceding paragraph, $A_3c^2/(1/M) = 0.06$, which must be included in the virial expansion for an accurate determination of molecular weight.

Three-Component Systems (Macroions)

The electric charges carried by a macroion (protein or polyelectrolyte) in an ionizing solvent, such as water, must be neutralized by an equal number of ions of opposite charge. Although these counterions are small enough to pass through the membrane, they must remain with the macroion to satisfy the condition of electroneutrality. Hence, for each macroion with charge Z, the osmotic pressure will not be a measure for $1/M$ but for $(1+Z)/M$. However, if an electrolyte consisting of small ions is added to the solution, the limiting law (eq. 1) becomes valid again. The presence of a sufficient amout of this supporting electrolyte also makes the solutions less nonideal, so that linear extrapolations of π/c to infinite dilution can be done with confidence.

The thermodynamics of solutions of electrically charged systems was treated by Scatchard (29). The osmotic pressure for a dilute solution of macroions, in the presence of a 1 : 1 electrolyte of small, membrane-permeating ions, is given by

$$\frac{\pi}{c} = \frac{RT}{M} + RTc\left[\frac{1000v_1}{M^2}\left(\frac{Z^2}{4m_3} + \frac{\beta_{22}}{2} - \frac{\beta_{23}^2 m_3}{4 + 2\beta_{33} m_3}\right)\right] \tag{15}$$

the expression in parentheses being the second virial coefficient A_2. The *molal* concentration of the added electrolyte is denoted as m_3, and v_1 is the specific volume of the solvent. Derivatives of the activity coefficients are $\beta_{22} = \delta \ln \beta_2/\delta m_2$, $\beta_{33} = \delta \ln \beta_3/\delta m_3$, and $\beta_{23} = \delta \ln \beta_2/\delta m_3$, the subscripts 2 and 3 referring to the macroion and the added electrolyte, respectively, M is the molecular weight of the macroion, and its meaning depends on the definition of c which, for instance, may include all the Z counterions per molecule.

The first term in parentheses in equation (15) is due to the Donnan effect as the presence of the macroion demands an unequal distribution of ions on opposite sides of the membrane. For solutions with low salt concentration and high macroion charge, this term can be dominant. The term β_{22}, a measure for the interaction between the macroions, contains a contribution from the excluded volume and, therefore, does not vanish when $Z = 0$. The third term in parentheses expresses the interactions between macroions and added electrolyte (β_{23}) and between the small ions (β_{33}). The Donnan term contains m_3 in the denominator and depends strongly on the salt concentration; β_{22} also depends on salt concentration, since the small ions have a screeing effect on the repulsive forces between macroions of the same charge.

Numerical values for the second virial coefficient and its constituent terms were determined for the macroion serum albumin by Scatchard and coworkers (30). They also estimated the extent of ion binding on globular proteins from the magnitude of A_2 at high salt concentrations (31).

Scatchard and coworkers (32) derived expressions for the third virial coefficient in their work on serum albumin.

Dimerization

Protein molecules have been observed to form dimeric and higher association products. In solutions in which the Donnan effect, being the main contributor to A_2 in equation (15), is kept small by a relatively high salt concentration, association will lead to a plot of π/c versus c with a negative slope. In the simplest case, where the solution behaves ideally except for dimer formation, the expression for the osmotic pressure is easily derived from the law of mass action, $K = c_{\mathrm{D}}/c_{\mathrm{M}}^2$, where c_{D} and c_{M} are the respective mass concentrations of dimer and monomer:

$$\frac{\pi}{cRT} = \frac{1}{2M}\left[1 + \frac{\sqrt{4Kc+1}-1}{2Kc}\right] \tag{16}$$

where M is the molecular weight of the monomer. Equations for higher association products have been derived by Steiner (33). Adams (34) has extended the treatment to include nonideal systems.

DATA HANDLING

Once the osmotic pressure has been determined in appropriate units for a series of polymer concentrations c, a plot of π/c versus c is prepared. The concentrations are best expressed as grams per cubic centimeter or grams per liter. At least four concentrations with roughly equal spacing should be used. For precise work, a larger number of concentrations may be needed, particularly when the plot has a steep slope and contributions from the third virial coefficient can be expected. The highest concentration of the series should exceed the lowest one by at least a factor of 3.

It is common practice to evaluate the plot of π/c versus c by simply drawing a straight line through the experimental points in accordance with the equation

$$\pi/c = RT/\bar{M}_n + RTA_2c \tag{17}$$

with the intercept I at $c = 0$ and with slope S. Then, $\bar{M}_n = RT/I$ and $A_2 = S/(RT)$. To yield M in conventional units (g/mol), the numerical value of the gas constant R must be adjusted to the units of π and c, as shown in Table 1.

A least-squares curve fit is superior to the graphic extrapolation. Since it

Table 1. Units of Measurement

Pressure	Concentration	Gas Constant R
Pascals	g/m^3	$8.3144\ m^3 \cdot Pa/mol\ K$
$Dynes/cm^2$	g/cm^3	$8.3144 \times 10^7\ cm \cdot dyne/mol\ K$
Atmospheres	g/cm^3	$82.057\ cm^3 \cdot atm/mol\ K$
Hydrostatic head (cm) (liquid with density ρ (g/cm^3) and $g = 980.64\ cm/s^2$)	g/cm^3	$8.4784 \times 10^4/\rho$

can be assumed that the absolute error of a pressure reading is the same for all pressures, the curve fitting must not be done on the π/c data but according to

$$\pi = a_0 + a_1 c + a_2 c^2 \tag{18}$$

For a good set of data, a_0 should be close to zero. Membrane asymmetry and shifting baselines contribute to the magnitude of a_0. Since a_0 has no physical meaning in the osmotic pressure equation, it may be assigned a value of zero in the curve-fitting procedure. The relationship of M with a_1 and of A_2 with a_2 is evident from a comparison of equations (17) and (18). If contributions from the third virial coefficient are likely, the term $a_3 c^3$ should be included in equation (18) for curve fitting.

Requisite computer programs for the fitting of power series are now widely available. More details of the statistical treatment of osmometry data and curve-fitting schemes without the use of a computer can be found in Reference 3.

When dealing with solutions of random coil polymers of high molecular weight in a good solvent, one has to expect contributions from the third virial coefficient. If the limited number of experimental points does not justify cubic curve fitting, it is recommended for better results to plot $(\pi/c)^{1/2}$ instead of π/c against c. Then, according to equation (13b), with the assumption that $\Gamma_3 = 0.25\Gamma_2^2$,

$$(\pi/c)^{1/2} = (RT/M)^{1/2}(1 + 0.5\Gamma_2 c) \tag{19}$$

If I is the intercept and S the slope of such a linear plot, then $\bar{M}_n = RT/I^2$ and $\Gamma_2 = 2S/I$.

This "square root" plot can also be applied (although it may not be needed) to polymers of lower molecular weight or to a less good solvent since, for random coil polymers, $\Gamma_3 = 0.25\Gamma_2^2$ is generally a better assumption than $\Gamma_3 = 0$, which is implied in the linear plot of π/c against c.

EXPERIMENTAL PRECAUTIONS

Several potential sources of error must be considered in the measurement of osmotic pressure.

Self-evident requirements for a functioning osmometer are leak-tight membrane and valves and adequate temperature control to make volume changes due to temperature fluctuations negligible. When a manual one-cell osmometer is used, control of the constant-temperature bath to 0.01°C should suffice, because the solvent surrounding the osmometer cell serves as a good damping device during a heating cycle. The capillaries of the osmometer should be at the same temperature as the cell.

Manual osmometers require capillaries of uniform bore. In small-bore capillaries liquid rise due to surface tension can be appreciable. Constancy of surface tension and contact angle must be assured for correct measurements. These effects become more important with liquids of higher surface tension. Aqueous solutions are generally not suitable as manometer liquids in capillary tubing.

The performance of the membrane has to be tested. Adequate solvent permeability can be recognized from the rate with which equilibrium is attained starting from some arbitrary pressure. If the approach to equilibrium is not exponential (cf. eq. 3), leaks can be suspected.

Some assurance is needed that the retention limit of the membrane is adequate for the molecular weight determination at hand. Testing for this limit can be done with polymer samples of relatively low, known molecular weight, preferably with narrow size distributions. If such molecular weight standards are not available, some well-characterized reference sample with a broad distribution extending below the presumed retention limit of the membrane can be used. The constancy of the pressure readings with time is an indicator for retentiveness, but a comparison of the measured osmotic pressure with the value that corresponds to the true molecular weight is more important, since solute permeation leads to osmotic pressures that are too low (or number average molecular weights that are too high). Homogeneous substances with molecular weights between 500 and 1000 can likewise be considered as test solutes for membranes with very low retention limits, although the structure of the molecule may also affect permeability (see section entitled "Membrane-Permeating Solutes"). Permeation limits of membranes tend to decrease with extended use, as pores are plugged by trapped polymer molecules, or because the membrane structure begins to collapse.

Several blank runs (solvent on both sides of the membrane) should be made with every freshly installed membrane. Pressure differences should decline with repeated rinsing, but often small "asymmetry" pressures persist. If these remain constant over the time required for a molecular weight determination, simply subtract the constant blank reading from the measured osmotic pressures. There is, however, no full assurance that the asymmetry pressure is not influenced by the presence of the solute. Membranes that show an asymmetry hydrostatic head much in excess of 0.1 cm should be discarded. More detailed discussions of membrane asymmetry can be found in Reference 3 or in the papers by Philipp (17) and by Wallach (35).

Aelenei (36), working with benzene–toluene as the solvent, reported that the membranes relaxed for 2–3 days after clamping in the osmometer, during which time inconstant osmotic pressures, were observed. This finding invites caution with freshly installed membranes.

Solvents should be degassed before use, particularly for measurements at elevated temperatures. It is recommended that solvents from the same batch be used in any given molecular weight determination, since small amounts of low molecular weight impurities at unequal concentrations on opposite sides of the membrane can give rise to significant transient osmotic pressures in a fast-responding instrument with a very retentive membrane.

OSMOMETERS

The performance of modern electronic osmometers is generally far superior to that of manual instruments. Of particular importance is the fast response of the former, which has shortened the time required for a measurement from hours to minutes. Points in favor of manual osmometers are low cost and simplicity of design. Figure 3, in a simplified diagram, illustrates the typical features of a well-designed small volume (2 mL) manual capillary osmometer (37) that can be used for static and dynamic measurements alike.

Early manual instruments were described by Fuoss and Mead (18) and by Zimm and Myerson (38). Improvements on these designs can be seen in the capillary osmometer of Hellfritz (39) and of Stabin and Immergut (40).

For work with aqueous solutions, a water-immiscible hydrocarbon, in contact with the aqueous solution, can be used as the manometer liquid in capillaries (41).

In principle, all manual osmometers can be used for static measurements. Their suitability for dynamic measurements depends on whether the membrane is firmly supported on both sides and whether a quick adjustment of pressure can be made in the course of the experiment.

A commercial osmometer (Wagner and Munz, Munich, FRG) consisting of a bulb of porous Vycor glass (pore size ~ 4 nm) with fused-on capillaries has been described by Elias (19) and by Elias and Ritscher (20). This osmometer can be used up to 300°C. Because of slow solvent transport through the relatively thick (0.1 cm) glass membrane, only dynamic measurements appear practical.

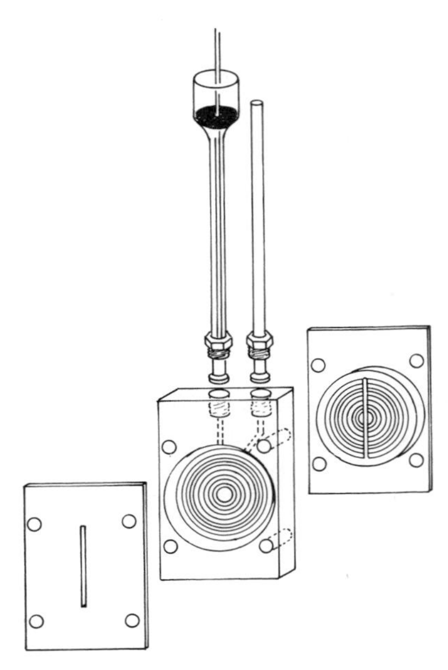

Figure 3. Manual one-cell osmometer of Bruss and Stross (simplified diagram; Ref. 37.) The osmometer body is made of stainless steel. The matched concentric lands and grooves in the body and end plates assure good support for the two membranes. The filling tube and measuring capillary are detachable. The push rod in the filling tube (for adjusting the meniscus in the reading capillary) is sealed with a pool of mercury. The assembled osmometer is placed in a solvent cup, which is thermostated in a water bath. A reference capillary (not shown) dips into the solvent. The sample volume is less than 2 mL.

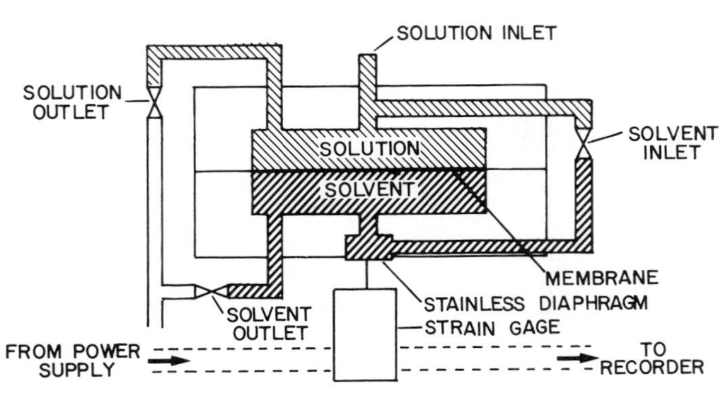

Figure 4. The Wescan recording membrane osmometer (schematic). (Reproduced with the permission of Dr. D. E. Burge, Wescan Instruments.)

In the 1960s, manual osmometers were superseded by self-balancing electronic instruments. These were precise, easy to handle and, most important, could give readings in times of the order of 10–30 minutes. Against these advantages stood considerably higher cost. Instruments of this kind were Hewlett-Packard's Mechrolab and the Shell osmometer (42).

The Mechrolab osmometer was based on the manual Hepp osmometer (43, 44). Solvent flow through the membrane was monitored by a photoelectric device focused on a tiny gas bubble in the solvent line. Movement of the bubble, via a servo loop, activated a leveling vessel (with pressure indicator) filled with solvent, to impose a zero-flow condition on the solvent. In the Shell osmometer (42) a thin flexible metal diaphragm in the bottom of the sample cell formed one plate of a capacitor. Through an oscillator and a servomechanism, the solvent level in a manometer tube was adjusted to restore the diaphragm to its undeflected position.

These instruments balanced quickly, but not instantaneously, because of the flexibility (slight ballooning) of the membrane, and it soon became apparent that osmometers equipped with a sensing diaphragm connected to a strain gage, without a servomechanism, reached equilibrium with comparable rapidity. Electronic osmometers of this design are simpler and less costly. They have completely replaced the self-balancing osmometers, which went out of production in 1972.

Two modern strain gage osmometers are the Wescan recording membrane osmometer (Wescan Instruments, Santa Clara, CA), shown schematically in Figure 4 and the Knauer membrane osmometer, manufactured in Germany (Wissenschaftlicher Gerätebau H. Knauer, Berlin; distributed by UIC, Joliet, IL). In the Wescan instrument, the solvent, contained in a cell that is tightly closed off by two valves, permeates upward through the membrane into the solution compartment, thereby generating a negative pressure, which causes a tiny upward displacement of the stainless steel diaphragm connected to the strain gage. The electrical output from the latter is proportional to the pressure difference across the membrane, and the signal is displayed on a recorder. Four pressure ranges extend up to a maximum of 100 cm H_2O. The required sample volume is stated as less than 0.5 mL (a larger amount is generally needed for rinsing of the compartment). A Wescan osmometer with a temperature range of 5–130°C is also available.

The distinguishing feature of the Knauer osmometer (Fig. 5) is the extremely small (50-μL) sample cell. With four rinses, as recommended, a total sample volume of only 0.25 mL is required. Six pressure ranges, up to 40 cm H_2O, are available. The maximum of the most sensitive range is 1.25 cm H_2O with a corresponding signal of 100 mV. The signal is automatically stored and recorded. The temperature range is stated as 25–125°C.

Both the Knauer and the Wescan osmometers can be used with aqueous and nonaqueous solutions, the sample cells being made of inert materials.

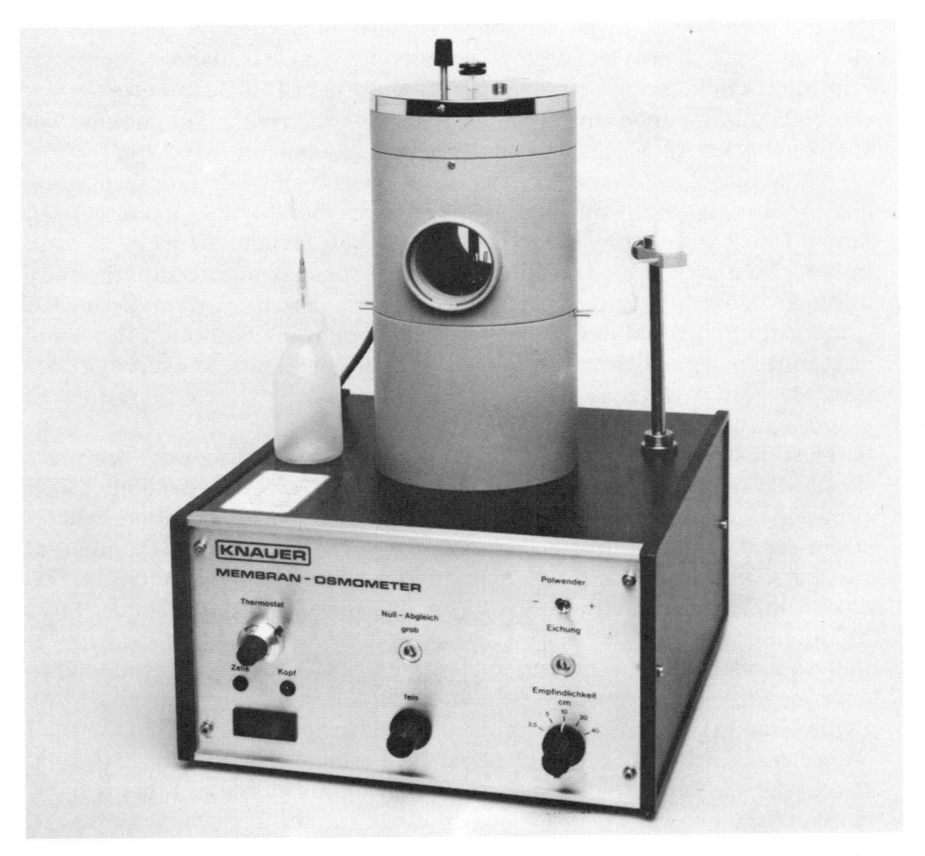

Figure 5. The Knauer membrane osmometer. (Photograph courtesy of UIC Inc., Joliet, IL.)

MEMBRANES

The quality of the membrane is of central importance in osmometry, and choosing the proper membrane requires some care. The use of commercial osmometer membranes is recommended for most applications.

Strictly semipermeable membranes are an idealization, but rigorous selectivity between solvent and solute molecules can be achieved if their molecular weights differ by a factor of about 100. Yet, membranes do not usually act as simple physical sieves. Widely used membranes are made of long chain organic polymers, such as cellulose or its derivatives, which swell in the osmotic solvent. These membranes may also act as a partitioning phase, into which molecules "dissolve." The swelling of cellulosic membranes often is not fully reversible, however, and a drying-out is usually to be avoided.

A membrane must be free from pinholes or other defects. Solvent permeation should be fast, yet retention of the solute is to be complete. A thin membrane and a uniform pore size help to reconcile these requirements. Practical solute retention extends down to a few thousand daltons, but lower retention limits have been reported. Vink (45), by acetylating cellophane membranes, was able to determine molecular weights below 1000 daltons. Very selective membranes may not be desirable for use with commercial polymers because solvent transport is slow, and impurities of low molecular weight cannot pass freely.

In the determination of very high (number average) molecular weights, the osmotic pressures will be small, and membranes having wider pores, which equilibrate faster, are preferable. Here, pressure asymmetry (see section above entitled "Experimental Precautions") becomes a special concern. Finally, membranes should be stable for an extended period of time. This is not a great problem at room temperature, where membranes can be usable for many months. Above 100°C, however, the cellulosic membranes tend to become brittle and impermable to the solvent, presumably because of oxidation. The use of antioxidants in the solvent is recommended. In a study of polyethylene, Schmieder (46) diluted the solvent *p*-xylene with 10–20% dimethylformamide, a swelling agent, which prolonged the life of the cellulose membranes to several months at 120°C.

Most widely used for aqueous solutions are collodion and cellulose acetate membranes. In the United States, the latter have been supplied by Carl Schleicher & Schuell (Keene, NH) under the designation B 20 for the most retentive member of the series (new designation, AC 61). A German supplier is Sartorius Membranfilter GmbH (Goettingen) as "Ultrafein allerfeinst" [superdense] membranes. Permeation limits can be expected to range from 5,000 to 20,000 daltons. Less retentive membranes are also available.

The same manufacturers supply regenerated cellulose membranes for use with nonaqueous solutions. Designations are RC 51 (formerly O 8) and "Ultracella allerfeinst," with retention limits similar to those stated above. American gel cellophane membranes of various porosities also have been widely used. The ultrafein superdense membranes, intended for work with aqueous solutions, were found to function well in benzene, with a reported retention limit of 2000 (47).

Membranes are usually supplied in a water–alcohol medium. They must be conditioned for water-immiscible solvents by soaking in an intermediate solvent, such as acetone, for a day before transfer to toluene, for instance. In-between soakings in water–acetone and acetone–toluene mixtures may be advisable to avoid "shocking" the membrane. The membrane is installed in the osmometer after equilibration in the final solvent.

For work at elevated temperatures with solvents such as xylene, trichlorobenzene, or decahydronaphthalene, solvent conditioning is followed by gradual heating, perhaps over several days, to the operating temperature. Cellulosic membranes tend to shrink in this process and sometimes deform. It may be advantageous to clamp the membranes in the osmometer at room temperature, and gradually heat the osmometer cell.

The proper treatment will depend on the membrane, and the reader should consult the manufacturers of electronic osmometers, who also supply membranes, for advice on conditioning and heating procedures. Much empirical information on membrane treatment and behavior has appeared in the literature. Vaughan (47) compared the behavior of cellulose acetate membranes. Cellulose acetate and collodion membranes were earlier studied by Philipp and Bjork (48). An extensive study of membrane behavior in aqueous and nonaqueous solvents has been made by Patat (49). Further accounts of membrane behavior can be found in Reference 3 and in Reference 14, which summarizes observations on membrane behavior before the mid-1960s. Of particular interest is the 1964 contribution by Chiang (13), who tabulated selectivity and permeability data for a number of membranes, with attention to high temperature osmometry.

Porous Vycor glass with 4-nm pore size already has been mentioned (19, 20) as a membrane suitable for osmometry up to 300°C. This membrane is too thick to be used in osmometers of conventional design. Immergut et al. (50) described the preparation of poly(trifluorochloroethylene) membranes, their outstanding features being inertness to solvent and stability at elevated temperatures (see also Ref. 3). Other materials used as membrane materials are poly(vinyl alcohol) (51), polyurethane (52), and poly(vinyl butyral) (50).

Anisotropic membranes, consisting of a very thin, highly selective layer of a material such as poly(vinyl acetate) on a thicker macroporous support should combine a low permeation limit with rapid solvent transport. Membranes of this kind are being supplied by Amicon Corporation (Danvers, MA). The use of these membranes has been mentioned by Burge (53), but a critical assessment of their performance in osmometry has not come to our attention.

SOLVENTS

In 1968 Armstrong (14) tested a number of commercial cellulosic membranes in a variety of solvents in self-balancing electronic osmometers. Partial results are reprinted in Table 2. We expect Armstrong's findings to be applicable to the modern electronic osmometers. They may also serve as a guide for use in manual osmometers. As stated in the preceding section, membranes O 8 and B 20 should correspond to RC 51 and AC 61 (products of Carl Schleicher &

Table 2. Solvent Performance[a] of Commercially Available Membranes in High-Speed Osmometers

Solvent	Temperature (°C)	Membrane Type					
		07	08[b]	B-19	B-20[c]	600-W[d]	450-W[d]
H_2O and buffered systems	5–25	1	1	3	3	1	1
Toluene	30–40	3	3	1	1	3	3
Benzene	20–25	2	3	1	1	3	3
Dimethylformamide	55–80	1	2	1	1	3	3
Dimethyl sulfoxide		None of these membranes will perform in this solvent					
Tetrahydrofuran	20–25	2	3	1	1	3	3
Cyclohexane	25–40	2	2	1	1	1	2
Dimethylacetamide	55–65	2	2	1	1	3	3
Acetone	20–25	2	3	1	1	3	3
Methyl ethyl ketone	20–25	2	3	1	1	3	3
Fluoroalcohols	25–40	2	3	1	1	3	3
Alcohols through C	20–35	2	2	2	2	2	2
Decalin	110–130	1	2	1	1	2	2
o-Dichlorobenzene	100–130	2	2	1	1	2	2
o-Chlorophenol	90–120	1	1	1	1	2	2
m-Cresol	90–110	1	1	1	1	2	2

[a] 1, Not recommended; 2, satisfactory; 3, excellent.
[b] New designation RC 51.
[c] New designation AC 61.
[d] American gel cellophane.
From Ref. 14.

Schuell). The American gel cellophane membranes 600-W and 450-W are supplied swollen in an aqueous solution, 600-W being the more retentive one. As can be seen from Table 2, cellulosic membranes are compatible with a variety of solvents, but none is rated "excellent" above 100°C. Additional information on membrane performance in various solvents can be found in the tables given in Reference 13.

Highly volatile solvents are not favored in osmometry, mostly because evaporative losses may cause concentration errors.

Mixed solvents have not been widely used, although they may offer the advantage of producing macromolecular solutions with small virial coefficients. Small initial differences of solvent composition on opposite sides of the membrane may cause high transient pressures, since membranes are often solvent selective. In addition to the osmotic equilibrium of the macromolecular solution, a compositional equilibrium of the solvent components must be attained. Palit et al. (54) reported on the use of acetone–methylcyclohexane mixtures in molecular weight determinations of polystyrene. The addition of dimethylformamide to *p*-xylene to prolong membrane life at high temperatures has already been mentioned (46).

NOTES ON APPLICATIONS

A large variety of natural and synthetic polymers had been successfully studied by osmometry before size exclusion chromatography became the foremost method for polymer characterization. Molecular weight determinations by osmometry require only that the concentration of the nonpermeating molecular species be known. Impurities such as dust or small amounts of gel in the solution affect the result only insofar as they introduce an error in the mass concentration of the sample. Molecular interactions, as reflected in the second virial coefficient, can be measured with great precision over a wide temperature range.

Osmometry, however, cannot deal well with highly polydisperse samples with an extensive low end in their molecular weight distribution. Large errors can arise if this distribution extends beyond the permeation limit of the membrane, since the number average molecular weight is heavily weighted by the low end.

Among the numerous studies on synthetic polymers, the osmometric characterization of polyethylene by Wagner and Verdier (55) may serve as a model for its care and thoroughness. Despite the low molecular weight (11,400) of one of the samples and the difficulties that accompany osmometry above 100°C, the results appear to be highly accurate.

Osmometry may be of particular usefulness in the study of electrically charged biopolymers. Apart from accurate determinations of the number average molecular weight in multicomponent systems, osmometry can provide reliable answers on intermolecular interactions (from the magnitude of the second virial coefficient), in particular on homo- and hetero association of proteins in solution [see, e.g., the review by Kelly and Kupke (6)]. Another interesting application of osmometry is the elucidation of the shape factor of the rodlike myosin molecule, described in a work cited earlier (23).

Osmometry is not well suited for studying the degradation of long chain molecules by random scission, since the low molecular weight fragments formed by this process are unlikely to be retained by the membrane.

It is however possible to determine the micellar weight of surfactants (56, 57), although the monomers, in equilibrium with the micelles, can diffuse through the membrane, as discussed in the next section.

MEMBRANE-PERMEATING SOLUTES

Thermodynamic equilibrium no longer exists when solute molecules permeate through the membrane, that is, when the membrane ceases to be semipermeable. Solute permeation can be readily recognized with a rapidly responding osmometer by transient behavior of the pressure. Typically, after an initial balancing period, the osmotic pressure declines with time. Figure 6 depicts the experimental pressure P as a function of time in an approach to equilibrium "from below." Curve A refers to a nonpermeating solute. Curves B to D illustrate permeation, the experimental pressure going through a maximum.

Extrapolation of the declining pressure to zero time gives a value $P(0)$ that marks the condition when all the solute molecules are still contained in the sample cell. Even under this condition, $P(0)$ is lower than π, the theoretical osmotic pressure which would be found if the membrane were semipermable. Substituting $P(0)$ in equation (1) results in too high a molecular weight. The inequality $P(0) < \pi$ is apparent if one considers that the experimental pressure just stops the net flow of material through the membrane. If the solvent flow into the cell is opposed by a solute flow *out* of the cell, this pressure must be less than for the flow of solvent alone. This argument also holds if the solute concentration in the sample cell has not yet been depleted by diffusion.

A quantitative formulation of the relationship between experimental and theoretical osmotic pressure, in terms of fluxes and forces acting on the solution components, has been given by Staverman (1, 2) with the help of nonequilibrium thermodynamics. The result is formally expressed as $P(0) = s\pi$, where s is called the reflection or Staverman coefficient, which is a measure for the selectivity of the membrane.

According to Vink (58), the time dependence of the measured osmotic

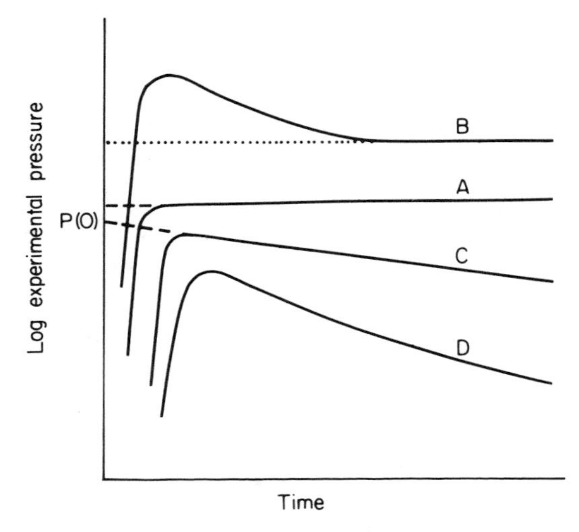

Figure 6. Pressure–time curves as obtained with a rapidly balancing osmometer. Curve A: nonpermeating solute; a constant value $P = \pi$ is attained. Curve B: the macromolecular solute is fully retained while an impurity of low molecular weight rapidly diffuses through the membrane. Curve C: the linear decrease of log P with time, after the initial balancing stage, indicates a homogeneous solute diffusing through the membrane. Curve D: a heterogeneous diffusing solute is indicated by a changing slope. With time a pressure plateau should be reached if a portion of the solute is retained by the membrane.

pressure for a homogeneous solute can be expressed as

$$P = s\pi \exp(-mt) + (p_0 - s\pi)\exp(-nt) \tag{20}$$

where m and n are the respective transport parameters for the solute and solvent, and p_0 is the hydrostatic pressure imposed at $t = 0$. Usually, $m \ll n$; thus after some initial period (of balancing of the osmometer), the second term in equation (20) becomes negligible and $\log P$ is a linear function of time. Equation (20) describes curve C in Figure 6. The exponential relationship does not hold unless the membrane volume is small compared to the cell volume (59, 60) and the concentration in the respective osmometer cells is uniform.

Neglecting the balancing term and assuming pressure additivity, we can write for a polydisperse solute with i components

$$P = \sum_i s_i \pi_i \exp(-m_i t) \tag{21}$$

The transport constants m_i depend on the membrane, the solvent, and the molecular weight, but possibly also on the structure of the solute. If calibration

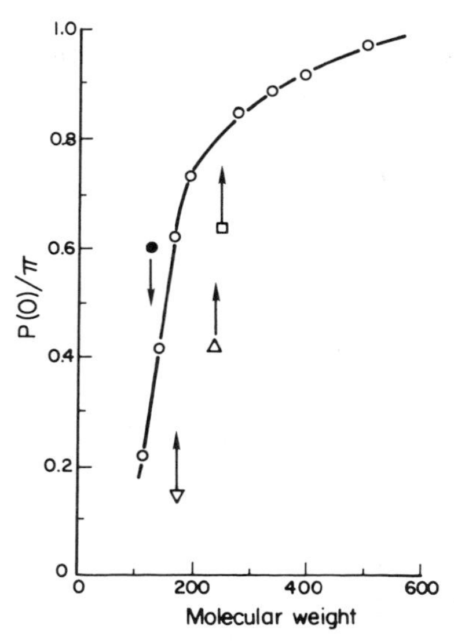

Figure 7. Staverman coefficients $s = P(0)/\pi$ obtained on a Shell osmometer, gel–cellophane 600 membrane with low-molecular weight solutes in toluene. ○, *n*-alkanes; ●, 2, 2, 5-trimethylhexane; ▽, anthracene; △, *n*-hexadecanol; □, 2-heptadecanone. (From Ref. 61.)

of the membrane with appropriate standards can establish the relationship between s_i and m_i, it is possible to deduce the molecular weight distributions of polymers from pressure–time curves. However, attempts to this end have been few (60) and unpromising compared with the much greater power of size exclusion chromatography. Also, permeability of membranes often changes with time, and the evaluation of sums of exponentials is generally precarious.

More modestly, a semiquantitative estimate of the concentration and size of a low molecular weight impurity in a polymer can sometimes be made by comparing the transient part of curves similar to B in Figure 6 (after subtraction of the constant plateau pressure) with s and m data for calibrating solutes in the appropriate molecular weight range (14, 61). A fast-responding osmometer is required for such measurements. However, the reflection coefficient of small molecules does not depend on molecular weight alone. Structure also plays a role, as illustrated in Figure 7 by a few examples from experiments with toluence solutions in a Shell self-balancing osmometer equipped with a cellulose acetate "superdense" membrane (61). Anthracene, *n*-hexadecanol, and 2-heptadecanone permeated more readily than the *n*-alkanes of corresponding molecular weight. The higher Staverman coefficient of 2, 2, 5-trimethylhexane suggests that the membrane offers greater resistance to branched hydrocarbons. Additivity of pressures according to equation (21) does not always seem to be assured, however. Thus, Elias and Männer (62), in their work with membrane-permeating poly(ethylene glycols) in manual osmometers, observed some anomalous effects.

Generally, an estimate of reflection coefficients, if only to assess the magnitude of error in a molecular weight determination with a leaking membrane, is highly desirable. Auxiliary measurements, such as diafiltration (63), can provide such information, but they are rarely practical. When working with a rapidly responding osmometer, it is therefore better to use calibrating substances in the appropriate molecular weight range to assess how the measured osmotic pressure differs from π when a decline of pressure with time is indicated. Such experiments also help to find the permeation limit of the membrane and enhance confidence in the molecular weight determinations.

The determination of the micellar weight of surfactants represents a special case: micelles with molecular weights above 10,000 can be retained by suitable osmometer membranes, but the monomer molecules, which are in a dynamic equilibrium with the micelles, are not retained. As the monomer molecules diffuse out through the membrane, their concentration is replenished by the continuous dissociation of micelles. This results in a significant decrease of osmotic pressure with time, which is characteristic for leaking membranes. Ionic surfactants which show this behavior are sodium dodecylsulfate (NaDDS) (56) and N-cetyltrimethylammonium bromide (CTAB) (57). To repress the dissociation of the micelles, it is then necessary to fill the solvent chamber likewise with surfactant solution having a concentration (c^*) slightly above the critical micelle concentration (cmc). Because the monomer concentration changes only very little above the cmc, the concentration gradient of monomer across the membrane always remains very small, even when the total surfactant concentrations on opposite sides of the membrane are very different. Consequently, diffusion through the membrane is effectively slowed down, and the micelle concentration decreases only very slowly with time. By extrapolating the pressure–time curves to $t = 0$, one obtains $P(0)$ for a given surfactant concentration. The equation for the experimental osmotic pressure is (56)

$$P(0) = sRTc/M + sRTA_2[c^2 + 2c(c^* - \text{cmc})] \tag{22}$$

where M is the micellar weight and c the mass concentration of the surfactant in the sample cell. By plotting P/c against c, an intercept of $sRT/M + 2sRTA_2(c^* - \text{cmc})$ is obtained. The second virial coefficient A_2 is given by the slope of the plot. Having determined the cmc by some appropriate method, one can then calculate M, provided the reflection coefficient s is known. An estimate of s can be made by repeating the measurements with a different surfactant concentration c^* on the solvent side of the membrane. Under the experimental conditions given in Reference 56 for NaDDS, s was found to be very close to unity. The micellar weight in 0.03 M sodium chloride solution at 21°C was 18,200, in good agreement with light-scattering data. Attwood and coworkers (57) estimated s with the help of an auxiliary diffusion and dialysis experiment. With s again near unity, the micellar weight of CTAB in 0.025 M KBr at 30°C was 105,000. The agreement with light-scattering results indicated that the micelles of these ionic surfactants have a narrow size distribution.

Some nonionic micelles (64) are seemingly stable enough to approach the behavior of nondissociating polymers, and their micellar weight can be determined in a rapidly responding osmometer in the conventional way. This behavior may be explained by a very low critical micelle concentration.

REFERENCES

1. A. J. Staverman, *Rec. Trav. Chim. Pays-Bas*, **70**, 344 (1951).
2. A. J. Staverman, *Rec. Trav. Chim. Pays-Bas*, **71**, 623 (1952).
3. R. U. Bonnar, M. Dimbat, and F. H. Stross, *Number-Average Molecular Weight*, Wiley-Interscience, New York, 1958.
4. M. P. Tombs and A. R. Peacocke, *The Osmotic Pressure of Biological Macromolecules*, Oxford University Press (Clarendon), London, New York, 1974.
5. D. W. Kupke, *Adv. Protein Chem.*, **15**, 57 (1960).
6. M. J. Kelly and D. W. Kupke, in *Physical Principles and Techniques of Protein Chemistry*, Part C, S. J. Leach, Ed., Academic Press, New York, 1973, p. 77.
7. W. R. Krigbaum and R. J. Roe, in *Treatise on Analytical Chemistry*, Part 1, Vol. 7, I. M. Kolthoff and P. J. Elving, Eds., Wiley-Interscience, New York, 1967, p. 4461.
8. W. Helbig, *Physikalisch-chemische Messverfahren*, Vol. 6, Akademische Verlagsgesellschaft, Leipzig, East Germany, 1973, Chap. 3.
9. R. D. Ulrich, in *Polymer Molecular Weights*, Dekker, New York, 1975, Chap. 2.
10. H. Vink, "Cellulose and Cellulose Derivatives," in *High Polymers*, Vol. 5, Part 4, N. M. Bikales and L. Segal, Eds., Wiley-Interscience, New York, 1971..
11. P. D. Jeffrey, in *Protein–Protein Interactions*, C. Frieden and L. W. Nichol, Eds., Wiley-Interscience, New York, 1981, Chap. 5.
12. C. Tanford, *Physical Chemistry of Macromolecules*, Wiley, New York, 1961.
13. R. Chiang, "Newer Methods of Polymer Characterization," in *Polymer Reviews*, Vol. 6, Bacon Ke, Ed., Wiley-Interscience, New York, 1964, Chap. 12.
14. J. L. Armstrong, "Characterization of Macromolecular Structure," NAS Conference Proceedings, April 5–7, 1967, Publication No. 1573, National Academy of Sciences, Washington DC, 1968.
15. H. Coll, *J. Polym. Sci., D*, **5**, 541 (1971).

16. D. B. Bruss and F. H. Stross, *J. Polymer Sci.*, **55**, 381 (1961).
17. H. J. Philipp, *J. Polym. Sci.*, **6**, 371 (1951).
18. R. M. Fuoss and D. J. Mead, *J. Phys. Chem.*, **47**, 59 (1943).
19. H.-G. Elias, *Chem.-Ing.-Tech.*, **33**, 359 (1961).
20. H.-G. Elias and T. Ritscher, *J. Polym. Sci.*, **28**, 648 (1958).
21. D. B. Bruss and F. H. Stross, *J. Polym. Sci., A*, **1**, 2439 (1963).
22. B. H. Zimm, *J. Chem. Phys.*, **14**, 164 (1946).
23. H. Portzehl, *Z. Naturforsch.*, **5b**, 75 (1950).
24. P. J. Flory and W. R. Krigbaum, *J. Chem. Phys.*, **18**, 1086 (1950).
25. Hiromi Yamakawa, *Modern Theory of Polymer Solutions*, Harper & Row, New York, 1971.
26. W. R. Krigbaum and P. J. Flory, *J. Am. Chem. Soc.*, **75**, 1775, 5254 (1953).
27. W. R. Krigbaum, *J. Am. Chem. Soc.*, **76**, 3758 (1954).
28. W. H. Stockmayer and E. F. Cassassa, *J. Chem. Phys.*, **20**, 1560 (1952).
29. G. Scatchard, *J. Am. Chem. Soc.*, **68**, 2315 (1946).
30. G. Scatchard, A. C. Batchelder, and A. Brown, *J. Am. Chem. Soc.*, **68**, 2320 (1946).
31. G. Scatchard, Y. V. Wu, and A. L. Shen, *J. Am. Chem. Soc.*, **81**, 6104 (1959).
32. G. Scatchard, A. C. Batchelder, A. Brown, and M. Zosa, *J. Am. Chem. Soc.*, **68**, 2610 (1946).
33. R. F. Steiner, *Arch. Biochem. Biophys.*, **49**, 400 (1954).
34. E. T. Adams, Jr., *Biochemistry*, **4**, 1655 (1965).
35. M. L. Wallach, *Polym. Prepr.*, **6**, 53 (1965).
36. N. Aelenei, *Eur. Polym. J.*, **17**, 533 (1981).
37. D. B. Bruss and F. H. Stross, *Anal. Chem.*, **32**, 1456 (1960).
38. B. H. Zimm and J. Myerson, *J. Am. Chem. Soc.*, **68**, 911 (1946).
39. H. Hellfritz, *Makromol. Chem.*, **7**, 184 (1951).
40. J. V. Stabin and E. H. Immergut, *J. Polym. Sci.*, **14**, 209 (1954).
41. G. S. Adair, in *A Laboratory Manual of Analytical Methods of Protein Chemistry*, Vol. 3, P. Alexander and R. J. Block, Eds., Pergamon Press, Elmsford, NY, 1961.
42. F. B. Rolfson and H. Coll, *Anal. Chem.*, **36**, 888 (1964).
43. O. Hepp, *Z. Ges. Exp. Med.*, **99**, 709 (1936).
44. R. H. Wagner, *Physical Methods of Organic Chemistry*, 2nd ed. Vol. 1, Part 1, A. Weissberger, Ed., Wiley-Interscience, New York, 1949, Chap. 11.
45. H. Vink, *J. Polym. Sci., A2*, **4**, 830 (1966).
46. W. Schmieder, *Kunststoffe*, **50**, 166 (1960).
47. M. F. Vaughan, *J. Polym. Sci.*, **33**, 417 (1958).
48. H. J. Philipp and C. F. Bjork, *J. Polym. Sci.*, **6**, 383 (1951).
49. F. Patat, *Makromol. Chem.*, **34**, 120 (1959).
50. E. H. Immergut, S. Rollin, A. Salkind, and H. Mark, *J. Polym. Sci.*, **12**, 439 (1954).

51. H. T. Hookway and R. Townsend, *J. Chem. Soc.*, 3190 (1952).
52. K. Ueberreiter, H. J. Ortman, and G. Sorge, *Makromol. Chem.*, **8**, 21 (1952).
53. D. E. Burge, *Am. Lab.*, **9**(6), 41 (1977).
54. S. R. Palit, G. Colombo, and H. Mark, *J. Polym. Sci.*, **6**, 295 (1951).
55. H. L. Wagner and P. H. Verdier, *J. Res. Natl. Bur. Stand.*, **83**, 179 (1978).
56. H. Coll, *J. Phys. Chem.*, **74**, 520 (1970).
57. D. Attwood, P. H. Elworthy, and S. B. Kayne, *J. Phys. Chem.*, **74**, 3529 (1970).
58. H. Vink, *Ark. Kem.*, **15**, 149 (1960).
59. M. Hoffmann and M. Unbehend, *Makromol. Chem.*, **88**, 256 (1965).
60. M. Hoffmann and M. Unbehend, *J. Polym. Sci., C*, **16**, 977 (1967).
61. H. Coll, *Makromol. Chem.*, **109**, 38 (1967).
62. H.-G. Elias and E. Männer, *Makromol. Chem.*, **40**, 207 (1960).
63. J. L. Talen and A. J. Staverman, *Trans. Faraday Soc.*, **61**, 2794 (1965).
64. H. Coll, *J. Am. Oil Chem. Soc.*, **46**, 593 (1969).

CHAPTER

5

LASER LIGHT SCATTERING

BENJAMIN CHU

Departments of Chemistry and of Materials Science and Engineering, State University of New York at Stony Brook, Long Island, New York

INTRODUCTION

Light scattering from dilute polymer solutions (1) has always been among the more difficult experiments in polymer characterization studies. The differential scattering cross section or the Rayleigh ratio R_θ (cm^{-1}), defined as $R_\theta = i_\theta r^2/(I_0 V)$, has magnitudes as small as $10^{-5}\ cm^{-1}$ for many ordinary liquids and dilute polymer solutions, where the scattered intensity (per unit solid angle), $i_\theta(r^2)$, is located at a distance r from the scattering volume V, I_0 is the incident beam intensity, and θ the scattering angle. Consequently, light-scattering experiments require special considerations:

1. in solution clarification, to eliminate the presence of dust particles which could contribute substantially to R_θ, especially at small scattering angles; and

2. in optical design, to reduce interface reflection and refraction effects. Small amounts of stray light could be inconsequential in other spectroscopic measurements. However, with a Rayleigh ratio of the order of 10^{-4}–$10^{-5}\ cm^{-1}$, unwanted stray light of only a few parts per million could become important. Recent advances in photon correlation spectroscopy (2) have also changed the design of a light-scattering spectrometer, requiring us to pay attention to coherence considerations (3). The main aim of this chapter is to provide

1. the basic equations used in the determination of molecular weight by measurements of angular distribution of absolute scattered intensity and in the determination of molecular weight distribution by combining light-scattering intensity and line width measurements;

2. discussions on the optical designs, including conflicting requirements for optimizing the signal-to-noise ratios in intensity and line width measurements;

3. methods of data analysis, allow access to the maximum amount of retrievable information based on appropriate mathematics; and

4. some guidelines for the beginners who are not familiar with light-scattering experiments.

The use of light-scattering intensity measurements for polymer solution characterization has been in existence for more than 40 years (4). Earlier achievements by Zimm (5) and others have been incorporated into a comprehensive book (1) edited by M. B. Huglin, who has subsequently updated the advances in another extensive review article entitled "Determination of Molecular Weight by Light Scattering" (6). I shall make use of Huglin's review (6) and book (1) as the main references for light-scattering intensity experiments and my own book (2) and review (7) and the excellent theory book by Berne and Pecora (8) as source material for dynamic light scattering.

Many aspects of laser light scattering have been discussed in a series of NATO ASIs on photon correlation and scattering techniques (9–12). Parallel to the NATO ASIs, there have also been conferences on photon correlation techniques in fluid mechanics (13). Light-scattering spectroscopy has been made into a powerful routine method for studying the dynamics of polymer solutions (14, 15) and melts (16). Instrumentation in photon correlation spectroscopy has been well developed (17), the only recent modification being the availability of dual delay time increments for a single-clipped correlator (18) and of commercial digital correlators with logarithmically spaced or variable delay time increments to increase the bandwidth of the measured time correlation function without increasing the available number of correlator channels. The development of a structurator (19, 20) for dynamic measurements, where drifts or slow fluctuations limit the precision of time correlation function determinations, represents another interesting alternative which requires further examination.

THEORY

Intensity of Scattered Light

Rayleigh Scattering

For a single particle of polarizability α, with a vertically polarized (x-direction) incident beam,

$$i_\theta = \frac{I_0 k^4 \alpha^2 \sin^2 \theta_1}{r^2} \tag{1}$$

where $k = 2\pi/\lambda_0$, with λ_0 being the incident wavelength in vacuo; θ_1 is the angle between the dipolar (x) axis and the line joining the dipole to the observer at a large distance r ($\gg$ dipole size) from the dipole. In a dilute solution with N identical particles per cubic centimeter in a solvent of refractive index n_0,

$$n^2 - n_0^2 = 4\pi N \alpha_{\text{ex}} \tag{2}$$

where the excess polarizability $\alpha_{\text{ex}} = \alpha - \alpha_0$ with α_0 being the solvent polarizability. Then,

$$\alpha_{\text{ex}} = \frac{n_0}{2\pi}\left(\frac{n - n_0}{C}\right)\frac{M}{N_{\text{A}}} \tag{3}$$

where we have taken $n + n_0 \sim 2n_0$ (or $2n$) and $C(= MN/N_{\text{A}})$ is the con- centimeter, and $H[\equiv k^4 n_0^2(\partial n/\partial C)_{\text{T,P}}^2/(4\pi^2 N_{\text{A}})]$ is an optical constant weight (g/mol) and Avogadro's number (mol^{-1}), respectively.

In the absence of particle interactions (i.e., at infinite dilution) and of particle interference, the excess scattered intensity from N single particle scattering per cubic centimeter $I_{\text{ex}}(\equiv i_{\theta,\text{ex}}N)$, using vertically ($x$ axis) polarized incident light and a scattering (y–z) plane perpendicular to the x-axis (such that $\sin\theta_1 = 1$), we have from equations (1)–(3):

$$\Delta R_\theta = \frac{I_{\text{ex}} r^2}{I_0} = \frac{k^4 n_0^2}{4\pi^2}\left(\frac{\partial n}{\partial C}\right)_{\text{T,P}}^2 \frac{CM}{N_{\text{A}}} = HCM \tag{4}$$

where $\Delta R_\theta[\equiv R_\theta(\text{solution}) - R_\theta(\text{solvent})]$ is the excess Rayleigh ratio in reciprocal centimeters, due to excess scattering of N solute particles per cubic centimeter, and $H[\equiv k^4 n_0^2(\partial n/\partial C)_{\text{T,P}}^2/(4\pi^2 N_{\text{A}})]$ is an optical constant (mol·cm^2/g^2). Equation (4) is the simplest basic expression for molecular weight determination by light-scattering intensity measurements; that is, it is valid for N identical particle scattering in the absence of interparticle interaction and intraparticle interference. Additional complications due to particle interactions (nonideality) and particle interference (angular dependence of scattered light) can be taken into account in actual experiments.

For polydisperse particles, each with the same H,

$$\Delta R_\theta = \sum_i \Delta R_{\theta,i} = H \sum_i C_i M_i \tag{5}$$

where the subscript i denotes species i. By definition,

$$M_w = \sum_i C_i M_i / \sum C_i \tag{6}$$

we then have ($\lim \theta \to 0$, $\lim C \to 0$)

$$\Delta R_\theta = HC\bar{M}_w \tag{7}$$

Thus, for polydisperse systems, light-scattering intensity experiments yield a weight average molecular weight $\bar{M}_w$, which differs from most other colligative property measurements, such as the osmotic pressure yielding a number average molecular weight $\bar{M}_n [\equiv \sum_i n_i M_i / \sum_i n_i \equiv \sum_i C_i / \sum_i (C_i/M_i)$, with n_i being the number of moles of species i].

If the incident beam is unpolarized, equation (4) becomes

$$\Delta R_{\theta,\mathrm{u}} = \frac{I_{\mathrm{ex}} r^2}{I_0(1+\cos\theta)} = \frac{k^4 n_0^2}{8\pi^2}\left(\frac{\partial n}{\partial C}\right)_{T,P}^2 \frac{CM}{N_\mathrm{A}} \tag{8}$$

which is seldom used with modern-day polarized laser light sources. Equation (8) is included for completeness and is used to show how the $(1+\cos^2\theta)/2$ term comes in.

Allowance for Particle Interaction and Intraparticle Interference

In the presence of interparticle interactions and intraparticle interference, equation (4) becomes

$$\frac{HC}{\Delta R_\theta} \cong \frac{1}{MP(\theta)} + 2A_2C \tag{9}$$

where A_2 is the second virial coefficient and $P(\theta)$ is the particle-scattering factor (or the static structure factor).

$$\lim_{\theta \to 0}\left(\frac{HC}{\Delta R_\theta}\right) = \frac{1}{M} + 2A_2C \tag{10}$$

$$\lim_{C \to 0}\left(\frac{HC}{\Delta R_\theta}\right) = \frac{1}{MP(\theta)} \tag{11}$$

$$\lim_{\substack{C \to 0 \\ \theta \to 0}}\left(\frac{HC}{\Delta R_\theta}\right) = \frac{1}{M} \tag{4}$$

Equations (4), (10), and (11) form the basis of a Zimm plot [$HC/\Delta R_\theta$ vs. $\sin^2(\theta/2) + b_\mathrm{s}C$, where b_s is an arbitrary constant making magnitudes of $\sin^2(\theta/2)$ and $b_\mathrm{s}C$ comparable]. From the Zimm plot, we can obtain

M, A_2, and R_g^2. In a polydisperse polymer solution,

$$\lim_{C\to 0} \Delta R_\theta = H\sum_i M_i C_i P_i(\theta) \tag{12}$$

If we consider particles with radius of gyration R_g comparable to the magnitude of the momentum transfer vector K,

$$P(K) \cong 1 - K^2 R_g^2/3 \tag{13}$$

where $K = (4\pi n_0/\lambda_0)\sin(\theta/2)$. At $KR_g \lesssim 1$, equation (12) becomes

$$\lim_{C\to 0} \Delta R_\theta = H\sum_i M_i C_i \left(\frac{1 - K^2 R_{g,i}^2}{3}\right) \tag{14}$$

or

$$\lim_{C\to 0} \frac{HC}{\Delta R_\theta} = \frac{1}{\bar{M}_w}\left(\frac{1 + K^2 \langle R_g^2 \rangle_z}{3}\right) \tag{15}$$

where $\langle R_g^2\rangle_z[\equiv \sum M_i C_i R_{g,i}^2/\sum M_i C_i]$ is the z-average square of the radius of gyration. We shall refer to $\langle R_g^2\rangle_z$ as R_g^2 and $\langle R_g^2\rangle_z^{1/2}$ as R_g for polydisperse polymers.

Allowance for Molecular Anisotropy

For anisotropic polymers in dilute solution, the Rayleigh ratio using vertically polarized incident and vertically polarized scattered light at $KR_g \lesssim 1$ has the form

$$\frac{HC}{\Delta R(K)} = \left(\frac{1 + \langle R_g^2\rangle_{\text{app}} K^2}{3}\right) M_{\text{app}}^{-1} + 2A_2 C \tag{16}$$

where the subscript app denotes apparent quantities.

$$\bar{M}_w = \frac{M_{\text{app}}}{1 + 4\delta^2/5} \tag{17}$$

and

$$\langle R_g^2\rangle = \frac{(1 + 4\delta^2/5)\langle R_g^2\rangle_{\text{app}}}{1 - 4\delta/5 + 4\delta^2/7} \tag{18}$$

where the molecular anisotropy δ can be determined from depolarized light scattering with

$$\lim_{\substack{C\to 0\\K\to 0}} \frac{R_{HV}}{HC} = \frac{3}{5}\delta^2 \bar{M}_w \tag{19}$$

and R_{HV} being the depolarized Rayleigh ratio using vertically polarized incident light and horizontally polarized scattered light.

Spectrum of Scattered Light

We shall consider only the self-beating mode in which stray light has been reduced to a negligible fraction of the scattered intensity. In the absence of optical mixing of scattered light with a local oscillator (either a fraction of the incident light or stray light from surface reflections), the intensity–intensity time correlation function $G^{(2)}(\tau, K)[\equiv \langle I(K,\tau)I(K,0)\rangle]$ is related to the first-order normalized electric field correlation function $g^{(1)}(\tau, K)[\equiv \langle E^*(K,\tau)E(K,0)\rangle / \langle E^*(K,0)E(K,0)\rangle]$

$$G^{(2)}(\tau, K) = A(1 + b|g^{(1)}(K,\tau)|^2) \tag{20}$$

where A is the baseline and is related to $\langle I(K,0)I(K,0)\rangle$; b is a spatial coherence factor depending on experimental conditions. Relations between the measured $G^{(2)}(K,\tau)$, the unnormalized net intensity–intensity time correlation function $[G^{(2)}(K,\tau) - A]/A = b|g^{(1)}(K,\tau)|^2$, and the normalized net intensity–intensity time correlation function $|g^{(1)}(K,\tau)|^2$ are shown schematically in Figure 1*a*, *b*, and *c*, respectively. Although the quantity b is often treated as an adjustable parameter; for dilute polymer solutions at finite concentrations, the b parameter depends not only on optical geometry, but also on polymer concentration. More specifically, $I(K)$ is the scattered intensity of polymer *solution*. Thus, $\langle I(K,0)I(K,0)\rangle = (I_2 + I_3)^2$, where I_2 and I_3 are the scattered intensity of the solvent and of the polymer solute, respectively. The intensity–intensity time correlation function of equation (20) can be expressed

$$G^{(2)}(K,\tau) = (I_2 + I_3)^2 \left\{ 1 + b\left[\left(\frac{I_2}{I_2 + I_3}\right)|g_2^{(1)}(K,\tau)| + \left(\frac{I_3}{I_2 + I_3}\right)|g_3^{(1)}(K,\tau)|\right]^2\right\} \tag{21}$$

where the subscripts 2 and 3 denote solvent and polymer, respectively. In a polymer solution, the solvent molecules are much smaller than polymer molecules. Therefore, the translational motions of the solvent molecules are much faster than those of the polymer molecules. Following the argument by

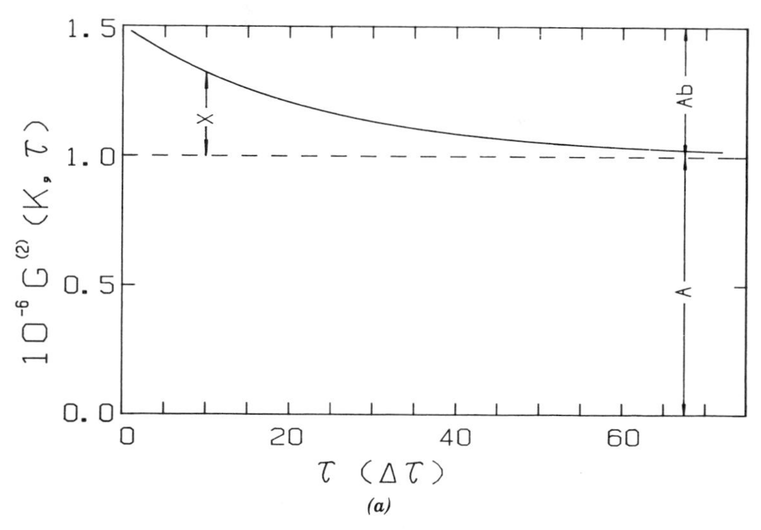

(a)

Figure 1. Schematic representations of time correlation functions $G^{(2)}(K, \tau) = A(1 + b|g^{(1)}(K, \tau)|^2$, where $\Delta\tau$ = delay time increment = 22 μs, N = total number of delay channels = 72. Delay time range varies from $\Delta\tau$ to $N\Delta\tau$. (*a*) Intensity–intensity time correlation function $G^{(2)}(K, \tau)$, denoted by solid curve:

$$x = Ab|g^{(1)}(K, 10\Delta\tau), \qquad A = 10^6, b = 0.5, \Gamma = 1000\,\mathrm{s}^{-1}$$

(*b*) Net intensity–intensity time correlation function $b|g^{(1)}(K, \tau)|^2$, denoted by solid curve:

$$b|g^{(1)}(K, \tau)|^2 = [G^{(2)}(K, \tau) - A]/A$$

and

$$\lim_{\tau \to 0} b|g^{(1)}(K, \tau)|^2 = b = 0.5;\ N = 72 \text{ channels}$$

By fitting the measured net intensity–intensity time correlation function, we can estimate the magnitude of b. (*c*) Normalized intensity–intensity time correlation function $|g^{(1)}(K, \tau)|^2$. *Note*: $\lim_{\tau \to 0}|g^{(1)}(k, \tau)|^2 = 1$; $N = 72$ channels. (*d*) Intensity–intensity time correlation function of a dilute polymer solution with solvent scattering I_2 comparable to polymer solution scattering I_3 (i.e., $I_2 \sim I_3$). To illustrate the scattering behavior, we took $I_2^2 = 10^5$ and $I_3^2 = 10^6$, yielding $(I_2 + I_3)^2 = 1.73 \times 10^6$. With $b = 0.5$, $\lim_{\tau \to 0} Ab|g^{(1)}(K, \tau)|^2 = 0.5 \times 1.73 \times 10^6 = 8.65 \times 10^5$. However, since contributions from solvent motions are very fast, the measured net intensity–intensity time correlation function can be represented by $bI_3^2|g_3^{(1)}(K, \tau)|^2$ [e.g., $Y = bI_3^2|g_3^{(1)}(K, 5\Delta\tau)|^2$, where the subscript 3 denotes polymer solute scattering]. (*e*) Net intensity–intensity time correlation function of a dilute polymer solution ($I_3 \gg I_2$) in the presence of large dust particles. To illustrate the scattering behavior, we took $I_3^2 = 10^6$ and $I_d^2 = 10^5$, yielding $(I_3 + I_d)^2 = 1.73 \times 10^6$, where the subscripts 2, 3, and d denote solvent, solute, and dust, respectively. We ignored other types of background, if any, and took $A = (I_3 + I_d)^2$. The solid curve is represented by $Ab|g^{(1)}(K, \tau)|^2$ with $\Gamma_d = 10\,\mathrm{s}^{-1}$ and $\Gamma_3 = 1000\,\mathrm{s}^{-1}$. Curve i, $I_3^2 b|g_3^{(1)}(K, \tau)|^2$; curve ii, $2I_3I_d b|g_3^{(1)}(K, \tau)|\,|g_d^{(1)}(K, \tau)|$; curve iii, $I_d^2 b|g_d^{(1)}(K, \tau)|^2$.

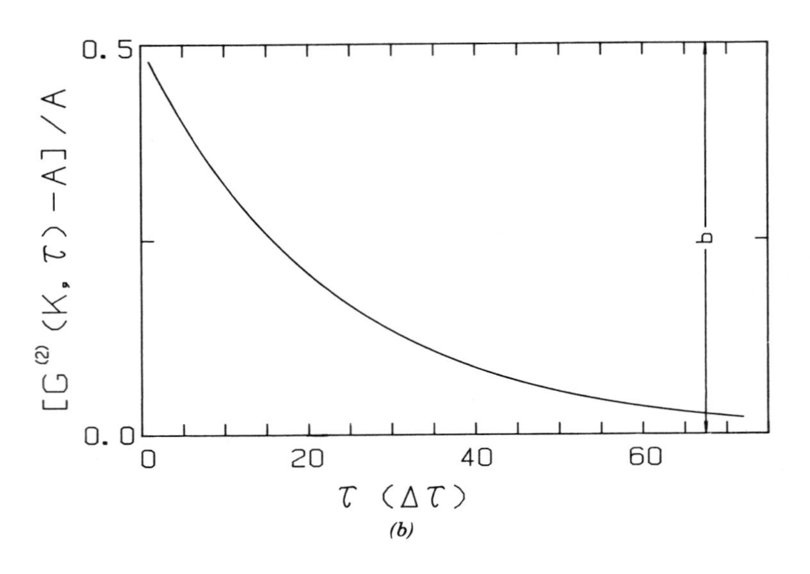

(b)

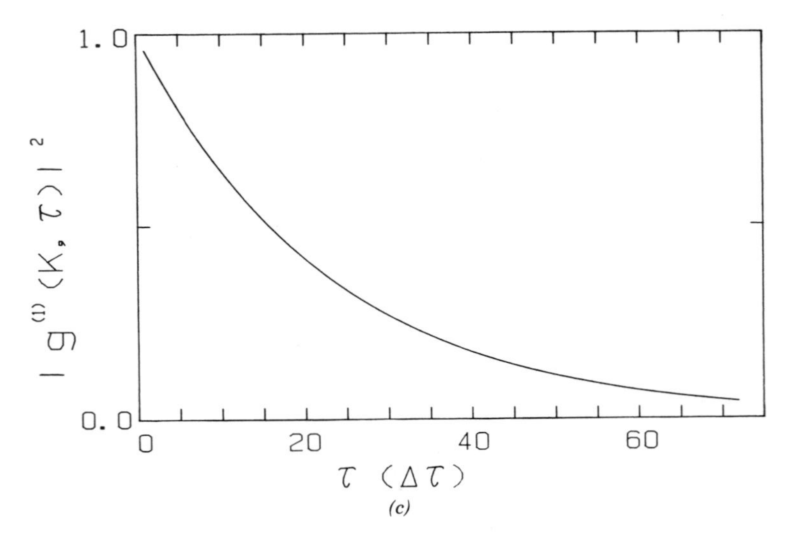

(c)

Figure 1. *(Continued)*

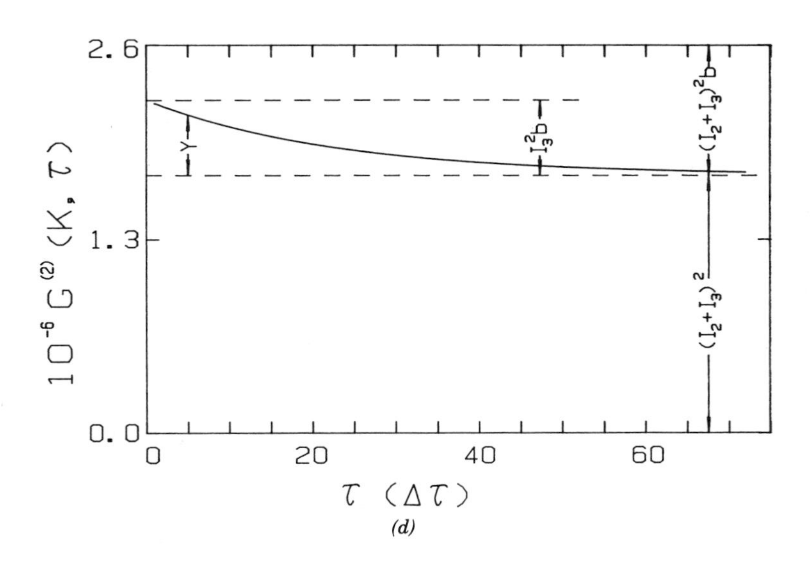

(d)

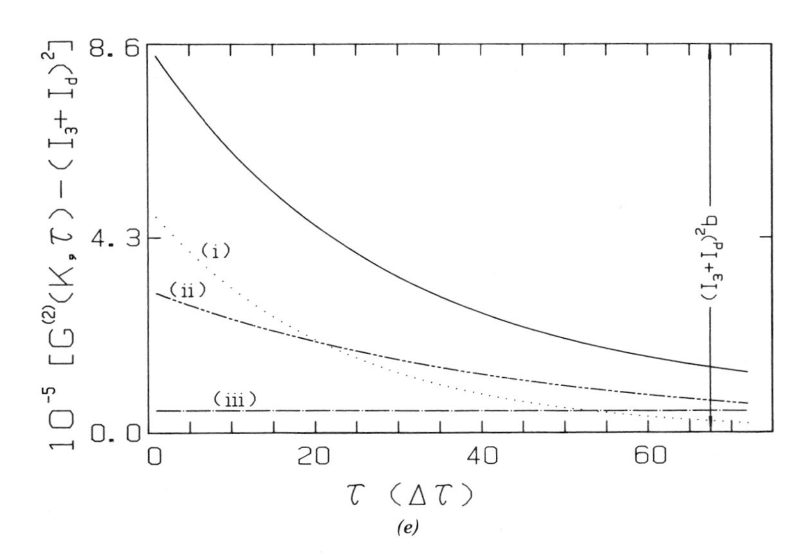

(e)

Figure 1. *(Continued)*

Tanaka (21), we have

$$G^{(2)}(K,\tau) \cong (I_2 + I_3)^2\left[1 + b\left(\frac{I_3}{I_2 + I_3}\right)^2 |g_3^{(1)}(K,\tau)|^2\right]$$
$$= (I_2 + I_3)^2[1 + b^*|g_3^{(1)}(K,\tau)|^2] \tag{22}$$

Equation (22) tells us that $g_2^{(1)}(K,\tau)$ for the solvent molecules becomes zero even with the first delay time increment ($\tau = \Delta\tau$) of the correlator setting, which is appropriate for measuring dynamical motions of polymers in solution—that is, $g_2^{(1)}(K,\tau \geqslant \Delta\tau) = 0$ as shown schematically in Figure 1*d*. It should be noted that with $g_2^{(1)}(K,\tau \geqslant \Delta\tau) = 0$, we have also dropped the cross-beating term between the solvent and the polymer. In fact, equation (22) tries to take advantage of the bandwidth limitation of present-day correlators. By realizing the fast decay contributions due to small solvent molecules and by appropriate setting of the correlator so that we can take $g_2^{(1)}(K,\tau) = 0$ over the delay time range of the measured time correlation function, we have an apparent change in the b parameter: $b^* = b[I_3/(I_2 + I_3)]^2$. Thus, in principle, if we calibrate the spectrometer using a polymer solution which scatters very strongly, such that I_3 is much greater than I_2, equation (22) becomes

$$G^{(2)}(K,\tau) \cong I_3^2(1 + b|g_3^{(1)}(K,\tau)|^2) \tag{23}$$

and the b parameter can be determined experimentally. Since the coherence factor b depends on optical geometry, it changes with scattering angle. For a dilute polymer solution, equation (22) not only flags an apparent drop in the beating efficiency ($b^* < b$) but also tells us the ratio of scattering intensity between the solvent and the solute (i.e., I_2/I_3), provided b is known precisely. Thus, if we know R_θ for the solvent, we can compute the excess Rayleigh ratio ΔR_θ for the polymer solute from dynamic light scattering. The approach above may be useful when $I_3 \sim I_2$. Nevertheless, it should be noted that when $I_3 \sim I_2$, precise dynamic light-scattering measurements are difficult to achieve.

In a dilute polymer solution, the presence of dust particles often renders both dynamic (line width) and static (intensity) light-scattering measurements ambiguous. Since dust particles are often much larger than polymer molecules, we may modify equation (21) and pay particular attention to its effect at small scattering angles when $I_d \sim I_3$:

$$G^{(2)}(K,\tau) = (I_2 + I_3(K) + I_d(K))^2\left[1 + b\left(\frac{I_2|g_2^{(1)}(K,\tau)|}{I_2 + I_3(K) + I_d(K)} + \frac{I_3(K)|g_3^{(1)}(K,\tau)|}{I_2 + I_3(K) + I_d(K)} + \frac{I_d(K)|g_d^{(1)}(K,\tau)|}{I_2 + I_3(K) + I_d(K)}\right)^2\right] \tag{24}$$

where the subscript d denotes dust particles. The terms $I_3(K)$ and $I_d(K)$ are used to emphasize the angular dependence of scattered intensity due to polymer solute molecules and dust particles. For convenience of discussion, if we take $I_d(K) \sim I_3(K) \gg I_2$, equation (24) is reduced to

$$G^{(2)}(K, \tau) \cong (I_3 + I_d)^2 \times \left\{ 1 + b\left[\frac{I_3}{I_3 + I_d} |g_3^{(1)}(K, \tau)| + \frac{I_d}{I_3 + I_d} |g_d^{(1)}(K, \tau)| \right]^2 \right\} \tag{25}$$

which has the same form as equation (21) except that the subscript 2 is now replaced by d. The physical meaning of $g_d^{(1)}(K, \tau)$ is, of course, quite different from that of $g_2^{(1)}(K, \tau)$ because dust particles are assumed to be large compared with polymer molecules. Again, we try to take advantage of the bandwidth limitation of the correlator ($\Delta\tau$ to $N\Delta\tau$, with $\Delta\tau$ and N being, respectively, the delay time increment and the total number of correlator channels). Figure 1*e* is a schematic representation of $G^{(2)}(K, \tau) - A$ in the presence of large dust particles with $I_d \sim I_3 \gg I_2$: we see that the dust particles, although large, are often still comparable in size to polymer molecules. Thus, $g_3^{(1)}(K, \tau)$ will be distorted by $g_d^{(1)}(K, \tau)$, particularly in the cross-beating contribution, dependent on the I_3/I_d ratio and the characteristic line widths associated with polymer molecules and dust particles. Finally at $N\Delta\tau$, $g_3^{(1)}(K, \tau = N\Delta\tau)$ could approach zero, but $g_d^{(1)}(K, \tau = N\Delta\tau)$ could still be finite, leading toward $G^{(2)}(K, \tau \sim N\Delta\tau) > A$. The last criterion may be used to estimate the problems related to the presence of dust particles. For precises determinations of $G^{(2)}(K, \tau)$, we usually do not accept $G^{(2)}(K, \tau \sim 4N\Delta\tau) - A \gtrsim 0.002A$.

INSTRUMENTATION

Light-scattering photometers, designed for static (intensity) measurements, have been in existence for more than 40 years. Many commercial instruments of earlier and current designs, including Sofica and Chromatix KMX-6, have been described by Huglin (1, 6). For dynamic (line width) measurements, an excellent introductory description of the apparatus has been presented by Ford (22) in a recent book on dynamic light scattering edited by R. Pecora. In this section we call to the attention of the uninitiated reader the importance of optical designs in trying to achieve some compromise which will permit both intensity and line width measurements over a range of scattering angles using essentially the same apparatus. It is assumed that the reader will try to first study the chapter by Ford (22) and that an instrument of a design comparable to those produced by Brookhaven (or Malvern) Instruments is available.

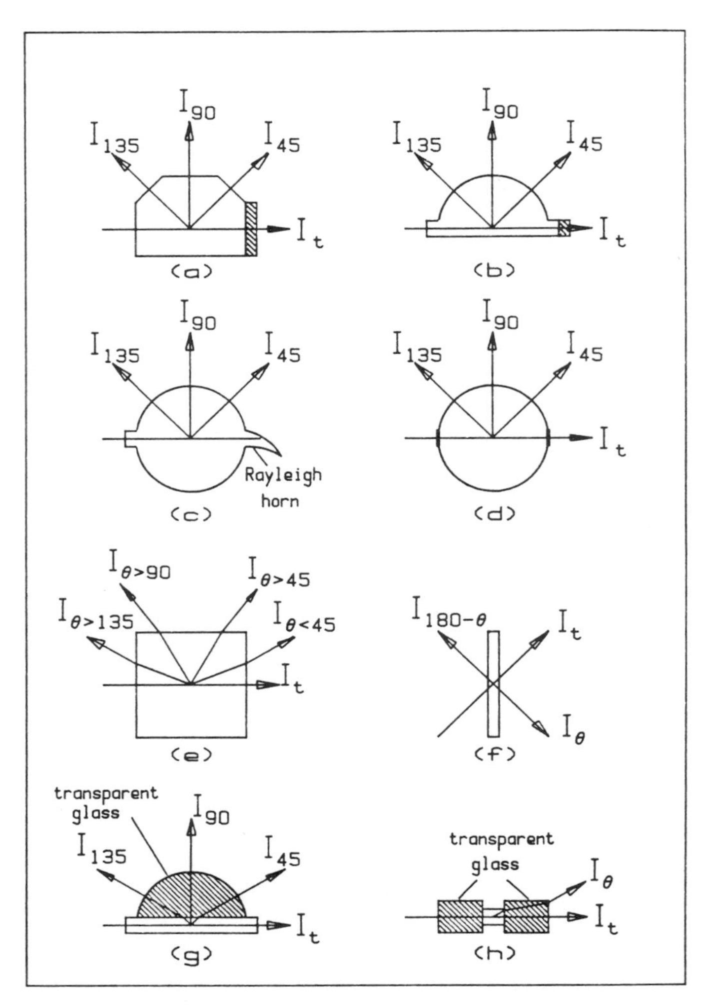

Figure 2. Typical light-scattering cells. Types *a–c* are often used without refractive-index-matching fluid; *d* is the most common configuration; with refractive-index-matching fluid, a cylindrical cell having an internal diameters of 8 mm can typically attain an angular range of $\sim 15° \leqslant \theta \leqslant 140°$. Types *e–h* require refraction corrections for scattering angle θ. (*a*) Semioctagonal cell with neutral density glass (shaded area) at exit of incident beam to reduce back-reflection at air–glass interface. (*b*) Semicylindrical cell with flat entrance window for the incident beam and neutral density glass (shaded area) at the exit window to reduce back reflection. (*c*) Cylindrical cell with flat entrance window and Rayleigh horn as a light trap for the incident beam. (*d*) Cylindrical light-scattering cell with flat entrance and exit windows. (*e*) Square light-scattering cell. (*f*) Flat light-scattering cell. (*g*) and (*h*) Light-scattering cells with small solution volumes suitable for flow measurements.

Optical Cell Designs

A general-purpose light-scattering spectrometer should be capable of making both static and dynamic measurements as a function of scattering angle θ. Typical light-scattering cells could have optical designs shown schematically in Figure 2. The designs in Figure 2*a*, *b*; and *c* emphasize a flat entrance window for the incident beam, the use of a Rayleigh horn or a piece of absorbing glass to eliminate or attenuate the back-reflection from the air–glass interface, and flat or cylindrical surfaces for angular distribution of scattering intensity. No refraction correction for the scattering angle θ is required. These cells are useful for light-scattering measurements when the cell is not immersed in a refractive-index-matching fluid. In Figure 2*d*, *e*, and *f*, the flat entrance and exit surfaces for the incident beam are retained; the cells, however, are usually immersed in a refractive-index-matching fluid. Figure 2*d* shows a modified cylindrical cell, which becomes advantageous only when the cell diameter is relatively small (< 10 mm). We have found cylindrical light-scattering cells with an inside diameter of 8 mm or more, without flat entrance and exit windows, to be quite satisfactory for most purposes. The angular range for a benzene solvent standard could agree to within 1% from $\sim 12°$ to 150° scattering angle with only a $\sin\theta$ volume correction. Typically, instead of precision diameter cylindrical cells (8.00 ± 0.01 mm i.d. and concentric to the same precision with wall thickness of 1 mm), we have used cylindrical vials of ~ 17 mm o.d. at $\sim 1/40$ the cost of a precision cell and have achieved an angular range of ~ 20–140° to $\pm 1\%$. Thus, for routine work, cylindrical light-scattering cells of slightly larger diameter ($\lesssim 15$ mm) together with the use of refractive-index-matching fluid are much cheaper and easier to use. For typical spectrophotometric (Beckman) cells having path lengths of 10 mm or less, see Figure 2*e* and *f*; these cells are useful for low angle light scattering or for solutions which scatter very strongly. Reduction of light path length (to as low as ~ 0.5 mm) by means of a thin cell (as shown in Fig. 2*f*) can be achieved readily. Noncylindrical cells usually have more limited angular ranges, including inaccessible scattering angles. For schematic configurations of two commercial flow light-scattering cells in the Dawn (Wyatt Technology) and KMX-6 (Chromatix) instruments, which are essentially for light-scattering intensity measurements, see Figure 2*g* and *h*. Finally, a flow prism light-scattering cell has been reported to provide both light-scattering intensity and line width measurements over a range of scattering angles, with simultaneous refractive index measurements to $\sim 10^{-7}$(23).

Detector Optical Geometry

Figure 3 shows a schematic diagram of a light-scattering spectrometer. In addition to cell design, as just discussed the essential design components in the

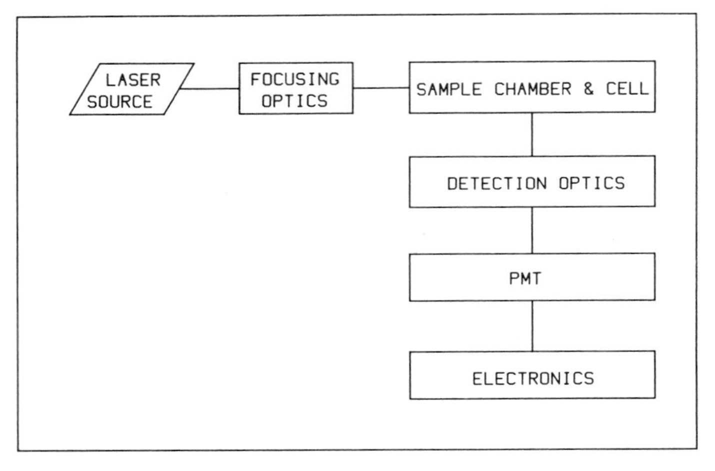

Figure 3. Block diagram of a light-scattering spectrometer (PMT = photomultiplier tube); see Figures 5 and 2 for detection optics and sample chamber, respectively. The intensity (static properties) is achieved by means of photon counting and the line width (dynamic properties) by means of photon correlation.

optical area are the focusing optics and the detection optics. A detailed account of an excellent spectrometer design has been reported by Cannell and his coworkers (24), and more elementary discussions on optics related to dynamic light scattering have been described elsewhere (2, 11, 22). In this section, we call attention to an often ignored consideration on the conflicts in requirements for optimizing light-scattering intensity and line width measurements.

Different designs of the detection optics are summarized schematically in Figure 4. To optimize the signal-to-noise ratio in light-scattering experiments, we want to have a decent scattering volume, not only to increase the total scattered intensity but also to avoid number fluctuations (2, 8). If the incident radiation is propagating in the z direction with polarization in the x direction, the coherence angle $(\Delta\theta)_{\text{coh}}$ in the $(y\text{–}z)$ scattering plane (2, 25) has the form

$$(\Delta\theta)_{\text{coh}} = \frac{0.5\lambda}{L_z \sin\theta + L_y \cos\theta} \tag{26}$$

In equation (26), we note that $(\Delta\theta)_{\text{coh}}$ is related to the scattering volume of dimensions L_x, L_y, and L_z; L_y can be made small without losing intensity by focusing the incident beam. Thus, in dynamic light scattering the incident laser beam is always focused, to increase the power density. One should use the

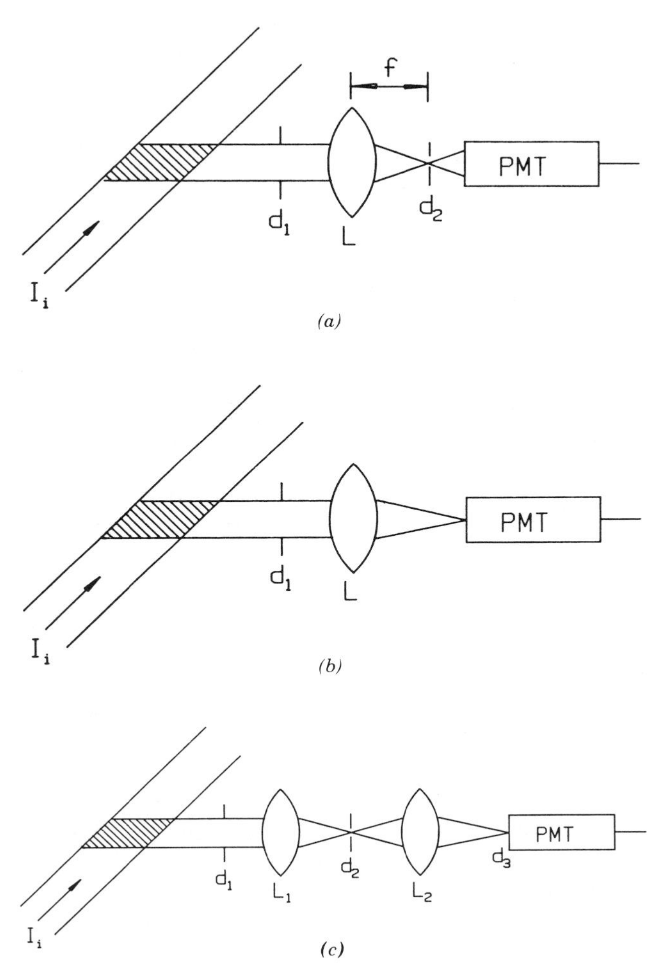

Figure 4. Five designs of detection optics: shaded area-scattering volume V. (*a*) The PMT must have a photocathode large enough to accept light with $\delta\theta = d_2/f$. In addition, I_i = incident beam, d_1 is a field stop, to control scattering volume V, L is a lens with focal length f, d_2 is an aperture stop, located at f from L, and $\delta\theta = d_2/f$. (*b*) Same as *a* except that the PMT has a small photocathode which can act as an aperture stop. This system is simpler than *a* but less flexible, since d_2 is fixed. (*c*) Same as *a* except that a second lens L_2 is used to transfer the image from d_2 to d_3. The distances between d_2L_2 and L_2d_3 are usually $2f$. The second lens is often needed for coolable PMTs, which have vacuum windows to isolate the PMTs from warm air. In such cases the PMTs can no longer be moved very close to d_2. (*d*) Alternative to *a*, where d_1 is an aperture stop with $\delta\theta \sim d_1/2f$ if $VL = Ld_2 \sim 2f$ and d_2 is a field stop, to control scattering volume V. (*e*) Similar to *d* except that d_1 has been moved to the front of the PMT; $\delta\theta$ is now controlled by d_2, d_1, and l. The diffraction from d_1 has been eliminated. This is the most versatile configuration, but it requires a relatively long working distance.

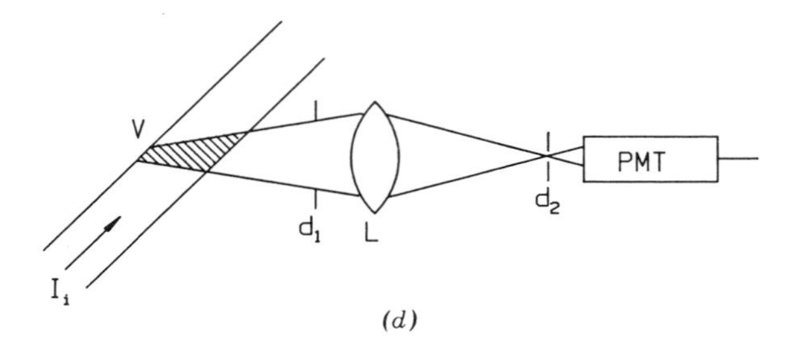

(d)

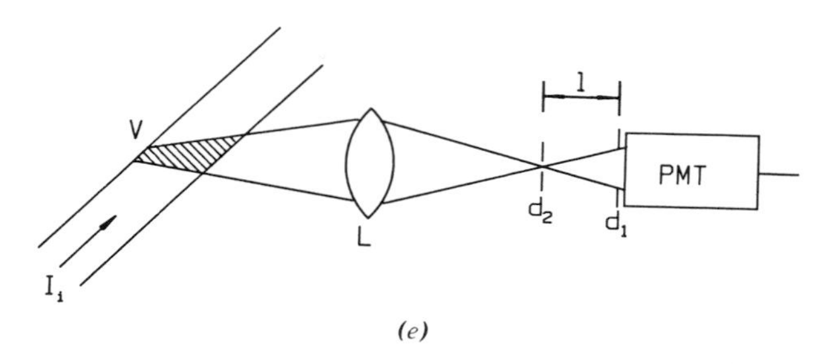

(e)

shortest focal length lens which is compatible with $\delta\theta$ uncertainties to achieve the highest power density. For example, with a laser beam diameter of ~ 1 mm and a focusing lens of $f = 20$ cm, $\delta\theta$ of the incident beam will be of the order of 0.3°. If we let the scattering volume be located near the waist of the incident beam, light-scattering intensity measurements down to a few degrees and line width measurements down to $\sim 10°$ could be measured without appreciable errors. Detailed equations for computing the beam profile and pinhole size needed to define an appropriate incident beam have been summarized in Reference 2. Based on Figure 4*a*, $L_z = d_1/\sin\theta$, where d_1 is the field stop. Then, equation (26) becomes

$$(\Delta\theta)_{\text{coh}} = \frac{0.5(\lambda/d_1)}{1 + (L_y/d_1)\cos\theta} \tag{27}$$

The value of $(\Delta\theta)_{\text{coh}}$ changes with scattering angle and becomes very small at $\theta < 90°$. For example, if $d_1 \sim L_y \sim 100\,\mu$m, $\lambda = 350$ nm, $(\Delta\theta)_{\text{coh}} \sim 0.1°$ at $\theta = 90°$ and $\sim 0.05°$ at small scattering angles. Thus, for dynamic light scattering, $\Delta\theta$ is limited to $\sim 0.1°$ even at $\theta \sim 90°$ using the optical geometry above, whereas for static light-scattering experiments, $\Delta\theta$ could be a few degrees at $\theta \sim 90°$, especially if $I(K)$ has no maxima or minima in the angular range under observation. A $\delta\theta$ variation by a factor of ~ 50 is often not feasible

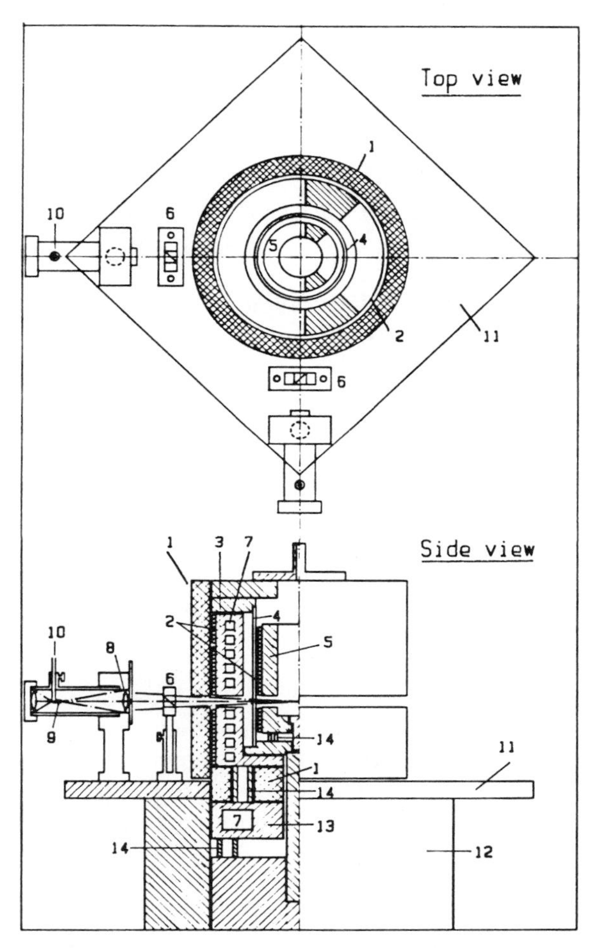

Figure 5. Schematic top and side views of a high temperature thermostat and detection system for a light-scattering spectrometer. 1, Silicone rubber insulation; 2, heating wires for the brass thermostats; 3, outer brass thermostat with fluid circulation facilities; 4, vacuum glass jacket for thermal isolation [the glass jacket is made of precision-polished glass of 2.25 in. o.d. with Kovar seals at both ends and machine centered to coincide with the center of rotation of the turntable (12) to 0.001 in.]; 5, inner brass thermostat, which has a separate temperature controller and thermometer and can accommodate light-scattering cells up to 27 mm o.d.; 6, Glan-Thompson polarizers; 7, fluid circulation paths; 8, lens; 9, field aperture; 10, optical fiber bundle; 11, rotating plate for multiple detectors; 12, RT-200 Klinger rotary table with 0.01° step size; 13, cooling plate to isolate the outer thermostat from the rotary table; 14, stainless steel standoffs for thermal isolation. (Reprinted with permission from *Macromolecules* (1987), reference 26. Copyright (1987) American Chemical Society.)

experimentally without physically changing the pinholes. We have on occasion used a $\delta\theta$ of a few tenths of a degree, which will provide sufficient scattering for intensity measurements even at large scattering angles ($\theta \sim 90°$) and a $\delta\theta$ sufficiently small that the coherence factor b is still reasonable at fairly small scattering angles ($\theta \sim 25°$). A working spectrometer, taking into account the consideration above, and capable of high temperature (up to $\sim 400°C$) light-scattering and refractive index measurements is shown schematically in Figure 5 (26).

Subsidiary Aspects: Solution Clarification, Refractive Index Increment, Calibration

Clarification of polymer solutions or colloidal suspensions represents one of the major obstacles to making light scattering a routine technique for molecular weight determinations. Different procedures have been devised to remove extraneous matter, such as dust particles, from the scattering medium. These include centrifugation followed by careful transfer of the clarified solution (or suspension) to the light-scattering cell and/or filtration of solution (or suspension) through ultrafine filters. Common sense and practice are indispensable in achieving success in this step. Although details have been described elsewhere[6], a few reminders are presented here for the uninitiated reader.

Filtration of solution (or suspension) directly into a dust-free light-scattering cell with exposure only to a dust-free (inert) atmosphere offers the best chance for success. Figure 6 is a schematic diagram of a high temperature polymer dissolution and solution clarification apparatus which has been used successfully for the light-scattering characterization of an alternating copolymer of ethylene and tetrafluoroethylene (PETFE) (26). The apparatus is constructed of glass, including the fine-grade sintered glass filter (f). Thus, the entire apparatus can be cleaned by pyrolysis. This capability is particularly important for polymers, such as PETFE, which have few known solvents, even at elevated temperatures ($> 230°C$). Solution preparation and clarification can be performed (1) under a filtered inert atmosphere without exposure to possible external contaminations, (2) at elevated temperatures, and (3) in an essentially closed system, to minimize solvent evaporation and consequent polymer concentration changes during solution preparation and clarification.

In Figure 6, chamber A is for polymer dissolution. Known amounts of polymer and filtered solvent as well as a small glass enclosed magnetic stirrer are placed in A at room temperature. The solution vessel A is connected to the precleaned filter chamber B and the dust-free light-scattering cell. After degassing and introduction of an inert gas, both stopcocks are closed to reduce solvent evaporation. The apparatus is heated in a temperature-controlled oven

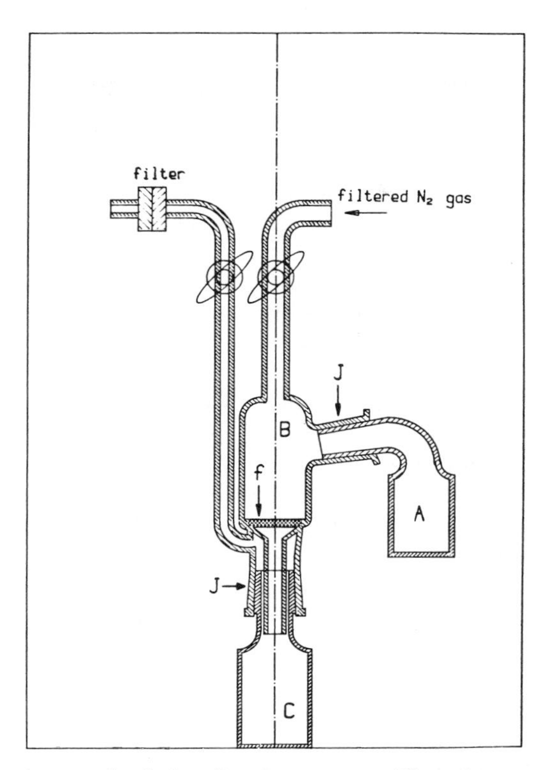

Figure 6. High temperature dissolution–filtration apparatus. The entire apparatus can be placed in a high temperature oven controlled at 250° ± 2°C. A, solution vessel where we can introduce known weights of the dried polymer and the solvent, as well as a small glass-enclosed magnetic stirrer; B, filter, connected with A and C by means of clear seal glass joints (J) (14/20, Wheaton Scientific); f, fine grade sintered glass filter; C, cylindrical light-scattering cell of 27 mm o.d. with a clear seal glass joint. (Reprinted with permission from *Macromolecules* (1987), reference 26. Copyright (1987) American Chemical Society.)

to the prescribed temperature, and the mixture in A is under gentle constant stirring until the polymer is dissolved. The polymer solution is transferred from A to B by turning the solution vessel A 180° using the clear seal glass joint J between A and B. With both stopcocks open, gentle excess pressure (~10 psi) is used to force the polymer solution directly into the precleaned dust-free light-scattering cell. Dust-free polymer solutions have been prepared in this manner with a high rate of success.

Figure 6 demonstrates one approach to a major practical difficulty in using light scattering as a routine technique, namely, variation in the details of solution clarification vary depending on the nature of the system under investigation. For example, aqueous solutions are often very difficult to clarify and require specific precautions (6, 27). Different types of filters with different pore size diameters are available (28). It is often necessary to find out whether the filter is blocking out particles of interest, and the procedure is not straightforward, depending on the nature of the system. Fortunately, photon correlation spectroscopy is able to (qualitatively) distinguish particle size distributions even without very precise time correlation function measurements and analysis. Thus, dynamic and static light-scattering measurements become built-in internal tests which permit us to set criteria for a proper solution clarification procedure. A filter having a 0.2-μm pore size diameter is suitable for most polymer solutions, except those with ultrahigh molecular weights ($\bar{M}_w \gtrsim 10^7$ g/mol).

For solutions of wormlike polymer chains or rigid rods, as well as large particle suspensions, clarification by filtration becomes more delicate. Now we need filter pore sizes a few times greater than the length of the longest polymer chain rather than its characteristic end-to-end distance. The appropriate filter pore size becomes quite large even for fairly small rigid rods. If we try to use pore diameters smaller than the polymer rod length, many polymer molecules will be blocked by the filter, reducing the initial polymer solution concentration in the clarified solution and changing the molecular weight distribution of the original polymer sample. In fact, for rodlike polymers in solution, centrifugation often is the preferred approach, provided polymer and solvent densities are comparable while the density of extraneous matter, such as dust particles, is higher. The centrifugal acceleration and time required to separate the extraneous matter (but not the dissolved polymer) from the solution should be tested out in separate experiments. Clarification by centrifugation often does not work well for solvents of high densities, such as CCl_4, where dust particles tend to float to the top. Under these circumstances, if we use a hypodermic needle and syringe to transfer the clarified polymer solution into a dust-free light-scattering cell, the syringe tip must travel through the dusty portion of the centrifuged polymer solution before it can reach the clarified portion, often resulting in unintended recontamination of the clarified solution.

The specific refractive index increment $(\partial n/\partial C)_{T,P}$ can be measured using a differential refractometer or a Rayleigh interferometer (6). Most $(\partial n/\partial C)_{T,P}$ values of typical polymer–solvent systems have been tabulated (16). On occasion, we need to determine $(\partial n/\partial C)_{T,P}$ values experimentally. Here, we shall introduce a modified differential refractometer consisting of any modern light-scattering spectrometer with a laser incident beam and a lens and Filer eyepiece combination as the detector. We modify the light-scattering cell, giving it an entrance window perpendicular to the incident laser beam and an exit window which interesects the incident beam at an angle θ, as shown-

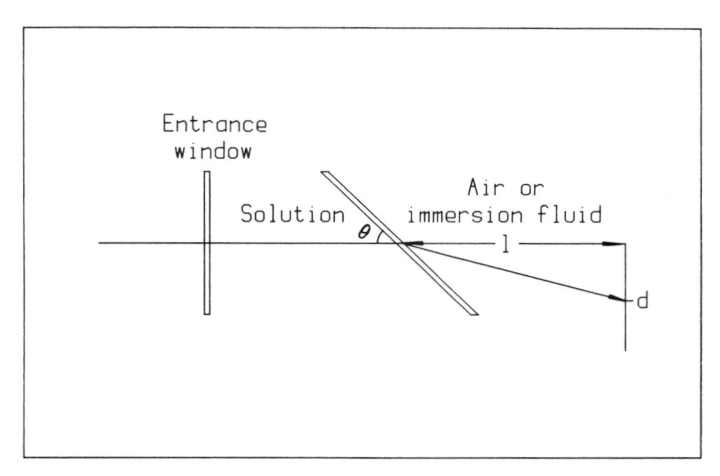

Figure 7. Refractive index increment by refraction measurements.

schematically in Figure 7. From the displacement d, measured at a distance l from the interface, we can compute the refractive index of the solution by means of Snell's law. The technique has been used successfully to determine $(\partial n/\partial C)_{T,P}$ of PETFE at 240°C.

Calibration standards of pure solvents, polymer solutions of known molecular weights, and/or colloidal suspensions have been used to obtain the absolute scattered intensity for a cylindrical light-scattering cell. We use

$$R_\theta^t = I_\theta^t \frac{R_\theta^{25}(\text{standard})}{I_\theta^{25}(\text{standard})} \exp(\gamma d^* C) \left[\frac{n_s^t}{n^{25}(\text{standard})} \right]^x \tag{28}$$

where R_θ^t, I_θ^t, R_θ^{25}(standard), I_θ^{25}(standard), n_s^t, n^{25}(standard), and x are, respectively, the Rayleigh ratio of the scattering medium at temperature t and scattering angle θ, the scattered intensity of the scattering medium at temperature t and scattering angle θ, the known Rayleigh ratio of reference standard at θ and 25°C, the scattered intensity of reference standard at θ and 25°C, the refractive index of the scattering medium at temperature t, the refractive index of the reference standard at 25°C, and a parameter which normally varies from 1 to 2 depending on detection optics geometry. The term $\exp(\gamma d^* C)$ corrects the effect of absorption, where C is the concentration in grams per cubic centimeter, d^* is the diameter of the light-scattering cell in centimeters, and γ is the extinction coefficient of the absorbing medium in square centimeters per gram. The value of γ should be determined in a separate experiment. For systems with strong forward scattering, one should also take into account the large forward acceptance cone of standard commercial spectrophotometers. However, these are special details. Suffice it to mention here that light scattering of absorbing solution or suspensions is feasible; but the scattered intensity should be corrected as if the light-scattering cell were infinitely thin. The $[n_s^t/n^{25}(\text{standard})]^x$ term is a refraction correction term.

In comparing the scattered intensity of a scattering medium with that of a reference standard, we need to use the same scattering volume. The parameter

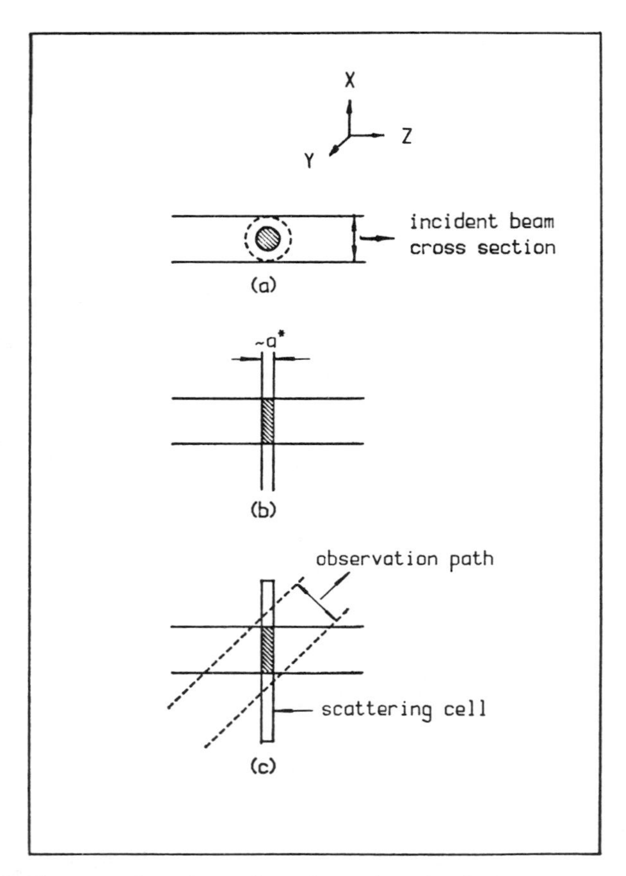

Figure 8. Refraction correction of scattering volume: n_s^t = refractive index of sample at temperature t. (*a*) Detection optics observes a scattering volume cross section smaller than the incident beam cross section: correction factor $= [n_s^t/n^{25}(\text{standard})]^2$. (*b*) Detection optics observes a scattering volume cross section with only one (horizontal) dimension controlled by the detection optics: correction factor $= n_s^t/n^{25}(\text{standard})$. (*c*) No refraction correction term for scattering volume if the detection optics observes a constant scattering volume.

x accounts for the detection optics variations. Suppose the detection optics of Figure 4 is set up in such a way that we view only a central portion of the incident beam as shown schematically in Figure 8*a*; then the x parameter has a value of 2. If we change the detection optics such that we always see the x–y directions of the incident beam, with only z having a refraction correction as shown in Figure 8*b*, then $x = 1$. Figure 8*b* can be achieved if we change d_2 in Figure 4*e* to a slit of width a^*. If we see the entire scattering volume, which is defined by the incident beam cross section in the x–y directions and by the scattering cell in the z direction, then $x = 0$. Refractive index correction for the scattering volume might be complex because the beam intensity profile is not uniform. It will become especially complicated if the detection optics define a cross section of the order of magnitude of the incident beam cross section. Then, depending on the refractive index of the scattering medium and that of the reference standard, x could vary from 1 to 2. In a similar manner, if we consider the detection optics of Figure 4*d*, the angular acceptance as defined by $\sim d_1/2f$ depends on the refractive index changes.

The main purpose of this more detailed discussion is to remind the uninitiated reader that precise measurements of Rayleigh ratio require attention to details. We need some knowledge of the light-scattering spectrometer optics to be able to determine those minor corrections properly. The most convenient scheme is to modify the detection optics of Figure 4*e* by replacing d_2 with a slit of width a^* and use $x = 1$ as shown schematically in Figure 8*b*.

METHODS OF DATA ANALYSIS

Intensity of Scattered Light

A detailed discussion of the treatment of light-scattering intensity data has been presented elsewhere (6). In dilute solutions studies, we are particularly interested in determining $\bar{M}_w$, R_g, and A_2 as discussed in connection with equations (1)–(8). The most popular approach is the Zimm plot. For large polymer coils, square root plots proposed by Berry (29) often yield a broader linear range in K^2 (and thus KR_g) than the Zimm plot. With

$$\frac{HC}{\Delta R_\theta} = \frac{1}{\bar{M}_w}\left[1 + \frac{16\pi^2}{3\lambda^2}\langle R_g^2(C)\rangle_z \sin^2\left(\frac{\theta}{2}\right) + 2A_2\bar{M}_w C\right] \tag{29}$$

and in a Zimm plot of $HC/\Delta R_\theta$ versus $\sin^2(\theta/2)$,

$$\langle R_g^2(C)\rangle_z = \frac{\text{initial slope}}{\text{intercept}}(1 + 2A_2\bar{M}_w C)\frac{3\lambda^2}{16\pi^2} \tag{30}$$

where $\langle R_g^2(C)\rangle^*[\equiv \langle R_g^2(C)\rangle/(1+2A_2\bar{M}_wC)]$ is defined as an apparent radius of gyration at concentration C. In a Berry plot,

$$\left(\frac{HC}{\Delta R_\theta}\right)^{1/2} = \left(\frac{1}{\bar{M}_w}\right)^{1/2}\left[1+\frac{8\pi^2}{3\lambda^2}\langle R_g^2(C)\rangle_z \sin^2\left(\frac{\theta}{2}\right)+A_2\bar{M}_wC\right] \tag{31}$$

A careful evaluation of the procedures using different ranges of KR_g is recommended for proper extrapolation to infinite dilution and determination of the initial slope if $KR_g > 1$. Computer programs have often been used to retrieve $\bar{M}_w$, A_2, and R_g using a Zimm plot, a Berry plot, or a nonlinear fitting procedure, including deviations from equation (29) or (31). The programs are often based on one of these equations, which are not valid at $KR_g > 1$ or high concentrations. Thus, by using linear least-squares fitting programs, we should be especially careful to test the invariance of $\bar{M}_w$, A_2, and R_g values by varying KR_g and concentration ranges. In a nonlinear fitting procedure, we also need to know the effects of contributions by higher order terms. Blind fitting of scattering data using unknown computer programs should be avoided.

Correlation Function Profile Analysis

Although we measure the intensity–intensity time correlation function $G^{(2)}(K,\tau)$, the quantity of interest is the first-order normalized electric field correlation function $g^{(1)}(K,\tau)$. The functions $G^{(2)}(K,\tau)$ and $g^{(1)}(K,\tau)$ are related by means of equation (20). Aside from effects due to large dust particles or small solvent molecules, our main interest is to determine experimentally $g^{(1)}(K,\tau)$ and to obtain information on the normalized characteristic line width distribution $G(\Gamma)$ using the Laplace integral equation:

$$g^{(1)}(K,\tau) = \int_0^\infty G(\Gamma)e^{-\Gamma\tau}\,d\Gamma \tag{32}$$

In a precise determination of $g^{(1)}(K,\tau)$, it is not advisable to consider A in equation (20) as an adjustable parameter. Furthermore, we can estimate whether the bandwidth of the correlator has been set properly by examining $g^{(1)}(K,\tau)$ as shown schematically in Figure 1*a*. This experimental test could be particularly important if the upper and lower bounds of $G(\Gamma)$ were unknown. In practice, the characteristic line width does not vary from zero to infinity but from the lower bound a_r to the upper bound b_r. Also, the origin of Γ may not necessarily come from translational motions only, especially when $KR_g > 1$. Thus, to determine a particle size distribution from $G(\Gamma)$, measurements should be made in the $KR_g(<1)$ range, where internal motions are not important. In the absence of internal motions, if we obtain a $g^{(1)}(K,\tau)$ starting from ~ 1 and decaying to ~ 0, we know that the delay time range for the correlator has been set properly to include dynamics of all the polymer molecules or particles of interest. This criterion is not easily satisfied, since the correlator does not have the zeroth channel. Since $g^{(1)}(K,\tau)\rightarrow 1$ only when $\tau\rightarrow 0$, extrapolation of $g^{(1)}(K,\tau)$ to $\tau\rightarrow 0$ is not trivial. Nevertheless, from data-fitting procedures, we can compute $g^{(1)}(K,\tau=0)$ and estimate whether we have missed any high-frequency components in $G(\Gamma)$. A similar approach can be specified for the slow components, by comparing $G^{(2)}(K,\tau=N\Delta\tau)$ with A of equation (20) where N is the last channel number of the correlator and $\Delta\tau$ is the delay time increment. The term $N\Delta\tau$ represents the longest delay time of the measured intensity–intensity time correlation function.

Laplace inversion of equation (32) is a difficult, ill-posed problem because of the bandwidth limitation of our correlators and the unavoidable experimental noise in $g^{(1)}(K,\tau)$. In practice, inspection on the upper and lower bounds of the experimentally measured $g^{(1)}(K,\tau)$, as discussed in the preceding paragraph, may prevent us from committing gross errors. However, it is quite difficult to determine exactly what the upper (b_r) and lower (a_r) bounds are. In fact, this uncertainty is associated with the ill-posed problem. Thus we do not expect to determine the details of the characteristic line width distribution function $G(K,\Gamma)$ from $g^{(1)}(K,\tau)$. Our aim is to be able to determine the first few moments of $G(K,\Gamma)$ in terms of

$$\bar{\Gamma}=\int G(\Gamma)\Gamma\,d\Gamma \tag{33}$$

$$\mu_i=\int(\Gamma-\bar{\Gamma})^i G(\Gamma)\,d\Gamma \tag{34}$$

with $\mu_2/\bar{\Gamma}^2(=\text{VAR})$ being the variance, that is, an approximation of $G(\Gamma)$, using some methods of regularization. The quality of the results now depends not only on experimental data but also on subtle methods of data analysis. The reader should be cautious in using computer programs with unfamiliar mathematical origins. Fortunately, in molecular weight distribution (MWD) determinations, most synthetic polymers exhibit a unimodal MWD. Then, the mathematical requirements for Laplace inversion of equation (32) become less stringent. Several approaches have been developed and used successfully. The algorithms have undergone extensive tests using simulated data as well as polymers of known MWDs.

The multiexponential singular value decomposition (MSVD) technique

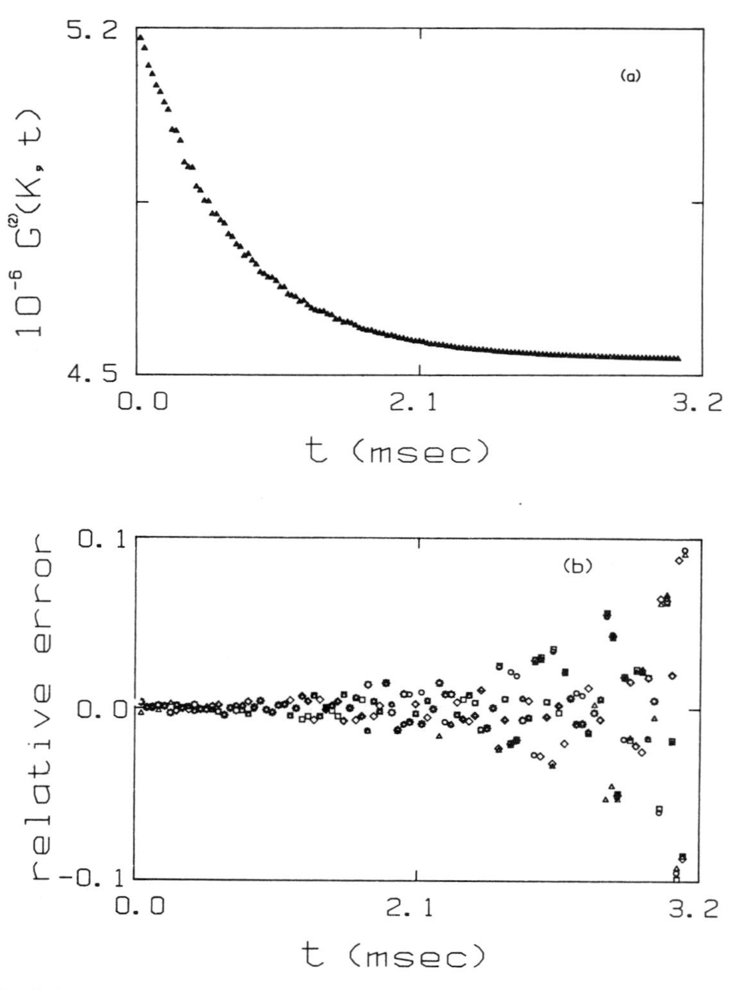

Figure 9. (*a*) A typical unnormalized intensity–intensity autocorrelation function 4.03 mg/mL PETFE ($\bar{M}_w = 5.4 \times 10^5$ g/mol) in diisobutyl adipate measured at $\theta = 30^\circ$ and 240°C using a delay time increment $\Delta\tau$ of 22.5 μs. (*b*) Relative deviation plots of the measured and the computed time correlation function using different methods of Laplace inversion as well as the method of cumulants (second-order). Relative deviation $= (b|g^{(1)}(t)|^2)_m - (b|g^{(1)}(t)|^2)_c/(b|g^{(1)}(t)|^2)_m$. (*c*) Normalized characteristic line width distribution for *a* using different methods of Laplace inversion. Results of computations are summarized in Table 1. Numerical values of $G(\Gamma)$ using different methods of data analysis are listed in Table 2. ▲, experimentally measured intensity–intensity time correlation function $G^{(2)}(K, t)$; ○, computed results based on MSVD; □, computed results based on our regularization method (RILIE); △, computed results based on CONTIN; ◇, computed results based on second-order cumulants expansion. (Reprinted with permission from *Macromolecules* (1987), reference 35. Copyright (1987) American Chemical Society.)

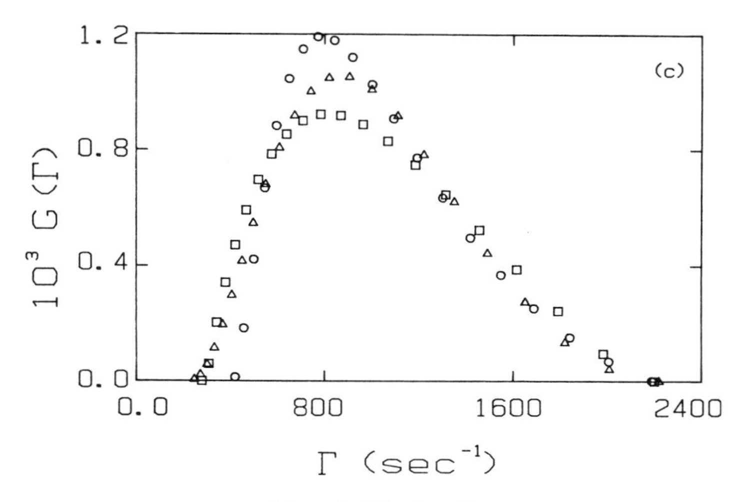

Figure 9. *(Continued)*

(13, 30), the regularized inversion of Laplace integral equation (RILIE) method (30), and Provencher's CONTIN program (13, 31) have been described elsewhere. Subtle details require careful study of updated computer programs. Fortunately, among the more popular approaches, including those developed by McWhirter and Pike (32, 33) and by Provencher (13, 31), the CONTIN and DISCRETE programs have been made readily available by Provencher with well-documented comments. In particular, the CONTIN program is capable of resolving multimodal characteristic line width distributions. Resolution in Laplace inversion is relatively poor even with very precise data; for example, with noises of less than 0.1% and an agreement between computed baseline (A in equation 20) and measured baseline [$\lim_{\tau \to \infty} G^{(2)}(K, \tau)$] to 0.1%. Thus, it is often advisable to confirm results of Laplace inversion using experimental data measured at different scattering angles and different concentrations, especially if we are concerned with the absolute magnitudes in the *estimates* of the upper and lower bounds in the determination of molecular weight distributions by means of laser light scattering. For unimodal characteristic line width distributions, the MSVD technique has been used successfully to characterize a variety of synthetic polymers such as poly(1,4-phenylene terephthalamide) (PPTA) (34) and an alternating copolymer of ethylene and tetrafluoroethylene (PETFE) (26, 35). Figure 9*a* shows a typical unnormalized intensity–intensity time correlation function. By using the three different methods of data analysis (MSVD, RILIE, and CONTIN), we can obtain deviation plots as shown in Figure 9*b* within experimental error limits. The resultant $G(\Gamma)$, as shown in Figure 9*c*, exhibits comparable magnitudes in

$\bar{\Gamma}(\sim 1.01 \pm 0.02\,\mathrm{s}^{-1})$, $\mu_2/\Gamma^2 \sim 0.10 \pm 0.01$ with ratio of $\bar{M}_z/\bar{M}_n{:}\bar{M}_w/\bar{M}_n \sim 2.1 \pm 0.1{:}1.3 \pm 0.1$, and $\bar{M}_w = 5.4 \times 10^5$ g/mol.

Figure 9 illustrates the order of magnitude of precision with which we can estimate MWD by means of laser light scattering and present-day algorithms. At $\mu_2/\bar{\Gamma}^2 \sim 0.05$–$0.3$, the cumulants method (37) provides an easier procedure to determine $\bar{\Gamma}$ and μ_2:

$$\ln|g^{(1)}(\Gamma, t)| = -\bar{\Gamma}t + (\tfrac{1}{2})\mu_2 t^2 + \cdots \tag{35}$$

In practice, it is difficult to determine cumulants to beyond the third order (skewness). For broad unimodal distributions, we find the MSVD and CONTIN techniques much more reliable. In any case, photon correlation spectroscopy (PCS) has an upper practical limit in determining the molecular weight distributions of synthetic polymers. If we approximate $G(K, \Gamma)$ by a set of linearly or logarithmically spaced discrete single exponentials, we have

$$G(K, \Gamma) = \sum_j I_j(K)\delta(\Gamma - \Gamma_j) \tag{36}$$

where the intensity weighting factor I_j is proportional to $M_jC_jP_j(K)$ as governed by equation (12), with the subscript j denoting the jth representative fraction and the particles being at infinite diluation ($A_2 = 0$). Thus, for $N_j[\equiv C_j/M_j]$ particles, I_j is proportional to M_j^2 and the light-scattering characteristic line width has an amplitude factor which is heavily weighted by the high molecular weight fractions. If we have two equal number representative fractions having a molecular weight ratio of 10, the intensity ratio, in the absence of intramolecular interference ($P(K) \sim 1$), is 100. In the characteristic line width distribution, it would be difficult to detect 1/100 of the intensity fraction. Conversely, for the high molecular weight fractions, light scattering is very sensitive and can easily detect small number fractions of polymers at high molecular weights. In this respect, laser light scattering is complementary to analytical techniques which emphasize detections based on the number of particles, since those methods will be difficult to use, especially to observe the presence of small amounts of large particles (i.e., high molecular weight fractions) in the presence of large amounts of small particles.

The intensity weighting of equation (12) does not imply that in the presence of a small amount of small particles, laser light scattering will yield an incorrect value for $\bar{M}_w$. On the contrary, by proper extrapolation to zero scattering angle and infinite dilution, the weight average molecular weight is defined precisely. In line width measurements, measurements of $G^{(2)}(K, \tau)$ are more delicate because of the bandwidth limitation of our present-day correlators. Nevertheless, if we consider the delay time range as discussed in the preceding sections and use appropriate methods of data analysis, it is quite easy to achieve reproducible and precise determinations of $\bar{\Gamma}$ and $\mu_2/\bar{\Gamma}^2$, especially for unimodal characteristic line width distributions. With care, one may be able to obtain the skewness. In terms of moments in the MWDs, ratios of $\bar{M}_z : \bar{M}_w : \bar{M}_n$ could be measured, provided the relation

$$D_T^0 = k_D M^{-\alpha_D} \tag{37}$$

is known, where D_T^0, k_D, and α_D are the translational diffusion coefficient at infinite dilution, a proportionality constant, and a scaling exponent. For polymer coils at the theta temperature, $\alpha_D = 0.5$.

Figure 10 shows typical plots of $G(D)$ versus D for a PETFE solution. At

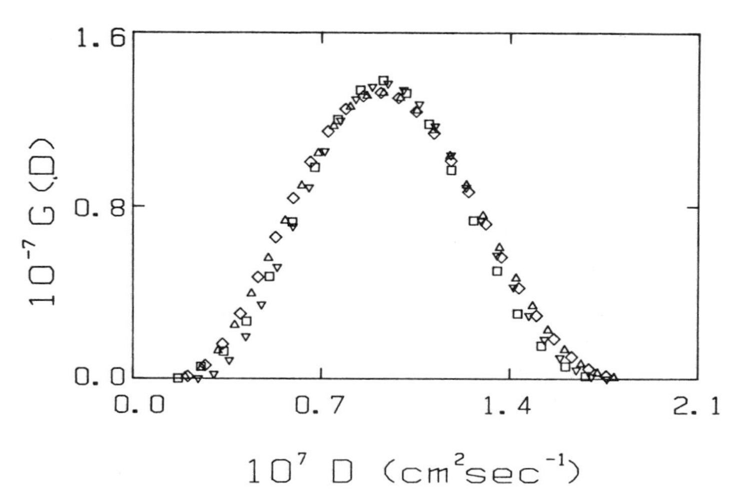

Figure 10. Plots of $G(D)$ versus $D(=\Gamma/K^2)$ for 3.50 mg/mL PETFE ($\bar{M}_w = 9.0 \times 10^5$ g/mol) in diisobutyl adipate at 240°C based on the CONTIN method of data analysis.

Notation	$\bar{D}$(cm²/sec)($\times 10^{-8}$)	$\mu_2/\bar{\Gamma}^2$
□ 30°	9.35	0.10
◇ 40°	9.12	0.096
△ 50°	9.25	0.10
▽ 60°	9.24	0.09

Overall $\bar{D} \simeq 9.24 \times 10^{-8}$ cm²/s

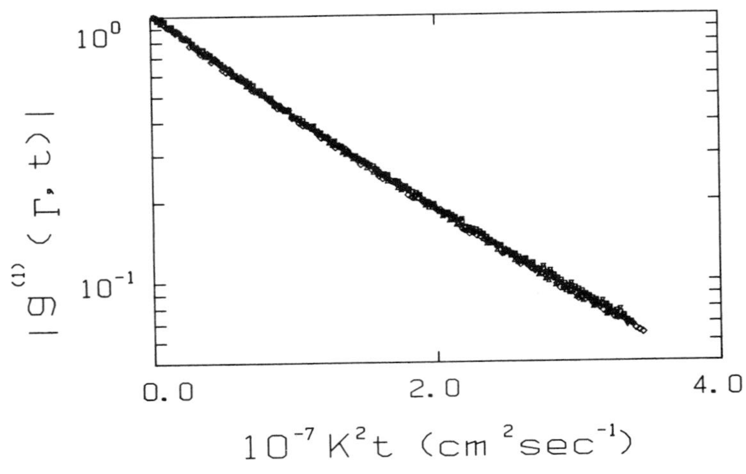

Figure 11. Scaling of $|g^{(1)}(\Gamma, t)|$ at different scattering angles by K^2 at constant concentration $C = 3.50$ mg/mL for PETFE ($\bar{M}_w = 9.0 \times 10^5$ g/mol) in diisobutyl adipate at 240°C. Same symbols as in Figure 10. (Reprinted with permission from *Macromolecules* (1987), reference 35. Copyright (1987) American Chemical Society.)

dilute concentrations, it is permissible to use

$$D_T = \bar{\Gamma}/K^2 = D_T^0(1 + k_d C) \tag{38}$$

where k_d is an experimentally determined average diffusion second viral coefficient. The subscript T is used to emphasize that we have avoided internal motions and have considered only translational diffusive motions by measuring the time correlation functions at small scattering angles ($KR_g < 1$). In Figure 10, the K^2 scaling for $\Gamma = DK^2$ is sufficient and can be shown to be appropriate experimentally without considering the complex Laplace inversion problem, as shown in Figure 11. In the presence of internal motions or intermolecular interactions, we can use

$$D = \Gamma/K^2 = D_T(1 + fR_g^2K^2) \tag{39}$$

and equation (38) to correct those effects. The details are best obtained from the original articles (34, 38).

Finally, laser light scattering permits us to estimate the MWD by making only one line width measurement at one scattering angle and one concentration, provided we know the magnitudes of equations (37)–(39). Thus, this noninvasive technique has the potential to become a powerful probe or monitor. Furthermore, it is permissible to use *difference* correlation functions to measure changes of molecular weight and polydispersity without any of the complications associated with the detailed analysis and complex mathematics. According to equations (35) and (37), and in the absence of internal motions and intermolecular interactions, the difference term can be represented by

$$\log|g^{(1)}_{M_1}(\Gamma, K^2t)| - \log|g^{(1)}_{M_2}(\Gamma, K^2t)| \cong -\frac{k_D}{2.303}[M_1^{-\alpha_D} - M_2^{-\alpha_D}]K^2t + \frac{1}{4.606}(\mu_{2,M_1} - \mu_{2,M_2})K^4t^2 \tag{40}$$

In equation (40), the first term on the right-hand side corresponds to the initial

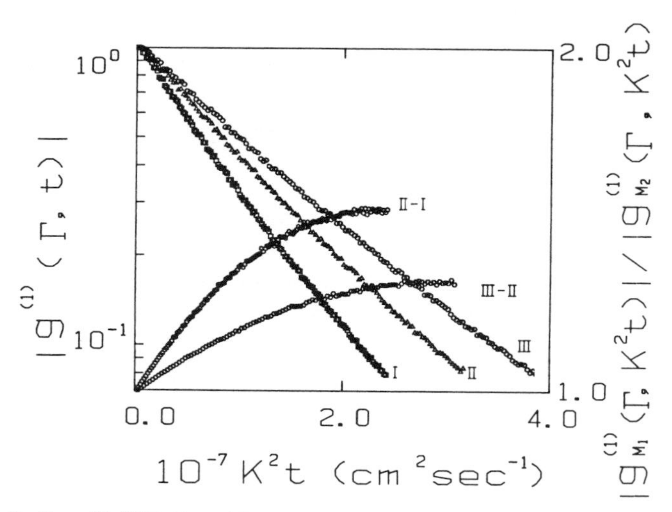

Figure 12. Scaling of $|g^{(1)}(\Gamma, t)|$ at different molecular weights; $\theta = 30°$. *Note*: Concentration has negligible effects.

PETFE	$\bar{M}_w$(g/mol)	C(mg/mL)	Notation
I	5.40×10^5	4.03×10^{-3}	□
II	9.00×10^5	3.50×10^{-3}	△
III	1.16×10^6	4.28×10^{-3}	○

$Y = \log|g^{(1)}(\Gamma, t)| - \log|g^{(1)}(\Gamma, t)|$

Slope	$\bar{M}_w$(g/mol)		
II − I 1.27×10^{-8} (cm²/s)	I = 5.62×10^5	5.40×10^5	5.3×10^5
III − II 4.97×10^{-9} (cm²/s)	III = 1.13×10^6	1.16×10^6	1.18×10^6

(Reprinted with permission from *Macromolecules* (1987), in reference 35. Copyright (1987) American Chemical Society.)

slope in a plot of $\log|g^{(1)}_{M_1}(\Gamma, K^2t)| - \log|g^{(1)}_{M_2}(\Gamma, K^2t)|$ versus K^2t as shown in Figure 12, with $Y[\equiv \log|g^{(1)}_{M_1}(\Gamma, K^2t)| - \log|g^{(1)}_{M_2}(\Gamma, K^2t)|]$ being the y axis. Thus, if we know that $\bar{M}_w$(PETFE II) = 9.00×10^5 g/mol we can determine PETFE I and III to be 5.62×10^5 and 1.13×10^6 g/mol using the initial slope and equation (40) with $D^0_T(\text{cm}^2/\text{s}) = 3.35 \times 10^{-4}\,\bar{M}_w^{-0.60}$ and $\bar{M}_w$ expressed in grams per mole. Similarly, we can determine $\mu_2/\bar{\Gamma}^2$ values (e.g., $\mu_2/\bar{\Gamma}^2$(II − I) yields 0.11 from equation 40), in agreement with more detailed computations.

SUMMARY

Laser light scattering can be used to characterize dilute polymer solutions in terms of macromolecular parameters such as $\bar{M}_w$, A_2, R_g, R_h, VAR, $G(\Gamma)$, and MWD. The technique is particularly powerful because of its ability to perform experiments at high temperatures in closed systems. Thus, even intractable polymers, such as PPTA and PETFE, have been characterized successfully. The procedures are as follows:

1. Light-scattering intensity measurements to determine $\bar{M}_w$, A_2, and R_g.
2. Light-scattering line width measurements to determine R_h, VAR, k_d, and $G(\Gamma)$, provided we are familiar with bandwidth and noise limitations of the measured time correlation function and appropriate methods of data analysis, including new developments (39, 40), which seem to improve our ability to resolve $G(\Gamma)$.
3. Knowledge of $D^0_T = k_D\bar{M}_w^{-\alpha_D}$ using different molecular weight fractions.
4. Estimates of MWD from $G(\Gamma)$ based on equation (37). In the molecular weight transformation from $G(\Gamma)$, we must take into account the intramolecular interference effect and intermolecular interactions if measurements were performed at $KR_g > 1$ and finite concentrations (k_dC not much less than 1). However, from an empirical approach, the approximation correction procedure represented by equations (38) and (39) is quite effective. Finally, a difference technique can be established to monitor changes in polymer molecular weight and polydispersity; this technique has the potential to develop into a remote sensing probe for the on-line monitoring of polymerization processes.

ACKNOWLEDGMENT

I gratefully acknowledge support of this work by the U.S. Army Research Office (DAAL0387K0136). I also thank Chi Wu for his help in doing the drawings.

REFERENCES

1. M. B. Huglin, Ed., *Light Scattering from Polymer Solutions*, Academic Press, New York, 1972.
2. B. Chu, *Laser Light Scattering*, Academic Press, New York, 1974.
3. H. S. Dhadwal and B. Chu, "Fiber optics in laser light scattering spectroscopy," *J. Collpid, Interface Sci.*, **115**, 561 (1987).
4. P. Debye, *J. Appl. Phys.*, **15**, 338 (1944).
5. B. H. Zimm, *J. Chem. Phys.* **16**, 1099 (1948).
6. M. B. Huglin, *Top. Curr. Chem.*, **73**, 141 (1977).
7. B. Chu, *Polym. J. (Japan)*, **17**, 225 (1985).
8. B. J. Berne and R. Pecora, *Dynamic Light Scattering*, Wiley-Interscience, New York, 1975.
9. H. Z. Cummins and E. R. Pike, Eds., *Photon Correlation and Light Beating Spectroscopy*, Plenum, New York, 1977.
10. H. Z. Cummins and E. R. Pike, Eds., *Photon Correlation Spectroscopy and Velocimetry*, Plenum, New York, 1977.
11. S. H. Chen, B. Chu, and R. Nossal, Eds., *Scattering Techniques Applied to Supramolecular and Nonequilibrium Systems*, Plenum, New York, 1981.
12. J. C. Earnshaw and M. W. Steer, Eds., *Application of Laser Light Scattering to the Study of Biological Motion*, Plenum, New York, 1983.
13. For example, see E. O. Schulz-DuBois, Ed., *Proceedings of the Fifth International Conference on Photon Correlation Techniques in Fluid Mechanics*, Springer-Verlag, New York Berlin, 1983.
14. *Dynamic Light Scattering: Applications of Photon Correlation Spectroscopy*, R. Pecora, Ed., Plenum, New York, 1985.
15. B. Chu, "Light scattering studies of polymer solution dynamics," *J. Polym. Sci., Polym. Symp.*, **73**, 137 (1985).
16. G. D. Patterson, *Adv. Polym. Sci.*, **48**, 125 (1983); *Annv. Rev. Mater. Sci.*, **13**, 219 (1983); also G. D. Patterson and P. J. Carroll, *J. Polym. Sci., Polym. Phys. Ed.*, **21**, 1897 (1983).
17. B. Chu, *Phys. Scripta*, **19**, 458 (1979).
18. K. M. Abbey, J. Shook, and B. Chu, in *The Application of Laser Light Scattering to the Study of Biological Motions*, J. C. Eranshaw and M. W. Steer, Eds., Plenum, New York, 1983, pp. 77–87.
19. E. O. Schulz-DuBois and I. Rehberg, *Appl. Phys.*, **24**, 323 (1981).
20. K. Schatzel, *Opt. Acta*, **30**, 155 (1983).
21. S.-T. Sun, I. Nishio, G. Swislow, and T. Tanaka, *J. Chem. Phys.*, **73**, 5971 (1980).
22. N. C. Ford, Jr., in *Dynamic Light Scattering, Applications of Photon Correlation Spectroscopy*, R. Pecora, Ed., Plenum, New York, and London, 1985, pp. 7–58.
23. U.S. Patent No. 4,565,446, Jan. 21, 1986.

24. H. R. Haller, C. Dester, and D. S. Cannell, *Rev. Sci. Instrum.*, **54**, 973 (1983).
25. H. S. Dhadwal and B. Chu, NASA Laser Light Scattering Workshop, Cleveland Ohio, 9/1988.
26. B. Chu and C. Wu, *Macromolecules*, **20**, 93 (1987).
27. G. Bernardi, *Makromol. Chem.*, **72**, 205 (1964).
28. H. I. Levine, R. J. Fiel, and F. W. Billmeyer, Jr., *Biopolymers*, **15**, 1267 (1976).
29. G. C. Berry, *J. Chem. Phys.*, **44**, 4550 (1966).
30. B. Chu, J. R. Ford, and H. S. Dhadwal, *Methods Enzymol.*, **117**, 256 (1985).
31. S. W. Provencher, *Biophys. J.*, **16**, 27 (1976); *J. Chem. Phys.*, **64**, 2772 (1976); *Makromol. Chem.*, **180**, 201 (1979).
32. J. G. McWhirter and E. R. Pike, *J. Phys. A*, **11**, 1729 (1978).
33. N. Ostrowsky, D. Sornette, P. Parker, and E. R. Pike, *Opt. Acta*, **28**, 1059 (1981).
34. B. Chu, Q. Ying, C. Wu, J. R. Ford, and H. S. Dhadwal, *Polymer*, **26**, 1408 (1985).
35. C. Wu, W. Buck, and B. Chu, *Macromolecules*, **20**, 98 (1987).
36. B. Chu and C. Wu., *Macromolecules*, **19**, 1285 (1986).
37. D. E. Koppel, *J. Chem. Phys.*, **57**, 4814 (1972).
38. B. Chu, C. Wu, and J. R. Ford, *J. Colloid Interface Sci.*, **105**, 473 (1985).
39. M. Bertero, P. Brianzi, E. R. Pike, G. de Villiers, K. H. Lau, and N. Ostrowsky, *J. Chem. Phys.*, **82**, 1551 (1985).
40. S. K. Livesey, P. Licinio, and M. Delaye, *J. Chem. Phys.*, **84**, 5102 (1986).

CHAPTER

6

NEUTRON AND X-RAY SCATTERING

RANDAL W. RICHARDS

Department of Pure and Applied Chemistry University of Strathclyde Glasgow, Scotland

INTRODUCTION

The essential features in the measurement of molecular weight by small angle X-ray or small angle neutron scattering (SAXS and SANS, respectively) have little distinction from the technique of light scattering. Only the fundamental process causing the scattering differs in each technique; in light scattering it is the variation in refractive index which is responsible for scattering, whereas X-ray scattering arises from fluctuations in electron density and neutron scattering relies on differences in scattering length density in the scattering system under examination. It is noted at the outset that the major advantage of SAXS and SANS is obtaining structural information over a range of length scales, not the accuracy of the molecular weights obtainable. Notwithstanding the classical use of light scattering to obtain radius of gyration (R_g) values for macromolecules in solution, for all except the very largest molecules it is the only structural quantity light scattering can provide. Moreover, for random coil polymers with relative molar masses less than 0.5×10^6 it is doubtful whether even R_g can be obtained (1). This differentiation between the techniques arises from the range of scattering vector $\mathbf{Q}$ probed by each (Table 1). As a rule of thumb, the length scales that can be investigated are $\approx 2\pi/\mathbf{Q}$. Consequently from Table 1 it is evident that light scattering is confined to global dimensions of macromolecules, whereas SAXS and especially SANS can provide information at much smaller dimensions. Currently, however, length of the largest scale on which SAXS and SANS can usefully provide data is limited.

Molecular weights are obtained from the scattering at $\mathbf{Q} = 0$, and this value has to be obtained by extrapolation of data from finite $\mathbf{Q}$ values. For light scattering this is not a long extrapolation, and therefore molecular weights will be intrinsically more accurate. Regrettably, light-scattering determination of molecular weight is confined to dilute solutions, thereby avoiding multiple

Table 1. Scattering Vector Ranges Probed

Method	Q(Å^{-1})
Light scattering	2.5×10^{-4}–2.9×10^{-3}[a]
SAXS	2×10^{-3}–0.1
SANS	10^{-3}–1

[a]Values quoted for a light wavelength of 632.8 nm; refractive index, 1.5; minimum angle, 10°; maximum angle, 170°.

scattering. In some situations in situ molecular weight information is required in bulk dense systems such as networks, crystalline polymers, or polymer blends. Evidently, light scattering will not be applicable here, and it will become apparent that the application of SAXS to such systems is of limited value for molecular weight determination because of the lack of available contrast (see below). It is in such situations that SANS is uniquely able to provide both molecular weight and structural information for individual molecules. Because of this and because small angle X-ray scattering has been exhaustively discussed in a recent publication (2), this chapter concentrates on SANS as a means of determining macromolecular molar mass. In view of the rarity of suitable diffractometers and the expense incurred in producing suitable neutron beams, it is not recommended that SANS be used a routine method of determining relative molar mass.

Four fundamental properties of the neutron make it an extremely versatile particle for the investigation of condensed matter.

1. Accessible wavelengths are between 1 and 20 Å, and a wide range of reciprocal space can be explored.
2. Neutron energies are of the same magnitude as excitation energies in condensed matter and permit the investigation of rotations, vibrations, and translations of molecules.
3. Nuclear scattering of neutrons is isotope dependent; therefore molecules can be labeled in part or in whole to enhance the neutron scattering from structural features. This becomes particularly important for hydrogen and its main isotope deuterium.
4. The absorption of neutrons is small for most atoms (unlike that for X rays), and therefore bulk specimens can be investigated.

Other properties which are of value but are not germane to the present discussion include the neutron's magnetic moment and the character of the available frequency range (viz., temporal properties in the NMR range at one end and in the range of photon correlation spectroscopy at the other should be accessible by neutron scattering). The interested reader can obtain more information on the application of such *inelastic* and *quasi-elastic* neutron scattering from the reviews by Maconnachie and Richards (3), Ullman (4), and Higgins (5). Discussion here is focused exclusively on *elastic* scattering; that is, there is no transfer of energy between neutron and molecule.

In the author's opinion, and without prejudicing the statements above, it is important to determine the relative molar mass from SANS data whenever possible. Although it may not be the most important quantity desired from the experiment, comparison of the value so obtained with that determined by classical techniques reveals whether individual macromolecules, aggregated structures, or degraded molecules are responsible for the scattering. In the remainder of this chapter we develop the theory of small angle neutron scattering, making explicit reference to molecular weight measurement. We also review experimental practice and give applications of molecular weight determination by SANS for the areas in which it has played an important or novel role.

THEORETICAL BACKGROUND

Neutron Scattering Theory

Usually interest in radiation scattering measurements is focused on the analysis of the scattered radiation intensity with respect to its angular variation (scattering vector) and energy. This variation is then interpretable in terms of the dynamic and structural properties of the scattering system. When only structural details are required, the energy analysis is unnecessary and the analysis for elastic scattering only is carried out. Figure 1 shows schematically the basic aspects of a scattering experiment with an incident beam of intensity I_0 and wave vector $\mathbf{k}_0$ is scattered by a sample through an angle 2θ with wave vector $\mathbf{k}$ into a detector. From the vector diagram accompanying the schematic view, the change in the wave vector on scattering is given by the scattering vector $\mathbf{Q}$

$$\mathbf{Q} = \mathbf{k}_0 - \mathbf{k}$$

From Figure 1 then

$$\mathbf{Q}^2 = \mathbf{k}_0^2 + \mathbf{k}^2 - 2\mathbf{k}_0 \cos 2\theta$$

For elastic scattering $|\mathbf{k}_0| = |\mathbf{k}| = 2\pi/\lambda$, where λ is the radiation wavelength in the scattering substance.

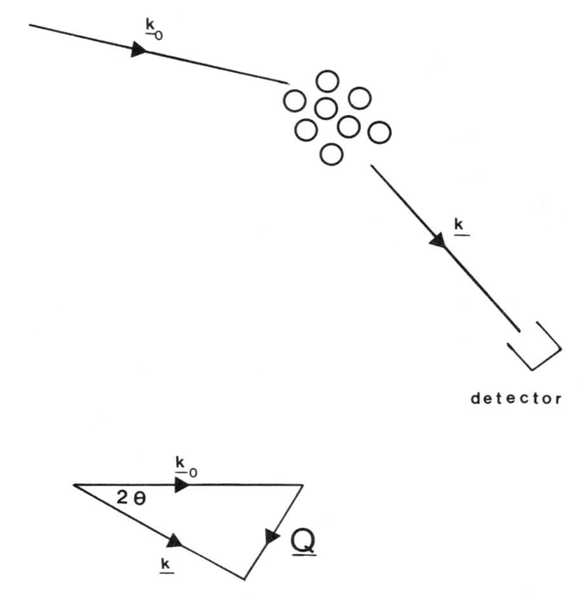

Figure 1. Schematic diagram of a scattering experiment and the associated vector diagram. Underscored symbols are vectors.

Hence,

$$Q = \mathbf{Q} = (4\pi/\lambda)\sin\theta$$

The incident neutron beam is assumed to be a well-collimated plane wave, whereas the spherically symmetric scattered wave is detected by the detector, which subtends an angle $\Delta\Omega$ at the sample because of its finite area. For large particles, scattered waves will originate from different parts of the particle. At $\mathbf{Q} = 0$ all these scattered waves are in phase; that is, we have constructive interference. As the magnitude of $\mathbf{Q}$ increases, destructive interference increases and the intensity of scattering decreases. The shape of this scattering distribution is determined by the pair correlation function of the scattering centers in the particle, and therefore analysis of the scattering distribution with respect to $\mathbf{Q}$ enables the particle size and shape to be obtained.

For an isolated nucleus, the scattered intensity in neutrons per square centimeter per second (flux) is given by

$$I = I_0\sigma \tag{1}$$

Table 2. Coherent Scattering Lengths, Incoherent Scattering Cross Sections, and Absorption Cross Sections for Some Elements

Element	b(cm)($\times 10^{12}$)[a]	σ_{inc}(cm^2)($\times 10^{24}$)	σ_A(cm^2)($\times 10^{24}$)[b]
^{1}H	-0.374	79.7	0.33
^{2}H	0.667	2.0	~ 0
C	0.665	~ 0	~ 0
N	0.936	0.46	1.88
O	0.58	~ 0	~ 0
Si	0.415	~ 0	0.16
Cl	0.958	5.9	33.6
B	0.535 + 0.021i	0.7	755
Cd	0.37 + 0.16i	—	2450

[a] For $\lambda = 1.0$ Å.
[b] For $\lambda = 1.8$ Å.
From Ref. 6.

where I_0 is the incident neutron flux and σ the total scattering cross section. The total scattering cross section has three components (we ignore magnetic scattering): coherent and incoherent scattering and neutron absorption processes. For common nuclei in macromolecules, absorption cross sections are negligible (see Table 2) and we ignore such processes. The distribution of the intensity scattered from an assembly of N nuclei is described by the probability that a neutron will be scattered into a solid angle $d\Omega$. This is the differential scattering cross section $d\sigma/d\Omega$ and for an assembly of N nuclei

$$I = KN(d\delta/d\Omega)\Delta\Omega \tag{2}$$

where K is a factor which includes the incident beam intensity, the neutron transmission of the specimen, and all geometrical parameters.

For purely elastic scattering, the application of fundamental neutron scattering theory leads to

$$d\delta/d\Omega = (m_n/2\pi\hbar N)^2 \left| \int V(r) \exp(i\mathbf{Q}\cdot\mathbf{r}) d^3\mathbf{r} \right|^2 \tag{3}$$

where m_n = neutron mass
$\hbar = h/2\pi$; h = Planck's constant
$V(\mathbf{r})$ = Fermi pseudopotential for neutron–nucleus vector separation of $\mathbf{r}$.

If only one nucleus is considered and the integral in equation (3) is confined to the nuclear volume, then a single constant value of $d\sigma/d\Omega$ independent of $\mathbf{Q}$ is obtained, since $|\mathbf{Qr}| \ll 1$.

Hence

$$-b = (m_n/2\pi\hbar)\int^{\text{nucleus}} V(\mathbf{r})d^3\mathbf{r} \tag{4}$$

The parameter b is the coherent scattering length and is the negative amplitude of the wave scattered by one nucleus. From equation (4) it will be appreciated that the scattering length is a nuclear property. Furthermore, the value of b varies for isotopes of the same element depending on the nuclear spin quantum number I. In addition, two different values of b can be taken for *many* nuclei of the *same* isotope depending on the interaction between nuclear and neutron spin (i.e., parallel or antiparallel). Thus for a scattering specimen of N nuclei, the resultant differential elastic scattering cross section is

$$d\delta/d\Omega = (1/N)\sum_{i,j=1}^{N} b_i b_j \exp(i\mathbf{Q}\cdot\mathbf{R}) \tag{5}$$

where $\mathbf{R} = \mathbf{r}_i - \mathbf{r}_j$. Since b_i and b_j depend on nuclear spin states and averaging over all possible spin states, we have

$$\langle b_i b_j \rangle = \bar{b}^2, \qquad i \neq j$$

$$\langle b_i b_j \rangle = \overline{b^2}, \qquad i = j$$

$$d\delta/d\Omega = (1/N)\left[\left|\sum_{\substack{i,j=1\\ i\neq j}}^{N} \bar{b}\exp(i\mathbf{Q}\cdot\mathbf{R})\right|^2 + \overline{b^2}\right] \tag{6}$$

Adding $\bar{b}^2$ to the first term in equation (6) and subtracting it from the second term gives

$$d\delta/d\Omega = (1/N)\left[\left|\sum_{i,j=1}^{N} \bar{b}\exp(i\mathbf{Q}\cdot\mathbf{R})\right|^2 + (\overline{b^2} - \bar{b}^2)\right] \tag{7}$$

Equation (7) explicitly shows the coherent and incoherent contributions to the differential scattering cross section. The second term on the right-hand side is the incoherent scattering; it is flat and isotropic and contains no structural information on the scattering sample. It is zero only for those nuclei with zero nuclear spin, and its magnitude can be calculated from tabulated values of incoherent scattering cross section σ_{inc}

$$\sigma_{inc} = 4\pi(\overline{b^2} - \bar{b}^2) \tag{8}$$

The first term on the right-hand side of equation (7) is the coherent scattering; it contains information on the phase of the scattered neutron intensity and thus provides the structural details of the scattering system. The differential scattering cross section is then

$$d\delta/d\Omega = (1/N)\left|\sum_{i,j=1}^{N} \bar{b}\exp(i\mathbf{Q}\cdot\mathbf{R})\right|^2 + \sigma_{inc}/4\pi \tag{9}$$

The incoherent scattering is a background signal, which must be subtracted from SANS data to leave the coherent scattering as the residual. Evidently from equation (9), the relative magnitudes of coherent and incoherent scattering are proportional to the values of b and σ. Values for these for some common nuclei in macromolecules are given in Table 2. More complete tables can be found in Bacon (7) and Kostorz (8). Entries in Table 2 of particular note are the large absorption cross section σ_A for boron and cadmium. Boron is commonly used as the active agent for neutron detection in neutron detectors as BF_3 gas, whereas cadmium is used for shielding purposes and in the construction of diaphragms or slits defining neutron beam geometry. Furthermore, the large difference between the values of b for the two common isotopes of hydrogen should be noted.

Small Angle Neutron Scattering

With reference to equation (9), as long as $\mathbf{Q}\cdot\mathbf{R} \ll 1$, then $\exp(i\mathbf{Q}\cdot\mathbf{R}) = 1$. Thus if we take 2 Å as an order-of-magnitude value for the length of a segment (monomer unit) in a macromolecule, the condition above will be valid for $\mathbf{Q} \leqslant 0.5\,\text{Å}^{-1}$. Consequently the scattering length b can be averaged over the volume of the segment to give a scattering length density $\rho(\mathbf{R})$, and the summation in equation (9) can be replaced by an integration over the whole sample volume.

$$d\delta/d\Omega = (1/N)\left|\int \rho(\mathbf{R})\exp(i\mathbf{Q}\cdot\mathbf{R})d^3\mathbf{R}\right|^2 + \sigma_{inc}/4\pi \tag{10}$$

Scattering length densities can be calculated from

$$\rho = \sum_i b_i N_A / M_m \bar{v} \tag{11}$$

Table 3. Scattering Length Densities ρ and Electron Densities ρ^x

Segment	Formula	Molar Mass	$\rho(cm^{-2})(\times 10^{10})$	$\rho^x(A^{-3})$
Ethylene	C_2H_4	28.05	-0.316	0.29
Ethylene-*d*4	C_2D_4	32.07	8.24	$\sim 0.28_6$
Styrene	C_8H_8	104.15	1.413	0.30
Styrene-*d*8	C_8D_8	112.19_9	6.50	0.30_6
Butadiene	C_4H_6	54.09	0.467	0.337
Butadiene-*d*6	C_4D_6	60.12	6.823	0.33_6
Methyl methacrylate	$C_5H_8O_2$	100.12	1.069	0.38_7
Methyl methacrylate-*d*8	$C_5D_8O_2$	108.17	7.03	$\sim 0.39_9$
Cyclohexane	C_6H_{12}	84.16	-0.24	0.26_7
Cyclohexane-*d*12	C_6D_{12}	96.23	~ 6.01	$\sim 0.26_4$
Water	H_2O	18.01	-0.56	0.33
Water-*d*2	D_2O	20.02	6.36	0.33_2
Toluene	C_7H_8	92.14	0.94	0.28
Toluene-*d*8	C_7D_8	100.19	~ 5.42	~ 0.29
Tetrahydrofuran	C_4H_6O	72.11	-0.246	0.213
Tetrahydrofuran-*d*8	C_4D_8O	80.16	$\sim 5.88_9$	~ 0.213

where b_i = scattering length of nucleus i in the segment
M_m = relative molar mass of the segment
$\bar{v}$ = partial specific volume for the polymer ($\approx$ 1/density)
N_A = Avogadro's number

Values of scattering length densities for some common polymers and solvents are given in Table 3.

To proceed further, we use a two-phase model for the scattering system (8); that is, we have N_p identical particles in the sample volume V. The particles (macromolecules) have a scattering length density ρ_p, whereas that of the medium in which they are dissolved or dispersed is ρ_m. Finally, the particles have random orientation with respect to each other and are sufficiently dilute that there is no interparticle interference scattering. Under these conditions equation (10) becomes

$$d\delta/d\Omega = (N_p(\rho_p - \rho_m)^2/N)\left|\int_V \exp(i\mathbf{Q}\cdot\mathbf{R})d^3\mathbf{R}\right|^2 + \sigma_{inc}/4\pi \qquad (12)$$

Since we have specified a dilute system, the integral in equation (12) is effectively restricted to pair correlations within one scattering particle, and this may be replaced by the single particle form factor $P(\mathbf{Q})$ since

$$P(\mathbf{Q}) = \left| 1/V_p \int_{V_p} \exp(i\mathbf{Q}\cdot\mathbf{R}) d^3\mathbf{R} \right|^2 \tag{13}$$

where V_p = particle volume and is included in equation (13) to ensure $P(\mathbf{Q}) = 1$ at $\mathbf{Q} = 0$. Hence replacing equation (13) into equation (12) we have

$$d\delta/d\Omega = (V_p^2 N_p/N)(\rho_p - \rho_m)^2 P(\mathbf{Q}) + \sigma_{inc}/4\pi \tag{14}$$

Consequently by replacing equation (14) into equation (2) the expression for the scattered neutron intensity distribution as a function of $\mathbf{Q}$ is

$$I(\mathbf{Q}) = K\Delta\Omega[V_p^2 N_p(\rho_p - \rho_m)^2 P(\mathbf{Q}) + \sigma_{inc}/4\pi \tag{15}$$

Determination of Molecular Weight

Inspection of equation (14) shows that the following conditions are the most favorable to its utilization:

1. Large value of $(\rho_p - \rho_m)^2$
2. Low value of background scattering $\sigma_{inc}/4\pi$

From Tables 2 and 3 we note that the first condition can be obtained by dissolving fully deuterated molecules in a hydrogenous solvent. The solvent can be macromolecular—as long as the conditions specified in arriving at equation (14) are fulfilled, the natures of solvent and solute are immaterial. This rules out phase-separated, aggregated structures for the moment, but these are discussed later. In principle it appears better to dissolve hydrogenous material as solute in a deuterated solvent. The contrast factor $(\rho_p - \rho_m)^2$ is the same, but the background scattering $\sigma_{inc}/4\pi$ would be much lower because of the small amount of hydrogenous material present. Although this is feasible when the solvent is a common low molecular weight liquid (i.e., when cost is not too penalizing), it is not advised for solid state solutions.

It was noted in one of the earliest papers (9) that where a deuterated host matrix was used there was considerable coherent scattering even in the absence of solute. This was attributed to the presence of microscopic voids and defects in the specimen. The deconvolution of two coherent scattering signals, one of unknown form, is not a task to be attempted lightly. Consequently, deuterated solutes are more often used in hydrogenous solvent systems, and the background scattering is removed by subtracting the scattering of an identical specimen without the presence of the labeled (deuterated) specimen.

Before equation (14) can be used, information is required on the following factors: (1) $K\Delta\Omega$, (2) $P(\mathbf{Q})$, (3) V_p and N_p.

The particle volume is given by

$$V_p = M\bar{v}/N_A \tag{16}$$

where M is the molecular weight of the particle.

The number of particles in the scattering volume is

$$N_p = cAtN_A/M \tag{17}$$

where c = concentration in mass [(unit vol)$^{-1}$]
A = cross-sectional area of the neutron beam
t = sample thickness

The single particle form factor $P(\mathbf{Q})$ is determined by the particle geometry. For the usual model of a macromolecule (i.e., a Gaussian coil), $P(\mathbf{Q})$ is given by the Debye equation (11, 12)

$$P(\mathbf{Q}) = \left(\frac{1}{Q^2R_g^2}\right)^2 [\exp(-Q^2R_g^2) - 1 + Q^2R_g^2] \tag{18}$$

For $QR_g \leqslant 1$, the Guinier limit, equation (18) reduces to

$$P(\mathbf{Q}) = 1 - Q^2R_g^2/3 \tag{19}$$

thus at $Q = 0$, $P(Q) = 1$.

The factor $K\Delta\Omega$ in more detailed form is

$$K\Delta\Omega = T_s I_0 e \Delta\Omega \tag{20}$$

where T_s = neutron transmission through the sample
e = detector efficiency factor

Sample transmissions are easily determined by measuring the straight-through intensity of the incident beam (attenuated if necessary), with and without the sample in the beam. The remaining parameters in equation (19) can be obtained by calibrating the diffractometer performance under *exactly* the same conditions used for measurements. This means that the neutron wavelength, **Q** range covered, and cross-sectional area of the beam must be identical in each case. The calibrant should be a substance which gives a flat scattered intensity over the whole range of **Q** used. This condition means that the calibrant should be a purely incoherent scatterer. Vanadium fulfills this condition; unfortunately, its incoherent scattering cross section is low ($\sigma = 4.97 \times 10^{-24}\,\text{cm}^2$). Water is more usually used ($\sigma_T \cong 250 \times 10^{-24}\,\text{cm}^2$ at $\lambda = 10\,\text{Å}$): a water speciment 1 mm thick scatters about half the incident neutron flux. For the long wavelength neutrons generally used in SANS, the exact fraction scattered is markedly dependent on the neutron wavelength (13). Hence for a planar water calibrant of thickness t_w and transmission T_w and assuming that the incoherent scattering is spherically isotropic, the incoherent scattered intensity I_w is given by

$$I_w = \frac{I_0 A t_w (1 - T_w) e \Delta\Omega}{4\pi} \tag{21}$$

This equation is valid only for $\lambda \geqslant 10.0\,\text{Å}$; for shorter wavelengths a proportion of inelastic scattering takes place which modifies the intensity calculated by equation (21) due to anisotropic scattering. A factor f can be included to accommodate this effect.

$$I_w = \frac{I_0 A t_w (1 - T_w) e A \Omega}{4\pi f} \tag{22}$$

Table 4 reproduces values of f reported by Jacrot and Zaccai (10). From equation (21)

$$I_0 e \Delta\Omega = \frac{4\pi f I_w}{A t_w (1 - T_w)} \tag{23}$$

Replacing equations (16), (17), (20), and (23) into equation (15), taking the limit

Table 4. Wavelength-Dependent Correction Factor f for the Incoherent Scattering of Water

Wavelength (Å)	f
1.0	0.461
1.45	0.521
4.2	0.680
5.0	0.752
5.68	0.794
8.0	0.909
10.0	1.00

as **Q** approaches 0, and subtracting the background scattering, we have

$$I(0) = \frac{4\pi t T_s I_w(\rho_p - \rho_m)^2 cM}{t_w(1 - T_w)N_A} \tag{24}$$

whence

$$\frac{1}{M} = \frac{4\pi t T_s(\rho_p - \rho_m)^2}{t_w(1 - T_w)N_A}\frac{cI_w}{I(0)} \tag{25}$$

It is standard practice in small angle neutron scattering to normalize all scattering data to the value obtained for the calibrant by dividing the scattered intensity by that for the calibrant at the same value of Q; this normalized intensity is $I_n(\mathbf{Q})$ and equals $I(Q)/I_w(Q)$. All the factors on the right-hand side of equation (25) are either measurable of calculable from tabulated data. For simplicity we can write

$$I_n(0) = K^*cM \tag{26}$$

$$1/M = K^*c/I_n(0) \tag{27}$$

where

$$K^* = \frac{4\pi f t(\rho_p - \rho_m)^2 T_s}{t_w(1 - T_w)N_A} \tag{28}$$

The derivation above is lengthy, and it is advisable to summarize here the procedure by which the molecular weight of a macromolecule can be evaluated by SANS. We start with a solution of deuterated molecule at low concentration in a hydrogenous solvent, which may itself be macromoleular. The scattered neutron intensity is obtained over a suitable range of scattering vector **Q**; the incoherent scattering from the specimen is also measured and subtracted to give the excess scattered intensity. Calibration and normalization data are obtained from the scattering from water, which is divided into the excess scattered for the specimen. This normalized intensity is then extrapolated to $\mathbf{Q} = 0$, and the molecular weight may be calculated from equation (27) using known parameters to calculate K^*. It is however advisable to make measurements over a finite concentration range to extrapolate to infinite dilution of the scattering species.

Extension to More Complex Cases

The discussion thus far has a great resemblance to derivations for light scattering and is applicable to dilute solutions of polymers dissolved in a low molecular weight solvent or mixtures of deuterated and hydrogenous homopolymers in which the labeled component is dilute. However, the possibility of altering the contrast $((\rho_p - \rho_m)^2)$ for otherwise identical molecules introduces additional features which are unexplorable by other techniques. For example, it permits the investigation of concentrated solutions by dispersing a small percentage of labeled material among the hydrogenous material, the whole mixture being then dissolved in a solvent. Such solutions now have three components: solvent *s*, deuterated polymer, and hydrogenous polymer. The differential coherent scattering cross section for this case can be written as the sum of a total scattering law, $S_T(\mathbf{Q})$, and the single chain structure factor $P(\mathbf{Q})$ as before. The total scattering law accounts for correlations between scattering segments on different molecules and for any scattering contrast between hydrogenous polymer and solvent and deuterated polymer and solvent. If the mole fraction of deuterated monomer units is X and the scattering length densities of the deuterated and hydrogenous polymer are ρ_p^D and ρ_p, respectively, while that of the solvent is ρ_s, the resultant scattering law is given by (14–17)

$$S(Q) = (X\rho_p^D + (1 - X)\rho_p - \rho_s)S_T(\mathbf{Q}) + X(1 - X)(\rho_p^D - \rho_p)^2 P(\mathbf{Q}) \quad (29)$$

and equation (14) becomes

$$I(\mathbf{Q}) = K\Delta\Omega V_p^2 N_p[(X\rho_p^D + (1 - X)\rho_p - \rho_s)S_T(\mathbf{Q}) + X(1 - X)(\rho_p^D - \rho_p)^2 P(\mathbf{Q}) + \sigma_{inc}/(4qV_p^2 N_p)] \quad (30)$$

where N_p is the total number of polymer molecules in the specimen. This equation can be used in two ways. First, if the deuterated and hydrogenous analogues of the solvent have scattering length densities of opposite sign, they can be mixed to give a solvent of mean scattering length density, ρ_s^m, which is chosen such that

$$\rho_s^m = X\rho_p^D + (1 - X)\rho_p \quad (31)$$

In this situation the first term inside the brackets of equation (30) disappears and we arrive at the same form of equation as in the simple case, modified by the presence of the quantity $X(1 - X)$.

Second, in the absence of solvent, the ρ_s term in equation (29) vanishes and measurements are made on two specimens with differing values of X; if the polymer specimens are incompressible, $S_T(\mathbf{Q})$ should vanish or certainly be very much less than $P(\mathbf{Q})$. Consequently by subtracting the values of $(I(\mathbf{Q})/X(1 - X))$ for each mixture which results after correcting for background, $S_T(\mathbf{Q})$ can be eliminated and the resultant data used to obtain R_g and molar mass values. If $S_T(\mathbf{Q})$ is indeed zero, there is no need to undertake measurements at two differing values of X; instead, the maximum value of $X(0.5)$ can be used to increase the intensity of scattering and therefore gain increased statistical accuracy in a shorter time (19). For mixtures of homopolymers, the use of high concentration labeling is strictly only valid when the deuterated and hydrogenous components are identical in molecular weight and in molecular weight distribution. The severity of this limitation is discussed later.

Nonideal Solutions and Polymer Mixtures

All the equations so far presented have presumed that thermodynamic ideality applies; that is, there are no excess thermodynamic functions arising from mixing of polymer with its matrix. Where nonideality occurs, it can be accommodated by a virial expansion of equation (14). After subtraction of the incoherent background and replacing for $K\Delta\Omega$, V_p^2, and N_p and incorporating the virial expansion, we have

$$I_n(\mathbf{Q}) = \frac{K^* c M P(\mathbf{Q})}{1 + 2A_2 Mc \cdots} \tag{32}$$

where $I_n(\mathbf{Q}) = I(\mathbf{Q})/I_w(\mathbf{Q})$.

If $P(\mathbf{Q})$ can be represented by the Debye equation at low Q (equation 19), then

$$I_n(\mathbf{Q}) = \frac{K^* c M(1 - Q^2 R_g^2/3)}{1 + 2A_2 Mc \cdots} \tag{33}$$

whence on restriction to the second virial coefficient only

$$\frac{K^* c}{I_n(\mathbf{Q})} = \frac{1}{M} \frac{1}{1 - Q^2 R_g^2/3}(1 + 2A_2 Mc) \tag{34}$$

for $QR_g < 1$, then $(1 - Q^2 R_g^2/3)^{-1}$ may be expanded by the binomial theorem to yield

$$\frac{K^* c}{I_n(\mathbf{Q})} = \frac{1}{M} \frac{1 + Q^2 R_g^2}{3}(1 + 2A_2 Mc) \tag{35}$$

The similarity of this equation to that used in classical intensity light scattering is evident. Consequently it is not surprising that the double extrapolation procedure of Zimm is used to evaluate molecular parameters from such data. For SANS this involves plotting $c/I_n(\mathbf{Q})$ as a function of $(Q^2 + kc)$, where k is arbitrarily chosen to give an open mesh of points. Extrapolation of the data to $Q = 0$ and $c = 0$ yields two lines, both of which have intercepts on the ordinate axis of $1/MK^*$. The slope of the $Q = 0$ line is $2A_2/k$, while that of the $c = 0$ line is $R_g^2/3K^*M$. Finally, in common with light scattering, the molecular weight obtained from SANS is a weight average value; that is, $M = \bar{M}_w$, while the radius of gyration obtained is a z average value.

EXPERIMENTAL APPARATUS AND ANALYSIS

Small Angle Neutron Scattering Diffractometers

SANS diffractometers differ from each other in the details of their construction, operation, accessible neutron wavelengths, and Q range available. However, they are sufficiently similar to each other for Figure 2 to serve as a schematic diagram applicable to all instruments. A "cold" neutron beam (i.e., a beam with a higher proportion of long wavelength neutrons, usually obtained from a nuclear reactor) is directed by a series of guides to a monochromator. The guides are usually parallel-sided rectangular tubes of nickel-plated glass. The monochromator may be a crystal oriented to the beam such that only one wavelength with a narrow distribution is reflected. More flexibility is obtained by using a velocity selector, consisting of a series of slotted neutron-absorbing disks, are arranged on a common axis such that slots are displaced helically from one another. By varying the rotational speed of the plates, one ensures that only a narrow wavelength range of neutrons has sufficient velocity to pass unimpeded through the monochromator. Typically the wavelength distribution $d\lambda/\lambda$ is $\sim 10\%$, but wider or narrower distributions can be achieved with an associated increase or reduction in neutron flux, respectively.

Another series of movable guides brings the neutron beam onto the sample position. After the sample position, the scattered neutrons pass down a flight tube to the detector. Both flight tube and guide tubes are evacuated to reduce air scattering. Sample positions are usually at ambient conditions and spacious

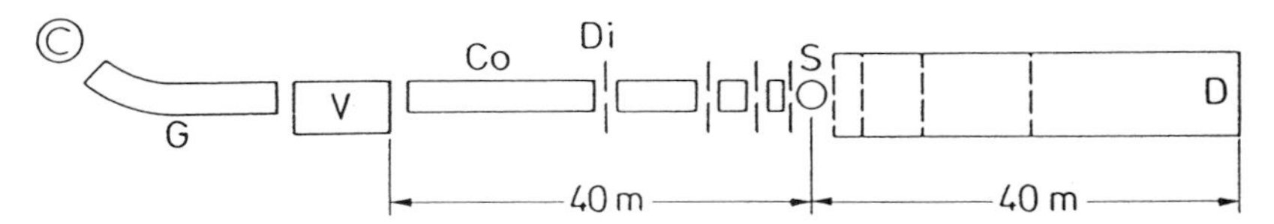

Figure 2. Essential components of a small angle neutron diffractometer. C, reactor core; G, neutron guide; V, velocity selector; Co, collimator, Di, diaphragm; S, sample position; D, detector in evacuated tube.

enough to accommodate a variety of experimental setups (e.g., automatic sample changers, magnets, furnaces). The ends of the guide and flight tube are therefore sealed with quartz windows, which transmit neutrons with little scattering except at very low Q.

The detector is generally a position-sensitive, two-dimensional detector and may be placed at distances selected by the user from the sample position, thus defining the accessible Q range. Such detectors consist of a series of discrete elements. For example, the detector on the D11 instrument at the Institut Laue-Langevin has an array of 64×64 cells, each having an area of $1\ \text{cm}^2$ (i.e., 4096 individual detector elements). The Q range accessible can be calculated from the distance L of the furthest element from the incident beam axis and the sample–detector distance D; then we have

$$Q_{\max} = (2\pi/\lambda)L/D$$

The detector is connected to a computer, which records the received counts and any counts on monitor detectors placed in the beam. On completion of a measurement, the operating system automatically records the data on disc memory and starts the next set of measurements.

Detailed descriptions of particular instruments may be obtained directly from the institutions. Appendix A gives addresses of some of these. The newest development is neutron beam production by means of spallation neutron sources, which produce a pulsed beam. At the time of writing (end 1985) the SANS diffractometer LOQ was about to go on line at ISIS, the spallation neutron source in the United Kingdom.

Sample and Background

Sample and background specimens can be enclosed between optical quality quartz windows. This is particularly convenient for solutions, since cells with a wide range of path lengths are commercially available. The sample thickness should be chosen to ensure that 50% of the incident neutron beam is scattered by both incoherent *and* coherent processes. This thickness maybe calculated from a Beer–Lambert type law:

$$I = I_0 \exp(-n\sigma_{\mathrm{T}} t) \tag{36}$$

where n is the number of scattering units (segments, etc.) per unit volume and t is the specimen thickness. For bulk polystyrene consisting mainly of hydrogenous material, this leads to thicknesses between 1 and 2 mm. For a polystyrene specimen with mainly deuterated polymer, the optimum thickness is ~ 1 cm.

Strictly, the background specimen should have the same isotopic composition as the sample but without the pair correlations, which give rise to the coherent scattering. For systems in which deuterated polymer is present only up to $\sim 10\%$ (w/w) in hydrogenous "solvent," the background incoherent scattering can be obtained with little error from the scattering of the "solvent" alone.

For higher concentrations or for systems in which the incoherent scattering is intrinsically low, it is advisable to incorporate low molecular weight deuterated material to match the isotopic composition of the sample. Another method of estimating the background scattering is to extend the Q range of measurements on the sample to high values ($\sim 0.5\,\text{Å}^{-1}$), where a flat signal devoid of coherent scattering contribution is obtained. The scattering in this range of Q is usually weak, and long counting times are required for statistical accuracy. Finally, there is the possibility of using polarized neutron scattering, in which the incoherent scattering is obtained directly from the proportion of the transmitted beam whose state of polarization has been altered by scattering.

The problem of properly accounting for background scattering is a serious one and often is the focus of much of the attention in SANS experiments. Maconnachie (20) has discussed the subject with particular relavence to high temperature measurements.

Data Analysis

On the assumption that the scattering from the sample is isotropic, the received counts on the detector are radially averaged about the incident beam direction. This produces a table of intensity as a function of Q. Background can then be subtracted, taking due account of variations in measurement duration and in thickness (and thus transmission) of the specimens. Normalization by division by the scattered intensity of water should produce a machine-independent, artifact-free set of data, which can be manipulated further in various ways at the discretion of the user. Typically for macromolecules the data can be plotted in a Zimm plot, which makes available data for several concentrations: that is, a plot of $c/I_n(\mathbf{Q})$ as a function of $Q^2 + kc$, where k is chosen to give an open mesh of points. Figure 3 shows a SANS Zimm plot for a polystyrene network swollen to equilibrium in a mixture of cyclohexane and cyclohexane-$d12$ at 35°C. Double extrapolation to $Q = 0$ and $c = 0$ gives a common intercept. Where only one concentration of labled material has been used, $c/I(Q)$ plotted as a function of Q^2 will produce a value of $1/MK^*$ from the $Q = 0$ intercept. The slope of the line will produce a radius of gyration, but this will pertain to the finite concentration c that was used, not to an infinite dilution value. In fact it will differ from the true R_g value according to the

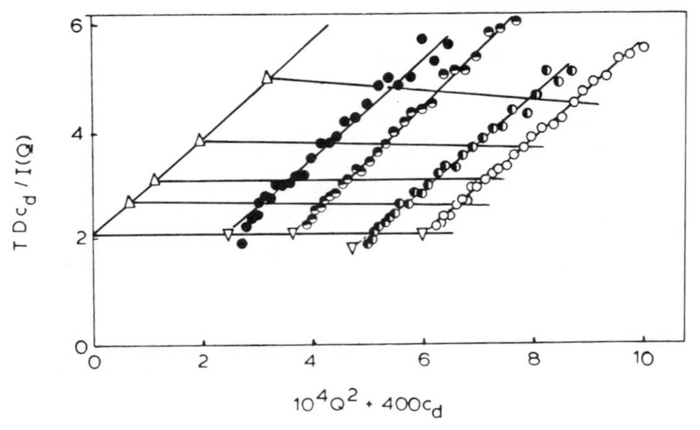

Figure 3. Full Zimm plot for a cross-linked network deuteropolystyrene in hydrogenous polystyrene at swelling equilibrium in cyclohexane ($3C_6H_{12}:1C_6D_{12}$) at 308 K. Volume fraction of polymer $\simeq 0.3$. $\triangle$, points obtained by extrapolation to zero concentration of deuterated polymer; ∇, points obtained by extrapolation to zero scattering vector.

relationship

$$(R_g^2)_{\text{true}} = (R_g^2)_{\text{meas}}(1 + 2A_2\bar{M}_w c) \tag{37}$$

Thus provided the product $A_2\bar{M}_w c \ll 1$, then $(R_g^2)_{\text{meas}} \approx (R_g^2)_{\text{true}}$.

Evidently with such large quantities of data, computer processing is essential. A particularly powerful and "user-friendly" software suite for this purpose has been written by Dr. R. E. Ghosh (21) of the Institut Laue-Langevin. This analysis also permits the user to delete, or reject on statistical grounds, regions on the detector in which the counts are excessively high due to anomalous scattering or reflections from obstructions in the neutron beam path.

APPLICATIONS OF SMALL ANGLE NEUTRON SCATTERING

It is not intended to review here all the areas in which SANS has been used. The great majority of studies have focused on configurational and morphological properties of polymer systems, and the determination of molecular weight has not been an overriding aspect of the work. There have been a number of reviews of the general application of SANS to polymers, and the interested reader should consult these (3, 4, 22–26). The applications discussed here either illustrate the use of SANS to determine relative molar mass or exemplify particular features.

Amorphous Polymers in Solid State and in Solution

The use of high concentrations of deuterated materials in SANS studies has been discussed by a number of workers. In addition to the work of Akcasu et al. and others referred to earlier (14–17), Tangari et al. (27) report results for bulk polystyrene. The SANS data were interpreted via "Zimm" plots assuming $A_2 = 0$, and the intercepts at $Q = 0$ for two different polymers were in the same ratio as molar masses determined by other methods. Similar results were obtained by Wignall et al. (28) and Boué et al. (29) at almost the same time as Tangari.

It was remarked earlier that applicability of the high concentration labeling method demands that matrix and labeled polymers have exactly equal molar mass and molar mass distribution. This aspect was further examined by Tangari and Summerfield (30, 31) using deliberately mismatched deuteropolystyrene and hydrogenous polystyrene. Boué et al. (29), who have made a very thorough study of the use of high concentration labeling in SANS studies of homopolymers, write down correction factors which should be applied if the deuterated and hydrogenous polymers differ substantially in molecular weight. For the correction of molecular weight, this is

$$\bar{M}_{w\text{meas}} = \bar{M}_{w\text{true}} \frac{X(1-X)(1+\Delta w)}{(1-X)(1+\Delta w)+X} \tag{38}$$

where $\bar{M}_w^{\mathrm{H}}/\bar{M}_w^{\mathrm{D}} = 1 + \Delta w$, the superscripts H and D referring to the hydrogenous and deutero polymer, respectively. Evidently if values of Δw are known, the measurement of $\bar{M}_w$ by SANS is redundant. These corrections are important, however, in ensuring that the correct value of R_g is obtained.

In the melt state, semicrystalline polymers adopt an isotropic molecular configuration. The papers by Crist et al. (32) and Kugler et al. (33) are typical in their approach to the determination of molecular weight for molten crystalline polymers.

Molecular weights of labeled molecular paths in swollen cross-linked networks have been reported by Richards and Davidson (34). Although the networks were dilute in labeled polymer, the presence of solvent meant that a pseudo-three-component system was obtained. Zero contrast between the cyclohexane solvent and the hydrogenous polystyrene was obtained by mixing hydrogenous and deuterated cyclohexane in the volume ratio of 3:1. For swollen gels, the molecular weights obtained were in good agreement with values obtained by gel permeation chromatography (GPC) before the labeled materials were cross-linked into a network. For dry, unswollen networks, problems were encountered with low Q scattering from voids. O'Reilly et al. (35) also remark on the presence of void scattering in their work on polymethyl methacrylates of differing tacticity. Nonetheless it was possible to obtain molar mass values in good agreement with known values.

The application of SANS for the evaluation of artifact-free values of molar mass and radius of gyration depends on random mixing between labeled and hydrogenous polymer, there being no difference in the thermodynamics of deuterated and hydrogenous versions of the same polymer. However, it is noteworthy that the theta temperature of deuteropolystyrene in cyclohexane is some 5°C lower than that for the hydrogenous polymer.

This effect arises from small differences in molar volume of deuterated and hydrogenous homologues. According to theoretical work of Buckingham and Hentschel (36), such differences gain significance only near a phase transition boundary. The consequence of this for semicrystalline polymers is discussed below. Preempting that discussion, we note that such differences lead to a segregation of deuterated polymer from the hydrogenous material. Such segregation has also been reported for *amorphous* poly(ethylene terephthalate) (PET) (37). Furthermore, as molecular weight determination from the SANS data showed, considerable transesterification had taken place during the sample preparation, leading to blocks of deuterated PET of molar mass between 1100 and 1600 (starting material molar mass = 23,000) being incorporated into the hydrogenous polymer.

Copolymers and Polymer Mixtures

Since copolymers consist of two chemically distinct monomer units that are highly unlikely to have the same scattering length density, it is evident that each component in a dilute solution will have a different contrast factor with the solvent. This situation is exactly analogous to that which prevails in the analysis of classical intensity light scattering from dilute copolymer solution (38). Apart from a change in symbols, the final equations obtained for SANS from copolymer solutions are identical in form to the light-scattering equations (39, 40). The important conclusion from this analysis is that both copolymer molar mass and radius of gyration obtained are *apparent* values, since they are markedly influenced by the scattering length density of the solvent. The variation of these two parameters with solvent scattering length density provides much information on copolymer configuration and compositional distribution. Since the scattering length desity of the solvent can be changed at will by merely mixing deuterated and hydrogenous versions in different proportions. This means that the variations in $\bar{M}_w$ and R_g can be obtained without significant change in the thermodynamics of the solution. It is this latter factor which has hindered such analysis, at least for R_g, using light scattering, since a change in specific refractive index increment necessitates a change in solvent. This contrast variation technique, first used in biological applications (18), can be exploited further in copolymers if one of the components of the copolymer is also deuterated. Thus by various combinations of deuterated/hydrogenous solvent or polymer, specific parts of the molecule can be highlighted.

For a copolymer of components A and B dissolved in a solvent S, the apparent molar mass M_{app} obtained from SANS is related to the true copolymer molar mass M_c and the molar mass of components A and B, which is M_A, and by the relation

$$M_{app} = \frac{K_A K_B}{K_C^2} M_C + \frac{K_A(K_A - K_B)}{K_C^2} w_A M_A + \frac{K_B(K_B - K_A)}{K_C^2} w_B M_B \tag{39}$$

where w_A, w_B = weight fractions of components A and B in the copolymer

$$\begin{aligned} K_A &= (\rho_A - \rho_S) \\ K_B &= (\rho_B - \rho_S) \\ K_C &= (\rho_C - \rho_S) \\ &= w_A K_A + w_B K_B \end{aligned}$$

with ρ_A, ρ_B, ρ_C, and ρ_S being the scattering length densities of polymer A, polymer B, the copolymer, and the solvent, respectively.

The apparent particle scattering function at $c = 0$ is given by

$$P(Q) = [w_A^2 K_A^2 P_A(\mathbf{Q}) + w_B^2 K_B^2 P_B(\mathbf{Q}) + 2 w_A w_B K_A K_B Q_{AB}(\mathbf{Q})] \tag{40}$$

where $P_A(\mathbf{Q})$ and $P_B(\mathbf{Q})$ are the particle scattering functions for the A and B parts and $Q_{AB}(\mathbf{Q})$ is a scattering function for cross correlation between the two components.

Use of these equations has so far been restricted to block copolymers. Ionescu et al. (40) have investigated R_g dependence in styrene–isoprene block copolymers in solution, and the results were in excellent agreement with theory. Edwards et al. (41) investigated the aggregation of styrene–methyl methacrylate block copolymers in dilute *p*-xylene solutions. By using a copolymer with a fully deuterated styrene block and hydrogenous and deuterated homologues of the *p*-xylene solvent, the investigators could examine each block individually by SANS. This was possible because of the almost identical values of the scattering length density for hydrogenous *p*-xylene and methyl methacrylate, thus resulting in zero contrast. Furthermore, there is essentially zero contrast between deuterostyrene and deutero-*p*-xylene. Consequently Edwards et al. (41) were able to determine unambiguously the aggregation number of the micellar structure formed at temperatures below 40°C.

In the solid state, block copolymers form a microphase-separated structure wherein domains of the minor component are dispersed with considerable long-range order in a matrix of the major component. The domain symmetry may be spherical, cylindrical, or lamellar, depending on the copolymer composition. The structure of these materials has been examined in detail by SANS (23), and some attempts have been made to ascertain the configuration of the domain forming block in the domain (42, 43). The problem is complicated by the regular domain morphology and the paracrystalline arrangement of the domains. This results in the total scattering law of equation (30) being a convolution of the scattering law for an isolated domain $\langle F_p(\mathbf{Q})\rangle^2$, and an interference function $A(Q)$, arising from the regular separation. Hence,

$$S_T(\mathbf{Q}) = \langle F_p(\mathbf{Q})\rangle^2 A(Q) \tag{41}$$

The form of $\langle F_p(\mathbf{Q})\rangle^2$ is known for regular morphologies, whereas $A(Q)$ is not known a priori. Although $A(Q)$ is not known, it dominates the low Q scattering from these materials, appearing as a series of Bragg peaks of considerable amplitude. These Bragg peaks fall in the region of Q used to determine R_g, hence influence the extrapolation to $Q = 0$ needed to determine the molar mass of the block. At higher Q, although $A(Q)$ may have been reduced to a value of 1, the scattering is still modulated, in principle, by $\langle F_p(\mathbf{Q})\rangle^2$, and nonlinear Zimm plots could result. Evidently, the prospects for obtaining R_g and $\bar{M}_w$ of the domain-forming blocks will be greatly enhanced if $S_T(\mathbf{Q})$ can be removed. Refer to equation (30) where we have a proportion of the block copolymer with one fully deuterated block mixed with its hydrogenous counterpart. In such case ρ_S is the scattering length density of the matrix polymer, and we note that if $X\rho_p^D + (1 - X)\rho_p = \rho_S$, the contribution of $S_T(\mathbf{Q})$ to the total scattering will be zero. This method has been used by Cohen et al. (44) to obtain the R_g and $\bar{M}_w$ of deuterobutadiene blocks in butadiene domains in styrene–butadiene block copolymers. Figure 4 show the success achieved by this method in removing both interference function and single domain scattering. This technique has also been used by Miller et al. (45, 46) in studies of polyurethane block copolymers and polyester block copolymers.

Mixtures of different polymers have been the subject of much activity. Small angle scattering studies have been used to ascertain the kinetics and mechanism of phase separation, and the application of SANS has been taken

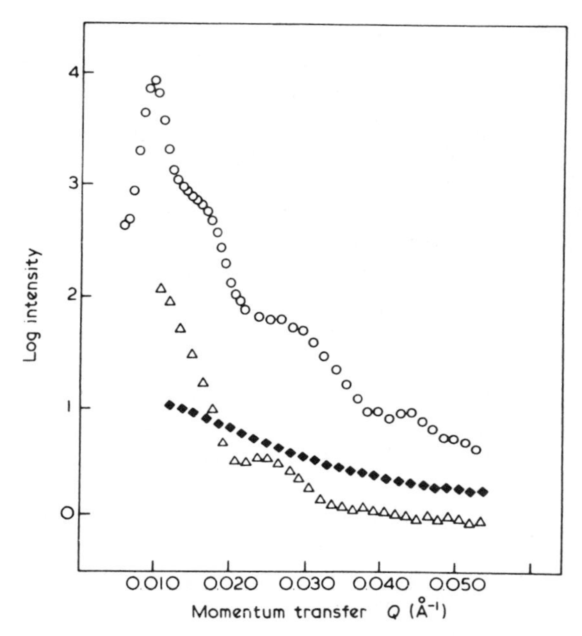

Figure 4. Small angle neutron scattering for block copolymers of styrene and butadiene. ○, deuterated butadiene blocks; △, fully hydrogenous copolymer; ◆, mixture of part deuterated and hydrogenous block copolymers, which is calculated to remove $S_T(\mathbf{Q})$ contribution. [Reproduced by permission from *Polymer*, **24**, 519 (1983).]

further to assess the nature of the molecular dispersion in the blends. From SANS-determined values of the molar mass of deuterated poly(methyl methacrylate) on deuterated polystyrene in blends with poly(vinylidene fluoride) and poly(vinyl methyl ether), respectively, Hadzioannou and Stein (47) were able to show that true molecular solutions prevailed in these amorphous blends. Much interest in such compatible blends lies in the nature and magnitude of the interactions between the blend constituents. It was shown independently by Stein and Hadzioannou (48) and Warner et al. (49) that values of the polymer–polymer interaction parameter can be obtained from the intercept at $Q = 0$ for finite concentrations of the labeled polymer. Interaction parameters and molecular weights of labeled components have also been reported for other polymer blends (50, 51).

Semicrystalline Polymers

Undoubtedly the most important area in which molar mass determination has made a major contribution is the investigation of the molecular trajectory in semicrystalline polymers. The possibility of segregation effects arising from differences in properties of deuterated and hydrogenous polymers was referred to earlier. This difference in properties is particularly evident in the different melting points for deuterated polyethylenes. In some of the earliest SANS experiments on polyethylene, strong evidence for segregation was obtained from the molecular weight of the labeled polymer (52, 53). This segregation and clustering was present to some extent in the polyethylene melt at 150°C ($T_m \simeq 114°C$), whereas measurements made on the solid semicrystalline polymer at 25°C produced molecular weight values suggesting that the clusters contained between 15 and 30 molecules. Evidently the occurrence of such clustering can invalidate a model evolved on the basis of data obtained from such systems. Subsequent work (54) overcame this effect by using a higher boil solvent to prepare mixtures of hydrogenous and deuterated polyethylene, and by rapid quenching of the subsequently hot compression-molded plaques.

Theoretical descriptions of the segregations in polyethylene have been attempted, notably the correlation network description of Schelten et al. (55, 56), which demonstrates that relatively few intermolecular contacts are required to produce erroneous values of molar mass and R_g. Wu and Wignall (57, 58) have capitalized on the segregation of polyethylene to investigate the incidence of melting and subsequent recrystallization in plastic deformation. Polyethylene specimens containing deuteropolyethylene (4%) were compressed at temperatures below the melting point, and SANS-determined molecular weights were compared to values obtained on specimens which had the same thermal history but were not compressed. In all cases the molar mass was reduced after compression, a result that was attributed to the occurrence of melting followed by a dispersive recrystallization of the original segregated deuteropolyethylene clusters. No evidence has been produced from SANS data for clustering in polypropylene (59).

McAlea et al. (60) have investigated the conformation of semicrystalline poly(ethylene terephthalate) by SANS. In contrast to the experience of Wu et al. (37) reported earlier, however, the values of molar mass obtained were in excellent agreement with values obtained by GPC. The authors attribute the absence of transesterification to the much lower pressing temperatures used to prepare SANS specimens. Finally we note that partially deuterated polyethylene (61) apparently does not suffer segregation but can produce measurable scattering suitable for the determination of R_g and $\bar{M}_w$, the low deuterium content notwithstanding.

COMPARISON WITH SMALL ANGLE X-RAY SCATTERING

In principle, all that has been written in the theory section above is applicable to small angle X-ray scattering measurements, and identical expressions can be derived. The major difference is the limited range of contrast factor available and the much smaller path lengths in SAXS due to the greater absorption. Furthermore, at intermediate values of scattering vector, the background scattering in SAXS due to thermal density fluctuation is not known with accuracy. Since this background is Q-dependent, it can have a crucial influence on the corrected data. This subject has been discussed by Koberstein et al. (62) and Roe (63). It was remarked earlier than SAXS arises from electron density fluctuations, and electron density ρ^{x} is calculated in a manner similar to that used for scattering length density fluctuations

$$\rho^{x} = \sum_{i} n_i^{e} N_A / M_m^{\bar{v}}$$

when n_i^{e} is the number of electrons in atom i. Values of ρ^{x} are included in Table 3, and it is immediately apparent that the contrast obtainable between deuterated and hydrogenous isomers is negligible, notwithstanding the finite density increase. Indeed, from Table 3 it would seem that there would be little X-ray contrast, and consequently little scattering, even for polymers in solution. However, canceling this factor is the much greater X-ray source "brightness" compared with neutron beam sources. For example, an experiment on a SAXS diffractometer on a synchrotron source may take up to 2 minutes; on the most powerful neutron source currently available, the same experiment may take 30 minutes to 1 hour. However, it must be stressed that SAXS experiments have to be performed on *dilute* solutions. If global molecular configuration and molar masses are required, measurements on bulk systems are impossible.

Sample thickness for SAXS is usually less than that for SANS and is calculable from a Beer–Lambert law

$$I = I_0 \exp(-\mu' t)$$

where μ' is the *linear* absorption coefficient for the scattering specimen. Values of *mass* absorption coefficients μ are tabulated for the elements as a function of X-ray wavelength (64). The relation between the two absorption coefficients is

$$\mu' = \mu d$$

where d is the specimen density, and optimum thickness is

$$t = 1/\mu'$$

Mass absorption coefficients for compounds are calculable as the weighted

Table 5. Optimum Thickness for 1.54-Å X Rays

Compound	t_{opt}(cm)
H_2O	0.1
C_6H_{12}	0.32
$CHCl_3$	0.007
C_8H_8 (polystyrene)	0.21
C_2H_4 (polyethylene)	0.19

sum of the elemental constituents

$$\mu = \sum w_i \mu_i$$

where w_i is the weight fraction of constituent i. Hence for an X-ray wavelength of 1.54 Å, the optimum sample thickness for a selection of compounds is shown in Table 5.

CONCLUSION

Small angle neutron scattering is a well-established technique for the measurement of macromolecular configuration. It has been shown that weight average molar masses can also be obtained from the same data by extrapolation to zero scattering vector. Given the limited accuracy and the expense involved in making measurements, SANS is not a technique for routine use. However, measurement of molar mass in parallel with configurational studies is vital to ascertain that the data pertain to the correct scattering particles. This has been especially true in the work on semicrystalline polyethylene and in detecting transesterification of poly(ethylene terephthalate). The variety of situations that can be examined using SANS (dilute solution, concentrated solution, bulk polymer, cross-linked networks, etc.) is due to the relative ease with which the contrast factor can be varied merely by deuteration of part or parts of the specimen. This facility is not available to SAXS.

The disadvantages of SANS are the relatively low intensity of neutron sources, necessitating long counting times, and the low spatial resolution of detectors. This last factor is not of great moment in molar mass determination by SANS.

APPENDIX

Addresses of Institutions with Small Angle Neutron Diffractometers

1. Office of the Scientific Secretary
   Institut Laue-Langevin
   BP156X
   38042 Grenoble-Cedex
   France

2. Mr. D. H. C. Harris
   Bldg. 436
   Atomic Energy Research Establishment Harwell
   Didcot
   Oxon OX11 ORA, UK

3. Dr. R. J. R. Bennett
   Neutron Division
   Bldg. R3
   Rutherford-Appleton Laboratory
   Chilton
   Didcot
   Oxon OX11 0QX, UK

4. Dr. G. D. Wignall
   National Center for Small Angle Scattering Research
   Oak Ridge National Laboratory
   Post Office Box X
   Oak Ridge, Tennessee 37830

5. Dr. J. R. D. Copley
   McMaster Nuclear Reactor
   McMaster University
   Hamilton
   Ontario L85 4K1, Canada

REFERENCES

1. P. Kratochvil, in *Light Scattering from Polymer Solutions*, M. B. Huglin, Ed. Academic Press, New York/London, 1972, Chap. 7.
2. O. Glatter and O. Kratky, *Small Angle X-ray Scattering*, Academic Press, New York/London, 1982.

3. A. Maconnachie and R. W. Richards, *Polymer*. **19**, 739 (1978).
4. R. Ullman, *Annu. Rev. Mater. Sci.*, **10**, 261 (1980).
5. J. S. Higgins, in *Developments in Polymer Characterisation*, 4th ed, J. V. Dawkins, Ed., Applied Science Publishers, Barking, 1982, Chap. 4.
6. Values from G. Kostorz and S. W. Lovesy, in *Treatise on Materials Science and Technology*, Vol. 15, *Neutron Scattering*, G. Korstorz, Ed., Academic Press, New York, 1979.
7. G. E. Bacon, *Neutron Diffraction*, Oxford University Press, London/New York, 1975.
8. G. Kostorz, in *Treatise on Materials Science and Technology*, Vol. 15, *Neutron Scattering*, G. Kostorz, Ed., Academic Press, New York, 1979.
9. J. P. Cotton, D. Decker, H. Benoit, B. Farnoux, J. S. Higgins, G. Jannink, R. Ober, C. Picot, and J. des Cloizeaux, *Macromolecules*, **7**, 863 (1974).
10. B. Jacrot and G. Zaccai, *Biopolymers*, **20**, 2413 (1981).
11. P. Debye, *Ann. Phys.* **46**, 809 (1915); *J. Phys. Colloid Chem.*, **51**, 98 (1947).
12. Expressions for the scattering functions of other particles can be found in A. Guinier and G. Fournet, *Small Angle Scattering of X-Rays*, Wiley, New York, 1955, and W. Burchard, in *Applied Fibre Science*, Vol. 1, F. Happey, Ed., Academic Press, New York/London, 1978, Chap. 10.
13. D. I. Garber and R. R. Kinsey, Ed., *Neutron Cross Sections*, Vol. II (BNL 325, 3rd ed.), Brookhaven National Laboratory, Upton, NY, 1976.
14. A. Z. Akcasu, G. C. Summerfield, S. N. Jahshan, C. C. Han, C. Y. Kim, and H. Yu, *J. Polym. Sci. Polym. Phys. Ed.*, **18**, 863 (1980).
15. S. N. Jahshan and G. C. Summerfield, *J. Polym. Sci., Polym. Phys. Ed.*, **18**, 1859 (1980).
16. S. N. Jahshan and G. C. Summerfield, *J. Polym. Sci., Polym. Phys. Ed.*, **18**, 2415 (1980).
17. H. Benoit, C. Picot, and M. Benmouna, *J. Polym. Sci., Polym. Phys. Ed.*, **22**, 1545 (1984).
18. This technique, called contrast variation, is extensively used in SANS investigations of biological materials. See, for example, B. Jacrot, *Annu. Rev. Biophys. Bioeng.*, **12**, 139 (1983).
19. C. Picot, in *Static and Dynamic Properties of the Polymeric Solid State*, R. A. Pethrick and R. W. Richards, Eds., Reidel, Dordrecht, 1982.
20. A. Maconnachie, *Polymer*, **25**, 1068 (1984).
21. R. E. Ghosh, Technical Report No. 81GH29T, Institut Lane-Langevin, Grenoble, France.
22. R. W. Richards, in *Developments in Polymer Characterisation*, 1st ed., J. V. Dawkins, Ed., Applied Science Publishers, Barking, 1978, Chap. 5.
23. R. W. Richards, *Adv. Polym. Sci.*, **71**, 1 (1985).
24. R. W. Richards, in *Developments in Polymer Characterisation*, 5th ed., J. V. Dawkins, Ed., Applied Science Publishers, Barking, 1986, Chap. 1.

25. D. G. H. Ballard and E. Janke, *Macromolecular Chemistry*, Vol. 2, Specialist Periodical Report, Royal Society of Chemistry, 1982.
26. J. S. Higgins, *Macromolecular Chemistry*, Vol. 3, Specialist Periodical Report, Royal Society of Chemistry, 1982.
27. C. Tangari, G. C. Summerfield, J. S. King, R. Berliner, and D. F. R. Mildner, *Macromolecules*, **13**, 1546 (1980).
28. G. D. Wignall, R. W. Hendricks, W. C. Koehler, J. S. Lin, M. P. Wai, E. L. Thomas, and R. S. Stein, *Polymer*, **22**, 886 (1981).
29. F. Boué, M. Nierlich, and L. Leibler, *Polymer*, **23**, 29 (1982).
30. C. Tangari, J. S. King, and G. C. Summerfield, *Macromolecules*, **15**, 132 (1982).
31. G. C. Summerfield, *J. Polym. Sci., Polym. Phys. Ed.*, **19**, 1011 (1981).
32. B. Crist, W. W. Graessley, and G. D. Wignall, *Polymer*, **23**, 1561 (1982).
33. J. Kugler, E. W. Fischer, M. Peuscher, and C. D. Eisenbach *Makromol. Chem.*, **184**, 2325 (1983).
34. R. W. Richards and N. S. Davidson, *Macromolecules*, to appear.
35. J. M. O'Reilly, D. M. Teegarden, and G. D. Wignall, *Macromolecules*, **18**, 2747 (1985).
36. A. D. Buckingham and H. G. E. Hentschel, *J. Polym. Sci. Polym. Phys. Ed.*, **18**, 853 (1980).
37. W. Wu, D. Wiswe, H. G. Zachmann, and K. Hahn, *Polymer*, **26**, 655 (1985).
38. D. Froelich and H. Benoit, in *Light Scattering from Polymer Solutions*, M. B. Huglin, Ed., Academic Press, New York/London, 1972, Chap. II.
39. L. Ionescu, C. Picot, R. Duplessix, H. Benoit, and J. P. Cotton, *J. Polym. Sci. Polym. Phys. Ed.*, **19**, 1019 (1981).
40. L. Ionescu, C. Picot, R. Duplessix, R. Duval, H. Benoit, J. P. Lingelser, and Y. Gallot, *J. Polym. Sci. Polym. Phys. Ed.*, **19**, 1033 (1981).
41. C. J. C. Edwards, R. W. Richards, and R. F. T. Stepto *Polymer*, **27**, 643 (1986).
42. R. W. Richards and J. L. Thomason, *Polymer* **22**, 581 (1981).
43. H. Hasegawa. T. Hashimoto, H. Kawai, T. P. Lodge, E. J. Amis, C. J. Glinka, and C. C. Han, *Macromolecules*, **18**, 67 (1985).
44. F. S. Bates, C. V. Berney, R. E. Cohen, and G. D. Wignall, *Polymer*, **24**, 519 (1983).
45. J. A. Miller, S. L. Cooper, C. C. Han, and G. Pruckmayr, *Macromolecules*, **17**, 1063 (1984).
46. J. A. Miller, J. M. McKenna, G. Pruckmayr, J. E. Epperson, and S. L. Cooper, *Macromolecules*, **18**, 1727 (1985).
47. G. Hadzioannou and R. S. Stein *Macromolecules*, **17**, 567 (1984).
48. R. S. Stein and G. Hadzioannou, *Macromolecules*, **17**, 1060 (1984).
49. M. Warner, J. S. Higgins, and A. J. Carter, *Macromolecules*, **16**, 1931 (1983).
50. J. Jelenic, R. G. Kirste, R. C. Oberthur, S. Schmitt-Strecker, and B. J. Schmitt, *Makromol. Chem.*, **185**, 129 (1984).

51. A. Maconnochie, R. P. Kambour, D. M. White, S. Rostami, and D. J. Walsh, *Macromolecules*, **17**, 2645 (1984).
52. J. Schelten, G. D. Wignall, and D. G. H. Ballard, *Polymer*, **15**, 685 (1974).
53. J. Schelten, G. D. Wignall, D. G. H. Ballard, and W. Schmatz, *Colloid Polym. Sci.*, **2523**, 749 (1974).
54. J. Schelten, D. G. H. Ballard, G. D. Wignall, G. Longman, and W. Schmatz, *Polymer*, **17**, 751 (1976).
55. J. Schelten, G. D. Wignall, D. G. H. Ballard, and G. W. Longman, *Polymer*, **18**, 111 (1977).
56. J. Schelten, A. Zinken, and D. G. H. Ballard, *Colloid Polym. Sci.*, **259**, 260 (1981).
57. G. D. Wignall and W. Wu, *Polym. Commun.*, **24**, 354 (1983).
58. W. Wu and G. D. Wignall, *Polymer*, **26**, 661 (1985).
59. D. G. H. Ballard, A. N. Burgess, A. Nevin, P. Cheshire, G. W. Longman, and J. Schelten, *Macromolecules*, **13**, 677 (1980).
60. K. P. McAlea, J. M. Schultz, K. H. Gardner, and G. D. Wignall, *Macromolecules*, **18**, 447 (1985).
61. J. D. Tanzer and B. Crist, *Macromolecules*, **18**, 1291 (1985).
62. J. Koberstein, B. Morra, and R. S. Stein, *J. Appl. Crystallogr.*, **13**, 34 (1980).
63. R. J. Roe, *J. Appl. Crystallogr.*, **15**, 182 (1982).
64. *International Tables for X-Ray Crystallography*, 2nd ed., Kynoch Press, Brimingham, 1965.

CHAPTER

7

ULTRACENTRIFUGATION

ANTHONY R. COOPER

Chemistry Department, Lockheed Palo Alto Research Laboratories Palo Alto, California

The history of polymer molecular weight characterization is closely linked the development of the ultracentrifuge. Svedberg and Fahreus (1) in 1925 developed a centrifuge with a force capability of 5000 g and determined the first definitive high molecular weight measurement, namely, 68,000 for hemoglobin. Improvements in the centrifugal field led to determination of the molecular weights of proteins from measurements of their rate of sedimentation (2). The difference between weight and number average molecular weights $\bar{M}_w$ and $\bar{M}_n$ was first pointed out when a polyester was characterized (3, 4), and the determination of molecular weight distribution (MWD) by ultracentrifugation was reported in 1934 (5). It became established as an accurate method for determination of MWD (6) following developments for correction of diffusion (7) and concentration (8). Svedberg's pioneering efforts were recognized by the award of the Nobel Prize for chemistry in 1926. Many books have been written about ultracentrifugation (9–20), and the literature contains numerous book chapters (21–25) and reviews (46–50), as well.

EQUIPMENT

Many designs for the ultracentrifuge have been described. The polymer sample, in solution, which is being characterized, is subjected to a force of about 400,000 times that of gravity by spinning the sample cell at speeds of up to 70,000 rpm. The Spinco model E (51), shown schematically in Figure 1, is the most widely used analytical ultracentrifuge. The rotor is suspended by a flexible shaft from a water-cooled drive mechanism consisting of an electric motor and a gearbox. Constant speed of the motor is obtained by a feedback loop from a photoelectric counter to the current in the motor. The rotor is enclosed inside an armored chamber, which can be raised or lowered. The chamber is evacuated, and temperature control is achieved from a balance between the refrigeration coils around the edge of the chamber and a heater in

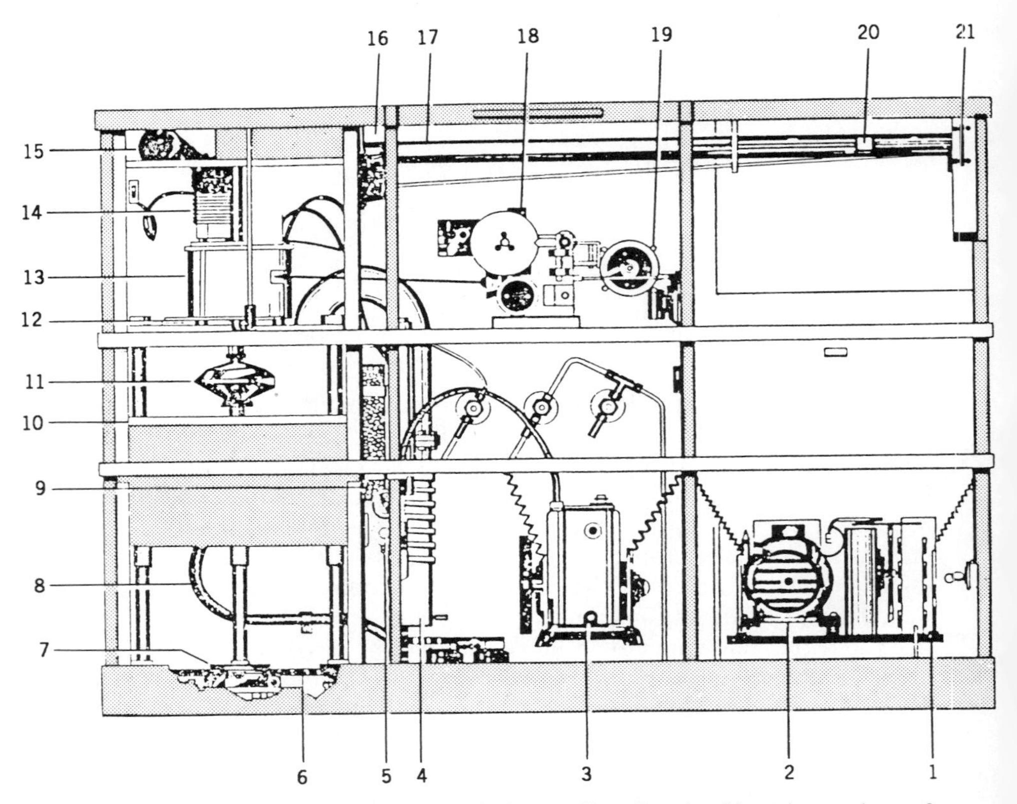

Figure 1. Schematic diagram of an analytical ultracentrifuge. Key: 1, refrigeration condenser; 2, refrigeration compressor; 3, mechanical vacuum pump; 4, diffusion pump; 5, Evapotrol; 6, chamber lift mechanism; 7, schlieren interference light; 8, capillary tube for refrigeration; 9, Drierite; 10, rotor chamber; 11, rotor; 12, drive oil; 13, rotor drive; 14, drive motor; 15, blower for drive motor; 16, plate shift mechanism; 17, optical tube; 18, differential gearbox; 19, synchronous motor; 20, viewer; 21, plate holder slot. (From Ref. 47.)

the base of the chamber. Generally temperature control is closely maintained in the range -10 to $+40\,^{\circ}$C, although special equipment has been designed for operation up to 130 °C.

Rotors are made from aluminum or titanium (the latter allows a 20% higher rotation speed), and they are available with two, four, or six cell cavities. Recently an eight-cell rotor has been reported (52) capable of rotation at 60,000 rpm and a controllable stroboscopic light source which enables the cells to be viewed and photographed. This considerably improves the throughput of samples. Cells are available in a variety of designs, and the selection of a

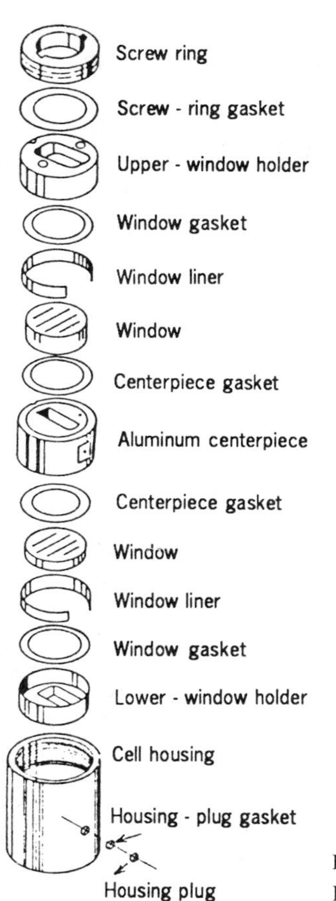

Figure 2. Schematic diagram of an ultracentrifuge cell (From Ref. 47.)

particular cell depends on the type of experiment being performed, the solvent and solute concentration, and the concentration detection system being employed. A simple single sector cell is shown in Figure 2. The sector-shaped channel, with a 2–4° sector angle, is machined into the centerpiece, and optically flat windows of quartz or sapphire are sealed with gaskets to the centerpiece. The sector shape ensures that sedimentation and diffusion occur in a radial direction. The sample is introduced through a small hole in the centerpiece. Other centerpiece designs are available; for example, double sectors allow simultaneous observation of solution and solvent. This is necessary with some optical systems such as the Rayleigh interference system. Centerpieces may also be made from other materials, such as filled epoxies or

Kel-F fluorocarbons. Advantages include low mass, easier sealing, and ease of manufacture. Disadvantages include lack of resistance to certain solvents and inferior high temperature properties. Cells may be obtained with optical path lengths between 1.5 and 30 mm. Window holders have different numbers and widths of slits depending on the cell type and detection system.

Detection systems are required to quantitate the amount of solute as a function of distance r from the spin axis during the experiment. The sample cell is placed in the rotor at the periphery and is interrogated by a light or UV beam at a certain point during its revolution. The detection of solute redistribution as sedimentation progresses may be detected by absorption or refractive index detection devices.

The light absorption scanning device (53, 54) detects optical transmission or absorbance as a function of radius. The solute is required to have an

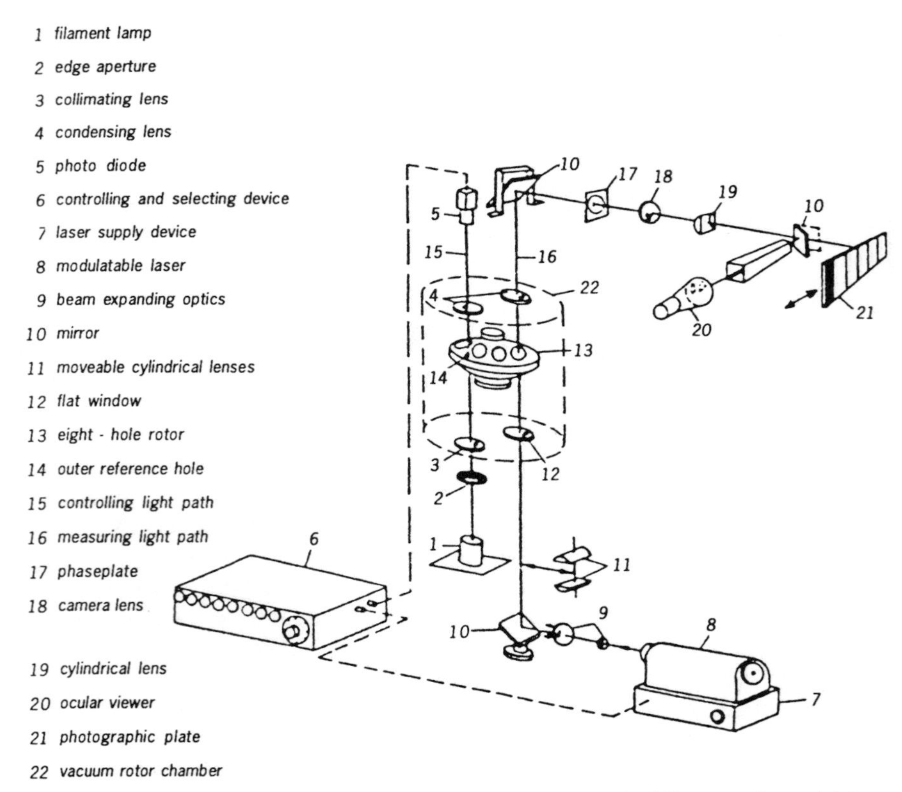

Figure 3. Optical arrangement of the combined interference and schilieren optics multiplexer for the analytical ultracentrifuge (From Ref. 52.)

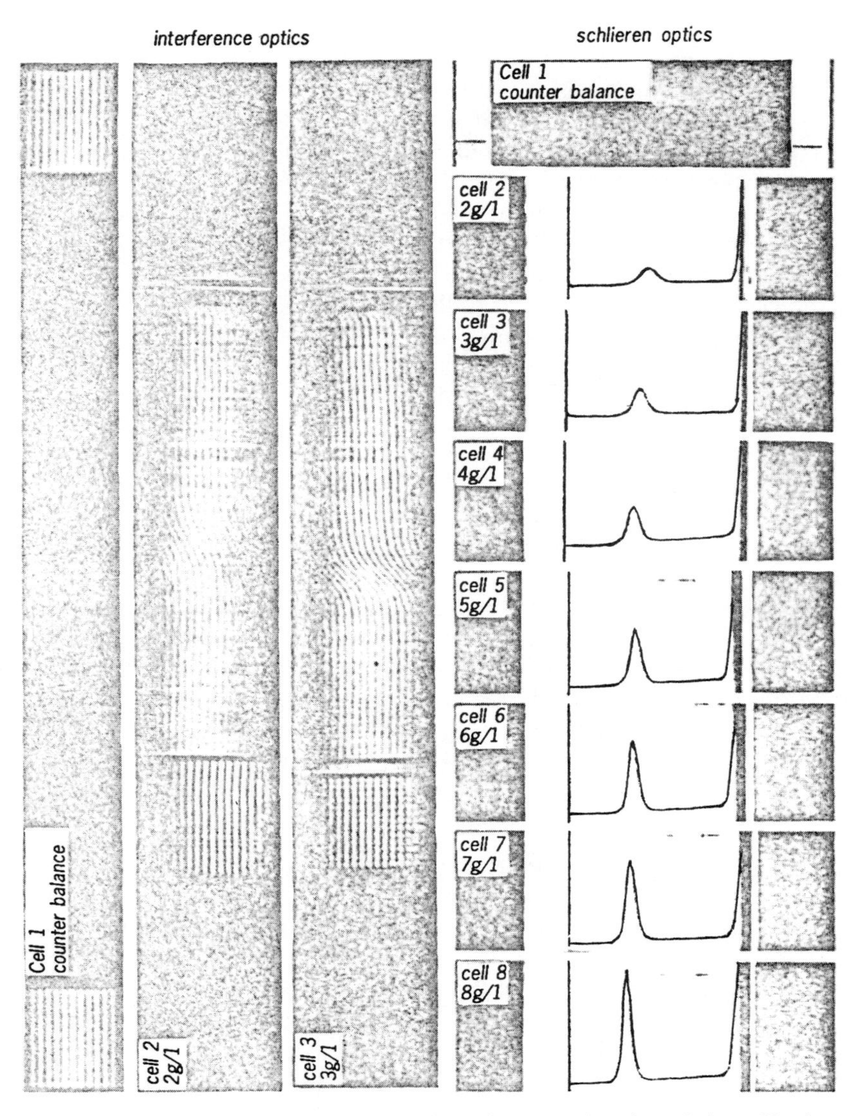

Figure 4. Interference and schlieren pictures taken after a running time of about 90 minutes during an eight-cell sedimentation run of polystyrene (MW 110,000) dissolved in toluene, at the maximum rotor speed of 60,000 rpm. (From Ref. 52.)

absorption band where the solvent is transparent. It is useful in that absorbance may be directly related to concentration and can be configured to permit simultaneous observation of many cells. A description of an improved UV detector has been reported (55) which provides a digital 500-channel intensity profile of the cell contents in real time. This system will also handle an eight-cell rotor.

Two systems based on detection of refractive index changes during sedimentation are in use: the schlieren and Rayleigh interference systems. In the former a light beam is deflected when passing through a concentration gradient which is proportional to the thickness of the solution in the optical axis and the refractive index gradient in the radial direction. The interference technique consists of passing two separate coherent light beams through the measuring and reference sections of the cell and bringing them together to form an interference pattern superimposed on the radial coordinate. The fringe displacement allows measurement of differences in concentration of solute with high precision.

Modifications to the Beckman instrument have been described (52) which incorporate combined interference and schlieren optics using multiplexing and a modulable laser. Figure 3 gives the optical arrangement, and the interference and schlieren patterns appear in Figure 4. An automatic photographic plate reading device has been reported (56) for digitizing Rayleigh interference patterns from a multislit interferometer (57, 58). The system was used to determine $\bar{M}_w$ of the standard polystyrene reference material (SRM 1478) as 37,400 by sedimentation equilibrium. The sample standard deviation was 0.7% (7 degree of freedom), and there was an expected limit of systematic error of 2%.

EXPERIMENTAL METHODS

Information about molecular weight and moleculer weight distribution can be achieved by two different experimental techniques. Sedimentation (or flotation) of a molecule occurs by the action of a centrifugal field and a density difference between the solute and solvent. When the rate of sedimentation, which depends on the mass and shape of the particles and the solution viscosity, is balanced by the rate of diffusion, caused by Brownian motion, the system is said to be in a state of equilibrium sedimentation. Both sedimentation rate measurements and sedimentation equilibrium measurements may be used to determine molecular weight.

Sedimentation Velocity Method (11, 15, 39, 46)

The attraction of using sedimentation velocity experiments is the relatively short time required, typically 2 hours. With automatic data reduction and multicelled rotors, the experimental time per sample is acceptable. On the other hand, the theory of the method is empirical, and two or sometimes three extrapolations are required. A considerable literature is available on the determination of MWD or the differential distribution of sedimentation coefficients $g(S)$. The latter may be transformed into MWD if the relation between S and M is known (15, 39, 43, 46, 59–62).

The Svedberg equation is

$$M = \frac{S_0}{D_0} \frac{RT}{(1 - \bar{v}_2 \rho_1)} \tag{1}$$

where S_0 and D_0 are the sedimentation coefficient and diffusion coefficient at infinite dilution, $\bar{v}_2$ is the partial specific volume of the solute, and ρ_1 is the density of the solvent (63).

To apply this equation to sedimentation coefficients measured by experiment, an extrapolation procedure must be employed. A typical extrapolation procedure employs an equation of the type

$$S = S_0(1 + k_s c + \cdots) \tag{2}$$

Sometimes concentration effects may be neglected if sensitive detectors are used—for example, a UV detector used with the polystyrene–cyclohexane system at 35 °C (64). To use refractometric detectors, it has been demonstrated that conducting the experiments below the θ temperature is useful (65, 66). Sufficiently high solute concentrations may be used for the detector to produce adequate signals; but importantly, the sedimentation coefficient is independent of concentration. Some restrictions apply to this approach, since high molecular weight materials are not soluble and soluble species may agglomerate. Methods of working at the θ temperature have been described (67) which allow the coefficients of the sedimentation and diffusion—molecular weight relationships to be obtained when monodisperse samples are not available. Values for D_0 and $\bar{v}_2$ must be obtained experimentally. An alternative approach to equation (1) is to use a semiempirical equation of the form

$$S_0 = kM^a$$

where k and a are constants for a given combination of polymer, solvent, and temperature. Relatively few values of k and a have been determined.

In the sedimentation velocity experiment, the data consist of a set of concentration boundaries between pure solvent and solution at various times as shown in Figure 5. The value of r at any given time is taken as the peak in the dc/dr trace or the position at half height of the c–r trace. Since the

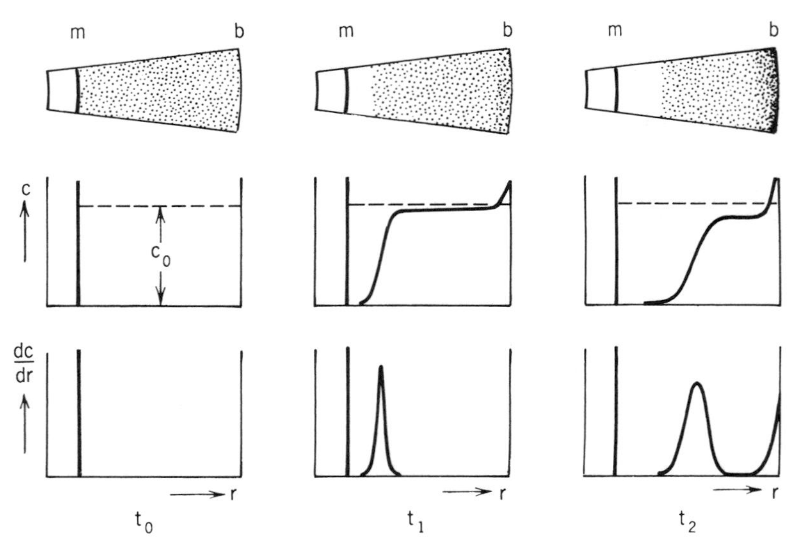

Figure 5. Schematic representation of the course of sedimentation in sectored Cells at times t_0, t_1 and t_2. (From Ref. 29.)

sedimentation coefficient can be expressed in the form

$$S = \frac{1}{\omega^2 r}\frac{dr}{dt} = \frac{1}{\omega^2}\frac{d \ln r}{dt} \tag{3}$$

a plot of ln r versus time will yield S from the initial slope $\omega^2 S$. The effect of radial dilution as the sample travels down the sector-shaped cell introduces a small curvature into the plot of ln r versus time. For monodisperse solutes the boundary is symmetrical and there is no problem in locating the peak value. However for polydisperse solutes the boundary becomes unsymmetrical because the larger molecules sediment faster but diffuse more slowly. Appropriate equations have been derived (68) to solve this problem.

The pressure exerted on the solution spinning in a conventional ultracentrifuge cell at 60,000 rpm is about 250 atm. Viscosity and density of the solvent are pressure dependent, as is the partial specific volume of the sample; hence the sedimentation coefficient is dependent on rotor speed. Although this set of properties may often be neglected for biopolymers in aqueous systems, it is important for flexible chain polymers in organic solvents. This effect also introduces curvature in the plots of ln r versus time, and there is no rigorous theory for conversion of pressure data to atmospheric pressure. An equation which is widely used (69) is

$$S_c = \frac{\ln r}{2\omega^2 t}[1 + K(r - 1)]^{-1} \tag{4}$$

where $K = \mu\rho M^2\omega^2/4$.

Unfortunately μ, which is supposed to be constant, has been shown to have a pressure dependence. Through lack of a better approach, equation (4) is used to correct sedimentation data for pressure effects. Correction for pressure and concentration effects when the polymer undergoes flotation (70) has been described. Other papers dealing with sedimentation that is dependent on concentration (71, 72) and on pressure (71) have appeared. Correction for diffusion (7) has been performed by extrapolating apparent distribution curves, corrected for pressure effects, to infinite time. The sedimentation velocity method cannot be regarded as an easy routine analysis for polydisperse synthetic polymers, but detailed studies of its use have been reported (12, 38, 59).

Molecular weights of proteins have been determined by a complete analysis of the sedimentation velocity experiment (73) by numerical integration of the Lamm differential equation.

$$\frac{dc}{dt} = \frac{1\partial}{r\partial r}\left[\frac{(D\partial c - S\omega^2 cr)r}{\partial r}\right] \tag{5}$$

Digital data acquisition was employed to obtained sedimentation profiles over a period of about 4 hours. The numerical evaluation took about 2 hours per run. Another approach (74) describes analysis of the early portion of the sedimentation velocity experiment using an approximate solution of the Lamm equation, to determine both S and S/D. Globular proteins with molecular weights of 10^4–10^6 were characterized using data obtained up to 50–80 minutes. Other treatments to solve the Lamm equation have been reported (75). By simultaneously determining sedimentation and diffusion coefficients, a new absolute method of molecular weight determination was obtained (76). Measurements have been reported which compare the precision of measurements, made using the polystyrene standards, for sedimentation and diffusion results with other methods (77). If the samples have narrow molecular weight distribution, then the results should be equal, however some differences were noted.

Johnston and Ogsten (78) pointed out that the concentration dependence of sedimentation coefficients causes the lack of proportionality of areas under the concentration gradient curves and the concentrations of the components. For the simple case of two components, one of which sediments much faster than the other, the faster component will sediment in the normal manner. For the slower component the sedimentation coefficient will be lower ahead of the first boundary with both components present than before the boundary where only the slower component is present. In extreme cases, for polydisperse systems or systems with highly concentration-dependent sedimentation coefficients, the expected broad sedimentation peak is not found and a much narrower peak is observed. Other discussions of the Johnston–Ogsten effect have appeared (79–81).

Branched polymers sediment faster than linear polymers of the same molecular weight because of the smaller hydrodynamic volume of the former. Branched polymers correspond to higher molecular weight species in the sedimentation velocity experiment but to a lower molecular weight in gel permeation chromatography. The distribution of long chain branching has been characterized using the combined results from these two experiments (82). Anomalous sedimentation behavior has been reported for branched polymers—for example, polymethyl methacrylate in *n*-butyl chloride (83). This was attributed to a lowering of the θ temperature by the branched structure.

Sedimentation of charge-carrying macromolecules involves some special considerations. The macroion will sediment faster than the counterions. The electric field set up will cause the macroion to move more slowly and the counterions to move faster; this is termed the primary charge effect. The counterions diffuse faster than the macroions, and the latter are pulled along by the former. Further complexities arise from the primary and secondary salt effects. For a discussion of polyelectrolyte sedimentation, see Eisenberg (84).

Equilibrium Sedimentation Method

In the equilibrium sedimentation approach, the ultracentrifuge is operated at much lower speeds (usually $< 20{,}000$ rpm) to establish a distribution of polymer along the radial direction of the cell. Equilibrium is established for each species when the sedimentation force is balanced by Brownian motion. The drawback to the method is the increased experimental time needed for the analysis: approximately 24 hours to 1 or 2 weeks. Several advantages are gained however. First, the method is based on sound thermodynamic principles, and absolute molecular weight distributions may be determined. Moreover, no calibration for the sedimentation velocity–molecular weight relationship is required, and pressure effects may be ignored because of the slower speeds involved. Other experimental benefits are gained over some alternate molecular weight techniques; for example, there is no need to remove dust particles, as is required in light-scattering techniques. For a monodisperse solute under ideal conditions in an incompressible solvent, the concentration C, at radius r, is derived as a function of the original solute concentration C_0, as follows (11)

$$C = \frac{\lambda M \exp(-\xi\lambda M) C_0}{1 - \exp(-\lambda M)} \tag{6}$$

where $\xi = (r^2 - a^2)/(b^2 - a^2)$ and a and b are the radii of the meniscus and the bottom of the cell.

Also, we can write

$$\lambda = \frac{(1 - \bar{v}\rho)\omega^2(b^2 - a^2)}{2RT}$$

For a polydisperse isopycnic solute with differential molecular weight distribution $f(M)$, the concentration profile in the cell is obtained by summing equation (6) over all M to give in integral form

$$C = C_0 \int_0^\infty \frac{\lambda M f(M) \exp(-\xi\lambda M) dM}{1 - \exp(-\lambda M)} \tag{7}$$

or in differential form

$$\frac{-dC}{d\xi} = C_0 \int_0^\infty \frac{\lambda^2 M^2 f(M) \exp(-\xi\lambda M) dM}{1 - \exp(-\lambda M)} \tag{8}$$

These equations may be used to generate equations for the various molecular weight averages.

The weight average molecular weight is given by

$$\bar{M}_w = \frac{1}{C_0\lambda} \int_0^1 \frac{dC}{d\xi} d\xi$$

Unfortunately, for averages higher than $\bar{M}_z$, determinations are difficult because of the increasing importance of measurements at the cell bottom. Difficulties also arise in determining the number average molecular weight $\bar{M}_n$, which can be evaluated only if the concentration falls to zero at some point in the cell. Increasing the rotor speed will achieve this, but conditions of solvent incompressibility must be maintained. Usually the application of equilibrium ultracentrifugation is limited therefore to the determination of $\bar{M}_w$ and $\bar{M}_z$. If ideal conditions do not prevail, then equations of the form

$$\frac{C_b - C_a}{\lambda C_0} = \bar{M}_w^* = \bar{M}_w - B_{ls}\bar{M}_w^2 C_0 + \cdots$$

are employed, where B_{ls} is the light-scattering second virial coefficient.

Since C_b and C_a, the equilibrium solution concentrations at the bottom and top of the solution column, are not measured directly, these values are obtained by extrapolation (56, 85). Results are then extrapolated to zero concentration and zero rotor speed if solvent compressibility has been encountered.

Results of comparisons of sedimentation equilibrium determinations of $\bar{M}_w$ have been compared with results from other methods (Table 1), and favorable agreement is observed. However the determination of accurate $\bar{M}_z$ values by other techniques is difficult, and similar comparisons cannot be made. The unique feature of sedimentation equilibrium is that at near θ conditions, absolute values for $\bar{M}_w$ and M_z may be obtained in a single experiment, and thus a measure of the variance of the molecular weight distribution may be derived.

Alternatively, data may be obtained at a fixed radius at various rotor speeds

TABLE 1. Weight Average Molecular Weights Determined by Sedimentation Equilibrium in the Ultracentrifuge (UC) and by Light Scattering (LS)

		$\bar{M}_w \times 10^{-3}$		
Sample	$\bar{M}_w/\bar{M}_n$	UC	LS	Ref.
Polycaprolactam	1.6	32.3	30	86
Polyethylene	1.4	126	130	86
Polyisobutylene	~1	75[a]	74[b]	87
Polymethylmethacrylate	~1	200[a]	204[b]	88
Polystyrene	~1	424	411	38
	~1	50.8	51	
	~1	1.07	1.08	
	~1	190	179	NBS SRM 705
	2.1	288	258	NBS SRM 706
	2.0	250	246C	
		37.4		56
				NBS SRM 1487

[a]Nonideal conditions.
[b]Viscosity average molecular weight.
[c]Gel permeation chromatography.

to calculate $\bar{M}_w$ and $\bar{M}_z$, and such values have been found (86) to agree with those obtained from fixed rotor speed experiments and measuring concentration profiles. Additionally, $\bar{M}_n$ was determined, and it compared favorably with osmometry results. Data obtained at various rotor speeds can be used to obtained results for polymers which exhibit heterogeneity in partial specific volume, second virial coefficient, or refractive index increment (88). Polyethylene has been characterized in terms of $\bar{M}_w$, $\bar{M}_z$ and $\bar{M}_{z+1}$ (89). Generally not enough averages are determined to accurately construct the molecular weight distribution curve by the method of moments.

The distribution of molecular weight may in principle be determined by solving the integral equation (8). This equation must be solved numerically, and it has been shown (90) that the solutions are inherently unstable after others had attempted solutions (91–93). More recent treatments have appeared (41, 91, 94–96). The Laplace transform method (91) is limited to unimodal distributions. Scholte's method (92) analyzes experimental data obtained at different rotor speeds and several different radii at each speed. Distributions of molecular weight were obtained from which the calculated values of $\bar{M}_n$, $\bar{M}_w$, $\bar{M}_z$, and $\bar{M}_{z+1}$ agreed well with those obtained from the normal sedimentation method.

A mixture of three anionically polymerized polystyrene samples was analyzed for molecular weight distribution by sedimentation equilibrium

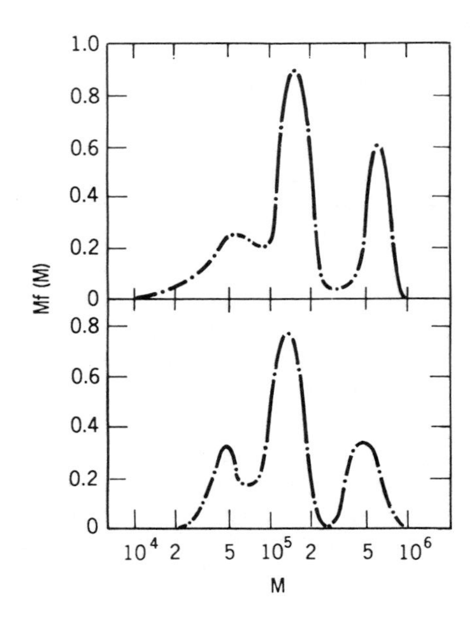

Figure 6. Plots of $MF(M)$ versus M for a trimodal blend of polystyrene. Upper curve: sedimentation velocity experiments. Lower curve: sedimentation equilibrium experiments using Scholte's linear programming method. (From Ref. 50.)

Table 2. Average Molecular Weights ($\times 10^{-3}$) of a Blend of Three Polystyrene Samples (65)

Method	$\bar{M}_n$	$\bar{M}_w$	$\bar{M}_z$
Calculated from components	102	204	385
From sedimentation equilibrium	106	202	362
From sedimentation velocity	101	205	365
From gel permeation chromatography	96	204	439

using seven different rotor speeds and also by gel permeation chromatography (65). Figure 6 shows the agreement between the sedimentation velocity and equilibrium sedimentation results. Table 2 shows the values for the molecular weight averages calculated from these two approaches and those obtained by gel permeation chromatography. Good agreement is observed between the results obtained and those calculated from the values for the individual components. The gel permeation chromatography data, however, are susceptible to calibration and band-broadening correction errors.

Applications to nonideal solutions have been described (95, 97–101). For a complete description of these correction methods see (References 50 and 102). One test involved characterizing the same material, a dextran, under θ and non-θ conditions (101). When the nonideal data were corrected, using the light-scattering second virial coefficient, good agreement was observed with the result obtained from the θ condition experiment (Fig. 7).

Fujita developed approximations to the sedimentation equilibrium equation for polydisperse nonideal solutions (103, 104), yielding an equation for $\bar{M}_w$ of similar form to that derived earlier (46) for monodisperse ideal solutions. Experimental tests of this approach were performed with two monodisperse polymers and this binary mixtures (105). It was reported that for nonideal solutions $\bar{M}_w$ values accurate to within $\pm 2\%$ could be achieved irrespective of the polydispersity value. Munk (106) has reported a method to determine $\bar{M}_w$, polydispersity, and second virial coefficient from a single sedimentation equilibrium run, if a unimodal Schultz–Zimm distribution can be assumed. The generalization of Fujita's equation for sedimentation equilibrium of nonideal polydisperse solutions (11) was reported by Suzuki (107) to give expressions for the higher molecular weight averages.

Experimental evaluations of methods to determine $\bar{M}_w$, $\bar{M}_z$, and the light-scattering second virial coefficient have been described (102). The application of nonlinear least-squares analysis to characterize nonideal self-associating macromolecular solutions has been reported (108, 109). It has been proposed that multiple regression analysis in conjunction with a simplex optimization

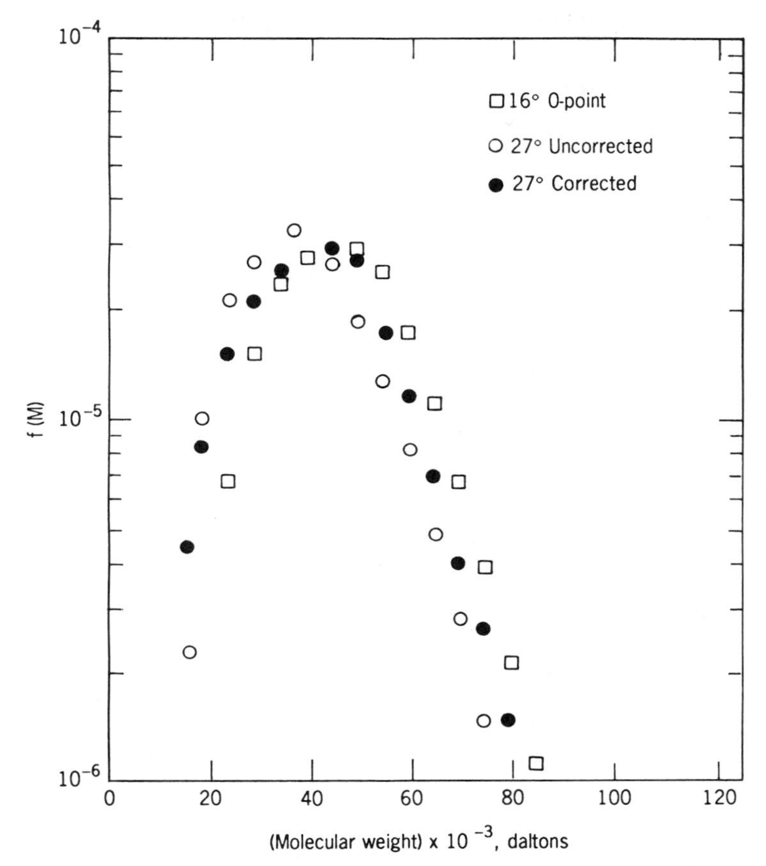

Figure 7. Molecular weight distributions of clinical dextran by Scholte's method. Experiments run at 16 and 27 °C. (From Ref. 101.)

routine (110), is more quantitative than the linear programming approach (111); these results were demonstrated using mixtures of known proteins.

Donnan effects have been investigated for equilibrium sedimentation experiments using cytochrome c in various ionic solutions (112). The use of salts with a partial specific volume of 1 (e.g., tetramethylammonium chloride) eliminates the Donnan effect (113).

Density Gradient Centrifugation Method

In the density gradient form of equilibrium centrifugation, the polymer is dissolved in a mixture of two miscible solvents which have different densities

(114, 115). This technique is useful for separating homopolymers, for characterizing tacticity of homopolymers, for determining the chemical composition distribution of copolymers, and for detecting homopolymers in copolymers. When the solution is centrifuged to equilibrium, a density gradient of solvent is established in the cell. A polymer will settle at the place where its effective buoyant density matches that of the solvent mixture. The solute band is Gaussian with a width proportional to the square root of the solute molecular weight in the ideal case (116). The centrifugation is carried out at intermediate to high speed, and an analysis requires many hours. The density gradient has a stabilizing effect, and some of the problems encountered in single solvent equilibrium centrifugation are eliminated.

The theory and practical techniques were developed around 1960 (116, 117). In biochemical applications, aqueous solutions of various salts have been widely employed to form density gradients. The theory has been developed for synthetic polymers (118–122). Practical difficulties involve the problem of achieving θ conditions and preferential sorption of one of the solvents by the polymer.

Applications to detect homopolymer impurity in graft copolymer (121) and polymethyl methacrylate in styrene–methacrylate copolymer (123) have been reported. In this case a bimodal distribution of solutes results, which is easily analyzed. The compositional heterogeneity of styrene–iodostyrene (122), styrene–butadiene (124), and styrene–methacrylate (123) copolymers has been determined. Such determinations require an assumption between the two variables, composition distribution, and molecular weight, before the compositional variance can be calculated. The density difference is dominant over molecular weight effects. The compositional distribution for different Kraton

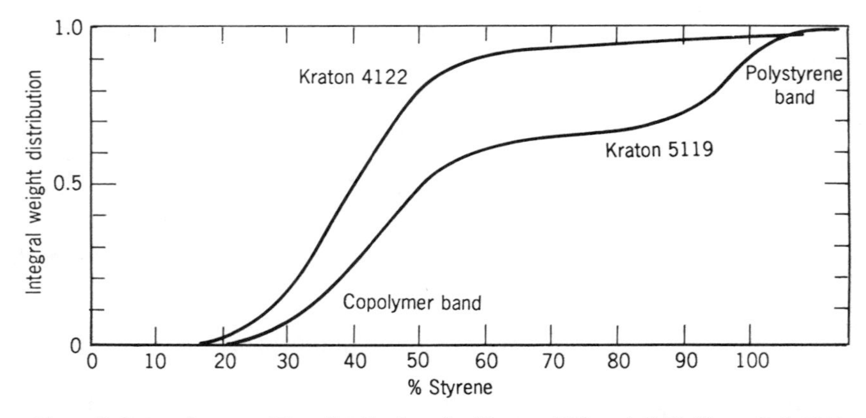

Figure 8. Integral composition distributions for Kraton 4112 and 5119. (From Ref. 124.)

samples (block copolymers of styrene and butadiene) is shown in Figure 8 (124).

Density gradient centrifugation has also been used to characterize the tacticity of poly(α-methylstyrene) (125) and the composition of nylon 6,6 (126). Many corrections are required when it is necessary to use good solvents, but some progress has been made (127, 128). Characterization of proteins by sedimentation equilibrium in density gradients has been reviewed (129). A method to determine the density gradient experimentally has been reported (130). Density gradient techniques are widely used in biochemistry (131). A theoretical analysis of the effects of solvation of macromolecules on sedimentation velocity in density gradients has been performed (132).

Approach to Equilibrium Method

The basic sedimentation equilibrium equation assumes that the equilibrium condition is one of no net flow anywhere in the cell. At the meniscus and at the base of the cell, this condition must hold at all times throughout the experiment. Archibald's analysis (133, 134) allows the values of $\bar{M}_w$ and the second virial coefficient to be determined from data obtained in the first 10–60 minutes of a run. Accurate concentration measurements at these points are difficult to make. Archibald proposed an extrapolation method based on experimental curves obtained at different times. Other mathematical techniques have been developed to obtain $\bar{M}_w$ from approach to equilibrium data (135–139).

Although principally used for monodisperse bipolymer characterization, approach to equilibrium method has also been applied to synthetic polymers. Its application to copolymers with molecular weight and compositional heterogeneity has been described (140). The Trautman method in a modified form (141) has been used to determine $\bar{M}_w$ for various synthetic and natural polyisoprenes (141). Nonlinear regression analysis has been applied (143) to approach to sedimentation equilibrium data to obtain estimates of molecular weight in one-tenth the time required for a complete sedimentation equilibrium experiment.

The meniscus depletion method is a subtechnique of the sedimentation equilibrium approach in which the concentration of solute at equilibrium at the meniscus–air interface is effectively zero (144). It was originally proposed to determine $\bar{M}_n$. Yphantis (145) proposed meniscus depletion methods, run under conditions that would ensure zero solute concentration at the meniscus. It was reinvestigated by Suzuki (146) using a high molecular weight polystyrene ($\bar{M}_w$ 2.7×10^7) with the objective of determining $\bar{M}_n$ of this material. Generally the concentration of solute at the bottom of the cell is too large for the optical system to be useful. Suzuki (146) used an extrapolation procedure to overcome this drawback and short columns (~ 2 mm) to keep the experimental time short Chervenka (147) developed a technique to use longer columns without long experimental times and applied the technique to ideal two component protein solutions. Budd (148) developed this approach to determine $\bar{M}_n$, $\bar{M}_w$, and $\bar{M}_z$ of the NBS standard polystyrene sample 706, and agreement with results from other methods was good. It was suggested that the method would be generally applicable as long as no significant amounts of low molecular weight material were present. This method has recently been applied to polyelectrolyte characterization (149).

Ultra-High Molecular Weight Characterization

The Svedberg method has been used to determine molecular weights up to 40×10^6 (150). The diffusion coefficients required for the calculation of molecular weight were obtained from intensity fluctuation spectroscopy (IFS) (38, 151). Molecular weight distributions were obtained by transforming the distribution of sedimentation coefficients obtained and correcting for concentration and diffusional broadening (152–154). The broadening of the boundary due to diffusion is proportional to the square root of time and that due to polydispersity is proportional to time (59); hence the diffusion dependence may be eliminated by extrapolation to infinite time (155). It appeared that a limit of 3×10^6 applies to the determination of MWD by sedimentation. Others (156) have used sedimentation velocity to investigate the molecular weight of polystyrene samples with molecular weights up to 1.2×10^7. Density gradient and sedimentation experiments have been used to establish the stability of polyacrylamide ($\bar{M}_w$ 8.4×10^6) in water over a period of 62 days (159).

Various ultracentrifuge techniques have been used to determine the high molecular weight of bacteriophages (Table 3). The use of combined spectro-

Table 3. Determination of High Molecular Weights by Various Ultracentrifugation Techniques

Sample	Method	$M \times 10^{-6}$	Ref.
T7 phage	Meniscus depletion	49.4	160
T7 DNA	CsCl density gradient	24.8	161
T5 DNA	CsCl density gradient	68.7	161
T4 DNA	CsCl density gradient	113	161
T7 phage	Svedberg equation	50.4	162
T5 phage	Svedberg equation	109.2	162
T4 phage	Svedberg equation	192.5	162

photometric and refractive index detectors in the Beckman model E analytical ultracentrifuge (163) has been employed to characterize bacteriophage $\phi 6$. Using experimentally determined turbidity and dn/dc values, the molecular weight was determined as 98.6×10^6 (164).

Latex particles may also be characterized using the ultracentrifuge (165, 166). The particles characterized were large enough to permit Mie scattering, and sedimentation coefficients were obtained, using a photoelectric scanner. The Stokes–Einstein relation was used to calculate particle radii.

Branching Characterization

Branching distribution characterization using the combined results of analytical techniques has been summarized (167). Using combined sedimentation velocity and diffusion results, the branching of polystyrene–divinylbenzene copolymers was achieved (168). Combined gel permeation chromatography and sedimentation results have also been used (82, 169) to elucidate details of branching distribution. Comparisons of sedimentation behavior for linear and four-branched star polystyrenes have been reported (170). Characterization of critically branched polycondensates and details of the calculation of the moments of the distribution have been described (171).

Copolymer Characterization

Archibald's method has been used to characterize block and random copolymers of methylmethacrylate and styrene (172). The extrapolation to zero concentration leads to an apparent molecular weight M_{app}, which is a function of the refractive index and density. By determining M_{app} in several solvents, the polydispersity parameters (the first and second moments of the z distribution function) and $\bar{M}_w$ may be obtained. The measurement of $\bar{M}_w$ for graft copolymers of starch and 1-amidoethylene in the range 0.19 to 3.7×10^6 has been reported (173). The buoyancy factors were calculated from plots of density versus copolymer concentration (174).

Preparative Ultracentrifugation

Fractionation or separation of polydisperse or mixtures of monodisperse polymers using ultracentrifugation in the preparative mode has been widely practiced. Equilibrium sedimentation has been performed in the preparative ultracentrifuge for many biopolymers. For example, a Beckman Spinco Airfuge operating at 5°C was used to characterize unpurified proteins (175), using a 3.8-cm diameter rotor spinning at 10,000 rpm at an angle of 18° to the axis of rotation, generating a force of 160,000g. After centrifugation for 24 hours, the sample was removed from the tube by a pipette in 10-μL aliquots and analyzed for protein content. This method used 1000-fold less sample than the analytical ultracentrifuge and was applicable over the molecular weight range 6,000–600,000. Later improvements to this method involved the development of a combined centrifuge tube and optical cell (176). At the conclusion of the centrifugation, the cell is transferred to an optically scanning spectrometer to determine concentration of solute versus position in the tube, and thus as a function of radial position. Similar equipment, used at higher rotational speeds (177) for 1–3 hours, allowed the rapid determination of sedimentation coefficients. These results were within a few percent of those obtained using the analytical ultracentrifuge.

The technique has been further improved for use in determining concentration gradients in small analytical ultracentrifuge tubes (178). Extreme sensitivity has been achieved using radiolabeled solutes. Simultaneous detection of species in a binary mixture has been achieved using differently labeled samples (viz., ^{14}C and ^{3}H) (179). To determine molecular weights from sedimentation coefficients, the diffusion coefficient must be found. Diffusion coefficients may be determined rapidly from intensity fluctuation spectroscopy (180, 181) or from ultracentrifugation measurements made using a synthetic boundary cell (182). For poor and intermediate solvents, the effect of polymer concentration is negligible. For good solvents and high molecular weight polymers, however, the effect of concentration cannot be ignored.

A new technique has been developed (183), combining sedimentation and low-angle laser light scattering (LALLS), to determine the MWD of high polymers. The polymer is separated by centrifugation in a Beckman L8-70 ultracentrifuge at 25,000–40,000 rpm (70,000–190,000g). After centrifugation, the tube contents are displaced by a dense liquid through a filter (0.4–1.0 μm) into a UV spectrophotometer and then through the LALLS instrument. The data together with a value for dn/dc allow $\bar{M}_w$ to be calculated at any point during the elution from the tube. This technique is particularly applicable to high molecular weight solutes because no large forces, which could cause degradation, are exerted on the polymer. Impurities which would interfere with the light-scattering data are sedimented before the polymer. The method has been used to study shear degradation of polyacrylamides (184).

Field Flow Fractionation

Field flow fractionation (FFF) is a broad-based technique (see Chapter 12), one variant of which uses a centrifugal field to effect a separation of molecules or particles in a flowing system. This is termed sedimentation field flow fractionation (SFFF) (185, 186). A dilute solution or suspension flows through a narrow, ribbonlike channel, which is located within a centrifuge. Systems

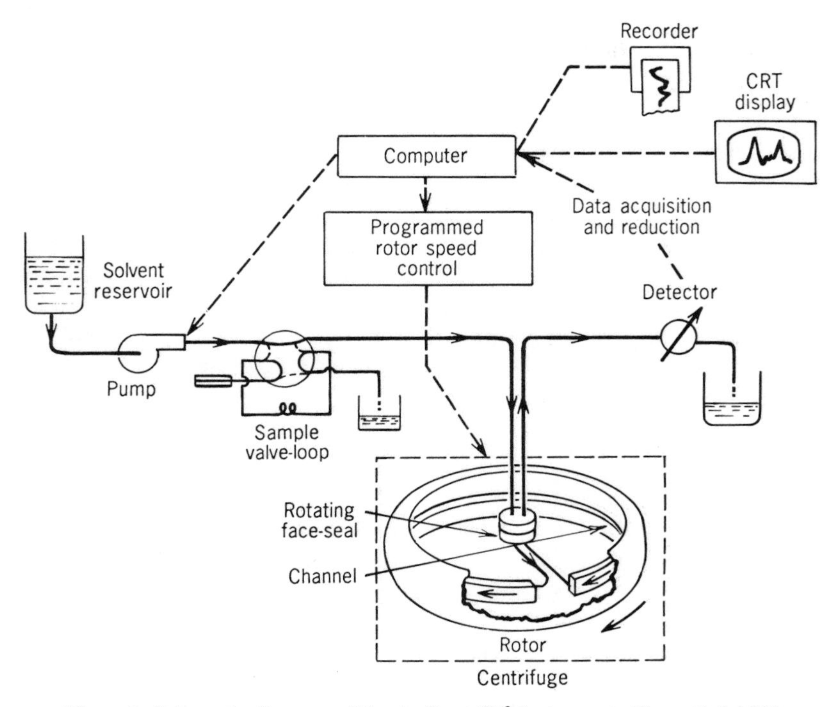

Figure 9. Schematic diagram of the du Pont SF^3 instrument. (From Ref. 189.)

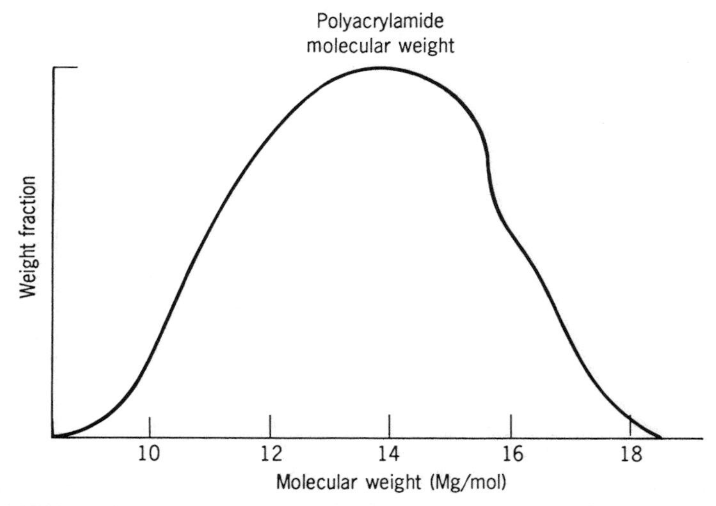

Figure 10. Molecular weight distribution of polyacrylamide determined by sedimentation field flow fractionation. (From Ref. 189.)

designed for aqueous and nonaqueous solvents have been developed (187). A unit marketed by du Pont, the SF^3 system (188), is shown schematically in Figure 9 (189). This instrument is designed to permit time-delayed exponential field decay (190, 191) to allow improved separations, and it uses higher centrifugal fields than the original equipment of Giddings. The instrument has been used to characterize high molecular weight polyacrylamide (188) as shown in Figure 10, as well as particulate systems. Dextran with molecular weight 5×10^7 has also been characterized (193). About 10,000g is required to characterize a molecular weight of 10^6; commercial equipment is now capable of operating at 30,000g (194, 195). The separation channel is a flat, ribbonlike space 250 μm thick, 25 mm wide, and 600 mm long, installed in a Sorvall high speed centrifuge. Another SFFF instrument (196) has been reported (197).

REFERENCES

1. T. Svedberg and R. Fahreus, *J. Am. Chem. Soc.*, **48**, 430 (1926).
2. T. Svedberg and J. B. Nichols, *J. Am. Chem. Soc.*, **49**, 2920 (1927).
3. E. O. Kraemer and W. D. Lansing, *J. Am. Chem. Soc.*, **55**, 4319 (1933).
4. E. O. Kraemer and W. D. Lansing, *J. Phys. Chem.*, **39**, 153 (1935).
5. R. Signer and H. Gross, *Helv. Chim. Acta*, **17**, 59, 335, 726 (1934).
6. H. W. McCormick, *J. Polym. Sci.*, **41**, 327 (1959).
7. J. W. Williams, R. L. Baldwin, W. M. Saunders, and P. G. Squire, *J. Am. Chem. Soc.*, **74**, 1542 (1952).
8. J. W. Williams and W. M. Saunders, *J. Phys. Chem.*, **58**, 854 (1954).
9. T. Svedberg and K. O. Pedersen, *The Ultracentrifuge*, Clarendon Press, Oxford, 1940.
10. H. K. Schachman, *Ultracentrifugation in Biochemistry*, Academic Press, New York, 1959.
11. H. Fujita, *Mathematical Theory of Sedimentation Analysis*, Academic Press, New York, 1962.
12. J. W. Williams, Ed. *Ultracentrifugal Analysis in Theory and Experiment*, Academic Press, New York, 1963.
13. T. J. Bowen, *An Introduction to Ultracentrifugation*, Wiley, London, 1970.
14. J. R. Cann, *Interacting Macromolecules: The Theory and Practice of Their Electrophoresis, Ultracentrifugation and Chromatography*, Academic Press, New York, 1970.
15. J. W. William, *Ultracentrifugation of Macromolecules*, Academic Press, New York, 1972.
16. J. S. McCall and B. J. Potter, *Ultracentrifugation*, Bailliere Tindall, London, 1973.
17. P. M. Lloyd, *Optical Methods in Ultracentrifugation, Electrophoresis and Diffusion*, Clarendon Press, Oxford, 1974.

18. J. Steensgaard, *Separation og Analysis of Macromolekyler ved s-zoneultra-centrifugering*, Arthus, 1974.
19. G. Batelier, *La Pratique de l'Ultracentrifugation Analytique*, Masson, Paris, 1979.
20. H. Fujita, *Foundations of Ultracentrifugal Analysis*, Wiley, London, 1975.
21. S. R. Rafikov, S. A. Pavlova, and I. I. Tverdokhlebova, *Determination of Molecular Weights and Polydispersity of High Polymers*, Israel Program for Scientific Translations Jerusalem, 1964, p. 142.
22. M. Wales, in "Characterization of Macromolecular Structure," publication 1573, National Academy of Sciences, Washington, DC, 1968, p. 343.
23. E. T. Adams, Jr., in "Characterization of Macromolecular Structure," publication 1573, National Academy of Sciences, Washington, DC, 1968, p. 84.
24. S. W. Provencher and W. Gobush, in "Characterization of Macromolecular Structure," publication 1573, National Academy of Sciences, Washington, DC, 1968, p. 143.
25. H. Eisenberg, in "Characterization of Macromolecular Structure," publication 1573, National Academy of Sciences, Washington, DC, 1968, p. 145.
26. P. C. Allen, E. A. Hill, and A. M. Stokes, *Plasma Proteins*, Blackwell, Oxford, 1978, p. 82.
27. T. G. Scholte, in *Polymer Molecular Weights*, Vol. 2, P. E. Slade, Jr., Ed., Dekker, New York, 1975, p. 501.
28. N. C. Billingham, *Molar Mass Measurements in Polymer Science*, Halsted (Division of Wiley), New York, 1977, p. 146.
29. H. G. Elias, *Macromolecules I*, 2nd ed., Plenum, New York, 1984, p. 329.
30. C. Tanford, *Physical Chemistry of Macromolecules*, Wiley, New York, 1961, p. 254.
31. F. W. Billmeyer, *Textbook of Polymer Science*, 3rd ed., Wiley, New york, 1984, p. 205.
32. D. J. Pollock and R. F. Kratz, in *Methods of Experimental Physics*, Vol. 16, Part A, R. A. Fava, Ed., Academic Press, New York, 1980, p. 65.
33. G. Reiss and P. Callot, in *Fractionation of Synthetic Polymers*, L. H. Tung, Ed., Dekker, New York, 1977, p. 476.
34. L.-O. Sundelof, in *Physical Chemistry of Colloids and Macromolecules*, B. Ranby, Ed., Blackwell, Oxford, 1987, p. 83.
35. A. J. Staverman, in *Physical Chemistry of Colloids and Macromolecules*, B. Ranby, Ed., Blackwell, Oxford, 1987, p. 88.
36. H. Suzuki, in *Physical Chemistry of Colloids and Macromolecules*, B. Ranby, Ed., Blackwell, Oxford, 1987, p. 101.
37. F. J. Bonner, in *Physical Chemistry of Colloids and Macromolecules*, B. Ranby, Ed., Blackwell, Oxford, 1987, p. 111.
38. R. Dietz, in *Industrial Polymers: Characterization by Molecular Weight*, J. H. S. Green and R. Dietz, Eds., Transcripta, London, 1973, p. 19.
39. E. T. Adams, Jr., P. J. Wan, D. A. Soucek, and G. H. Barlow, *Adv. Chem. Ser.*, **125**, 235 (1973).

40. A. H. Pekar, P. J. Wan, and E. T. Adams, Jr., *Adv. Chem. Ser.*, **125**, 260 (1973).
41. M. Gehatia and D. R. Wiff, *Adv. Chem. Ser.*, **125**, 216 (1973).
42. C. H. Chervenka, "A Manual of Methods for the Analytical Ultracentrifuge," Spinco Division of Beckman Instruments Inc., Palo Alto, CA, 1969.
43. H. W. McCormick, in *Polymer Fractionation*, M. J. R. Cantow, Ed., Academic Press, New York, 1967, p. 251.
44. J. E. Blair, *J. Polym. Sci., C*, **8**, 287 (1965).
45. D. R. Wiff and M. Gehatia, *J. Polym. Sci., Polym. Symp.*, **43**, 219 (1973).
46. J. W. Williams, K. E. van Holde, R. L. Baldwin, and H. Fujita, *Chem. Rev.*, **58**, 715 (1958).
47. R. C. Williams, Jr., and D. A. Yphantis, in *Encyclopedia of Polymer Science and Technology*, N. M. Bikales, Ed., Wiley-Interscience, New York, 1971, Vol. 14, p. 97.
48. J. F. Johnson, M. J. R. Cantow, and R. S. Porter, in *Encyclopedia of Polymer Science and Technology*, N. M. Bikales, Ed., Wiley-Interscience, New York, 1967, Vol. 7, p. 231.
49. L. H. Tung, in *Encyclopedia of Polymer Science and Engineering*, J. I. Kroschwitz, Ed., Wiley-Interscience, New York, 1987, 2nd ed., Vol. 7, p. 298.
50. E. T. Adams, Jr., W. E. Ferguson, P. J. Wan, J. L. Sarquis, and B. M. Escott, *Separ. Sci.*, **10**, 175 (1975).
51. Beckman Instruments, Inc., Palo Alto, CA.
52. W. Machtle and U. Klodwig, *Makromol. Chem.*, **180**, 2507 (1979).
53. H. K. Schachman and S. J. Edelstein, *Biochemistry*, **5**, 2681 (1966).
54. S. J. Edelstein and H. K. Schachman, *J. Biol. Chem.*, **242**, 306 (1967).
55. W. Machtle, in *Preprints of Short Communications, IUPAC MAKRO*, Mainz, September 1979, Vol. II, p. 731.
56. F. W. Wang and F. L. McCrackin, *Polymer*, **24**, 1541 (1983).
57. H. Svensson, *Acta Chem. Scand.*, **5**, 1301 (1951).
58. I. H. Billick and R. J. Bowen, *J. Phys. Chem.*, **69**, 4024 (1965).
59. R. L. Baldwin and K. E. van Holde, *Fortschr. Hochpolym-Forsch.*, **1**, 451 (1960).
60. M. Wales and S. J. Rehfeld, *J. Polym. Sci.*, **62**, 179 (1962).
61. J.-P. Merle and A. Sarko, *Macromolecules*, **5**, 132 (1972).
62. T. Bluhm and A. Sarko, *Macromolecules*, **6**, 578 (1973).
63. H. Elmgren, *J. Polym. Sci., Polym. Lett. Ed.*, **20**, 57 (1982).
64. L. H. Tung and J. R. Runyon, *J. Appl. Polym. Sci.*, **17**, 1589 (1973).
65. T. G. Scholte, *Eur. Polym. J.*, **6**, 51 (1970).
66. C. M. L. Atkinson and R. Dietz, *Br. Polym. J.*, **6**, 133 (1974).
67. M. Okabe, *J. Polym. Sci., Polym. Lett. Ed.*, **22**, 477 (1984).
68. R. J. Goldberg, *J. Phys. Chem.*, **57**, 194 (1953).
69. J. E. Blair and J. W. Williams, *J. Phys. Chem.*, **68**, 161 (1964).
70. J. W. A. van den Berg and P. LeGrand, *Eur. Polym. J.*, **18**, 51 (1982).

71. B. Nystrom and L.-O. Sundelof, *Chem. Scripta*, **10**, 16 (1976).
72. A. J. Rowe, *Bipolymer*, **16**, 2595 (1977).
73. C. Urbanke, B. Ziegler, and K. Stieglitz, *Fresenius Z. Anal. Chem.*, **301**, 139 (1980).
74. L. A. Holladay, *Biophys. Chem.*, **11**, 303 (1980).
75. R. Wohlschiess, K. F. Elgert, H. J. Cantow, and S. Bantle, *Ber Bunsenges. Phys. Chem.*, **83**, 367 (1979).
76. R. Wohlschiess, K. F. Elgert, and H.-J. Cantow, *Angew. Makromol. Chem.*, **74**, 323 (1978).
77. V. Petrus, B. Porsch, B. Nystrom, and L.-O. Sundelof, *Makromol. Chem.*, **183**, 1279 (1982).
78. J. P. Johnston and A. G. Ogsten, *Trans. Faraday Soc.*, **42**, 789 (1946).
79. F. A. H. Peters and A. J. Staverman, *Proc. K. Ned. Akad. Wet.*, **B85**, 273 (1982).
80. F. A. H. Peters and A. J. Staverman, *Proc. K. Ned. Akad. Wet.*, **B86**, 89 (1983).
81. A. J. Staverman, in *Physical Chemistry of Colloids and Macromolecules*, B. Ranby, Ed., Blackwell, Oxford, 1987, p. 89.
82. L. H. Tung, *J. Polym. Sci., A2*, **7**, 47 (1969).
83. M. Wales and H. Coll, paper presented at the Conference on Advances in Ultracentrifugal Analysis, New York, February 1968.
84. H. Eisenberg, *Biophys. Chem.*, **5**, 243 (1976).
85. E. G. Richards, D. C. Teller, and H. K. Schachman, *Biochemistry*, **7**, 1054 (1968).
86. T. G. Scholte, *J. Polym. Sci., A2*, **6**, 91 (1968).
87. D. A. Albright and J. W. Williams, *J. Phys. Chem.*, **71**, 2780 (1967).
88. T. Kotaka, N. Donkai, and H. Inagaki, *J. Polym. Sci., A2*, **9**, 1379 (1971).
89. A. Kotera, N. Iso, A. Senuma, and T. Hamada, *J. Polym. Sci., A2*, **5**, 277 (1967).
90. D. A. Lee, *J. Polym. Sci.*, **8**, 1039 (1970).
91. T. H. Donnelly, *J. Phys. Chem.*, **70**, 1862 (1966).
92. T. G. Scholte, *J. Polym. Sci., A2*, **6**, 11 (1968).
93. S. W. Provencher, *J. Chem. Phys.*, **46**, 3229 (1967).
94. M. Gehatia and D. R. Wiff, *Eur. Polym. J.*, **8**, 585 (1972).
95. E. T. Adams, P. J. Wan, D. A. Soucek, and G. H. Barlow, *Adv. Chem. Ser.*, **125**, 235 (1973).
96. D. R. Wiff and M. T. Gehatia, *Biophys. Chem.*, **5**, 199 (1976).
97. M. Gehatia and D. R. Wiff, *J. Polym. Sci., A2*, **8**, 2039 (1970).
98. M. Gehatia, *Polym. Prepr.*, **12**, 875 (1971).
99. D. R. Wiff and M. Gehatia, *J. Macromol. Sci. Phys.*, **B6**, 287 (1972).
100. M. Gehatia and D. R. Wiff, *J. Chem. Phys.*, **57**, 1070 (1972).
101. D. A. Soucek and E. T. Adams, Jr., *J. Colloid Interface Sci.*, **55**, 571 (1976).
102. P. J. Wan and E. T. Adams, Jr., *Biophys. Chem.*, **5**, 207 (1976).
103. H. Fujita, *J. Phys. Chem.*, **63**, 1326 (1059).
104. H. Fujita, *J. Phys. Chem.*, **73**, 1759 (1969).
105. H. Uliyama, N. Tagata, and M. Kurata, *J. Phys. Chem.*, **73**, 1448 (1969).

106. P. Munk, *Macromoleculers*, **13**, 1215 (1980).
107. H. Suzuki, *Bull. Inst. Chem. Res. Kyoto Univ.*, **56**, 89 (1978).
108. M. L. Johnson, J. J. Correia, D. A. Yphantis, and H. R. Halvorson, *Biophys. J.*, **36**, 575 (1981).
109. M. L. Johnson and S. G. Frasier, *Methods Enzymol.*, **117**, 301 (1985).
110. S. Nakai and F. Van de Voort, *J. Dairy Res.*, **46**, 283 (1979).
111. T. G. Scholte, *Ann. New York Acad. Sci.*, **164**, 156 (1969).
112. M. Syvanen and H. K. Schachman, *Bipolymers*, **17**, 943 (1978).
113. R. Zicardi and V. Schumaker, *Biopolymers*, **10**, 1701 (1961).
114. J. J. Hermans and H. A. Ende, in *Newer Methods of Polymer Characterization*, B. Ke, Ed., Wiley-Interscience, New York, 1964.
115. H. W. McCormick, in *Polymer Fractionation*, M. J. R. Cantow, Ed., Academic Press, New York, 1967, Chap. C.2.
116. M. Meselson, F. W. Stahl, and J. Vinograd, *Proc. Natl. Acad. Sci. U.S.*, **43**, 581 (1957).
117. J. E. Hearst and J. Vinograd, *Proc. Natl. Acad. Sci. U.S.*, **47**, 999 (1961).
118. J. J. Hermans and H. A. Ende, *J. Polym. Sci.*, **C1**, 161 (1963).
119. J. J. Hermans and H. A. Ende, *J. Polym. Sci.*, **C4**, 519 (1963).
120. J. J. Hermans, *J. Chem. Phys.*, **38**, 597 (1963).
121. H. A. Ende and V. T. Stannett, *J. Polym. Sci.*, **A2**, 4047 (1964).
122. H. A. Ende and J. J. Hermans, *J. Polym. Sci.*, **A2**, 4053 (1964).
123. A. Nakazawa and J. J. Hermans, *J. Polym. Sci., A2*, **9**, 1871 (1971).
124. C. J. Stacy, *J. Appl. Polym. Sci.*, **21**, 2231 (1977).
125. J. M. G. Cowie and P. M. Toporowski, *Eur. Polym. J.*, **5**, 493 (1969).
126. J. J. Burke and T. A. Orofino, *J. Polym. Sci., A2*, **7**, 1 (1969).
127. J. Dayantis and H. Benoit, *J. Chim. Phys.*, **61**, 773 (1964).
128. M. Jacob, J. Dayantis, and H. Benoit, *J. Chim. Phys.*, **62**, 73 (1965).
129. J. B. Ifft, in *Analytical Methods of Protein Chemistry*, Vol. 5, P. Alexander and H. P. Lundgren, Eds., Pergamon Press, Elmsford, NY, 1969, p. 151.
130. P. Munk, *Macromoleculers*, **15**, 500 (1982).
131. R. Hinton and M. Dobrota, *Density Gradient Centrifugation*, North-Holland, Amsterdam, 1976.
132. P. Nieuwenhuysen, *Biopolymers*, **18**, 277 (1979).
133. W. J. Archibald, *J. Phys. Colloid Chem.*, **51**, 1204 (1947).
134. W. J. Archibald, *J. Appl. Phys.*, **18**, 362 (1947).
135. R. L. Baldwin, *Biochem. J.*, **44**, 644 (1953).
136. H. Gutfreund and A. G. Ogsten, *Biochem. J.*, **44**, 163 (1949).
137. G. Kegeles, S. M. Klainer, and W. J. Salem, *J. Phys. Chem.*, **61**, 1286 (1957).
138. R. Trautman, *J. Phys. Chem.*, **60**, 1211 (1956).
139. R. Trautman and C. F. Compton, *J. Am. Chem. Soc.*, **81**, 4036 (1959).

140. T. Kotaka, N. Donkai, H. Ohnuma, and H. Inagaki, *J. Polym. Sci., A2*, **6**, 1803 (1968).
141. H. C. Nielsen, G. E. Babcock, and F. R. Senti, *Arch. Biochem. Biophys.*, **96**, 252 (1962).
142. C. L. Swanson, M. E. Carr, and H. C. Niesen, *J. Polym. Mater.*, **3**, 211 (1986).
143. L. A. Holladay, *Biophys. Chem.*, **10**, 183 (1979).
144. W. O. Lansing and E. O. Kraemer, *J. Am. Chem. Soc.*, **57**, 1369 (1935).
145. D. A..Yphantis, *Biochemistry*, **3**, 297 (1964).
146. H. Suzuki, *Br. Polym. J.*, **11**, 91 (1979).
147. C. H. Chervenka, *Anal. Biochem.*, **34**, 24 (1970).
148. P. M. Budd, *J. Polym. Sci., B, Polym. Phys. Ed.*, **26**, 1143 (1988).
149. P. M. Budd, *Br. Polym. J.*, **20**, 33 (1988).
150. B. Appelt and G. Meyerhoff, *Macromolecules*, **13**, 657 (1980).
151. P. N. Pusey, in *Industrial Polymers: Characterization by Molecular Weight*, Transcripta, London, 1973, p. 26.
152. H.-J. Cantow, *Makromol. Chem.*, **30**, 169 (1959).
153. H. W. McCormick, *J. Polym. Sci.*, **36**, 341 (1959).
154. C. W. Pyun and M. Fixman, *J. Chem. Phys.*, **41**, 937 (1965).
155. A. F. V. Ericsson, *Acta. Chem. Scand.*, **10**, 360 (1965).
156. T. Fujimoto and M. Nagasawa, *Polym. J.*, **7**, 397 (1975).
157. W. Machtle, *Makromol. Chem.*, **185**, 1025 (1984).
158. W. Machtle, *Colloid Polym. Sci.*, **262**, 270 (1984).
159. W. Machtle, *Makromol. Chem.*, **183**, 2515 (1982).
160. F. C. Bancroft and D. Freifelder, *J. Mol. Biol.*, **54**, 537. (1970).
161. C. W. Schmid and J. E. Hearst, *Biopolymers*, **10**, 1901 (1971).
162. S. B. Dubin, in *Methods in Enzymology*, Vol. 26, C. H. W. Hirs and S. N. Timasheft, Eds., Academic Press, New York, 1972, p. 119.
163. S. A. Berkowitz and L. A. Day, *Biochemistry*, **19**, 1969 (1980).
164. L. A. Day and L. Mindich, *Virology*, **103**, 376 (1980).
165. J. W. A. Averink, H. Reerink, J. Boerma, and W. J. M. Jaspers, *J. Colloid Interface Sci.*, **21**, 66 (1966).
166. H.-J. Cantow, *Makromol. Chem.*, **70**, 130 (1964).
167. Z. Dobkowski, in *Applied Polymer Analysis and Characterization: Recent Developments in Techniques, Instrumentation and Problem Solving*, J. W. Mitchell, Ed., Macmillan, New York, 1987, p. 341.
168. H. Matsuda, I. Yamada, and S. Kuroiwa, *Polym. J.*, **8**, 415 (1976).
169. L. H. Tung, *J. Polym. Sci., A2*, **9**, 759 (1971).
170. B. Nystrom and J. Roots, *Polymer*, **21**, 183 (1980).
171. C. G. Leonis, H. Suzuki, and M. Gordon, *Makromol. Chem.*, **178**, 2867 (1977).

172. T. Kotaka, N. Donkai, H. Ohnuma, and H. Inagaki, *J. Polym. Sci., A2*, **6**, 1803 (1968).
173. J. J. Meister, M. L. Sha, and E. G. Richards, *J. Appl. Polym. Sci.*, **33**, 1873 (1987).
174. H. K. Eisenberg and E. F. Casassa, *J. Polym. Sci.*, **49**, 29 (1960).
175. R. J. Pollet, B. A. Haase, and M. L. Standaert, *J. Biol. Chem.*, **254**, 30 (1979).
176. A. K. Attri and A. P. Minton, *Anal. Biochem.*, **133**, 142 (1983).
177. A. K. Attri and A. P. Minton, *Anal. Biochem.*, **136**, 407 (1983).
178. A. K. Attri and A. P. Minton, *Anal. Biochem.*, **152**, 319 (1986).
179. A. K. Attri and A. P. Minton, *Anal. Biochem.*, **162**, 409 (1987).
180. M. E. M. McDonnell and A. M. Jamieson, *J. Macromol. Sci. Phys.*, **13**, 67 (1977).
181. S. W. Provencher, J. Hendrix, L. De Maeyer, and N. Paulussen, *J. Chem. Phys.*, **69**, 4273 (1978).
182. T. M. Aminabhavi and P. Munk, *Macromolecules*, **12**, 1194 (1979).
183. G. Holzwarth, L. Soni, and D. N. Schultz, *Macromolecules*, **19**, 422 (1986).
184. G. Holzwarth, L. Soni, D. N. Schultz, and J. Bock, *Polym. Mater. Sci. Eng.*, **55**, 511 (1986).
185. J. C. Giddings, M. N. Myers, K. D. Caldwell, and S. R. Fisher, in *Methods of Biochemical Analysis*, Vol. 26, D. Glick, Ed., Wiley, New York, 1980, p. 79.
186. J. C. Giddings and K. D. Caldwell, in *Physical Methods in Chemistry* Vol. IIIB, B. W. Rossiter, Ed., Wiley, New York, 1989.
187. K. D. Caldwell, G. Karaiskakis, M. N. Myers, and J. C. Giddings, *J. Pharm. Sci.*, **70**, 1350 (1981).
188. J. J. Kirkland, C. H. Dilks, and W. W. Yau, *J. Chromatogr.*, **255**, 255 (1983).
189. Du Pont Company, Materials Characterization Systems, Wilmington, Del.
190. W. W. Yau and J. J. Kirkland, *Separ. Sci. Technol.*, **16**, 577 (1981).
191. J. J. Kirkland, S. W. Rementer, and W. W. Yau, *Anal. Chem.*, **53**, 1730 (1981).
192. J. J. Kirkland, W. W. Yau, and W. A. Doerner, *Anal. Chem.*, **52**, 1944 (1980).
193. L. E. Oppenheimer and T. H. Maury, *J. Chromatogr.*, **298**, 217 (1984).
194. G. B. Levy, *Am. Lab.*, June 1987, p. 84.
195. R. L. Blaine, *Res. Dev.*, September, 1987, p. 78.
196. FFFractionation Inc., Salt Lake City, UT.
197. J. C. Giddings, *Chem. Eng. News*, Oct. 10, 1988, p. 34.

CHAPTER

8

VISCOMETRIC DETERMINATION OF MOLECULAR WEIGHT

KENJI KAMIDE and MASATOSHI SAITO

Fundamental Research Laboratory of Fibers and Fiber Forming Polymers Asahi Chemical Industry, Company Ltd. Hacchonawatecho, Takatsuki, Osaka, Japan

INTRODUCTION

The method for the determinating molecular weight by viscometry, first applied by Staudinger and Freudenberger (1) to cellulose and cellulose derivatives, played an important role in establishing the concept of macromolecules. Now this is one of the most familiar methods, being widely used in fundamental research in polymer science as well as in industrial fields. The viscometric method is classified as an indirect determination method and has various advantages over direct methods. For example, viscosity is easy to measure, hence saves time, and uses relatively inexpensive apparatus compared with other methods, such as light scattering and sedimentation. Rigorous preparation of the solution for measurements is unnecessary, contrary to the case of light-scattering measurements. On the other hand, the theoretical foundation of the viscosity of a polymer solution has not been completely established, even for linear polymers in the dilute region, and there are some problems still to be solved. Consequently, some uncertainties remain in the determination of molecular weight using the relationship between the limiting viscosity number $[\eta]$ and the molecular weight. Another noteworthy point is that the molecular weight determined by this method is not an intrinsic value of the given samples but depends on the parameter in the viscosity equation used. Nevertheless methods involving viscosity probably will continue to be of paramount importance in the characterization of polymer molecules.

In this chapter we introduce theories on the viscosity of linear polymers in dilute solution and the methods to determine the molecular weight using the molecular weight dependence of the limiting viscosity number.

VISCOSITY OF DILUTE POLYMER SOLUTIONS

Viscosity

Most polymer solutions exhibit a significant viscous nature. When an external force is exerted on these solutions, an attractive force works between any two arbitrarily chosen points in the solution.

Figure 1 illustrates laminar flow, which has only one component of velocity $\mathbf{v}$ proportional to coordinate z, that is,

$$\mathbf{v} = \begin{pmatrix} 0 \\ gz \\ 0 \end{pmatrix} \tag{1}$$

where g is the velocity gradient. In the flow we choose arbitrarily the adjacent cubes I and II, whose centers of mass are a distance dz apart. Cube I applies the attractive force on cube II in the positive y-axis direction. The reaction of cube II to the opposite side on cube I results in a shear stress σ_s in plane dA (Fig. 1). The stress σ_s is related to the gradient of the fluid velocity to coordinate z as follows:

$$\sigma_s = \eta\, dv/dz \qquad (= \eta g) \tag{2}$$

where η is referred to as the viscosity or viscosity coefficient. A fluid for which η is independent of the velocity gradient at a given temperature is called a Newtonian fluid. The viscosity of polymer or colloidal solutions generally

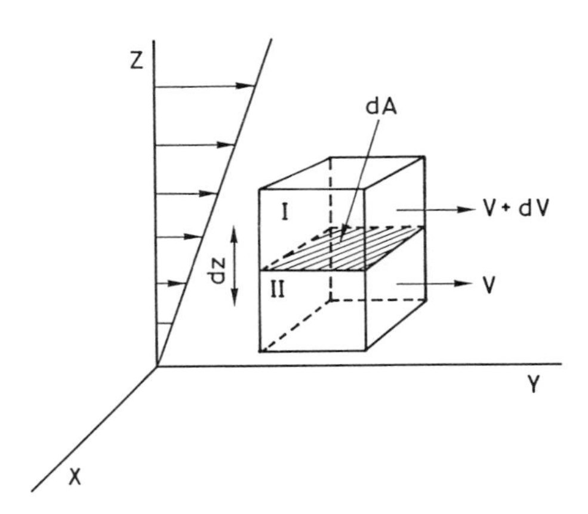

Figure 1. Two adjacent cubes in laminar flow.

indicates significant velocity gradient dependence, and in most cases η decreases with an increase of velocity gradient, especially in the high gradient region.

The viscosity of a polymer solution also depends on the concentration c of the polymer; in the region of low concentration, η is expressed in the expansion form

$$\eta = \eta_0\{1 + [\eta]c + Kc^2 + \text{higher term of } c^3\} \tag{3}$$

where η_0 is the viscosity of the solvent; $[\eta]$ is the limiting viscosity number (first designated by Kraemer (2) as intrinsic viscosity), which physically represents the increase in viscosity of the solution due to introduction of the polymer molecules; and K is a parameter independent of concentration and sometimes dependent on g. The concentration c is usually given as grams of solute per cubic centimeter (g/cm^3). In the analysis of the experimental data, the ratio η/η_0 is called the relative viscosity η_r, and the ratio $(\eta - \eta_0)/\eta_0$ is the specific viscosity η_{sp}.

For the purpose of determining the molecular weight of polymers, the relationship between η or $[\eta]$ and molecular weight is required. If solvent is present for the polymer concerned, $[\eta]$, extrapolated to zero shear rate for the case of relatively high molecular weight polymers, is preferred to η, taking into account of the accuracy of the data and ease of measurements. In undiluted polymers and in highly concentrated solutions, we measure η at constant temperature; then greater attention is paid to the shear rate dependence of η.

Theory of Viscosity of Dilute Polymer Solutions

Molecular Weight Dependence of the Limiting Viscosity Number

In the case of polymer solutions, the ratio η_{sp}/c exhibits significant concentration dependence, so that $[\eta]$ defined in equation (4) is frequently used as an index of the viscosity.

$$[\eta] = \lim_{c \to 0} (\eta_{sp}/c) \tag{4}$$

A theory on the viscosity of dilute colloidal suspensions was proposed in 1906 and 1911 by Einstein (3,4), who derived the equation for $[\eta]$ of colloidal solutions as a function of the specific volume of the solute v as follows (see Appendix A):

$$[\eta] = 2.5v \tag{5}$$

Many assumptions were made in the course of deriving equation (5): for example, the colloidal particles were considered to be rigid spheres moving in a continuous medium of the solvent, and the velocity of the solvent on the surface of the particles was taken to be zero (i.e., the surface of the particle was fully draining). The extension of Einstein's theory to semidilute colloidal solutions has been worked out by Guth and Simha (5) and Gold (6), and third term $14.1v^2$ was added to the right-hand side of equation (5). The viscosity of solutions including non-spherical colloidal particles was derived for ellipsoids by Jeffery (7) and for rods and disks in the framework of Einstein's treatment, by Simha (8).

Einstein's equation successfully explained the viscosity of dilute colloidal solutions of glass spheres, fungi spores, and sulfur (9, 10). However the value of $[\eta]$ of typical polymer solutions was several hundred times larger than that calculated through equation (5), and Einstein's equation must be modified to be applicable to polymer solutions.

On the bases of the rigid rod model for a polymer molecule in solvent, Staudinger (1, 11–13a) generalized Einstein's equation assuming that $[\eta]$ was proportional to the volume swept out by the suspended particles. The proposed viscosity equation, which is called Staudinger's law, is

$$[\eta] = K''_m M \tag{6}$$

where K''_m is the characteristic parameter predominantly determined by given homologous polymer groups. However, experimental data collected did not always agree with this law. For example, Signer (13b) obtained in the ultracentrifuge average molecular weights of 3×10^4, 8×10^4, and 3×10^5 for polystyrene samples for which Staudinger had estimated 1.5×10^4, 3.1×10^4, and 1.35×10^5. Rather, equation (6) is correct for polymers of very low molecular weight.

After a semiempirical modification of Staudinger's law made independently by Mark (14), Houwink (15), and Sakurada (16) (MHS), it became possible to express the viscosity equation in a more generalized form than equation (6), as

$$[\eta] = K_m M^{\mathsf{a}} \tag{7}$$

where K_m is a parameter determined by the combination of polymer and solvent at constant temperature. The power a of the given polymer–solvent combination at constant temperature remains constant over a wide range of molecular weights.

Many attempts to provide a theoretical foundation for the molecular weight dependence of $[\eta]$ for linear polymer solutions were made between 1932 and 1946 by Huggins (17–19), Hermans (20), Kuhn (21–23), Kraemer

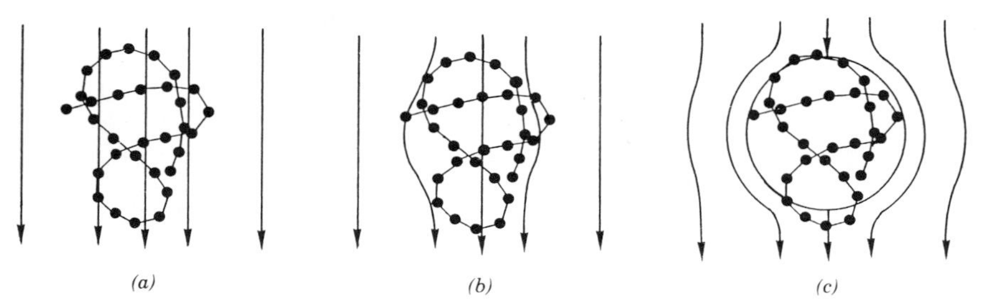

Figure 2. Polymer molecules in the flow of a solvent: (*a*) perfectly free draining, (*b*) partially free draining, (*c*) perfectly nondraining.

(24), and Debye (25). Debye (25) (1946) derived $[\eta]$ as a function of the average square end-to-end distance of polymer molecules in solution $\langle R^2 \rangle_0$ as follows:

$$[\eta] = \tfrac{1}{36}(\zeta/m\eta_0)\langle R^2 \rangle_0 \tag{8}$$

where ζ is the coefficient of friction between a segment and a solvent, and m is the molecular weight of a segment. In equation (8), Debye assumed that the distribution function of the segments around the center of mass of the polymer chain was Gaussian and that solvent molecules could pass freely through the polymer chain (perfectly free draining) as shown in Figure 2*a*. For Gaussian chains, $\langle R^2 \rangle_0$ is proportional to the molecular weight, resulting in $[\eta] \propto M$, which reduces to Staudinger's law. In real polymer solutions, solvent molecules near or inside the polymer chain region seem to be subjected to a disturbance, and the assumption of perfectly free drainage is not always acceptable. On the other hand, $[n]$ corresponding to the perfectly nondraining state is expected to be proportional to $M^{1/2}$ from Einstein's equation (eq. 5) (26, 27), if a polymer chain is regarded as a sphere with diameter $\langle R^2 \rangle_0^{1/2}$ and specific volume proportional to $\langle R^2 \rangle_0^{3/2}/M$.

The formulation of $[\eta]$ under condition of partial free drainage (Fig. 2*b*) was derived independently by Debye and Büche (28) (DB) and Kirkwood and Riseman (29) (KR) in 1948. The models used in these theories and the final formulas are summarized in Table 1. Appendix B gives the derivation of the KR equation.

In the DB theory, the parameter x (representing the shielding effect) ranges from 0 at free draining state to infinite at perfect nondrainage. The power **a** in the MHS equation, which is related to the parameter x (eq. 9), changes from 0.5 to 1 corresponding to $x = 0$ to ∞, respectively.

$$\mathbf{a} = \tfrac{1}{2} + \tfrac{1}{4}(x/\phi)(d\phi/dx) \tag{9}$$

Table 1. Comparison of the Debye–Büche (BD), Kirwood–Riseman (KR), and Kurata–Yamakawa (KY) Theories

Theory (Ref.)	Model	Deriving Method	Theoretical Expression for $[\eta]$
BD	Chain segment disperses in sphere of hydrodynamic diameter R_s; solvent molecule can pass through the sphere	Rigorous solution	$[\eta] = (4/3)\pi R_s^3 \phi(x)/M$ $\phi(x) = \dfrac{0.025[1 + (3/x^3) - (3/x)\coth x]}{1 + (10/x^2)[1 + (3/x^2) - (3/x)\coth x]}$
KR	Pearl necklace model (link length a'); hydrodynamic interaction between pearls is considered using Oseen's method	Approximate solution	$[\eta] = \dfrac{\zeta N_A a'^2 n F_0(X)}{36\eta_0 m}$ $F_0(X) = (6/\pi)\sum \dfrac{1}{k^2(1 + X/k^{1/2})}$ $X = \dfrac{\zeta n^{1/2}}{(6\pi^3)\eta_0 a'}$
KY	Pearl necklace model, based on the KR theory; excluded volume effect is taken into account	Approximate solution	$[\eta] = \dfrac{\zeta N_A a'^2 n F(X)}{36\eta_0 m}$ $F(X) = F_0[1 + p(X)z - \cdots]$, where p is a function of X: see text

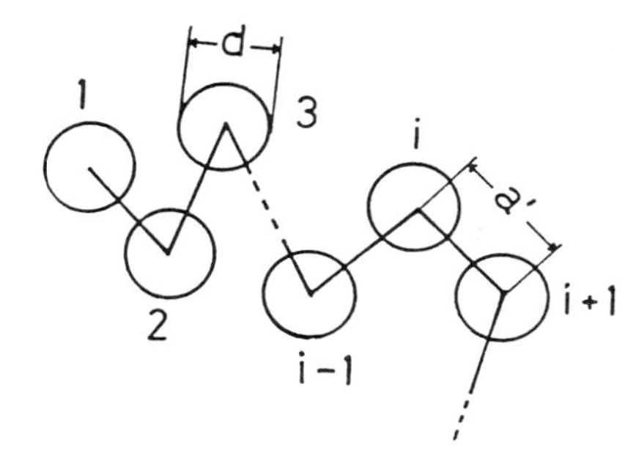

Figure 3. Schematic representation of the pearl necklace model.

Using the pearl necklace model (Fig. 3), Kirkwood and Riseman took into account the hydrodynamic interaction between polymer segments using Oseen's procedure, and the exponent **a** in the MHS equation was related to the draining parameter X as follows:

$$\mathbf{a} = 1 + (\tfrac{1}{2})[d \ln F_0(X)/d \ln X] \tag{10}$$

where

$$X = \zeta n^{1/2}/\{(6\pi^3)^{1/2} a' \eta_0\} \tag{11}$$

and a' and n are the length of a link and the number of segments, respectively, in the pearl necklace model. Table 2 lists the exponent **a** values calculated through equations (9) and (10). Both parameters, x in the DB theory and X in the KR theory, increase with an increase of molecular weight and consequently the exponent **a** decreases. However the **a** value experimentally obtained for typical polymer solutions remains constant over a wide range of molecular weights, and the DB and KR theories cannot rigorously explain this experimental fact.

In addition, these two theories rely on the assumption that even in a good solvent, the polymeric chain behaves as a random coil. But the segments of the polymer chain are not able to occupy the same place at the same time (physical excluded volume effect), and in a good solvent, the polymeric chain expands as a result of strong polymer–solvent interaction (solvent effect) in the equilibrium state. From these effects, the hypothesis of a random coil configuration, irrespective of the kind of the solvent, is not always realized. The important role of these two effects, which are referred to simply as the excluded volume effect, in determining the chain configuration in polymer solutions was first suggested by Flory (30, 31). Later, in the 1950s, studies were made by Grimley (32), Krigbaum (33), James (34), Zimm (35), Fixman (36), and Kurata et al. (37, 38).

Table 2. Numerical Values of Viscosity Functions in the Debye–Büche, Kirkwood–Riseman, and Kurata–Yamakawa Theories

	Debye–Büche[a]		Kirwood–Riseman[b]			Kurata–Yamakawa[c]		
x	$\phi(x)$	a	X	$XF_0(X)$	a	X	$XF_0(X)$	a
0	0.000	1.000	0	0.000	1.000	0	0.000	1.000
1	0.0947	0.973	0.1	0.092	0.963	0.1	0.073	0.963
2	0.327	0.910	0.3	0.242	0.905	0.3	0.192	0.905
3	0.600	0.839	0.5	0.358	0.862	0.5	0.284	0.862
4	0.857	0.778	1.0	0.564	0.791	1.0	0.447	0.791
5	1.07	0.731	1.5	0.700	0.746	2.0	0.634	0.715
6	1.25	0.693	2.5	0.875	0.691	3.0	0.742	0.674
7	1.40	0.664	3.5	0.984	0.66	4.0	0.812	0.647
8	1.52	0.642	4.0	1.024	0.647	5.0	0.864	0.632
9	1.62	0.625	5.0	1.090	0.632	10.0	0.999	0.583
10	1.72	0.611	10	1.260	0.583	20.0	1.110	0.552
20	2.10	0.549	50	1.486	0.524	50.0	1.178	0.524
∞	2.50	0.500	∞	1.588	0.500	∞	1.259	0.500

[a] Reference 28.
[b] Reference 29.
[c] Reference 38.

Flory (39) derived the following equations for $[\eta]$, based on the KR theory and taking into consideration the excluded volume effect

$$[\eta] = (\pi/6)^{3/2} N_A X' F(X') \langle R^2 \rangle^{3/2} / M \tag{12}$$

and

$$X' = \zeta n / \{(6\pi^3 \langle R^2 \rangle)^{1/2} \eta_0\} \tag{13}$$

where N_A is Avogadro's number, $F(X')$ the function derived by KR theory, and $\langle R^2 \rangle^{1/2}$ the expanded end-to-end distance of the polymer chain in the solution. At the limit of $X \to \infty$, equation (12) can be expressed as

$$[\eta] = \Phi \langle R^2 \rangle^{3/2} / M \tag{14}$$

where

$$\Phi = (\pi/6)^{3/2} N_A [X' F(X')]_{x \to \infty} \tag{15}$$

Flory and Fox (FF) (40) set forth the assumptions that the molecular weight and solvent dependence of $[\eta]$ are more closely related to the excluded volume effect than to the draining effect and that Φ is independent of the molecular weight.

The parameter representing the expansion of the polymer chain (expansion coefficient) is defined as

$$\alpha = \langle R^2 \rangle^{1/2} / \langle R^2 \rangle_0^{1/2} \tag{16}$$

Substitution of α into equation (14) gives

$$[\eta] = KM^{1/2}\alpha^3 \tag{17}$$

where

$$K = \Phi(\langle R^2 \rangle_0 / M)^{3/2} \tag{18}$$

When Φ and the ratio $\langle R^2 \rangle_0/M$ are independent of M, Flory's constant K is invariable. Considering the swelling of an isolated polymer coil in which an equilibrium is attained between osmotic forces favoring the dilution of polymer segments and rubberlike elasticity opposing coil expansion, Flory (39) obtained the relation between α and M, so-called fifth power-law type of equation:

$$\alpha^5 - \alpha^3 = C(\tfrac{1}{2} - \chi)M^{1/2} \tag{19}$$

The constant C is inversely proportional to the molar volume of the solvent, and χ is the polymer–solvent interaction parameter in the Flory–Huggins (40–44) theory. In the region of very high molecular weight, α^5 is expected to be proportional to $M^{1/2}$ from equation (19), and $[\eta]$ in equation (17) tends to increase proportionately to $M^{4/5}$. A criticism to this theory has been made by Krigbaum and Carpenter (45), who have indicated that contrary to the prediction of Flory and Fox, the hydrodynamic radius of a polymer coil increases less rapidly than the statistical radius as the excluded volume increases.

Kurata and Yamakawa (38) (KY) developed a theory of $[\eta]$ on the basis of the KR scheme, taking account of the influence of the excluded volume effect on the chain configuration. The KY formulation of $[\eta]$ is given in Table 1, where $F_0(X)$ and X are the same as those in KR theory and the function $p(X)$ and the excluded parameter z are given as follows:

$$p(X) = \frac{F_1'(X) + F_1''(X)}{F_0(X)} \tag{20}$$

$$F_1'(X) = \left(\frac{4}{\pi^2}\right) \sum_{k=1}^{\infty} \left\{ \frac{1}{k^2} \frac{1}{1 + Xk^{-1/2}} \times \left[\frac{1 + 2\pi^{1/2}C(2\pi k)}{(\pi k)^{1/2}} + \frac{3\pi^{1/2}S(2\pi k)}{2(\pi k)^{3/2}} \right] \right\} \tag{21}$$

$$F_1''(X) = \frac{2.493}{\pi^2} \sum_{k=1}^{\infty} \left\{ \frac{(1/k^{5/2})X}{[1 + Xk^{-1/2}]^2} \right\} \tag{22}$$

$$C(2\pi k) = \int_0^{2\pi k} \frac{\cos t}{(2\pi t)^{1/2}} dt \tag{23}$$

$$S(2\pi k) = \int_0^{2\pi k} \frac{\sin t}{(2\pi t)^{1/2}} dt \tag{24}$$

$$z = (\tfrac{3}{2}\pi a'^2)^{3/2} \beta n^{1/2} \tag{25}$$

where β is the binary cluster integral.

At the asymptotic limit of $X \to 0$, and $X \to \infty$, $[\eta]$ of KY theory reduces to the following equations:

$$[\eta]_{x \to 0} = (N_A \zeta / 6\eta_0 M_0) \langle S^2 \rangle_0 \alpha_s^2 \tag{26}$$

$$[\eta]_{x \to \infty} = \pi^{3/2} N_A [X F_0(X)]_{x \to \infty} (\langle S^2 \rangle_0^{3/2} / M) \alpha_s^{2.43} \tag{27}$$

Here α_s is the expansion factor for the radius of gyration, defined as follows:

$$\alpha_s = \langle S^2 \rangle^{1/2} / \langle S^2 \rangle_0^{1/2} \tag{28}$$

In KY theory, $[\eta]_{x \to \infty}$ is proportional to $\alpha_s^{2.43}$, which is different from the Flory–Fox theory (eq 17). This means that the expansion of a polymer coil due to the volume effect does not occur uniformly, but more in the skin part than in the core part.

The radius of gyration calculated by perturbation theory using the cluster expansion method is expressed as a function of z in the form (46),

$$\langle S^2 \rangle = \langle S^2 \rangle_0 (1 + 1.276z - 2.082z^2 + \cdots) \tag{29}$$

Consequently, α_s^3 is calculated approximately from a Fixman (47) (F)-type equation as

$$\alpha_s^3 \simeq 1 + 1.91z \tag{30}$$

If we put $\langle R^2 \rangle$, Φ, and α_s as

$$\langle R^2 \rangle \simeq 6 \langle S^2 \rangle \tag{31}$$

$$\Phi = N_A \left(\frac{\pi}{6} \right)^{3/2} [X F_0(X)]_{x \to \infty} \frac{1 + 1.55z}{1 + 1.91z} \tag{32}$$

and

$$\alpha_s \simeq \alpha \tag{33}$$

then $[\eta]$ in KY theory is represented in a form analogous to equation (14) as

$$[\eta] = \Phi(\langle R^2 \rangle / M)^{3/2} M^{1/2} \alpha^3 \tag{34}$$

The value of Φ in equation (32) is not constant, but depends on the excluded volume effect parameter z. Using the limiting value of $[XF_0(X)]_{x \to \infty}$, Φ is obtained as 2.87×10^{23}.

The exponent **a** in the MHS equation is obtained as a function of X in the KY theory (eq. 35), as listed in Table 2.

$$\mathbf{a} = \tfrac{1}{2} + \tfrac{1}{2}[d \log XF_0(X)/d \log X] \tag{35}$$

Note that all theories of $[\eta]$ mentioned above assume the polymeric chain segments in the unperturbed state to obey a Gaussian distribution function around the center of mass of the polymer. For extremely inflexible or semiflexible polymers, however, this assumption does not remain valid. The wormlike chain, first proposed by Kratky and Porod (48) as a model for chain molecules with large equilibrium rigidity, is conventionally expressed in terms of the persistence length q. The relationship between q and $\langle S^2 \rangle^{1/2}$ for a wormlike chain was derived by Benoit and Doty (49) (BD). Equations for $[\eta]$ and the sedimentation coefficient **s** of the same chain were obtained by Hearst (50), Peterlin (51), Hearst and Stockmayer (52), Yamakawa and Stockmayer (53), Ullman (54, 55), and Yamakawa and Fujii (56, 57) (YF). The models used in their theories are a little different from ours, as illustrated in Figure 4.

Neglecting the draining term in Kirkwood–Riseman integral equation, Yamakawa and Fujii (57) derived an expression of $[\eta]$ for unperturbed continuous wormlike cylinders (Fig. 4*c*), which leads to

$$[\eta] = \Phi'(2q/M_L)^{3/2} M^{1/2} \tag{36}$$

where Φ' is a function of q, contour length L, and the diameter of the cylinder d and is calculated numerically in Reference 57. The term M_L is equal to M/L and is referred to as a shift factor.

Yamakawa and Fujii (56) proposed a method to evaluate M_L in conjunction with $[\eta]$ and **s**. Kamide et al. (58, 59) and Saito (60) analyzed the literature data for $[\eta]$ of solutions of cellulose and its derivatives, assuming M_L and d to be equal to the ratio of the molecular weight to the length of a pyranose ring and the hydrodynamic diameter of the pearl necklace model, respectively. The q values evaluated by the YF theory were smaller than those by the BD theory,

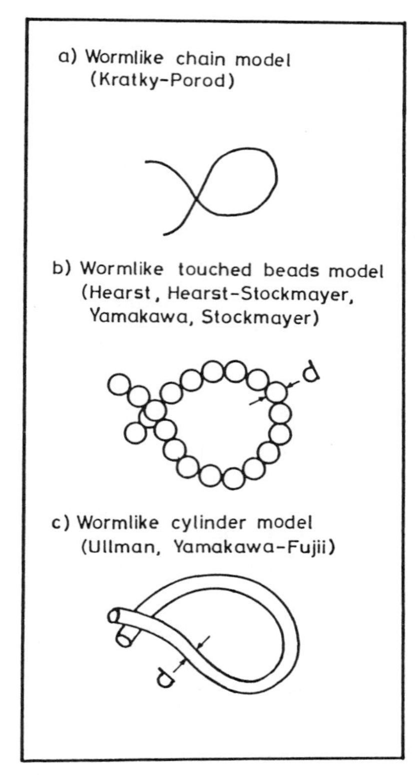

Figure 4. Schematic representation of three wormlike chain models: d is the diameter of a wormlike touched bead (b) or a wormlike cylinder (c).

and this discrepancy was attributed to the neglect of the partial free draining effect in the YF theory.

Theoretical Relationship Between the K_m Parameter and the Exponent **a** *in the MHS Equation*

Kawai and Kamide (61) pointed out that if the FF theory has general applicability, the following relationship must be established between the constant K_m and the exponent **a** in the MHS equation

$$\begin{aligned} -\log K_m + \tfrac{3}{2}\log\{\tfrac{4}{3}[1/(\mathbf{a} - 0.5) - 10/3]^{-1} + \} \\ = -\log K + (\mathbf{a} - 1/2)\log M_0 \qquad \text{(FF)} \end{aligned} \tag{37}$$

where M_0 is a molecular weight representative of the molecular weight range covered by the limiting viscosity number–molecular weight relationship. This equation furnishes the "single-sample method" for estimating the parameters in the MHS equation. When the values for K and M_0 are known for a given polymer, the parameter K_m and the exponent **a** for the polymer in a given solvent can be easily computed from the measurements of $[\eta]$ and molecular weight for only one sample, without recourse to the more laborious methods usually used.

For chains of poor flexibility, however, equation (37) does not hold, owing to the neglect of the partial free draining effect.

Estimation of the consequences of excluding the volume effect and the draining effect from viscosity data become possible through the use of the following approximation method adopted by Kawai and Kamide (61). Kurata and Yamakawa (38) pointed out that the countercontributions of these two effects make the Φ values approximately independent of M (i.e., the FF theory is applicable) over wide ranges in so far as the draining parameter X takes a value larger than ~ 10; as X decreases below 10, however, the balance inclines toward the draining effect and the decrease in the Φ values will become significant. Along this line of argument, Kawai and Kamide expressed $[\eta]$, to a satisfactory approximation, in the form

$$[\eta] = K\{XF(X)/(X_0F(X_0))\}M^{1/2}\alpha^3 \tag{38}$$

Then the exponent **a** in the MHS equation can be expressed as

$$\mathbf{a} = 0.5 + \mathbf{a}_1 + \Delta \tag{39}$$

with

$$\mathbf{a}_1 = d \log \alpha_s / d \log M \tag{40}$$

and

$$\Delta = \nu(X) - \nu(X_0) \tag{41}$$

where

$$\nu(X) = \tfrac{1}{2}[d \log XF(X)/d \log X] \tag{42}$$

For convenience, and bearing in mind that the FF theory retains its validity to a satisfactory approximation in the range $X = 20 \sim \infty$, the value 50 has been chosen for X_0, giving a standard beyond which the deviation from the FF theory due to the draining effect arises. Strictly speaking, the value for X_0 must be determined case by case, by

$$\nu(X_0) = \epsilon[3 - n(X_0)] \tag{43}$$

The quantities appearing in this equation are also defined in the KY theory.

Owing to the counteractivity of the volume effect and the draining effect on the molecular weight dependency of $[\eta]$, however, use of the more exact X_0 value does not alter the result to any great extent.

Making use of equations (37), (41), and (42), we obtain in place of equation (37)

$$\frac{K_m}{(Kf)} = \frac{\{\frac{4}{3}[1/(\mathbf{a} - 0.5 - \Delta) - \frac{10}{3}]^{-1} + \mathbf{a}\}^{3/2}}{M_0^{a-0.5-\Delta}} \tag{44}$$

where $f = XF(X)/X_0F(X_0)$. In arriving at this result, Kawai and Kamide employed a fifth-power-law type of equation for α (equation 19). In view of the nonuniform expansion character of an actual polymer chain in a good solvent, equation (19) is not rigorously correct.

An alternative expression of α_s was derived as a function of the excluded parameter z by Kurata, Stockmayer, and Roig (62) (KSR), adopting an ellipsoid having its principal axes proportional to the radius of gyration with fixed end-to-end distance, in the third-power-law form, as

$$\alpha_s^3 - \alpha_s = \{(1 + 1/(3\alpha_s))^{-3/2}\}(4/3)^{5/2}(3/(2\pi))^{3/2}(\beta/a'^3)N^{1/2} \tag{45}$$

The combinations of equations (17) and (45), and equations (27) and (45) give the following equations (63)

$$\begin{aligned} &-\log K_m + (3/2)\log\{1 + (4/3)[(\mathbf{a} - 0.5)^{-1} - 2]^{-1}\} \\ &\quad = -\log K + (\mathbf{a} - 0.5)\log M_0 \quad \text{(FF-KSR)} \end{aligned} \tag{46}$$

and

$$\begin{aligned} &-\log K_m + (2.43/2)\log\{1 + (4/2.43)[(\mathbf{a} - 0.5)^{-1} - 6/2.43]^{-1}\} \\ &\quad = -\log K + (\mathbf{a} - 0.5)\log M_0 \quad \text{(KY-KSR)} \end{aligned} \tag{47}$$

respectively. In the course of deriving equations (46) and (47), the approximation of $\alpha \simeq \alpha_s$ (eq. 33) is used.

On the basis of the relations between α and z derived by Fixman (47) (eq. 30), Ueda and Kajitani (64) (UK), and Shiokawa and Oyama (65) (SO), the theoretical equations between K_m and the exponent **a** were derived by Kamide et al. (66, 67).

The combinations of the treatments of FF theory and Fixman equation (FF-F), and KY theory to Fixman equation (KY-F) led, respectively, to (66)

$$\begin{aligned} &-\log K_m + \log\{1 + 2[(\mathbf{a} - 0.5)^{-1} - 2]^{-1}\} \\ &\quad = -\log K + (\mathbf{a} - 0.5)\log M_0 \quad \text{(FF-F)} \end{aligned} \tag{48}$$

and

$$- \log K_m + (2.43/2) \log \{1 + (6/2.42)[(\mathbf{a} - 0.5)^{-1} - 6/2.43]^{-1}\}$$
$$= - \log K + (\mathbf{a} - 0.5) \log M_0 \quad \text{(KY-F)} \tag{49}$$

Similarly, the combination of the treatments of Flory and Fox and Ueda and Kajitani (FF-UK) leads to (67)

$$- \log K_m + \log \{0.983 + (0.983/0.4)[(\mathbf{a} - 0.5)^{-1} - 2.5]^{-1}\}$$
$$= - \log K + (\mathbf{a} - 0.5) \log M_0 \quad \text{(FF-UK)} \tag{50}$$

and the combination of Flory–Fox and Shiokawa–Oyama (FF-SO) leads to

$$- \log K_m + \log \{1 + 2.5[(\mathbf{a} - 0.5)^{-1} - 2.5]^{-1}\}$$
$$= - \log K + (\mathbf{a} - 0.5) \log M_0 \quad \text{(FF-SO)} \tag{51}$$

The relationships between $-\log K_m$ and **a** as calculated from equations (37) and (46)–(51) are illustrated in Figure 5 (67), assuming $K = 1 \times 10^{-3}$ and $M_0 = 1 \times 10^5$. Reference to Figure 5 shows that the difference between the FF-UK and FF-SO treatments with respect to $-\log K_m \sim \mathbf{a}$ relations is negligible, and both treatments are almost identical to that of KY-KSR.

As Kamide and Moore (68a) were the first to indicate, $\partial \log K_m / \partial \mathbf{a}$ is in practice always positive. Therefore, theoretical predictions of $K_m \sim \mathbf{a}$ relations become inconsistent with experimental results at least for values of **a** greater than $\mathbf{a}^*$, where $\mathbf{a}^*$ is the value of **a** at which $\partial \log K_m / \partial \mathbf{a} = 0$. Such values of $\mathbf{a}^*$

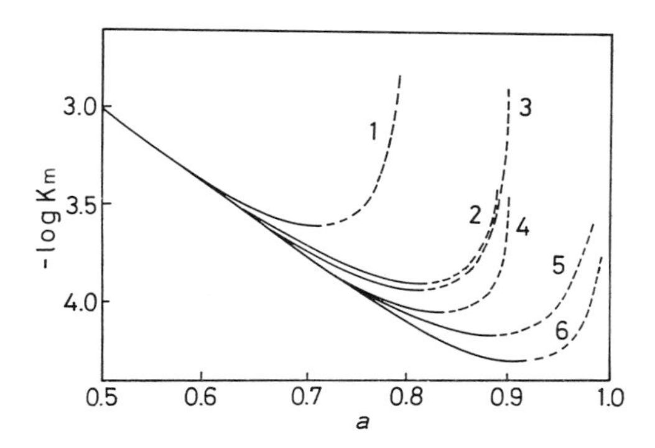

Figure 5. Some theoretical relationships between $-\log K_m$ and **a** (67). Curve 1, FF; 2, FF-UK and FF-SO; 3, KY-KSR; 4, KY-F; 5, FF-KSR; 6, FF-F. Values of parameters used in the theoretical curves are $K = 1 \times 10^{-3}$ and $M_0 = 1 \times 10^5$.

can be obtained (67) from the FF treatment as

$$\{1-2(\mathbf{a}^*-0.5)\}\{1-(10/3)(\mathbf{a}^*-0.5)\}=2/\log M_0 \tag{52}$$

from combination of the KY and KSR treatments as

$$\{1-(6/2.43)(\mathbf{a}^*-0.5)\}\{1-(2/2.43)(\mathbf{a}^*-0.5)\}=2/\log M \tag{53}$$

from combination of the FF and F theories as

$$\mathbf{a}^*=1-1/\log M_0 \tag{54}$$

from combination of the FF and Inagaki–Suzuki–Kurata (68b) (ISK) treatments, as

$$\mathbf{a}^*=0.905-0.81/\log M_0 \qquad \text{(FF-ISK)} \tag{55}$$

from combination of the FF and KSR theories as

$$(\mathbf{a}^*-0.5)^5\{[(\mathbf{a}^*-0.5)^{-1}-2]^2+(4/3)[(\mathbf{a}^*-0.5)^{-1}-2]\}=2/\log M_0 \tag{56}$$

and from combination of the treatments of FF and UK or FF and SO as

$$\mathbf{a}^*=0.900-1/\log M_0 \tag{57}$$

It can be shown from equations (52)–(57) that the magnitude of $\mathbf{a}^*$ depends on M_0, as illustrated in Figure 6.

The parameter M_0 is determined by the maximum and minimum molecular weight (M_2 and M_1, respectively) between which the MHS equation holds. The exponent $\mathbf{a}$ is expressed in terms of M_1 and M_2 as

$$\mathbf{a}=\log\frac{([\eta]_1/[\eta]_2)}{\log(M_1/M_2)} \tag{58}$$

where $[\eta]_1$ and $[\eta]_2$ are the limiting viscosity numbers calculated through equation (17) to which M_1 and M_2 are substituted as M, respectively. Putting $\Delta=0$ in equation (39), then $\mathbf{a}$ is expressed as

$$\mathbf{a}=\tfrac{1}{2}+3d\log\alpha_s/d\log M \tag{59}$$

The second term in the right-hand side of equation (59) is calculated in each

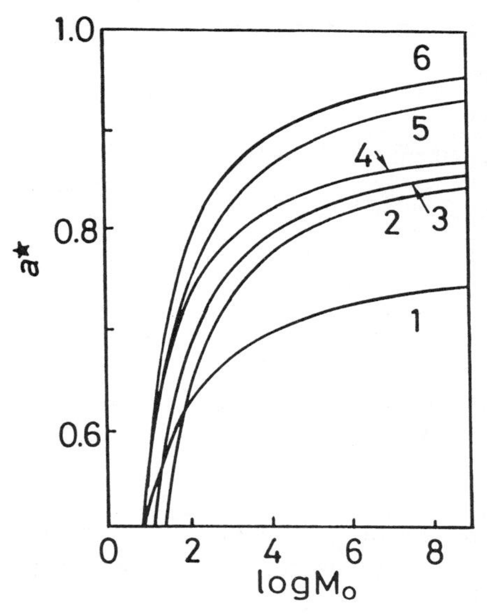

Figure 6. Dependence (67) of **a*** on log M_0 for $K = 1 \times 10^{-3}$. Curve 1, FF; 2, FF-ISK; 3, FF-UK and FF-SO; 4, KY-F; 5, FF-KSR; 6, FF-F.

case for the theories of α_s. For Fixman's theory of α_s (eq. 30), M_0 is obtained analytically as follows (61, 69, 70).

Equation 30 is simply rewritten as a function of M in the form

$$\alpha_s^3 - 1 = CM^{1/2} \tag{30'}$$

Substituting Equations (58) and (30′) into Equation (59) and replacing M by M_0 in Equation (59), we obtain

$$M_0 = \frac{2\xi}{C(1 - 2\xi)} \tag{60}$$

with

$$\xi = \frac{\log(1 + CM_1^{1/2}) - \log(1 + CM_2^{1/2})}{\log M_1 - \log M_2} \tag{61}$$

Equation (60) can be readily rewritten in the form

$$M_0^{1/2} = (\mathbf{a} - \tfrac{1}{2})(1 - \mathbf{a})^{-1} M_2^{1/2} (M_1/M_2)^{(\mathbf{a} - 1/2)} \times [\{1 - (M_1 M_2)^{(1-\mathbf{a})}\}\{1 - (M_1 M_2)^{(\mathbf{a} - 1/2)}\} - 1] \tag{62}$$

Using equation (62), we can calculate M_0 from **a**, M_1, and M_2.

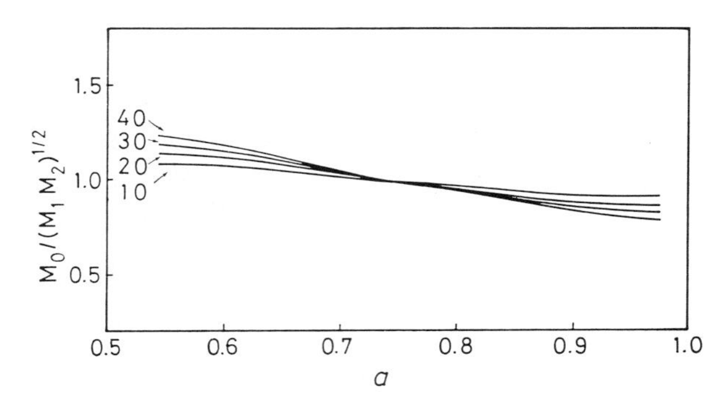

Figure 7. Ratio of M_0 to $(M_1M_2)^{1/2}$ plotted as a function of **a** for various M_1/M_2 values, indicated by the numbers on the curves.

Figure 7 plots $M_0/(M_1M_2)^{1/2}$ against **a** for various M_1/M_2 values. When $\mathbf{a} = 0.75$ in equation (62), we obtain the following identical equation, irrespective of M_1 and M_2.

$$M_0 = (M_1/M_2)^{1/2} \tag{63}$$

In the range $0.55 \leqslant \mathbf{a} \leqslant 0.95$ and $M_1/M_2 \leqslant 40$, in which almost all experimental conditions lie, the ratio $M_0/(M_1M_2)^{1/2}$ is always less than 1.25; in other words, $\log M_0$ is obtained within an accuracy of ± 0.1. Accordingly, the relative error of $(\frac{1}{2}) \log (M_1/M_2)$ against $\log M_0$ is considered to be less than 2% at $M_0 = 1 \times 10^5$ and 2.5% at $M_0 = 1 \times 10^4$ under the usual experimental conditions.

A more generalized equation for M_0, which is obtained by taking into consideration the non-Gaussian nature of the polymer in the solvent, was derived by Kamide and Saito (70). Kamide et al. (69) calculated the parameter M_0 numerically in the equation of FF and FF-KSR, using the experimental data for the Flory temperature and polymer–solvent interaction parameter for a polyisobutylene and polystyrene solution. They found that M_0 can be replaced with $(M_1M_2)^{1/2}$ within an accuracy of $\pm 20\%$, irrespective of the theories and the solvent power.

The value of **a*** can be considered to be a parameter representing the upper limit of applicability of the treatment from which **a*** is derived for comparable M_0. Inspection of Figure 6 also shows us that in the range of M_0 from 1×10^5 to 1×10^6, the limit of applicability of the treatments or combination of them as defined by a^* increases in the following order (87)

$$\text{FF} \ll \text{KY-KSR} < \text{FF-UK} \simeq \text{FF-SO} < \text{KY-F} < \text{FF-KSR} < \text{FF-F}$$

It is clear from Figure 6 that the FF-UK and FF-SO theories prove to be of limited applicability to experimental data in comparison with the FF-F treatment, which offers the most simple and straightforward expression. Furthermore, these treatments described above cannot account for data on the solutions of semiflexible macromolecules, because for such polymer solutions the experimental value of **a** often exceeds **a*** predicted by the FF-F treatment. To avoid confusion, it is necessary to notice that even in the $\mathbf{a} < \mathbf{a}^*$ range the theoretical relation between K_m and **a** differs for different treatments.

The K_m–**a** relations predicted by the theoretical treatments mentioned above have been examined by Kamide and Kataoka (67) using experimental data for atactic polypropylene (71, 72), poly(hexene-1 sulfone) (73), polyvinyl acetate (74), poly(methyl methacrylate) (75), polystyrene (76, 77), and polyacrylonitrile (78, 79) (curves 1–6 in Fig. 8). The most appropriate theory so

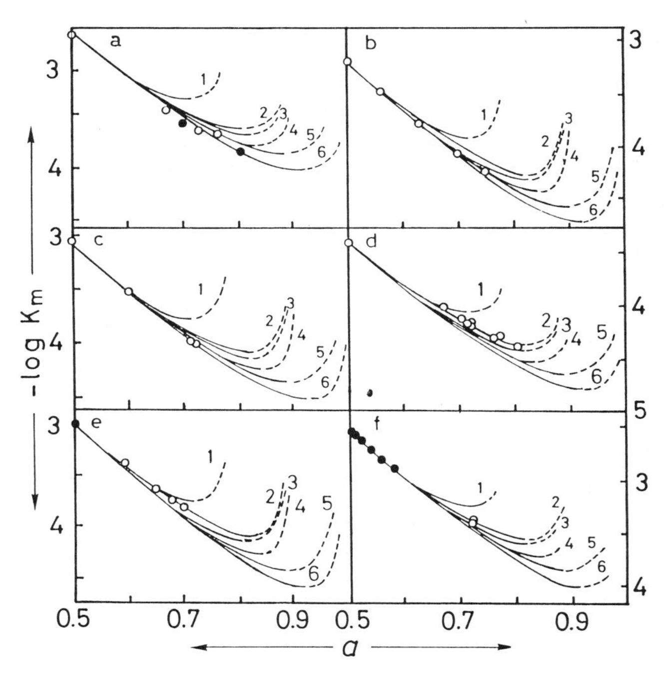

Figure 8. Relationships between $-\log K_m$ and **a** for various polymer solutions (67). Curves are numbered as in Figure 5. (*a*) Attactic polypropylene solutions; ○, data of Danusso and Moraglio (71) at 30°C; ●, Kinsinger and Hughes (72) at 25°C. (*b*) Polyhexene-1-sulfone solutions at 20.5 (Θ solvent) and 25°C: ○, Bates and Ivin (73). (*c*) Poly(vinyl acetate) solutions at 30°C: ○, Matsumoto and Ohyanagi (74). (*d*) Poly(methyl methacrylate) solutions at 30°C: ○, Cohn–Ginsberg et al. (75). (*e*) Polystyrene–toluene–methanol solutions at 25°C: ○, Bawn et al. (76); ●, Bianchi and Magnasco (77). (*f*) Polyacrylonitrile solutions : ○, Kamide et al. (78); ●, Kamide et al. (79).

determined differs for the different polymers, leading to the conclusion that the FF-UK and FF-SO treatments are of limited applicability.

By the way, equations (37) and (46)–(51) afford us the unperturbed chain dimension A, which is defined as $A = (\langle R^2 \rangle_0/M)^{1/2}$, from the plot of K_m versus **a**, assuming a Φ value. Other methods to estimate A by the use of viscosity data are described by Kamide et al. (80–82) and by Stockmayer and Fixman (83).

THE MARK–HOUWINK–SAKURADA EQUATION

Using MHS to Find Molecular Weight

The Mark–Houwink–Sakurada (MHS) equation offers a convenient means of determining the molecular weight of a polymer which is soluble in a solvent:

$$[\eta] = K_m M^{\mathbf{a}} \tag{64}$$

For many linear polymer–solvent systems, the MHS equation parameters K_m and **a** are constant over a wide range of molecular weights.

In Figure 9, the MHS equation is plotted against weight average molecular weight $\bar{M}_w$ of polystyrene (PS), with a relatively narrow molecular weight distribution in cyclohexane at 34.5°C and in benzene at 25 and 30°C (84).

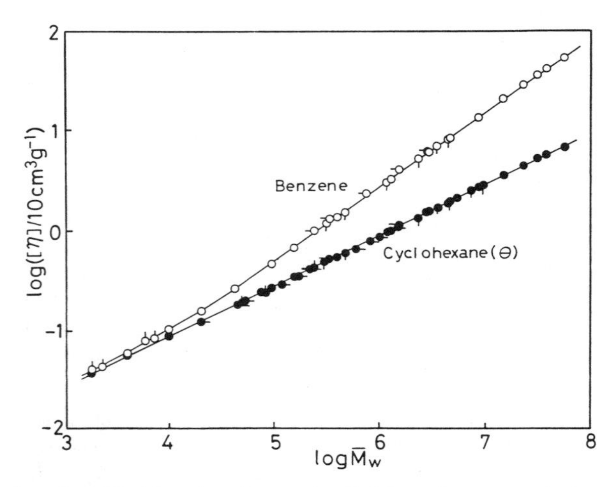

Figure 9. Log–log plot of limiting viscosity number $[\eta]$ for polystyrene solutions against weight average molecular weight $\bar{M}_w$ (84): ○, benzene at 25 and 30°C; ●, cyclohexane at Θ temperature; ○, ●, data of Einaga et al. (84); ○, ●, Altares et al. (85), ●–, Berry (86); ○, ●, Fukuda et al. (87); –○, –●, Yamamoto et al. (88).

Table 3. Mark–Houwink–Sakurada Parameters K_m and a for Polystyrene at the Flory Θ or near θ-conditions

Solvent	K_m (cm^3/g)($\times 10^2$)	a	Temperature (°C)
1-Chloro-*n*-decane	8.71	0.49	8.5
Cycloheptane	7.45	0.51	19.0
Cyclooctane	5.88	0.53	20.5
Cyclopentane	7.26	0.51	20.5
1-Chloro-*n*-undecane	8.10	0.50	32.8
Cyclohexane	8.37	0.50	34.5
Diethyl malonate	8.15	0.49	34.5
Diethyl oxalate	8.15	0.46	58.2
1-Chloro-*n*-dodecane	7.24	0.50	58.6
Dimethyl succinate	8.65	0.48	67.6
Methylcyclohexane	7.61	0.50	68.0
Ethylcyclohexane	6.08	0.53	75.0

From Ref. 89.

These convinsing experimental results show that for polystyrene–solvent systems, the parameters K_m and **a** exhibit constant values over four decades of $\bar{M}_w$. Cyclohexane at 34.5°C is known as a Flory θ solvent, and the exponent **a** in this solvent is almost 0.5. Table 3 summarizes K_m and **a** in the MHS equation, which was established using $\bar{M}_w$ determined by means of a GPC method (89), for PS solvents at or near the state. The value of K_m depends significantly on the kind of solvent used, ranging from 0.61×10^{-2} to 8.71×10^{-2} cm^3/g, and **a** of given systems is approximately 0.5. In benzene at 25 and 30°C the exponent **a** in the region of $\bar{M}_w \geqslant 10^4$ is 0.75; in good solvent the exponent **a** of PS with molecular weight higher than about 10^4, reported previously (90), lies in the range of 0.5 to 0.8. In the range of $\bar{M}_w \geqslant 10^4$, a significant upward discrepancy from the straight line was observed, implying that the unperturbed chain dimension is larger than those observed at the θ point. It should be noted that the MHS equation, established using relatively high molecular weight polymers, is not always applicable in the low molecular weight region. The critical molecular weight at which MHS equation deviates from its linearity depends on the polymer–solvent combination. Figure 10 exemplifies the case for poly(methyl methacrylate)–benzene system at 30°C (91), in which the exponent **a** value changes from 0.76 to 0.5 at a molecular weight $\sim 10^5$.

The effect of polymer chain branching on the MHS equation is demonstrated in Figure 11 (92, 93). Segmented polyurethane employed here was polymerized with methylene bis(4-phenyl isocyanate), polycaprolactone, and

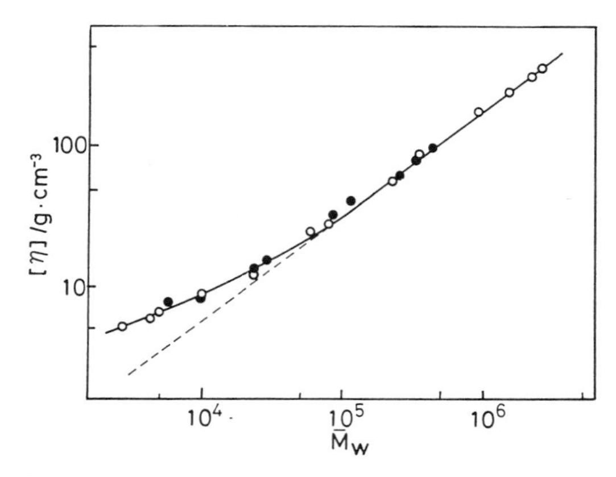

Figure 10. The Mark–Houwink–Sakurada relation for poly(methyl methacrylate) in benzene at 30°C (91): ○, Cohn-Ginsberg et al. (75); ●, Dondos and Benoit (91).

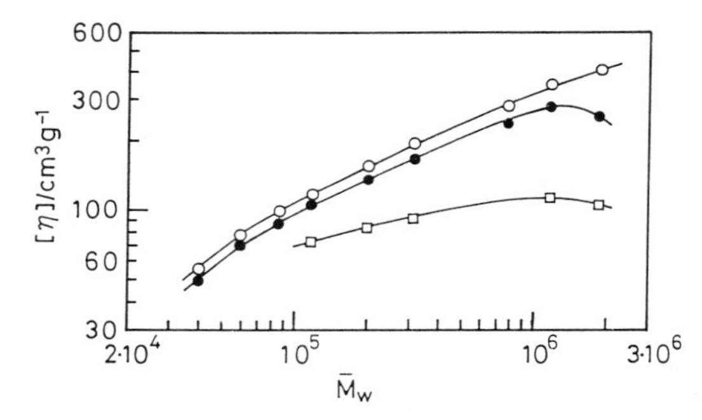

Figure 11. Log–log plots of limiting viscosity number $[\eta]$ versus weight average molecular weight $\bar{M}_w$ for a polyurethane fraction in dimethylacetamide (○), dimenthlformamide (●), and dimethylsulfoxide (□) (92, 93).

2-aminoethanol and was successively precipitation fractionated using N,N-dimethylacetamide (DMAc) as a good solvent and a mixture of ethyl ether and n-heptane as nonsolvent. With an increase of $\bar{M}_w$ larger than 1×10^5, the degree of branching per molecule increases monotonically, resulting in the apparent upward curvature of the log–log plot of $[\eta]$ and $\bar{M}_w$ in all solvents. This indicates that the molecular weight of the branched polymer cannot be unconditionally determined by using the MHS equation for the linear polymer to the sample involved and for a branched polymer with a different number of branches per molecule.

One of the factors determining the reliability of the MHS equation is the uniformity of the molecular weight distribution of the polymer samples used to establish it.

The parameters in the MHS equation of given polymer–solvent systems, such as systems consisting of cellulose and its derivatives plus solvent, do not always coincide in the literature (94–98). For example, the exponent **a** in the MHS equation of cellulose trinitrate–acetone (94), cellulose diacetate–acetone (95), and cellulose triacetate (CTA)–trichloromethane (98) (TCM) has values scattered from 0.67 to 1, 0.75 to ~1, and 0.65 to 1.0, respectively. These differences in exponent **a**, especially for the MHS equation of CTA–solvent systems, were explained, to a first approximation, by the wide molecular weight distribution of samples employed (98). In particular, the systematic molecular weight dependence of the molecular weight distribution of the samples yields incorrectly high **a** values.

Table 4 lists reliable and recommended values for MHS equations for linear polymer–solvent systems, all of which were established using $\bar{M}_w$ or viscosity average molecular weight $\bar{M}_v$ (defined in the next section) and selected on the basis of the second edition of *Polymer Handbook* (see Ref. 90).

Viscosity Average Molecular Weight

The molecular weight, of a polydisperse polymer determined with the aid of the MHS equation is usually referred to as the viscosity average molecular weight $\bar{M}_v$, defined as follows.

We suppose that the MHS equation of the ith monodisperse polymeric chain with molecular weight M_i is written as

$$[\eta]_i = K'_{\mathrm{m}} M_i^{\mathbf{a}} \tag{65}$$

where K'_{m} is the K_{m} for a monodisperse polymer. When the polydisperse polymer contains monodisperse molecules with molecular weight M_i in the weight ratio $W(M_i)$, then $[\eta]$ of the polydisperse polymer is given by superposition of $[\eta]_i$, as

$$[\eta] = K'_{\mathrm{m}} \sum M_i^{\mathbf{a}} W(M_i) \tag{66}$$

Here we define $\bar{M}_v$ as

$$\bar{M}_v = \{\sum M_i^{\mathbf{a}} W(M_i)\}^{1/\mathbf{a}} \tag{67}$$

The number average and weight average molecular weight $\bar{M}_n$ and $\bar{M}_w$ are

Table 4. The Parameter K_m and the exponent a in the Mark–Houwink–Sakurada Equation for Polymer Solutions

Polymer	Solvent	Temperature (°C)	K_m (cm^3/g) ($\times 10^3$)	a	Number of Samples	Molecular Weight Range ($\times 10^{-4}$)	Ref.
Polybutadiene	Benzene	30	8.5	0.78	4	15–50	90, 99
	Cyclohexane	30	11.2	0.75	4	15–50	90, 99
Polychloroprene	Butanone	25	113	0.5	7	15–300	90, 100
	Toluene	30	20.2	0.73	12	14–580	90, 101
Polyethylene	Trichlorobenzene	135	71.1	0.67	8	0.4–90	102
(high density)	Decalin	135	62	0.7	11	2–104	90, 103, 104
	Diphenyl ether	161	295	0.5	6	2–104	90, 103
Polypropylene (atactic)	Decalin	135	11.0	0.80	6	2–62	90, 105
Polyacrylonitrile	Dimethylformamide	20	46.6	0.71	36	7–170	78
		25	52	0.69	7	5–52	106
(atactic)	Dimethyl sulfoxide	25	153	0.60	4	11–52	107
	85% Aqueous ethylene carbonate	25	256	0.49	7	5–52	106
	55% HNO_3	25	342	0.50	7	5–52	
	67% HNO_3	25	122	0.62	7	5–52	
(high meso)	Dimethyl sulfoxide	25	204	0.58	8	7–78	107
	57% HNO_3	25	363	0.50	7	7–78	
	67% HNO_3	25	217	0.57	8	7–78	
Acrylonitrile–methylacrylate copolymer	Dimethylformamide	25	21.3	0.74	9	5–53	108
	51% HNO_3	25	152	0.50	9	5–53	
	80% HNO_3	25	62	0.68	9	5–53	
Poly(methyl-acrylate)	Butanone	20	3.5	0.81	13	6–240	90, 109

Poly(methyl methacrylate)	Acetone	25	5.3	0.73	7	2–78	90, 110
	Benzene	25	5.5	0.76	11	2–740	
	4-Heptanone	33.8	48	0.50	5	1–172	90, 111
Poly(vinyl chloride)	Cyclohexanone	25	13.8	0.78	28	1–12	90, 112
	Tetrahydrofuran	25	16.3	0.77	23	2–30	90, 113
Polystyrene (linear atactic)	Benzene	25	7.8	0.75	12	40–6000	84
	Butanone	25	39	0.58	16	1–180	90, 114
	Cyclohexane	34.5	88.	0.50	16	1–6000	84
	Toluene	25	10.5	0.73	6	16–100	90, 115
Polyimino(1-oxo-hexamethylene) (nylon 6)	Conc. H_2SO_4	20–35	3.32[a]	0.78	—[b]	0.1–10	116
Polyoxydimethyl-silylene	Bromocyclohexane	29.0	74	0.50	5	3.3–106	90, 117
Cellulose	Cadoxen	25	38.5	0.76	9	1–95	97, 118
	6 wt% Aqueous LiOH	25	27.8	0.79	6	4–19	119
	Iron sodium tartrate	25	53.1	0.78	10	0.6–64	120
Cellulose triacetate (2.92)[c]	Dimethylacetamide	25	26.4	0.75	10	6–69	98
	Trifluoroacetic acid	25	39.6	0.71	8	6–69	
	Acetone	25	28.9	0.76	9	6–64	
	Trichloromethane	25	45.4	0.65	7	6–69	
	Dichloromethane	20	24.7	0.70	7	6–69	
Cellulose diacetate (2.46)[c]	Dimethylacetamide	25	13.4	0.82	5	5–27	121
	Acetone	25	133	0.62	9	6–27	122
Carboxymethyl cellulose (1)[c]	NaCl (limit of infinite ion strength)	25	1.9	0.6	4	14–106	129
Amylose	Dimethyl sulfoxide	25	1.25	0.87	9	22–310	130
Amylose triacetate	Nitromethane	22.5	8.5	0.73	12	14–310	131

[a] Established using viscosity average molecular weight.
[b] Obtained by analysis of literature data.
[c] Degree of substitution.

given in equations (68) and (69), respectively.

$$\bar{M}_n = \frac{1}{\{\sum W(M_i)/M_i\}} \tag{68}$$

$$\bar{M}_w = \sum M_i W(M_i) \tag{69}$$

Substitution of $\bar{M}_v$ into equation (67) gives an expression for $[\eta]$ as

$$[\eta] = K'_{\mathrm{m}} \bar{M}_v^{\mathbf{a}} \tag{70}$$

If the parameters K'_{m} and **a** for monodisperse polymers are provided, $\bar{M}_v$ values are calculated from the $[\eta]$ values experimentally determined. Note that $\bar{M}_v$ depends markedly on the value of exponent **a**; that is, $\bar{M}_v$ is influenced by solvent and temperature condition used for the MHS equation. From equations (67) and (69), $\bar{M}_v$ becomes equal to $\bar{M}_w$ when $\mathbf{a} = 1$. Usually the value of **a** for linear polymer ranges from 0.5 to 1; then $\bar{M}_v$ lies in the range of

$$\bar{M}_n < \bar{M}_v \leqslant \bar{M}_w \tag{71}$$

Equation (71) can be derived more generally by a use of Jensen's inequality. We provide the function $\psi(x)$, which is a continuous and real function of x in the region of $x > 0$. If the second-order differentiation of ψ on x is possible, the following relation (Jensen's inequality) holds

$$\psi\left(\sum p_i q_i / \sum p_i\right) \gtreqless \sum p_i \psi(q_i) / \sum p_i \tag{72}$$

corresponding to the relation

$$d^2\psi(x) \gtreqless 0 \tag{73}$$

When we put $\psi = x^{\kappa/\omega}$, $p_i = W(M_i)$, and $x = M_i^{\omega}$ in equations (72) and (73), these two equations become

$$\left\{\frac{\sum M_i^{\kappa} W(M_i)}{\sum W(M_i)}\right\}^{1/\kappa} \lesseqgtr \left\{\frac{\sum M_i^{\omega} W(M_i)}{\sum M_i}\right\}^{1/\omega} \tag{74}$$

corresponding to

$$\kappa \lesseqgtr \omega \tag{75}$$

Putting $\kappa = 1$ and $\omega = -1$ into equations (74) and (75), we obtain the relation between $\bar{M}_n$, $\bar{M}_v$, and $\bar{M}_w$

$$\bar{M}_n < \bar{M}_v \leqslant \bar{M}_w \tag{71}$$

corresponding to

$$-1 < \mathbf{a} \leqslant 1 \tag{76}$$

Considering the range of the exponent **a** that has been obtained experimentally, $\bar{M}_v$ is far closer to $\bar{M}_w$ than $\bar{M}_n$.

Effect of Polymer Polydispersity on the MHS Equation

The parameter K_{m} is not independent of the polydispersity of the samples used to established the MHS equation. Thus in determining molecular weights with high accuracy, we must pay attention to the effect of polydispersity on the MHS equation.

We consider $[\eta]$ of the polymer with a Schulz–Zimm (SZ) type molecular weight distribution function

$$f_w(M) = \{b^{h+1}/\Gamma(h+1)\} \exp(-bM)M^h \tag{77}$$

where

$$h^{-1} = \bar{M}_w/\bar{M}_n - 1 \tag{78}$$

Also, b is the normalization factor of the SZ function and Γ is the gamma function.

In equation (66), $[\eta]$ is expressed in continuous form with respect to the molecular weight as

$$[\eta] = K'_{\mathrm{m}} \int_0^\infty M^{\mathbf{a}} f_w(M) dM$$
$$= K'_{\mathrm{m}} \frac{h + \mathbf{a} + 1}{b^{\mathbf{a}} \Gamma(h+1)} \tag{79}$$

The hypothetical $[\eta]$ for a monodisperse polymer which has the same molecular weight as $\bar{M}_w[=(h+1)/b]$ and $\bar{M}_n(=h/b)$, can be written

$$[\eta]_w = K'_{\mathrm{m}} \left\{\frac{(h+1)}{b}\right\}^{\mathbf{a}} \tag{80}$$

$$[\eta]_n = K'_{\mathrm{m}} \left(\frac{h}{b}\right)^{\mathrm{a}} \tag{81}$$

The viscosity equation for polydisperse polymers can be expressed as a function of $\bar{M}_w$ and $\bar{M}_n$ as

$$[\eta] = K'_{\mathrm{m}} \frac{\Gamma(h + \mathbf{a} + 1)}{[(h+1)^{\mathbf{a}}\Gamma(h+1)]} \bar{M}_w^{\mathbf{a}} \tag{82}$$

$$= K'_{\mathrm{m}} \frac{\Gamma(h + \mathbf{a} + 1)}{h^{\mathbf{a}}\Gamma(h+1)} \bar{M}_n^{\mathbf{a}} \tag{83}$$

Similarly, when the Wesslau-type molecular weight distribution function is employed, the limiting viscosity number is represented as

$$[\eta] = K'_{\mathrm{m}}(\bar{M}_w/\bar{M}_n)^{\mathbf{a}(\mathbf{a}-1)/2} \bar{M}_w^{\mathbf{a}} \tag{84}$$

$$= K'_{\mathrm{m}}(\bar{M}_w/\bar{M}_n)^{\mathbf{a}(\mathbf{a}+1)/2} \bar{M}_n^{\mathbf{a}} \tag{85}$$

Equations (82)–(85) enable us to convert the MHS equation obtained using polydisperse polymers to that for the monodisperse case.

In the case of $\mathbf{a} = 1$, equations (82) and (85) reduce to the single form,

$$[\eta] = K'_{\mathrm{m}} \bar{M}_w \tag{86}$$

The parameter K_{m} coincides with K'_{m} for polymer–solvent systems with $\mathbf{a} = 1$, irrespective of the $\bar{M}_w/\bar{M}_n$ ratio.

The effect of polydispersity on the MHS equation in the case of SZ distribution is illustrated in Figure 12. In this figure values for $[\eta]$ in equations (82) and (83), with $K'_{\mathrm{m}} = 10^{-4}$ and $\mathbf{a} = 0.8$, are plotted against $\bar{M}_w$ and $\bar{M}_n$ for various polydispersity index $\bar{M}_w/\bar{M}_n$ values. When $\bar{M}_w/\bar{M}_n$ increases, $[\eta]$ versus $\bar{M}_w$ at constant molecular weight decreases slightly monotonically. Contrary to this, $[\eta]$ versus $\bar{M}_n$ increases significantly with an increase of $\bar{M}_w/\bar{M}_n$ and is very sensitive to the molecular weight distribution of the samples used for establishment of MHS equation.

When we use the MHS equation based on $\bar{M}_n$ to determine $\bar{M}_n$, the difference between $\bar{M}_w/\bar{M}_n$ of the sample to be measured and that of the polymers used to establish the MHS equation must carefully be taken into account. For example, consider a sample with an SZ distribution function of $\bar{M}_w/\bar{M}_n = 5$ and $[\eta]$ of the sample in a solvent measured to be unity. The MHS equation determined using the polymer samples with $\bar{M}_w/\bar{M}_n = 2$ is

$$[\eta] = 1.67 \times 10^{-4} \bar{M}_n^{0.8} \tag{87}$$

which is chosen conveniently to correspond to the broken line with $\bar{M}_w/\bar{M}_n = 2$

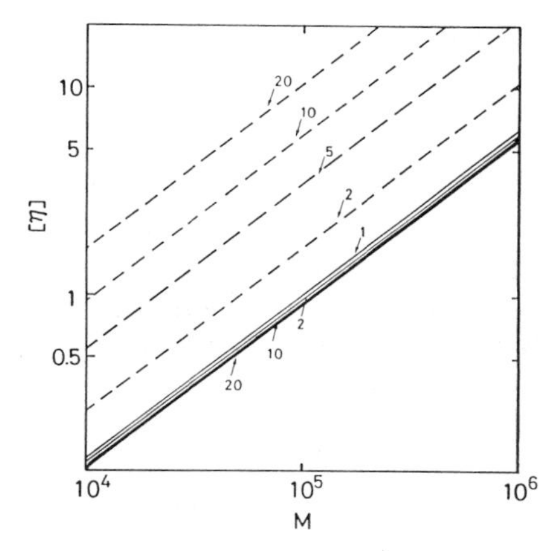

Figure 12. The effect of polymolecularity on the Mark–Houwink–Sakurada relation with constant $K_m = 1 \times 10^{-4}$ and $\mathbf{a} = 0.8$. The molecular weight distribution function is the Schulz–Zimm type. ——, weight average molecular weight $\bar{M}_w$; ···, number average molecular weight $\bar{M}_n$. Indicated number denotes the ratio $\bar{M}_w/\bar{M}_n$.

in Figure 12. If we put $[\eta] = 1$ into equation (87), we erroneously obtain 5.24×10^4 as $\bar{M}_n$ of this sample. By comparing equation (83), into which $\bar{M}_w/\bar{M}_n = 2$ and $\mathbf{a} = 0.8$ are substituted, to equation (87), we obtain the correct value of constant the K'_m in equation (83) (in this case $K'_m = 1 \times 10^{-4}$). Substituting $\bar{M}_w/\bar{M}_n = 5$, $\mathbf{a} = 0.8$, and $K'_m = 1 \times 10^{-4}$ into equation (83) gives

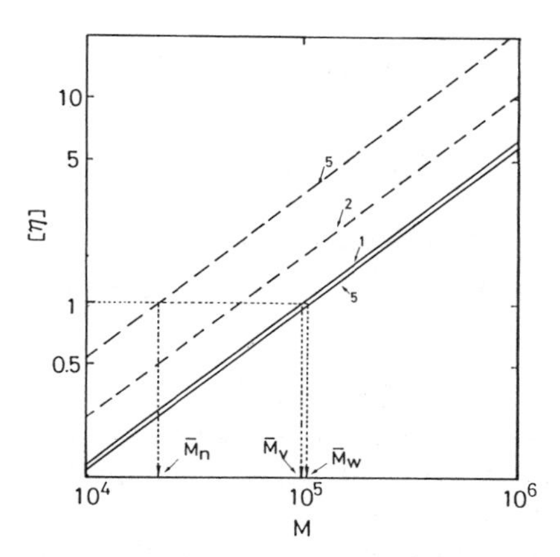

Figure 13. Method for estimating various kinds of average molecular weight using the MHS equation. $\bar{M}_n$, number average molecular weight; $\bar{M}_v$, viscosity average molecular weight, $\bar{M}_w$, weight average molecular weight. The lines and associated numbers have the same meaning as in Figure 12.

the following MHS equation,

$$[\eta] = 3.417 \times 10^{-4} \bar{M}_n^{0.8} \tag{87'}$$

which is demonstrated as the broken line with $\bar{M}_w/\bar{M}_n = 5$ in Figure 13. When $[\eta] = 1$, equation (87) gives the correct $\bar{M}_n$ value of the sample as 2.15×10^4. Here $\bar{M}_v$ is estimated from the MHS equation for monodisperse polymers, namely $[\eta] = 1 \times 10^{-4} M^{0.8}$, to be 1×10^5. On the other hand, substitution of $\bar{M}_w/\bar{M}_n = 2$, $\mathbf{a} = 0.8$, and $[\eta] = 1$ into equation (82) gives for this sample the incorrect $\bar{M}_w$ value of 1.04×10^5. Using similar calculation procedures in the case of $\bar{M}_n$ mentioned above, the correct $\bar{M}_w$ value is finally obtained, namely 1.08×10^5, which is very close to 1.04×10^5. Using the MHS equation against $\bar{M}_w$, the polydisperse nature of the polymers is practically negligible.

Some attempts (132–134) have been made to determine degree of polymolecularity using the difference of MHS equations in various solvents and temperatures.

Suppose that the MHS equations in two different kinds of solvent (or at two temperatures) are given as

$$[\eta] = K_{\text{mr}} \bar{M}_v^{\text{r}} \tag{88}$$

$$[\eta] = K_{\text{ms}} \bar{M}_v^{\text{s}} \tag{88'}$$

A parameter $\delta(r, s)$ of the polymolecularity is defined as

$$\delta(r, s) = \frac{\bar{M}_v(r)}{\bar{M}_v(s)} - 1 \tag{89}$$

where $\bar{M}_v(r)$ and $\bar{M}_v(s)$ are $\bar{M}_v$ of a polymer, determined by means of equations (88) and (88′), respectively. The parameter $\delta(r, s)$ is theoretically related to the parameter h in equation (78) for various types of molecular weight distribution function as listed in Table 5 (135). Equation (89) predicts that the two values of $\bar{M}_v$ calculated through the MHS equation in two solvents will give us the polymolecularity of the sample. However, $\delta(r, s)$ values experimentally determined for actual polymer are sometimes remarkably different from those calculated through the theoretical relation in Table 5. Typical examples (135) are listed in Table 6, which compares experimentally determined $\delta(r, s)$ (δ_{ex}) for poly(isopropyl acetate) and polystyrene and $\delta(r, s)$ calculated for SZ (δ_{SZ}) and log-Gaussian (δ_{LG}) distributions. In every combination of polymer–solvent pair, δ_{ex} is far larger than δ_{SZ} and δ_{LG}. In addition, for polypropyrene with various $\bar{M}_w/\bar{M}_n$ in t-decalin (at 135°C) and in α-chloronaphthalene (139.2°C), δ_{ex} almost failed to correlate to $\bar{M}_w/\bar{M}_n$ of the samples. These experimental

Table 5. Relationship Between h and δ for Several Distribution Types

Distribution Type	Distribution Function $f(M)$	$h^{-1} = \bar{M}_w/M_n - 1$	$\delta(r, s)$
Rectangular distribution	$1/a$, upper limit; $\bar{M}_w + a/2$ lower limit; $\bar{M}_w - a/2$	$a^2/3\bar{M}_w^2$	$(r-s)(a^2/6)\bar{M}_w^2$
Gaussian distribution	$\dfrac{\exp\{(M-\bar{M}_w)^2/2\sigma^2\}}{\sigma(2\pi)^{1/2}}$	$\sigma^2/\bar{M}_w^2$	$(r-s)(\sigma^2/2)\bar{M}_w^2$
Zimm distribution	$\dfrac{y^{z+1}}{\Gamma(z+1)}\exp(-yM)M^z$	$1/z$	$\dfrac{\Gamma(z+r+1)^{1/r}}{\Gamma(z+s+1)^{1/s}}$
Log–Gaussian	$w(t) = [(2\pi)^{-1}\exp\{-t^2/2\}$	$\exp(\gamma^2/2\} - 1$	$\Gamma(z+1)^{1/r-1/s}$ $\exp(\gamma^2/2)^{1/2(r-s)} - 1$

Table 6. Comparison of Viscometric Polydispersity Index of Experimental and Calculated Values

Polymer	h^{-1}	Solvent Pair	Difference $r - s$	$\delta(r, s)_{ex}$	$\delta(r, s)_{SZ}$	$\delta(r, s)_{LG}$
Polyisopropyl acetate	4.5	Benzene–acetone	0.71–0.69	0.13	0.008	0.015
Polystyrene	3.1	Tetralin–butanone	0.57–0.58	0.59	0.054	0.128
	3.1	Toluene–butanone	0.69–0.58	0.41	0.030	0.081
	3.1	Tetralin–toluene	0.75–0.69	0.13	0.023	0.043

From Ref. 135.

data indicate that the viscometric method is not suitable for determining the polymolecularity of polymers.

SHEAR RATE DEPENDENCE OF THE LIMITING VISCOSITY NUMBER

Many polymer solutions, especially elongated and/or stiff chain polymers, such as DNA and cellulose derivatives, exhibit significant non-Newtonian behavior even in the dilute region. For the purpose of determinating molecular weight using the limiting viscosity number, the shear rate dependence of $[\eta]$ has been determined. This shear rate dependence is attributed to two predominant factors: (1) the orientation of the polymeric chain by the flow and the resistance accompanying the rapid conformational change of the segments in the solvent (136). These two effects essentially depend on the hydrodynamic interaction between the polymer chain and the solvent molecules. Consequently, shear rate effect on $[\eta]$ is concerned with concentration, molecular weight or molecular weight distribution, polymer flexibility, and solvent power.

Several theoretical attempts to elucidate the shear rate dependence of $[\eta]$ were made in the framework of the theory of Rouse and Zimm (137, 138). This is based on the ideal spring–bread model, introducing the internal viscosity concept (138, 140) and hydrodynamic interaction in the deformed coil (141), and using an equivalent ellipsoid model with the excluded volume effect (142, 143). To date, no comprehensive explanation combining the orientation and resistance arising from the conformational change of segments has been given.

On the other hand, such general features of the non-Newtonian $[\eta]$ of flexible linear macromolecules as the effects of molecular weight of the polymer, solvent power, and solvent viscosity on the shear rate dependence of the limiting viscosity number have been investigated experimentally (144, 145).

Figure 14 shows non-Newtonian effect on the limiting viscosity number for polystyrene (PS) with $M_w = 6.20 \times 10^6$ at θ and near-θ temperatures (145). Figure 14 plots the ratio of $[\eta]$ at shear rate $\dot{\gamma}$ ($[\eta]_{\dot{\gamma}}$) to $[\eta]$ at zero shear rate against a parameter β', which is related to shear rate thus:

$$\beta' = (M[\eta]\eta_0/RT)\dot{\gamma} \tag{90}$$

At θ or near-θ temperature, the ratio $[\eta]_{\dot{\gamma}}/[\eta]$ of PS is a unique function of β'. In the β' region lower than $\log \beta' \sim -0.5$, $[\eta]_{\dot{\gamma}}$ is almost independent of η. The ratio $[\eta]_{\dot{\gamma}}/[\eta]$ decreases monotonically with an increase of β' in the range of $\log \beta' > -0.5$, indicating that the non-Newtonian effect does exist, even at the θ temperature. In a good solvent $[\eta]_{\dot{\gamma}}/[\eta]$ is not always a universal function of β', but the curve of $[\eta]_{\dot{\gamma}}/[\eta]$ versus β' decreases more rapidly with increasing

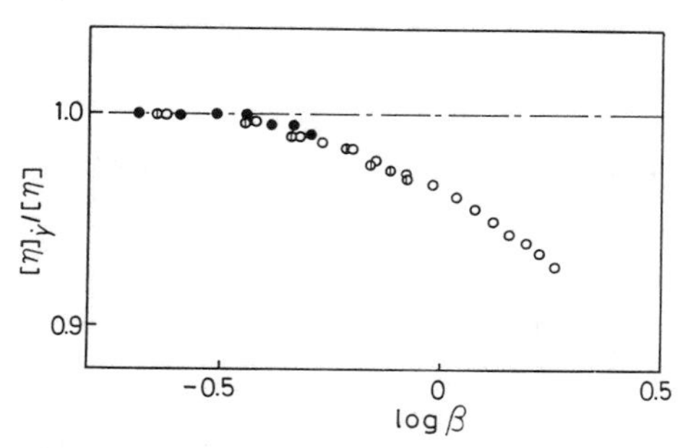

Figure 14. The non-Newtonian limiting viscosity number behavior of polystyrene at and near θ conditions (145). ○, data of polystyrene in dioctyl phthalate at 40°C; ●, cyclcohexane at 34.8°C; ⦶, mixed decalin at 25°C.

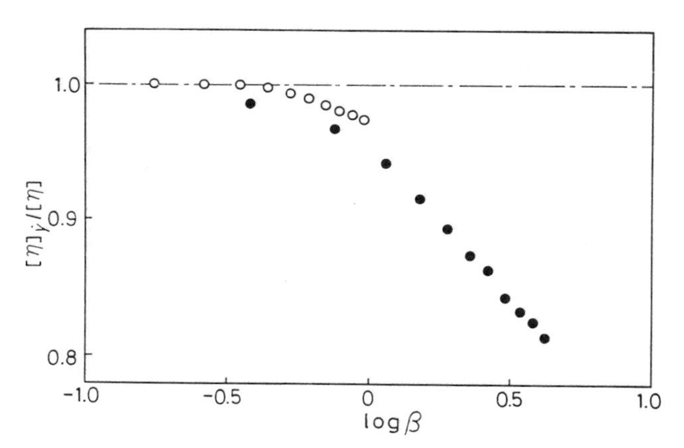

Figure 15. Molecular weight dependence of non-Newtonian limiting viscosity number for polystyrene samples with $\bar{M}_w = 3.16 \times 10^6$ (○) and 7.14×10^6 (●) in benzene at 30°C (145).

solvent power. The molecular weight effect on the relation between $[\eta]_{\dot{\gamma}}/[\eta]$ and $\log \beta'$ is demonstrated in Figure 15 (145). Comparison of the data on two samples in benzene shows that the non-Newtonian effect is larger for the higher molecular weight sample, which obviously has a higher excluded volume effect.

The polydispersity of the sample also influences the shear rate dependence of the limiting viscosity number, and the drop in $[\eta]_{\dot{\gamma}}/[\eta]$ with β' for a polydisperse sample is larger (or earlier) than that for a monodisperse sample (145).

METHODS FOR THE EXPERIMENTAL DETERMINATION OF VISCOSITY

Methods for Measuring Viscosity

The viscosity of polymer solutions is generally measured by capillary or rotational viscometers. For the high accuracy determination of the viscosity of polymer solutions, a capillary-type viscometer (see, e.g., Fig. 16) is frequently used.

The viscosity η of the solution which flows in the capillary of length L' is proportional to the fourth power of the diameter d' of the capillary as follows:

$$\eta = \pi \Delta P \, d'^4 t/(8QL') \tag{91}$$

Here ΔP is the pressure difference between the head of the fluid and the end of the capillary and Q is the flux of the solution passing through capillary in time t. Equation (91), known as the Hagen–Poiseuille law (146, 147), is derived by solving the Stokes equation under the assumption of slow and steady flow without external force, except for the gravity. When ΔP is replaced by $\rho g'h$, where ρ is the density of the solution, g' the gravitational acceleration, and h the height of the fluid, equation (91) is rewritten as

$$\eta = \pi \rho g' h d'^4/(8QL') \tag{92}$$

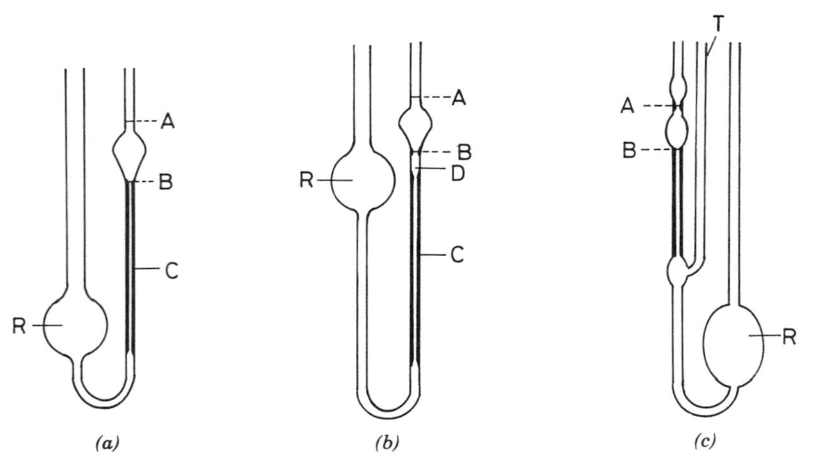

Figure 16. Typical suspension-type capillary viscometers: (*a*) Ostwald type, (*b*) modified Ostwald type, (*c*) Ubbelohde type.

Using a capillary viscometer like the one shown in Figure 16*a*, it is possible to measure the time for the fluid to fall from the upper fiducial mark at A to the lower one at B. Using equation (92), the ratio of the viscosity of solution 1 to that of solution 2 is expressed as

$$\eta_1/\eta_2 = \rho_1 t_1/\rho_2 t_2 \tag{93}$$

The subscripts to ρ and t correspond to the kinds of solution. When solution 1 is the dilute polymer solution and 2 is solvent, the density of solution 1 is assumed to be equal to that of solvent 2. Then equation (93) has the simpler form of

$$\eta_1/\eta_2 = t_1/t_2 \tag{94}$$

and left-hand side of equation (94) gives relative viscosity η_r.

For dilute non-Newtonian solutions, η_r is given in the more complex form of

$$\eta_r = \frac{4t_2}{3/t_1 + 1/t_2 \dfrac{d(1/t_2)}{d(1/t_1)}}(\rho_2/\rho_1) \tag{95}$$

In the actual capillary type viscometer, equations (94) and (95) are not realized without modifications. One deviation from the Hagen–Poiseuille law arises from the additional resistances at the end of the capillary, where the flow diverges or undergoes sudden enlargement (the end effect) and turbulent flow occurs (the Couette correction). In the capillary, the viscous fluid falls with a definite velocity, such that the pressure difference ΔP is not equal to the static pressure $\rho g'h$ (the velocity correction). Moreover, at both menisci of the solution in the viscometer, the effect of the surface tension S on η should be taken into account (the surface tension correction). The final form of η considering the foregoing corrections is (148)

$$\eta = \frac{\pi d'^4}{8Q(L' + L_e + \lambda')}\left\{\rho g'ht - \int_0^t (2S/r_H - 2S/r_L)dt\right\} - \frac{m'Q}{8\pi(L' + L_e + \lambda')t} \tag{96}$$

Here L_e is equivalent length of hypothetical tube, which has the same diameter d' as the capillary. The increase of resistance arising from turbulent flow is expressed in terms of the increase of the capillary length λ'. The radii of the curvature of the higher and lower menisci at time t are r_H and r_L, respectively;

m' is a constant, which is determined by the form of the tube at the end of the capillary.

To minimize the above-mentioned corrections, some improvements were made. Figure 16*b* shows the modified Ostwald viscometer. The reservoir R is set at a higher position than in the case of prototype Ostwald device (Fig. 16*a*) to lower the head drop of the two menisci. The diffuser channel D provided between B and the top of the capillary helps to suppress the occurrence of turbulent flow. The Ubbelohde-type viscometer (Fig. 16*c*) has a larger reservoir R, than the other two viscometers, usually 30–100 mL, which makes it possible to dilute the solution by adding the solvent through tube 1.

In the course of the conversion of the potential energy of the fluid due to its position to kinetic energy, some fraction of the energy thermally disperses in the stream, resulting in a rise in the temperature of the fluid. In this sense, the viscosity measurement is not strictly an isothermal process. The temperature increase ΔT of the incompressible fluid from the frictional force is represented as

$$\Delta T = \rho \Delta P/(JC') \tag{97}$$

where J is the mechanical equivalent of heat ($= 4.1855$ J/cal) and C' is the specific heat of the fluid. Under the condition of $\Delta P = 500$ mm H_2O, $\rho = 0.9$, and $C = 0.5$, ΔT is only 0.02°C. In the case of polystyrene in benzene, the average rate $(1/[\eta])(d[\eta]/dT)$ for molecular weights larger than 2×10^5 is about 1%. The error of η caused from the temperature increase is in the order of 0.01% (148), which is practically negligible.

To obtain η of the polymer solution with an accuracy of 0.1%, the temperature around the capillary must be regulated at least within ± 0.1°C. Commercial electronic circulators, which have recently become available, allow the temperature of the thermal bath to be easily controlled within an accuracy of ± 0.01°C or less.

In measurements of viscosity at high temperature, some polymers, such as polyethylene (PE) and polypropyrene (PP), inevitably suffer from oxidation. To avoid this possibility, nitrogen gas is used for the atmosphere in the capillary viscometer. Alternatively, oxidation-preventing agents (e.g., 2, 6, ditertiary-butyl-*p*-cresol for PE and PP) are added to the polymer solution (149).

Analysis of Viscosity Data

Specific viscosity η_{sp} represents the increase of the viscosity due to presence of solute and is represented by a polynomial approximation in dilute solution, neglecting c^3 and higher terms, as follows:

$$\eta_{sp}/c = [\eta] + k'[\eta]^2 c \tag{98}$$

A plot of η_{sp}/c versus c is called a Huggins plot (150, 151). The coefficient k', referred to as the Huggins constant, is obtained from the slope of the Huggins plot. It is commonly independent of temperature and solvent and characterizes the polymer–polymer interaction in the solvent. The probability that a polymer molecule will have contact with others increases in poor solvent, with the result that k' takes a higher value. For the polymers in a Flory θ solvent, k' is about 0.5 or slightly higher; in good solvent k' frequently lies in the 0.2–0.5 range.

Another formula which represents the concentration dependence of viscosity was proposed by Kraemer (2) as follows:

$$(\ln \eta_r)/c = [\eta] + k''[\eta]^2 c \tag{99}$$

where k'' is independent of c and is related to the Huggins constant by

$$k'' = k' - \tfrac{1}{2} \tag{100}$$

The $[\eta]$ is obtained by extrapolation of either η_{sp}/c or $(\ln \eta_r)/c$ to zero concentration that is,

$$[\eta]_H = \lim_{c \to 0} (\eta_{sp}/c) \tag{101}$$

$$[\eta]_K = \lim_{c \to 0} (\ln \eta_r)/c \tag{102}$$

where $[\eta]_H$ and $[\eta]_K$ are the limiting viscosity numbers obtained from Huggins and Kraemer plots, respectively. Inspection of equation (100) indicates that the slope of the Kraemer plot is smaller than that of Huggins, so we can obtain $[\eta]$ at higher accuracy by the former method than the latter.

By definition, η_r is related to η_{sp} as

$$\eta_r = 1 + \eta_{sp} \tag{103}$$

If η_{sp} is sufficiently small compared with unity, the logarithm of η_r in the equation above can be expressed by the Taylor series

$$\frac{\ln \eta_r}{c} = \frac{\eta_{sp}}{c} - \frac{\eta_{sp}(\eta_{sp}/c)}{2} + \frac{\eta_{sp}^2(\eta_{sp}/c)}{3} - \cdots \tag{104}$$

Table 7. Comparison of Calculated Values of ln η_r and the Summation

$$\sum_{i=1}^{j} [(-1)^{i+1}/i]\eta_{sp}^{i}$$

η_r	$\ln \eta_r$	η_{sp}	η_{sp}^2	η_{sp}^3	η_{sp}^4	η_{sp}^5	η_{sp}^6
1.2	0.1823	0.200	0.180	0.1827	0.1823	0.1824	0.1824
1.3	0.2624	0.300	0.255	0.2640	0.2620	0.2615	0.2614
1.4	0.3365	0.400	0.320	0.3413	0.3347	0.3368	0.3361
1.6	0.4700	0.600	0.420	0.4840	0.4552	0.4690	0.4621
1.8	0.5878	0.800	0.480	0.6507	0.5483	0.6138	0.5801

From Ref. 152.

At the limit of $c \to 0$, η_{sp} approaches zero, and equation (104) reduces to

$$\lim_{c \to 0} (\ln \eta_r)/c = \lim_{c \to 0} (\eta_{sp}/c) \tag{105}$$

Accordingly, $[\eta]_H$ should theoretically coincide with $[\eta]_K$.

Table 7 indicates the values of $\ln \eta_r$ in equation (104) in the range of $\eta_r = 1.2$–1.8[152]. The third through eighth columns in this table list the values of the summation $\sum_{i=1}^{j}(-1)^{i+1}\eta_{sp}^{i}/i$ from equation (104) with $j = 1$–6. Inspection

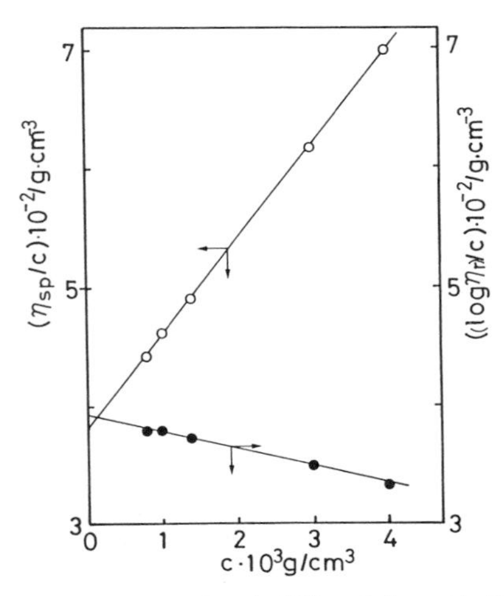

Figure 17. Concentration c dependence of η_{sp}/c (○) and $(\log \eta_r)/c$ (●) for polyisobutylene in methylcyclohexane (152).

Table 8. Comparison of the Limiting Viscosity Numbers Determined from the Huggins Plot $[\eta]_H$ and from the Kraemer Plot $[\eta]_K$ and of the Coefficients k' and k''

Polymer	Sample number	Solvent	Temperature (°C)	$[\eta]$ (cm^3/g)		Coefficient	
				$[\eta]_H$	$[\eta]_K$	k'	k''
Polyurea	PTH7	99.7% HCOOH + 1% HCOONa	25.0	17.96	17.96	0.33	0.33
	PTH7	96% H_2SO_4	25.0	28.22	28.22	0.39	0.37
	PH14	98% HCOOH + 2% KCl	25.0	57.9	56.2	0.08	0.22
	PH20	90% H_2SO_4	25.0	17.49	17.32	0.143	0.225
	PTH7	$HCCl_2COOH$	46.0	19.2	21.9	1.85	0.84
	PH17	$HCCl_2COOH$	46.0	29.11	32.28	1.146	0.575
Polyisobutylene	B100	Methylcyclohexane	25.0	380.6	393.4	0.548	0.404
	B100	*n*-Hexane	25.0	293	309	0.61	0.410
	B100	Benzole	24.4	120.1	125.9	0.94	0.576
	B100	Carbone tetrachloride–butanone	25.0	120	138.1	1.60	0.70
	B100	Cyclohexane–butanone	25.0	116	143.8	2.44	0.816

Table 9. Experimental Equations Representing the Concentration Dependence of $[\eta]$ and the Relation of the Huggins Coefficient k' to the Constant k in Each Equation

Author	Formula	k and k'	Ref.
Arrhenius	$\log \eta_r = [\eta]c$	$k' = 0.5$	153
Fikentscher–Mark	$\eta_{sp} = \dfrac{[\eta]c}{1 - k[\eta]c}$	$k' = k$	154
Staudinger–Heuer	$\log(\eta_{sp}/c) = \log[\eta] + kc$	$k' = k/[\eta]$	155
Baker	$\eta_r = \left\{\dfrac{(1 + [\eta]c)}{k}\right\}^k$	$k' = (k-1)/2$	156

of Table 7 shows that if η_r is 1.2 or less, terms higher than the third one are negligible within an accuracy of 10%.

Figure 17 plots of η_{sp}/c and $\ln \eta_r/c$ against c for the polyisobutylene–methylcyclohexane system (152). Both plots are expressed well by straight lines. The intercept of the ordinate axis gives $[\eta]_H$ for the plot of η_{sp}/c versus c and $[\eta]_K$ for that of $(\ln \eta_r)/c$ versus c. In the case of this polymer–solvent system, $[\eta]_K$ is a little larger than $[\eta]_H$, but the difference is only about 3.6%, in comparison with the value of $[\eta]_H$.

Table 8 summarizes the experimental results on polyurea and polyisobutylene in various solvents (152), including the results of Figure 17. Values of $[\eta]_H$ and $[\eta]_K$ for these polymer–solvent systems coincide within 10% error. However the relation between k' and k'' (eq. 91) does not always hold experimentally.

Many kinds of experimental equation representing the concentration dependence of the viscosity of polymer solutions have been proposed (153–156), four are listed in Table 9. Each equation in the table is a special case or equivalent to the Huggins equation (eq. 98), because the Huggins coefficient k' can be suitably defined to make each proposed equation equal to the Huggins equation, as shown in the third column of the table. The Huggins plot is generally used now because it has such wide applicability to experimental data on the concentration dependence of the viscosity of polymer solutions, except for the cases described below.

The extrapolation methods to zero concentration cited above rest on the assumption that the linear relationship between η_{sp}/c or $(\ln \eta_r)/c$ and c holds even in very dilute region such as $c \ll 0.1$ g/cm^3. But significant, usually upward, discrepancies from the straight lines in Huggins plots were observed for polymer solutions—for example, poly(vinyl chloride) (157) and polystyrene (158) (Fig. 18) in various solvents and isotactic polypropylene in decalin

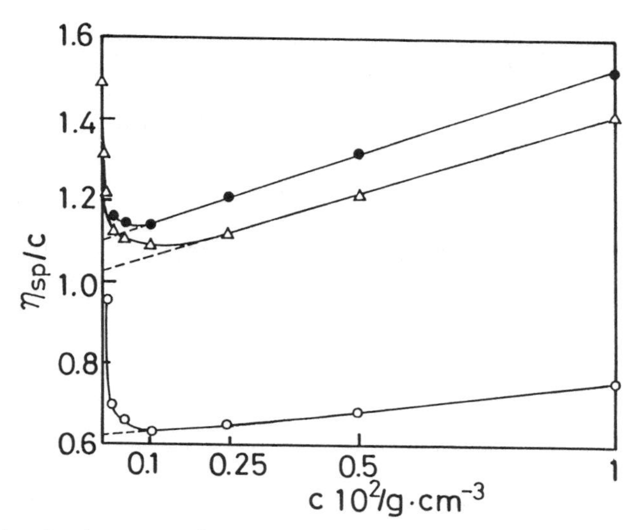

Figure 18. Plot of η_{sp}/c versus c at low concentration for polystyrene in 2-butanone (○), toluene (△), and benzene (●) (158).

(149). Many interpretations have been proposed for these extraordinary observations, which are summarized in three categories (159):

1. When a capillary-type viscometer is used in measurements of viscosity, the polymer molecules are adsorbed to the surface of the capillary, resulting in a smaller cross-sectional area of the capillary with an apparent increase in η_{sp}/c.
2. The thermodynamic expansion of the polymer molecule is greater in the very dilute region than in a moderately dilute state.
3. Several polymer molecules in a moderately dilute solution are associated with each other and dissolve monodispersely only in a very dilute region.

A satisfactory explanation for these phenomena has not been given, but it is possible that all three factors just enumerated operate to a greater or lesser degree in this region of very low concentration.

To determine $[\eta]$ of a polymer solution from Huggins or Kraemer plots, it is insufficient to measure η_{sp} at one concentration. To reduce the amount of samples and the time required for measurements, however, some methods (160–162) for estimating $[\eta]$ from viscosity data at only one concentration have been considered. These are known generically as "one-point" methods.

For solutions whose viscosity can be determined accurately at sufficiently low concentration, when the value of η_{sp} is very small (e.g., $\eta_{sp} < 0.05$), η_{sp}/c can be regarded as equal to $[\eta]$.

When η_{sp} is smaller than 0.6, $[\eta]$ can be expressed as (160)

$$[\eta] = \frac{\eta_{sp}/c}{1 + k_\eta \eta_{sp}} \tag{106}$$

Here k_η is the constant determined by the combination of polymer and solvent.

Combining the equations of Baker (Table 9) and Kraemer (eq. 99), Sakurada (161) derived a relation between $[\eta]$ and η_r which is useful in the range of $\eta_{sp} < 5$, such that

$$[\eta] = \frac{k(\eta_r^{1/k} - 1)}{c} \tag{107}$$

where k is the constant in Baker's equation and relates to Huggins's coefficient k' through the equation indicated in Table 9.

Solomon and Gottesman (162) defined two parameters Ξ and Λ, which are functions of η_{sp} and η_r, as follows

$$\Xi = \eta_{sp} - \ln \eta_r \tag{108}$$

$$\Lambda = \eta_{sp} \ln \eta_r \tag{109}$$

The constant k'' in Kraemer's equation (eq. 99) is expressed using Ξ and Λ as follows

$$k'' = \frac{1}{\ln \eta_r} - \frac{1}{2^{1/2}\Xi} \tag{110}$$

or

$$k'' = \frac{1}{\ln \eta_r} - \frac{1}{\Lambda^{1/2}} \tag{111}$$

Substitution of equation (110) or (111) into equation (99) gives a method for determining $[\eta]$ from experimental data on η_{sp} and η_r at one concentration.

In every one-concentration method, the linearity of η_{sp}/c and/or $\ln \eta_r/c$ versus c is assumed over the whole range of c from the relatively dilute region to $c = 0$, as in the case of the extrapolation methods. But for polymer solutions in which extraordinary upward discrepancies are observed, viscosity measurements must be carried out in a concentration region higher than that at which a significant discrepancy from the Huggins plot occurs.

REFERENCES

1. H. Staudinger and H. Freudenberger, *Ber. Dtsch. Chem. Ges.*, **63**, 2331 (1930).
2. E. O. Kraemer, *Ind. Eng. Chem.*, **30**, 1200 (1938).
3. A. Einstein, *Ann. Phys.*, **19**, 289 (1906).
4. A. Einstein, *Ann. Phys.*, **34**, 591 (1911).
5. E. Guth and R. Simha, *Kolloid Z.*, **1**, 266 (1936).
6. O. Gold, Dissertation, Vienna, 1937.
7. G. B. Jeffery, *Proc. R. Soc. London*, **A102**, 163 (192).
8. R. Simha, *J. Phys. Chem.*, **44**, 25 (1940).
9. F. Eirich, *Kolloid Z.*, **74**, 276 (1936).
10. F. Eirich, *Kolloid Z.*, **81**, 7 (1937).
11. H. Staudinger and R. Nozu, *Ber. Dtsch. Chem. Ges.*, **63**, 721 (1930).
12. H. Staudinger and E. Ochiai, *Z. Phys. Chem.*, **A158**, 51 (1931).

13a. H. Staudinger and F. Staiger, *Ber. Dtsch. Chem.*, **68**, 707 (1935).

13b. R. Signer, *Trans. Faraday Soc.*, **32**, 296 (1936).

14. H. Mark, *Der Feste Körper*, Hirgel, Leipzig, 1983, p. 103.
15. R. Houwink, *J. Prakt. Chem.*, **155**, 241 (1940).
16. I. Sakurada, *Kasenkouenshyu*, **6**, 177 (1941).
17. M. L. Huggins, *J. Phys. Chem.*, **42**, 911 (1938).
18. M. L. Huggins, *J. Phys. Chem.*, **43**, 439 (1939).
19. M. L. Huggins, *J. Appl. Chem.*, **10**, 700 (1939).
20. J. J. Hermans, *Physica*, **10**, 777 (1943).
21. W. Kuhn, *Z. Phys. Chem.*, **A161**, 1 (1932).
22. W. Kuhn, *Kolloid Z.*, **62**, 269 (1932).
23. W. Kuhn and H. Kuhn, *Helv. Chim. Acta*, **26**, 1324 (1943).
24. H. A. Kraemer, *J. Chem. Phys.*, **14**, 415 (1946).
25. P. Debye, *J. Chem. Phys.*, **14**, 415 (1946).
26. W. Kuhn, *Kolloid Z.*, **68**, 2 (1934).
27. H. Fikentscher and H. Mark, *Kolloid Z.*, **49**, 185 (1929).
28. P. Debye and A. M. Büche, *J. Chem. Phys.*, **16**, 573 (1948).
29. J. G. Kirikwood and J. Riseman, *J. Chem. Phys.*, **16**, 565 (1948).
30. P. J. Flory, *J. Chem. Phys.*, **17**, 134 (1949).
31. P. J. Flory and W. R. Krigbaum, *J. Chem. Phys.*, **18**, 1086 (1959).
32. T. B. Grimley, *J. Chem. Phys.*, **21**, 185 (1953).
33. W. R. Krigbaum, *J. Chem. Phys.*, **23**, 2113 (1955).
34. H. M. James, *J. Chem. Phys.*, **21**, 1628 (1953).
35. B. H. Zimm, *J. Chem. Phys.*, **21**, 1716 (1953).
36. M. Fixman, *J. Chem. Phys.*, **23**, 1656 (1955).

37. M. Kurata, H. Yamakawa, and E. Teramoto, *J. Chem. Phys.*, **28**, 785 (1958).
38. M. Kurata and H. Yamakawa, *J. Chem. Phys.*, **29**, 311 (1958).
39. P. J. Flory, *Principles of Polymer Chemistry*, Cornell University Press, Ithaca, NY, 1953.
40. P. J. Flory and T. G. Fox, *J. Am. Chem. Soc.*, **73**, 1904 (1951).
41. M. L. Huggins, *J. Phys. Chem.*, **46**, 151 (1942).
42. M. L. Huggins, *Ann. N.Y. Acad. Sci.*, **43**, 1 (1942).
43. M. L. Huggins, *J. Am. Chem. Soc.*, **64**, 1712 (1942).
44. P. J. Flory, *J. Chem. Phys.*, **10**, 51 (1942).
45. W. R. Krigbaum and D. K. Carpenter, *J. Phys. Chem.*, **59**, 1166 (1955).
46. See, for example, H. Yamakawa, *Modern Theory of Polymer Solutions*, Harper & Row, New York, 1971, Ch. 3.
47. M. Fixman, *J. Chem. Phys.*, **36**, 3123 (1962).
48. O. Kratky and G. Porod, *Re. Trav. Chim. Pays-Bas*, **68**, 1106 (1949).
49. H. Benoit and P. M. Doty, *J. Phys. Chem.*, **57**, 958 (1953).
50. J. E. Hearst, *J. Chem. Phys.*, **40**, 1506 (1964).
51. A. Peterlin, *J. Polym. Sci.*, **8**, 173 (1952).
52. J. E. Hearst and W. H. Stockmayer, *J. Chem. Phys.*, **37**, 1425 (1962).
53. H. Yamakawa and W. H, Stockmayer, *J. Chem. Phys.*, **57**, 2843 (1972).
54. R. Ullman, *J. Chem. Phys.*, **49**, 5486 (1968).
55. R. Ullman, *J. Chem. Phys.*, **53**, 1734 (1979).
56. H. Yamakawa and M. Fujii, *Macromolecules*, **6**, 407 (1973).
57. H. Yamakawa and M. Fujii, *Macromolecules*, **7**, 128 (1974).
58. K. Kamide, M. Saito, and H. Suzuki, *Makromol. Chem., Rapid Commun.*, **4**, 33 (1983).
59. K. Kamide and M. Saito, *Eur. Polym. J.*, **19**, 507 (1983).
60. M. Saito, *Polym. J.*, **15**, 213 (1983).
61. T. Kawai and K. Kamide, *J. Polym. Sci.*, **54**, 343 (1961).
62. M. Kurata, W. Stockmayer, and A. Roig, *J. Chem. Phys.*, **33**, 151 (1960).
63. K. Kamide and T. Kawai, *Chem. High Polym. (Tokyo)*, **20**, 512 (1963).
64. M. Ueda and K. Kajitani, *Makromol. Chem.*, **109**, 22 (1967).
65. K. Shiokawa and T. Oyama, *Rep. Prog. Polym. Phys. Jpn.*, **11**, 77 (1968).
66. K. Kamide and W. R. Moore, *Chem. High Polym. (Tokyo)*, **21**, 694 (1964).
67. K. Kamide and K. Kataoka, *Makromol. Chem.*, **128**, 217 (1969).
68a. K. Kamide and W. R. Moore, *Makromol. Chem.*, **80**, 225 (1964).
68b. H. Inagaki, H. Suzuki, and M. Kurata, *J. Polym. Sci.*, **C15**, 409 (1966).
69. K. Kamide, Y. Inamoto, and G. Livingstone, *Chem. High Polym. (Tokyo)*, **23**, 1 (1966).
70. K. Kamide and M. Saito, *Eur. Polym. J.*, **18**, 661 (1982).

71. F. Danusso and G. Moraglio, *Atti Accad. Naz. Lincei, Rend. Cl., fis. Mat. Nat.*, **25**, 509 (1958).
72. J. B. Kinsinger and R. E. Hughes, *J. Phys. Chem.*, **63**, 2002 (1959).
73. T. W. Bates and K. J. Ivin, *Polymer (London)*, **8**, 263 (1967).
74. M. Matsumoto and S. Ohyanagi, *Chem. High Polym. (Tokyo)*, **17**, 1 (1960).
75. E. Cohn-Ginsberg, T. G. Fox, and H. F. Mason, *Polymer (London)*, **3**, 97 (1962).
76. C. E. Bawn, T. B. Grimley, and M. A. Wajid, *Trans. Faraday Soc.*, **46**, 1112 (1950).
77. U. Bianchi and V. Magnasco, *J. Polym. Sci.*, **41**, 177 (1959).
78. K. Kamide, H. Kobayashi, Y. Miyazaki, and C. Nakayama, *Chem. High Polym. (Tokyo)*, **24**, 679 (1967).
79. K. Kamide, K. Fujii, and H. Kobayashi, *Makromol. Chem.*, **117**, 190 (1968).
80. K. Kamide and W. R. Moore, *J. Polym. Sci., Polym. Lett.*, **2**, 1029 (1978).
81. K. Kamide and Y. Miyazaki, *Polym. J.*, **10**, 409 (1978).
82. K. Kamide and M. Saito, *Eur. Polym. J.*, **17**, 1049 (1981).
83. W. H. Stockmayer and M. Fixman, *J. Polym. Sci.*, **C1**, 137 (1963).
84. Y. Einaga, Y. Miyaki, and H. Fujita, *J. Polym. Sci., Polym. Phys. Ed.*, **17**, 2103 (1979).
85. T. Altares, D. P. Wyman, and V. R. Alten, *J. Polym. Sci.*, **A2**, 4533 (1964).
86. G. C. Berry, *J. Chem. Phys.*, **46**, 1338 (1967).
87. M. Fukuda, M. Fukutomi, Y. Kato, and H. Yamakawa, *J. Polym. Sci., Polym. Phys. Ed.*, **12**, 871 (1974).
88. A. Yamamoto, M. Fujii, G. Tanaka, and H. Yamakawa, *Polym. J.*, **2**, 799 (1971).
89. J. W. Mays, N. Hadjichristidis, and L. J. Fetters, *Macromolecules*, **18**, 2231 (1985).
90. See, for example, M. Kurata, T. Tsunashima, M. Iwama, and K. Kamada, in *Polymer Handbook*, 2nd ed., J. Brandrup and E. H. Immergut, Eds., Wiley New York, 1975, Chap. 4.
91. A. Dondos and H. Benoit, *Polymer*, **18**, 1161 (1977).
92. Y. Miyazaki and K. Kamide, *Kohbunshi Ronbunshyu (Tokyo)*, **44**, 1 (1987).
93. K. Kamide, A. Kiguch, and Y. Miyazaki, *Polym. J.*, **18**, 919 (1986).
94. W. G. Harland, *J. Text. Inst.*, **46**, T483 (1955).
95. R. Cumberbirch and W. G. Harland, *J. Text. Inst.*, **49**, T664 (1958).
96. R. Cumberbirch and W. G. Harland, *J. Text. Inst.*, **49**, T679 (1958).
97. K. Kamide, Y. Miyazaki, and T. Abe, *Makromol. Chem.*, **180**, 2081 (1979).
98. D. Henley, *Ark. Kem.*, **18**, 327 (1961).
99. H. Fujita, N. Takeguchi, K. Kawahara, M. Abe, H. Utiyama, and M. Kurata, paper given at the 12th Polymer Symposium, Nagoya, Japan, November 1963.
100. K. Kawahara, T. Norisuye, and H. Fujita, *J. Chem. Phys.*, **49**, 4339 (1968).
101. W. H. Beattie and C. Booth, *J. Appl. Polym. Sci.*, **7**, 507 (1963).
102. A. Peyrouset, R. Precher, R. Panaris, and H. Benoit, *J. Appl. Polym. Sci.*, **19**, 1363 (1975).

103. R. Chian, *J. Phys. Chem.*, **69**, 1645 (1965).
104. R. Chian, *J. Polym. Sci.*, **24**, 333 (1957).
105. J. B. Kinsinger and R. E. Hughes, *J. Phys. Chem.*, **63**, 2002 (1959).
106. K. Kamide, Y. Miyazaki, and H. Kobayashi, *Polym. J.*, **17**, 607 (1985).
107. K. Kamide, H. Yamazaki, and Y. Miyazaki, *Polym. J.*, **18**, (1986).
108. K. Kamide, Y. Miyazaki, and H. Kobayashi, *Polym. J.*, **14**, 591 (1982).
109. S. Kause, *Dilute Solution Properties of Acrylic and Methacrylic Polymers*, Part I, Revision, Rohm & Haas Co., Philadelphia, February 1961.
110. H. J. Cantow and G. V. Schulz, *Z. Phys. Chem. (Frankfurt)*, **2**, 117 (1957).
111. T. G. Fox, *Polymer*, **3**, 111 (1962).
112. M. Bohdanecky, K. Solc, P. Kratochvil, M. Kolinsky, M. Pyska, D. Lim, *J. Polym. Sci., A2*, **5**, 343 (1967).
113. M. Freeman and P. P. Manning, *J. Polym. Sci., A2*, **2**, 2017 (1964).
114. C. Rossi, *Chimica delle Macromolecole* (September 1961); Consiglio Nazionalle delle Ricerche, Rome, 1963, p. 153.
115. Breitenbach, H. Gabler, and O. F. Olaj, *Makromol. Chem.*, **81**, 32 (1964).
116. K. Kamide and Y. Miyazaki, *Kohbunshi Ronbunshyu*, **35**, 467 (1978).
117. G. V. Schulz and A. Haug, *Z. Phys. Chem. (Frankfurt)*, **34**, 328 (1962).
118. W. Brown and R. Wirkström, *Eur. Polym. J.*, **1**, 1 (1965).
119. K. Kamide and M. Saito, *Polym. J.*, **18**, 569 (1986).
120. L. Valtassari, *Makromol. Chem.*, **150**, 117 (1971).
121. K. Kamide and M. Saito, *Polym. J.*, **14**, 517 (1982).
122. K. Kamide, T. Terakawa, and Y. Miyazaki, *Polym. J.*, **11**, 285 (1979).
123. M. Saito, *Polym. J.*, **15**, 249 (1983).
124. K. Kamide, M. Saito, and T. Abe, *Polym. J.*, **13**, 421 (1981).
125. G. V. Schulz and E. Penzel, *Makromol. Chem.*, **112**, 260 (1968).
126. V. P. Shanbhag, *Ark. Kem.*, **29**, 1 (1968).
127. W. Brown, D. Henley, and J. Öhman, *Makromol. Chem.*, **64**, 49 (1963).
128. R. S. Manley, *Ark. Kem.* **9**, 519 (1956).
129. W. Brown and D. Henley, *Makromol. Chem.*, **79**, 68 (1964).
130. J. M. G. Cowie, *Makromol. Chem.*, **42**, 230 (1961).
131. J. M. G. Cowie, *J. Polym. Sci.*, **49**, 455 (1961).
132. H. L. Frish and J. L. Lundeberg, *J. Polym. Sci.*, **37**, 123 (1959).
133. J. L. Lundberg, M. Y. Hellman, and H. L. Frisch, *J. Polym. Sci.*, **46**, 3 (1969).
134. P. F. Onyon, *Nature*, **183**, 1670 (1959).
135. K. Kamide, *J. Soc. Fiber Sci. Tecnol. Jpn.*, **17**, 1159 (1961).
136. M. Bohdanecky and J. Kovár, *Viscosity of Polymer Solutions*, Elsevier, Amsterdam/New York, 1982, Chap. 4.
137. P. E. Rouse, *J. Chem. Phys.*, **21**, 1272 (1955).

138. B. H. Zimm, *J. Chem. Phys.*, **24**, 269 (1956).
139. W. Kuhn and H. Kuhn, *Helv. Chim. Acta*, **29**, 830 (1946).
140. R. Cerf, *Fortschr. Hochpolym. Forsch.*, **1**, 382 (1959).
141. A. Peterlin, *J. Chem. Phys.*, **33**, 1799 (1960).
142. Y. Chikahisa and T. Fujiki, *J. Phys. Soc. Jpn.*, **19**, 2188 (1964).
143. Y. Chikahisa, *J. Phys. Soc. Jpn.*, **21**, 2324 (1966).
144. T. Kotaka, H. Suzuki, and H. Inagaki, *J. Phys. Chem.*, **45**, 2770 (1966).
145. H. Suzuki, T. Kotaka, and H. Inagaki, *J. Chem. Phys.*, **51**, 1279 (1969).
146. E. Hagenbach, *Prog. Ann.*, **109**, 385 (1960).
147. J. M. L. M. Poiseuille, *Mém. Savants Étrang.*, **9**, 433 (1946).
148. H. Mizutani, *Measurements on Degree of Polymerization*, Experiments in Polymer Science Series No. 6, Society of Polymer Science of Japan, Kyoritu Shuppan, Ed., Tokyo, 1957, Chap. 2.
149. K. Kamide, *Chem. High Polym. (Tokyo)*, **21**, 152 (1964).
150. M. L. Huggins, *J. Am. Chem. Soc.*, **64**, 2716 (1942).
151. M. L. Huggins, *Ind. Eng. Chem.*, **35**, 980 (1943).
152. F. Ibrahim and H. G. Elias, *Makromol. Chem.*, **76**, 1 (1964).
153. S. F. Arrhenius, *Z. Phys. Chem.*, **1**, 285 (1887).
154. H. Sataudinger and W. Heuer, *Z. Phys. Chem.*, **A171**, 129 (1934).
155. H. Fikentscher and H. Mark, *Kolloid Z.*, **49**, 135 (1930).
156. F. Baker, *J. Chem. Soc. (London)*, **103**, 1653 (1913).
157. M. Takeda and E. Tsuruta, *Bull. Chem. Soc. Jpn.*, **25**, 80 (1952).
158. D. J. Streeker and R. F. Boyer, *J. Polym. Sci.*, **14**, 5 (1954).
159. M. Kaneko and N. Kuwamura, *High Polym. Jpn.*, **6**, 324 (1957).
160. G. V. Schulz and F. Blaschk, *J. Prakt. Chem.*, **158**, 130 (1941).
161. I. Sakurada, *Chem. High Polym. (Tokyo)*, **2**, 253 (1945).
162. O. F. Solomon and B. S. Gottesman, *Makromol. Chem.*, **127**, 153 (1969).

APPENDIX

A. Einstein's Viscosity Equation

Assumptions employed by Einstein in the course of the derivation of equation (5) are summarized as follows.

1. The solute is in the form of rigid spherical particles
2. The solvent molecules are regarded as a continuous field, and there exists no specific interaction between solvent molecules and solute particles.

3. The solution is so dilute that hydrodynamic interaction between particles can be neglected.
4. The solution is incompressible.
5. Flow is sufficiently slow and the Stokes equation is valid for this fluid.
6. The velocity of the flow on the surface of these particles is zero (i.e., perfectly nondraining).

The simplest derivation of Einstein's equation among the many proposed may be that performed by Ishihara (1), who used Oseen's formula to represent the perturbational flow due to an introduction of the particles. In this appendix we deduce Einstein's equation along the lines of Ishihara's work.

We consider the particles in a simple laminar shear flow (Couette flow) for which velocity, in the absence of the particles, is expressed as

$$\mathbf{v}^0 = \begin{pmatrix} 0 \\ gx \\ 0 \end{pmatrix} \tag{A-1}$$

The field $\mathbf{v}^0$ may be divided in two parts: rotational flow $\mathbf{v}_r$ and deformational flow $\mathbf{v}'$, that is,

$$\mathbf{v}^0 = \mathbf{v}_r + \mathbf{v}' \tag{A-2}$$

where

$$\mathbf{v}_r = \begin{pmatrix} -gy/2 \\ gx/2 \\ 0 \end{pmatrix} \tag{A-3}$$

and

$$\mathbf{v}' = \begin{pmatrix} gy/2 \\ gx/2 \\ 0 \end{pmatrix} \tag{A-4}$$

A particle with radius a placed at the origin of the Cartesian coordinate rotates along the z axis with an angular velocity $g/2$. An increase of viscosity in this solution arises from the deformational flow. However, an introduction of a particle with finite volume brings about an additional local disturbance, and this effect can be treated by replacing the sphere with the equivalent forces in the fluid.

To reproduce the perturbational flow due to the presence of the particle at the origin, we introduce a doublet force at points infinitely near the origin.

According to Oseen, a positive force at $(x=0, y=-\epsilon, z=0)$ and a negative force at $(x=0, y=+\epsilon, z=0)$ gives rise to velocity such that

$$\mathbf{v}_{\mathrm{I}} = A \begin{pmatrix} -y/r^3 - 3x^2y/r^5 \\ x/r^3 - 3xy^2/r^5 \\ -3xyz/r^5 \end{pmatrix} \tag{A-5}$$

and

$$\mathbf{v}_{\mathrm{II}} = A \begin{pmatrix} y/r^3 - 3x^2y/r^5 \\ -x/r3 - 3xy^2/r^5 \\ -3xyz/r^5 \end{pmatrix} \tag{A-6}$$

respectively. Here A is an undetermined constant relating to the strength of the doublet and r is the distance from the origin.

By the requirement to represent correctly the initial unperturbed velocity, the third velocity field $\mathbf{v}_{\mathrm{III}}$, which is derived from the potential ϕ through equation (A-7), is added to the two velocity fields above.

$$\mathbf{v}_{\mathrm{III}} = \nabla\phi \tag{A-7}$$

The potential ϕ in the form

$$\phi = \frac{B\partial^2(1/r)}{\partial x\, \partial y} \tag{A-8}$$

yields

$$\mathbf{v}_{\mathrm{III}} = B \begin{pmatrix} 3y/r^5 - 15x^2y/r^7 \\ 3x/r^5 - 15xy^2/r^7 \\ -15xyz/r^7 \end{pmatrix} \tag{A-9}$$

Superposing these three velocity fields, we choose a suitable relation between constant A and B at $r = a$, as

$$B = -2Aa^2/5 \tag{A-10}$$

Then, on the surface of the sphere, the perturbed velocity $\mathbf{v}'$ becomes

$$\mathbf{v}' = \frac{-6A}{5a^3} \begin{pmatrix} y \\ x \\ 0 \end{pmatrix} \tag{A-11}$$

Comparing this with equation (A-4), we conclude that the choice of A should be such that

$$-6A/5a^3 = g/2 \tag{A-12}$$

At large distance from the origin, the effect of the sphere should vanish and the third velocity field $\mathbf{v}_{\text{III}}$ is relatively small compared to $\mathbf{v}_{\text{I}}$ and $\mathbf{v}_{\text{II}}$. Thus, a combination of the fields $\mathbf{v}_{\text{I}}$ and $\mathbf{v}_{\text{II}}$ yields the perturbation

$$\mathbf{v}' = \frac{5ga^3}{2}\begin{pmatrix} x^2y/r^5 \\ xy^2/r^5 \\ xyz/r^5 \end{pmatrix} \tag{A-13}$$

Energy dispersion per unit volume and unit time in this solution ($= \eta g^2$) is attributed to the frictional force in the original solvent flow ($\eta_0 g^2$) and to that due to the particles with number density n. Neglecting hydrodynamic interaction between the particles, we simply superpose the energy dispersion arising from individual particles. Then the energy equation is

$$\eta g^2 = \eta_0 g^2 + 2n \iint (5ga^3/2)(xy^2/r^5)(\eta_0 g)\, dx\, dy \tag{A-14}$$

Carrying the integration over all the space of the fluid, we obtain the specific viscosity η_{sp} as

$$\eta_{\text{sp}} = (5n/2)(4\pi a^3/3) \tag{A-15}$$

where $4\pi a^3 n/3$ is the product of concentration c (g/cm^3) and the specific volume v of the solute. Finally we arrive at the Einstein's equation

$$[\eta] = 2.5v \tag{5}$$

B. The Kirkwood–Riseman (KR) Theory

First we consider a polymer molecule consisting of a number of segments n in laminar flow for which the velocity field is expressed as by equation (A-1). The most important assumption in the KY theory is that the segment is hydrodynamically regarded as a material point, with the result that the effect of drainage on the surface of the segment is ignored, contrary to the case of Einstein's theory.

The origin of the coordinate is set at the center of mass of this molecule, and the radius vector of the ith segment is $\mathbf{r}_i$. In the low velocity field, a frictional force $\mathbf{F}_i$ at the location of the ith segment is expected to be proportional to the velocity difference of the segment $\mathbf{u}_i$ and the solvent $\mathbf{v}_i$, such that

$$\mathbf{F}_i = \zeta(\mathbf{u}_i - \mathbf{v}_i) \tag{B-1}$$

Here ζ is a frictional constant.

On the other hand, the flow field at the ith segment is perturbed as a result of the frictional forces exerted by other segments. The perturbed velocity $\mathbf{v}'_j$ at the ith segment, which arises from the frictional force $\mathbf{F}_j$ due to the jth segment, is derived by Oseen as

$$\mathbf{v}'_j = \mathbf{T}_{ij}\mathbf{F}_j \tag{B-2}$$

and

$$\mathbf{T}_{ij} = \tfrac{1}{8}\pi\eta_0\left[\frac{\mathbf{I}}{r_{ij}} + \frac{\mathbf{r}_{ij}\mathbf{r}_{ij}}{r_{ij}^3}\right] \tag{B-3}$$

where $\mathbf{I}$ is a unit tensor, $\mathbf{r}_{ij} = \mathbf{r}_i - \mathbf{r}_j$, r_{ij} is the absolute value of vector $\mathbf{r}_{ij}$. The Oseen tensor $\mathbf{T}_{ij}$ is a symmetric tensor.

The sum of the initial unperturbed velocity and the perturbed velocity due to all other segments is $\mathbf{v}_i$, which is given as

$$\mathbf{v}_i = \mathbf{v}_i^0 + \zeta \sum_{j=1}^{n}{}' \, \mathbf{T}_{ij}(\mathbf{u}_i - \mathbf{v}_j) \tag{B-4}$$

Then the frictional force $\mathbf{F}_i$ is

$$\mathbf{F}_i = -\zeta(\mathbf{v}_i - \mathbf{u}_i) - \zeta \sum_{j=1}^{n}{}' \, \mathbf{T}_{ij}\mathbf{F}_j \tag{B-5}$$

When we neglect the deformation of the chain molecules in the flow field, all the segments rotate around the origin with the same angular velocity $g/2$. Similar to the case of the spherical particles mentioned in Appendix A, the velocity difference $\mathbf{v}_i^0 - \mathbf{u}_i$ is deduced as

$$\mathbf{v}_i^0 - \mathbf{u}_i = \frac{g}{2}\begin{pmatrix} y_i \\ x_i \\ 0 \end{pmatrix} \tag{B-6}$$

The energy dispersion per unit volume and unit time due to the presence of a molecule in an instantaneous conformation is

$$-\sum_{i=1}^{n} \mathbf{F}_i \mathbf{v}_i^0$$

where $\mathbf{F}_i$ works in the opposite direction to the flow. Neglecting hydrodynamic interactions between molecules and averaging $\sum \mathbf{F}_i \mathbf{v}_i^0$ over all the possible configurations of a molecule, we obtain the viscosity of the solution which contains N molecules per unit volume in the form

$$\eta g^2 = \eta_0 g^2 - Ng \sum_{i=1}^{n} \langle \mathbf{F}_i \mathbf{v}_i^0 \rangle \tag{B-7}$$

Specific viscosity is

$$\eta_{\text{sp}} = -N_{\text{A}}/\eta_0 Mg \sum_{i=1}^{n} \langle \mathbf{F}_i \mathbf{v}_i^0 \rangle \tag{B-8}$$

where N_{A} is Avogadro's number and M is the molecular weight of the polymer.

Using equations (A-1), (B-5), and (B-6), the vector inner product on the right-hand side of equation (B-8) may be rewritten as

$$\langle \mathbf{F}_i \mathbf{v}_i^0 \rangle = -\zeta (g/2) \langle x_i x_i \rangle - \zeta \sum_{j=1}^{n}{}' \langle (\mathbf{T}_{ij} \mathbf{F}_j)_y x_i \rangle \tag{B-9}$$

Kirkwood and Riseman introduced an approximation on the Oseen tensor, the so-called preaverage assumption, such that

$$\langle (\mathbf{T}_{ij} \mathbf{F}_j)_y x_i \rangle = \langle (\langle \mathbf{T}_{ij} \rangle \mathbf{F}_j)_y x_i \rangle \tag{B-10}$$

When the radial distribution function of the segments in the polymer is assumed to be Gaussian and the number of segments is sufficiently large, $\langle \mathbf{T}_{ij} \rangle$ and $\langle x_i x_i \rangle$ are readily calculated in the form

$$\langle \mathbf{T}_{ij} \rangle = \frac{I}{(6\pi^3)^{1/2} |i-j|^{1/2} a' \eta_0} \tag{B-11}$$

and

$$\langle x_i x_i \rangle = (a'^2/9n)[3i^2 - 3ni + n^2] \tag{B-12}$$

where a' is bond length.

If we assume that $\mathbf{F}_j$ is a function with the same form as $\mathbf{F}_i$ and the summation $\sum'$ in equation (B-9) can be converted to a definite integral, then equation (B-9) becomes the second kind of Fredholm integral equation.

Here we define a function $\phi(x, x)$ as

$$\phi(x, x) = -(18/na'^2\zeta)\langle \mathbf{F}_i \mathbf{v}_i^0 \rangle \tag{B-13}$$

Then equation (B-9) is reduced to

$$\phi(x, x) = f(x, x) - \lambda \int_{-1}^{1} \frac{\phi(t, x)}{|x - t|^{1/2}} dt \tag{B-14}$$

where

$$x = \frac{2i}{n} - 1 \tag{B-15}$$

$$f(x, x) = \frac{3x^2 + 1}{4} \tag{B-16}$$

and

$$\lambda = \frac{\zeta}{(6\pi^3)^{1/2} \eta_0 a'} n^{1/2} \tag{B-17}$$

Using the function $\phi(x, x)$, the limiting viscosity number may be rewritten in the form

$$[\eta] = (N_A \zeta a'^2 / 36\eta_0 m) n F(\lambda) \tag{B-18}$$

$$F(\lambda) = \int_{-1}^{1} \phi(x; \lambda)\, dx \tag{B-19}$$

where m is the molecular weight of a segment.

The integral equation (eq. B-14) is solved with an aid of the Fourier expansion method. Fourier series of the functions ϕ and f are

$$\phi(x, x) = \sum_{-\infty}^{\infty} \phi_k e^{i\pi kx} \tag{B-20}$$

and

$$f(x, x) = \sum_{-\infty}^{\infty} f_k e^{i\pi kx} \tag{B-21}$$

Equation (B-14) leads to a set of linear equations for the Fourier coefficients ϕ_k and f_k, such as

$$\phi_k + \lambda \sum_{-\infty}^{\infty} \alpha_{ks}\phi_s = f_k \qquad (k = 0, \pm 1, \pm 2, \ldots) \tag{B-22}$$

where

$$\alpha_{ks} = \tfrac{1}{2}\int_{-1}^{1} \frac{\exp(i\pi(st - kx))}{|x - t|^{1/2}}\, dx\, dx \tag{B-23}$$

The matrix component α_{ks} is given by

$$\alpha_{00} = \frac{8 \cdot 2^{1/2}}{3}$$

$$\alpha_{k0} = (-1)^{k+1} 4\pi^{1/2} \frac{S(2\pi|k|)}{(2\pi|k|)^{3/2}}$$

$$\alpha_{kk} = 4\pi^{1/2}\left[\frac{C(2\pi|k|)}{(2\pi|k|)^{1/2}} + \frac{S(2\pi|k|)}{2(2\pi|k|)^{3/2}}\right]$$

$$\alpha_{ks} = \frac{2(-1)^{s-k}}{\pi^{1/2}(s-k)}\left[\frac{kS(2\pi|k|)}{|k|(2\pi|k|)^{1/2}} - \frac{sS(2\pi|s|)}{|s|(2\pi|s|)^{1/2}}\right]$$

where

$$C(x) = \int_0^x \frac{\cos t}{(2\pi t)^{1/2}}\, dt$$

and

$$S(x) = \int_0^x \frac{\sin t}{(2\pi t)^{1/2}}\, dt$$

Following Kirkwood and Riseman, here we employ an approximation for α_{sk} which is valid for large $|k|$ and $|s|$:

$$\alpha_{sk} = (2/|k|)^{1/2}\delta_{ks} \qquad \text{for} \quad k \neq 0 \tag{B-24}$$

Then we readily obtain $\phi(x; \lambda)$ in the form

$$\phi(x; \lambda) = \sum_k \frac{f_k(x)\exp(i\pi kx)}{1 + \lambda(2/|k|)^{1/2}} \tag{B-25}$$

with

$$f_k(x) = \frac{3}{4\pi^2 k^2} \exp(-i\pi k x) + (-1)^k \frac{3ix}{4\pi k} \tag{B-26}$$

Finally $[\eta]$ in equation (B-18) becomes

$$[\eta] = \frac{N_A \zeta a'^2}{36\eta_0 m} nF(X) \tag{B-27}$$

where

$$F(X) = \frac{6}{\pi^2} \sum_k \frac{1}{k^2(1 + X/k^{1/2})} \tag{B-28}$$

where

$$X = 2^{1/2}\lambda \tag{B-29}$$

Note: An attempt to take into account the draining effect on the surface of the segments was made by Edwards and Oliver (2) in the course of the derivation of the sedimentation coefficient. The problems involved in the preaveraging of the Oseen tensor are the subject of an excellent review by Yamakawa in Reference 46. An improvement on the solution of equation (B-14) was carried out by Kirkwood, Zwanzig, and Plock (3).

REFERENCES

1. A. Ishihara, *Adv. Polym. Sci.*, **5**, 531 (1968).
2. S. F. Edwards and M. A. Oliver, *J. Phys. A, Gen. Phys.*, **4**, 1 (1971).
3. J. G. Kirkwood, R. W. Zwanzig, and R. Plock, *J. Chem. Phys.*, **23**, 213 (1955).

CHAPTER

9

FRACTIONATION METHODS FOR THE DETERMINATION OF MOLECULAR WEIGHT DISTRIBUTION

KENJI KAMIDE and SHIGENOBU MATSUDA

Fundamental Research Laboratory of Fibers and Fiber Forming Polymers, Asahi Chemical Industry Company Ltd., Hacchonawate Cho, Takatsuki, Osaka, Japan

INTRODUCTION

Since Schulz's first paper (1) in 1936, molecular weight fractionation by solubility difference has been extensively employed as a principal method for evaluation of the molecular weight distribution (MWD) of macromolecules. Figure 1 shows a steplike cumulative weight fraction curve (a) and its smooth cumulative weight fraction curve (b) of the polyisobutylene–petroleum ether–acetone system constructed by Schulz (1) from fractionation data. Schulz obtained the MWD curve (c) by differentiating curve (b). The figure also shows the cumulative weight fraction curve (d) calculated by use of equation (1) derived from the theory of probability of radical polymerization (1).

$$\int_0^x X\,dn = (1 - \alpha^x) + X\alpha^x \ln \alpha \qquad (1)$$

where X is the degree of polymerization and α is probability of polymerization (in this case, $\alpha = 0.9982$), calculated from the maximum X of curve (c) ($\bar{X} = 550$). Agreement of curves (b) and (d) is not satisfactory.

The rapid development of gel permeation chromatography (GPC) from the mid-1950s through the mid-1960s—for example, separation of dextran by use of dextran gel (2); of polystyrene (PS) by use of PS soft-type gel (3), and of PS by use of PS hard-type gel (4)—has led to a decrease in the frequency of use of the fractionation method as a means of in-plant process control. An exact evaluation of MWD by GPC is possible only in two cases: (1) when nearly monodisperse and well-characterized polymer samples are available as standard materials for calibrating the GPC, and (2) when the hydrodynamic volume universal calibration plot is strictly justified for the polymer type of interest (5).

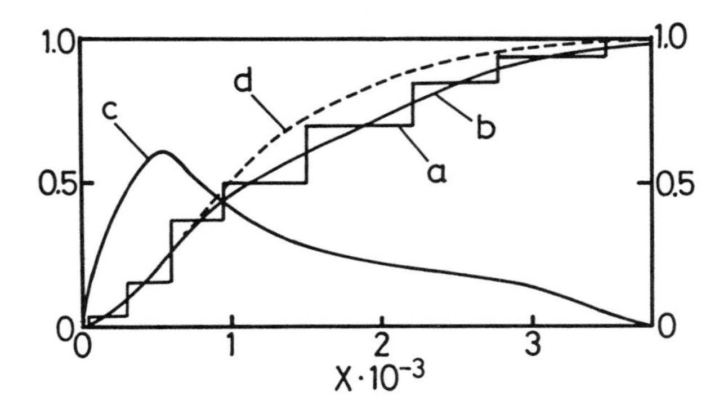

Figure 1. Steplike cumulative weight fraction curve (a), its smooth cumulative weight fraction curve (b) of the polyisobutylene–petroleum ether–acetone system constructed by Schulz in 1936 (1), molecular weight distribution curve (MWD) (c) obtained by differentiating curve (b) and cumulative weight fraction curve (d) calculated by use of equation (1).

Studies (6–10) on the application limit of this universal plot revealed that the plot was not universal when a second separation mechanism exists (i.e., in addition to steric exclusion separation) and when polymer samples have different molecular geometries and rigidities. It seems sufficient to note that the efficiency of GPC fractionation cannot be thoroughly studied without the aid of a complete theory, which has not yet been established (5).

Fractionation by solubility difference has several merits, including inexpensive equipment, wide choice of solvents, relatively small amount of solvent needed to obtain the samples (as compared with GPC and column fractionation), and definite theoretical background (5). Some people hold the opinion that analytical use of fractionation by solubility difference is now out of date and meaningless. However, the importance of analytical successive fractionation by solubility difference should never diminish, as is evident from the discussion above; even in the future this classical method has, we believe, a good chance of continued use in a research laboratory for analytical purposes (5).

The solubility of a polymer is, in principle, expressed in terms of the solution theory. Up to now, numerous reviews and papers (11–42) have discussed successive precipitation and successive solution fractionation (SPF and SSF) on the basis of the Flory–Huggins–Fujishiro theory. Nevertheless, although many studies have been published, the reliability of fractionation methods, including both SPF and SSF, has not always been completely demonstrated, nor are the operating conditions always completely established. For example, in the studies just cited, except for the work of Tung (18), Kamide and his coworkers (20, 21, 30–44), and Koningsveld and Staverman (22–28), hypothetical fractions were carried out either at constant value for the ratio of polymer-lean to polymer-rich phase R, and for a given value of the partition coefficient σ (see eqs. 8 and 24), or at constant initial polymer concentration (expressed as polymer volume fraction) v_p^0, and for a given value of the polymer–solvent interaction parameter χ. In the former case, v_p^0 as well as the amount of the fraction (i.e., the weight fraction of polymer) which is in the polymer-rich phase (in the case of SPF, ρ_p) or in the polymer-lean phase (in the case of SSF, ρ_s) necessarily varied in a very complicated manner, and the values of v_p^0 were not given in the original papers.

In the latter case, ρ_p (or ρ_s) was not kept constant throughout the fractionation. In addition, the calculation was on occasion done with the use of approximate equations. The practical fractionation experiment is usually undertaken under conditions of constant v_p^0 and constant ρ_p (or ρ_s). Hence, simulation fractionation should be performed under the above-mentioned conditions, bearing in mind the value of easy comparison of the experimental data with simulations. It was only after many years, however, that the principal mechanism underlying the fractionation technique became well understood, through the use of electronic computers.

In a fractionation experiment, first, a solution of the polymer in a single solvent is cooled (or nonsolvent added) to allow liquid–liquid separation into two phases. The polymer in the polymer-lean phase is precipitated by addition of a nonsolvent or by evaporation of the solvent (the first fraction). A large amount of solvent is added to the polymer-rich phase to yield a dilute solution, and the temperature of the solution is lowered (or nonsolvent is added to the

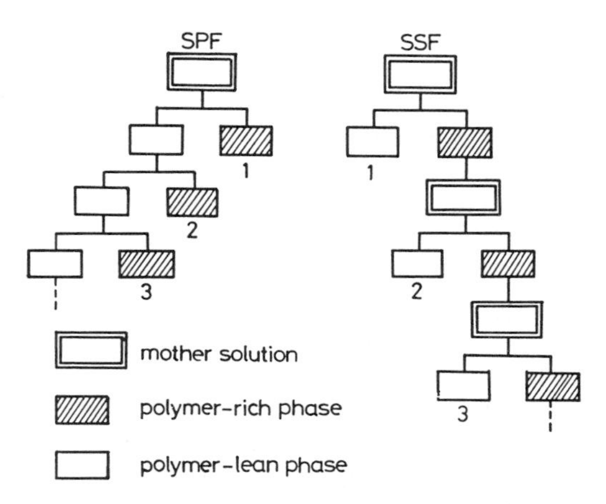

Figure 2. Schematic representation of successive precipitation fractionation (SPF) and successive solution fractionation (SSF); numbers 1–3 denote fraction numbers.

solution), again to produce further phase separation. This procedure is repeated to give more fractions. This method for the fractionation of macromolecules by coacervation is named successive solution fractionation (SSF) in this chapter. In the successive precipitation method, the polymer-rich phase is separated as the fraction, and the principal difference between SPF and SSF is schematically demonstrated in Figure 2. In 1954 Matsumoto and Ohyanagi (14) predicted that SSF was a good candidate for the preparation of sharp fractions. Since 1972, Kamide et al. (34–44) have studied SSF experimentally and theoretically in a systematic manner.

SPF as a method of molecular weight fractionation by solubility difference has been widely used to study polymerization mechanisms; consequently, a large amount of fractionation data has been accumulated (35, 44). We regret to say that insufficient attention was paid to the analytical procedure and to the accuracy of the results obtained. If the reliability of MWD curves constructed from SPF data is made clear, these literature data are expected to play a very important role in determining polymerization mechanisms. In this sense, it is very important to shed light on the reliability difference in relation to the operating conditions of the fractionation and to the analytical procedure adopted.

The extent of uncertainty inherent in this method can be clarified in the following sophisticated manner. A computer-simulated fractionation may be performed, on the basis of a rigorously valid theory, for a given polymer of predetermined MWD under defined operating conditions [v_p^0, ρ_p (or ρ_s), and total number of fractions n_t]. The MWD of the original polymer evaluated from the fractionation data is then compared with the true MWD, which is known in advance.

THEORY OF FRACTIONATION

Quasi-Binary Systems Consisting of Multicomponent Polymers in a Single Solvent

Theoretical studies following the line of the foregoing discussion were performed by Matsumoto and Ohyanagi (MO) (14) for constant σ (see eq. 8), Booth and Beason (BB) (17), Tung (18), Koningsveld et al. (22–28), and Kamide et al. (20, 21, 30–44). The details are summarized in Table 1. Among them, these investigators, Kamide and his coworkers established the modern theory of phase separation and of fractionation by solubility difference, where the original Flory–Huggins (FH) theory was modified by taking into account

Table 1. Theoretical Studies of the Accuracy of the Fractionation Method by Solubility Difference

	Reporter						
	Matsumoto and Ohyanagi	Booth and Beason	Tung	Kamide et al.		Koningsveld et al.	
Theory	FH[a]	FH	FH	Midified FH[b]		Modified FH	
Procedure	SPF[c] and SSF[d]	SPF	SPF	SPF	SSF	SPF	SSF
Original polymer	$\bar{X}_n = 500$[e]	$\bar{X}_n = 50$	$\bar{X}_w = 50$[f]	$\bar{X}_w = 300$–3000 $\bar{X}_w/\bar{X}_n = 1.05$–$5$ SZ[g] and W[h]	$\bar{X}_w = 300$ $\bar{X}_w/\bar{X}_n = 2$ SZ	$\bar{X}_w = 100$–1500 $\bar{X}_w/\bar{X}_n = 1.01$–$2$ LN[i] and GE[j]	$\bar{X}_w = 100$–1500 $\bar{X}_w/\bar{X}_n = 1.01$–$2$ LN and GE
Operating conditions	$R = 10^2$[k] $n_t = 8$[m]	$R = 10^2$ $n_t = 5$	$v_p^0 = 0.5$ and 2%[l] $n_t = 10$	$v_p^0 = 0.01$–1% $n_t = 4$–20 $p_1 = 0$ and 0.5[n]	$v_p^0 = 1$ and 5% $n_t = 15$ $p_1 = 0$ and 0.5	$v_p^0 = 1\%$ $n_t = 20$ $p_1 = 0$ and 0.6	$v_p^0 = 1\%$ $n_t = 20$ $p_1 = 0$ and 0.6
Number of fractionation runs	2	1	2	17	3	7	5
Analytical methods	Schulz	Schulz	Schulz	Schulz and Kamide Kamide	Schulz and Kamide Kamide	Schulz	Schulz
Evaluation	Qualitative	Qualitative	Qualitative	E[o] and E'[p]	E and E'	Qualitative	Qualitative

[a] Flory–Huggins.
[b] The concentration dependence of the polymer–solvent interaction parameter expressed by p_1 was taken into account for the original FH theory.
[c] Successive precipitation fractionation.
[d] Successive solution fractionation.
[e] Number average degree of polymerization.
[f] Weight average degree of polymerization.
[g] Schulz–Zimm distribution.
[h] Wesslau distribution.
[i] Log-normal distribution.
[j] Generalized exponential distribution.
[k] Volume ratio of polymer-lean phase to polymer-rich phase.
[l] initial polymer volume fraction.
[m] Total number of fractions in a run.
[n] p_1 is defined in equation (2).
[o] E is defined by equation (48).
[p] E' is defined by equation (56).

the dependence on concentration and molecular weight of the polymer–solvent interaction parameter χ_i (41, 43, 45):

$$\chi_i = \chi_{00}\left(1 + \frac{k'}{X_i}\right)\left(1 + \sum_{j=1}^{n} p_j v_p^j\right) \qquad (i = 1, \ldots, m) \tag{2}$$

where χ_{00} is a parameter independent of the polymer volume fraction v_p and the molar volume ratio of the ith component of the polymer to the solvent X_i; k' and p_j are parameters dependent on molecular weight and concentration, respectively; χ_{00} and k' are functions of temperature and probably of pressure; and m is the total number of polymer components, all belonging to the same chemical homologue. Kamide et al. (36, 37, 41, 43, 46) confirmed experimentally that equation (2) for $p_1 \neq 0$ with $p_j = 0$ $(j \geqslant 2)$ is accurate enough to represent the two-phase equilibrium of a multicomponent polymer–single solvent system and disclosed in detail the effects of the k' and p_1 parameters on the characteristics of the coexistence curve. In other words, although the χ parameter containing terms higher than v_p^2 can be observed by membrane osmometry, vapor pressure (see, e.g., Fig. 1 of Ref. 39), cloud point curve, and critical solution points measurements (note that in the former two methods p_2 cannot be accurately evaluated), these higher terms do not significantly contribute to the phase separation characteristics, including the coexistence curve, within experimental certainty.

The well-known Gibbs conditions for the phase equilibrium of a quasi-binary solution at constant temperature and pressure are:

$$\Delta\mu_{0(1)} = \Delta\mu_{0(2)} \tag{3}$$

$$\Delta\mu_{x_i(1)} = \Delta\mu_{x_i(2)} \qquad (i = 1, \ldots, m) \tag{4}$$

Here the subscripts (1) and (2) denote the polymer-lean and -rich phases, respectively. The chemical potential of the solvent $\Delta\mu_0$ can be rewritten as (45)

$$\Delta\mu_0 = \tilde{R}T\left\{\ln v_0 + \left(1 - \frac{1}{\bar{X}_n}\right)v_p + \chi_{00}\left(1 + \frac{k'}{\bar{X}_n}\right)\left(1 + \sum_{j=1}^{n} p_j v_p^j\right)v_p^2\right\} \tag{5}$$

where $\tilde{R}$ is the gas constant, T the Kelvin temperature, $\bar{X}_n$ the number average X_i, and v_0 the solvent volume fraction $(= 1 - v_p)$.

The chemical potential of X_i-mer, $\Delta\mu_{X_i}$ is derived from equation (4) when these materials satisfy the Gibbs–Duhem equation. The results are (45):

$$\Delta\mu_{xi} = \tilde{R}T\Bigg[\ln v_{X_i} - (X_i - 1) + X_i\left(1 - \frac{1}{\bar{X}_n}\right)v_p + X_i(1 - v_p)^2\chi_{00}\Bigg[\left(1 + \frac{k'}{\bar{X}_n}\right)\left\{1 + \sum_{j=1}^{n}\frac{p_j}{j+1}\left(\sum_{q=1}^{j}(q+1)v_p^q\right)\right\} + k'\left(\frac{1}{X_i} - \frac{k'}{\bar{X}_n}\right)\left\{\frac{1}{1-v_p} + \sum_{j=1}^{n}\frac{p_j}{j+1}\left(\sum_{q=1}^{j}\frac{v_p^q}{1-v_p}\right)\right\}\Bigg]\Bigg] \tag{6}$$

Combination of equations (3)–(6) yields (45)

$$\chi_{00} = \frac{\ln\dfrac{1-v_{p(1)}}{1-v_{p(2)}} + (v_{p(1)} - v_{p(2)}) - \left(\dfrac{v_{p(1)}}{\bar{X}_{n(1)}} - \dfrac{v_{p(2)}}{\bar{X}_{n(2)}}\right)}{\left[(v_{p(2)}^2 - v_{p(1)}^2) + k'\left(\dfrac{v_{p(2)}^2}{\bar{X}_{n(2)}} - \dfrac{v_{p(1)}^2}{\bar{X}_{n(1)}}\right) + \sum_{j=1}^{n} p_j\left\{(v_{p(2)}^{j+2} - v_{p(2)}^{j+2}) + k'\left(\dfrac{v_{p(2)}^{j+2}}{\bar{X}_{n(2)}} - \dfrac{v_{p(1)}^{j+2}}{\bar{X}_{n(1)}}\right)\right\}\right]} \tag{7}$$

and

$$\sigma_i \equiv \frac{1}{X_i}\ln\frac{v_{x_i(2)}}{v_{x_i(1)}} = \sigma_0 + \frac{\sigma_{01}}{X_i} \tag{8}$$

with

$$\sigma_0 = (v_{p(1)} - v_{p(2)}) - \left(\frac{v_{p(1)}}{\bar{X}_{n(1)}} - \frac{v_{p(2)}}{\bar{X}_{n(2)}}\right) - \chi_{00}\Bigg[2(v_{p(1)} - v_{p(2)}) - (v_{p(1)}^2 - v_{p(2)}^2) + \sum_{j=1}^{n} p_j\left\{\frac{j+2}{j+1}(v_{p(1)}^{j+1} - v_{p(2)}^{j+1}) - (v_{p(1)}^{j+2} - v_{p(2)}^{j+2})\right\}\Bigg] - \chi_{00}k'\Bigg[\left(\frac{v_{p(1)}}{\bar{X}_{n(1)}} - \frac{v_{p(2)}}{\bar{X}_{n(2)}}\right) - \left(\frac{v_{p(1)}^2}{\bar{X}_{n(1)}} - \frac{v_{p(2)}^2}{\bar{X}_{n(2)}}\right) - \sum_{j=1}^{n} p_j\left\{\left(\frac{v_{p(1)}^{j+1}}{\bar{X}_{n(1)}} - \frac{v_{p(2)}^{j+1}}{\bar{X}_{n(2)}}\right) - \left(\frac{v_{p(1)}^{j+2}}{\bar{X}_{n(1)}} - \frac{v_{p(2)}^{j+2}}{\bar{X}_{n(2)}}\right)\right\}\Bigg] \tag{9}$$

and

$$\sigma_{01} = k'\left\{(v_{p(1)} - v_{p(2)}) + \sum_{j=1}^{n}\frac{p_j}{j+1}(v_{p(1)}^{j+2} - v_{p(2)}^{j+2})\right\} \tag{10}$$

When $k' = 0$, equation (2) reduces to

$$\chi = \chi_0 \left(1 + \sum_{j=1}^{n} p_j v_p^j \right) \tag{11}$$

The normalized molecular weight distribution of the original polymer $g_0(X_i)$ is a summation of the relative amounts of X_i-mer separated into polymer-lean and -rich phase $g_{(1)}(X_i)$ and $g_{(2)}(X_i)$:

$$g_0(X_i) = g_{(1)}(X_i) + g_{(2)}(X_i) \tag{12}$$

According to the definition, ρ_s and ρ_p are given by equations (13a) and (13b):

$$\rho_s = \sum_{i=1}^{m} g_{(1)}(X_i) \tag{13a}$$

$$\rho_p = \sum_{i=1}^{m} g_{(2)}(X_i) \tag{13b}$$

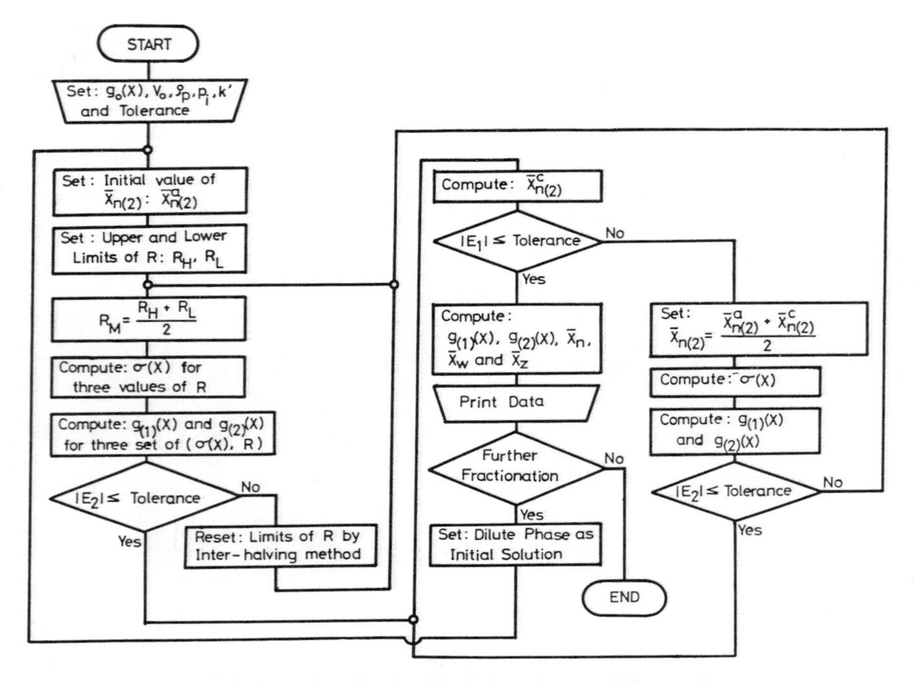

Figure 3. Flowchart for the simulation of a quasi-binary system.

Using σ, R, and $g_0(X_i)$, $g_{(1)}(X_i)$, and $g_{(2)}(X_i)$ are expressed as:

$$g_{(1)}(X_i) = \frac{R}{R + \exp(\sigma X_i)} g_0(X_i) \tag{14a}$$

$$g_{(2)}(X_i) = \frac{\exp(\sigma X_i)}{R + \exp(\sigma X_i)} g_0(X_i) \tag{14b}$$

The phase separation and SPF and SSF were undertaken according to a mathematical simulation procedure established by modifying the Kamide–Sugamiya method (33, 34). The computer simulation was performed by IBM-370 or FACOM-M360. The flowchart of the simulation is presented in Figure 3 (41).

Figure 4 (*a* and *b*) illustrates the MWD curves of some typical fractions obtained by SPF for a 0.94% solution of PS in methylcyclohexane (MCH) as well as those obtained by hypothetical SPF calculation from a 0.94% solution ($\rho_p = 1/18$) (46). Also shown (Fig. 4*c*, *d*) are the MWD curves of fractions isolated by SSF for a 0.94% solution of PS in MCH together with those from the computer simulation with $p_1 = 0.7$ from a 0.94% solution ($\rho_s = 1/23$) (46). These plots indicate that the characteristic features observed for SPF and SSF calculations by Kamide et al. are also maintained qualitatively in practical

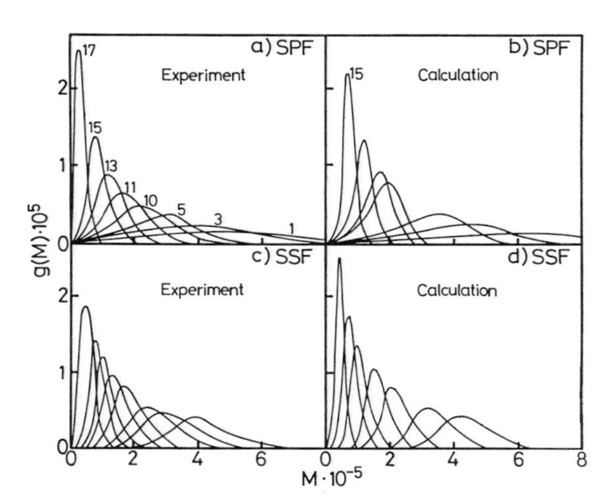

Figure 4. Normalized molecular weight distribution curves of fractions separated by SPF and SSF from 0.94 vol% solution: (*a*) and (*c*) experiments for polystyrene in methylcyclohexane, (*b*) and (*d*) theoretical calculations $p_1 = 0.7$, $\rho_p = 1/18$, and $\rho_s = 1/23$. Numbers on curves denote fraction numbers.

experiments: namely, the fractions separated by SSF contain the lower molecular weight region to a lesser extent, suggesting that the "tail effect" can be almost neglected.

Quasi-Ternary Systems Consisting of Multicomponent Polymers in a Binary Solvent Mixture

The phase separation phenomena of monodisperse polymer–solvent–nonsolvent systems had been studied by Flory (47), Scott (13, 48), Tompa (49), Nakagaki and Sunada (50), Krigbaum and Carpenter (51), and Suh and Liou (52). Even for this relatively simple case, the two-phase equilibrium calculation is far from rigorous and is based on several crude assumptions: (1) a solvent mixture can be approximated as a "single solvent" [Flory (47), Scott (13, 48)], (2) polymer molecular weight is infinite [Scott (13, 48), Nakagaki and Sunada (50)], (3) among three thermodynamic interaction parameters χ_{12}, χ_{13}, and χ_{23} [originally defined in the Flory–Huggins theory (53): 1, solvent 1; 2, solvent 2; 3, polymer), $\chi_{12}=\chi_{13}$ and $\chi_{23}=0$ hold [Tompa (49), Nakagaki and Sunada (50)], and (4) polymer does not exist in the polymer-lean phase (Krigbaum and Carpenter (51), Shu and Liou (52)]. Very recently, Kamide, Matsuda, and Miyazaki (54, 55) proposed a rigorous theory of phase equilibrium of quasi-ternary systems [solvent 1 (good solvent) + solvent 2 (poor solvent) + multicomponent polymers], where χ_{12}, χ_{13}, and χ_{23} were assumed to be concentration dependent:

$$\chi_{12}=\chi_{12}^{0}\left(1+\sum_{s=1}^{n_s} p_{12,s}v_{\mathrm{p}}^{s}\right) \tag{15}$$

$$\chi_{13}=\chi_{13}^{0}\left(1+\sum_{q=1}^{n_q} p_{13,q}v_{\mathrm{p}}^{q}\right) \tag{16}$$

$$\chi_{23}=\chi_{23}^{0}\left(1+\sum_{r=1}^{n_r} p_{23,r}v_{\mathrm{p}}^{r}\right) \tag{17}$$

where χ_{12}^0, χ_{13}^0, and χ_{23}^0 are parameters independent of concentration and degree of polymerization and depend only on the temperature; $p_{12,s}$, $p_{13,q}$, and $p_{23,r}$ are independent of X_i and the temperature.

The conditions of two-phase equilibrium of the quasi-ternary system at constant temperature and pressure are given by equations (18)–(20):

$$\Delta\mu_{1(1)}=\Delta\mu_{1(2)} \tag{18}$$

$$\Delta\mu_{2(1)} = \Delta\mu_{2(2)} \tag{19}$$

$$\Delta\mu_{x_i(1)} = \Delta\mu_{x_i(2)} \qquad (i = 1, \ldots, m) \tag{20}$$

The chemical potentials of solvents 1 and 2 and the X_i-mer ($\Delta\mu_1$, $\Delta\mu_2$, and $\Delta\mu_{X_i}$, respectively) were given by following equations (55):

$$\begin{aligned}\Delta\mu_1 = \tilde{R}T\Bigg[&\ln v_1 - \left(1 - \frac{1}{\bar{X}_n}\right)v_\mathrm{p} \\ &+ \chi_{12}^0\left\{v_2(1-v_1) + \sum_{s=1}^{n_s} p_{12,s}v_2 v_\mathrm{p}^s(1-(s+1)v_1)\right\} \\ &+ \chi_{13}^0 v_\mathrm{p}(1-v_1)\left(1 + \sum_{q=1}^{n_q} p_{13,q}v_\mathrm{p}^q\right) \\ &+ \chi_{23}^0 v_2 v_\mathrm{p}\left\{1 + \sum_{r=1}^{n_r} p_{23,r}\left(\frac{r}{r+1}\right)\frac{v_\mathrm{p}^r}{(1-v_2)^{r+1}}\left(\frac{1-(v_1+v_\mathrm{p})^{r+1}}{v_2}\right)\right\}\Bigg]\end{aligned} \tag{21}$$

$$\begin{aligned}\Delta\mu_2 = \tilde{R}T\Bigg[&\ln v_2 - \left(1 - \frac{1}{\bar{X}_n}\right)v_\mathrm{p} \\ &+ \chi_{12}^0\left\{v_1(1-v_2) + \sum_{s=1}^{n_s} p_{12,s}v_1 v_\mathrm{p}^s(1-(s+1)v_2)\right\} \\ &+ \chi_{23}^0 v_\mathrm{p}(1-v_2)\left(1 + \sum_{r=1}^{n_r} p_{23,r}v_\mathrm{p}^r\right) \\ &+ \chi_{13}^0 v_1 v_\mathrm{p}\left\{1 + \sum_{q=1}^{n_q} p_{13,q}\left(\frac{q}{q+1}\right)\frac{v_\mathrm{p}^q}{(1-v_1)^{q+1}}\left(\frac{1-(v_2+v_\mathrm{p})^{q+1}}{v_1}\right)\right\}\Bigg]\end{aligned} \tag{22}$$

$$\begin{aligned}\Delta\mu_{x^i} = \tilde{R}T\Bigg[&\ln v_{x_i} - (X_i - 1) - X_i\left(1 - \frac{1}{\bar{X}_n}\right)v_\mathrm{p} \\ &- X_i\chi_{12}^0 v_1 v_2\left\{1 - \sum_{s=1}^{n_s} p_{12,s}s v_\mathrm{p}^{s-1}\left(1 - \frac{s+1}{s}v_\mathrm{p}\right)\right\} \\ &- X_i\chi_{13}^0 v_1\Bigg\{1 - v_\mathrm{p}\left(1 + \sum_{q=1}^{n_q} p_{13,q}v_\mathrm{p}^q\right) \\ &+ \sum_{q=1}^{n_q} p_{13,q}\left(1 - \left(\frac{q}{q+1}\right)\frac{v_\mathrm{p}}{1-v_1}\right)\left(\frac{v_\mathrm{p}}{1-v_1}\right)^q\left(\frac{1-(v_2+v_\mathrm{p})^{q+1}}{v_1}\right)\Bigg\}\end{aligned}$$

$$- X_i \chi_{23}^0 v_2 \left\{ 1 - v_p \left(1 + \sum_{r=1}^{n_r} p_{23,r} v_p^r \right) \right.$$

$$\left. + \sum_{r=1}^{n_r} p_{23,r} \left(1 - \left(\frac{r}{r+1} \right) \frac{v_p}{1 - v_2} \right) \left(\frac{v_p}{1 - v_2} \right)^r \left(\frac{1 - (v_1 + v_p)^{r+1}}{v_2} \right) \right\} \Bigg]$$

$$(i = 1, \ldots, m) \quad (23)$$

where v_1 and v_2 are the volume fractions of solvents 1 and 2, respectively. We assumed that the molar volume of solvent 1 is the same as that of solvent 2, that solvents 1 and 2, and the polymer are volumetrically additive, and that the densities of solvents 1 and 2 and the polymer are the same. These assumptions are not absolutely necessary and do not limit the theory (54, 55).

The partition coefficient σ is defined by:

$$\sigma = \frac{1}{X_i} \ln \frac{v_{x_i(2)}}{v_{x_i(1)}} \quad (24)$$

Combination of equations (20), (23), and (24) gives (55):

$$\sigma = (v_{p(2)} - v_{p(1)}) + \left(\frac{v_{p(2)}}{\bar{X}_{n(2)}} - \frac{v_{p(1)}}{\bar{X}_{n(1)}} \right)$$

$$+ \chi_{12}^0 \Bigg[(v_{1(2)} v_{2(2)} - v_{1(1)} v_{2(1)})$$

$$- \sum_{s=1}^{n_s} p_{12,s} s \left\{ v_{1(2)} v_{2(2)} v_{p(2)}^{s-1} \left(1 - \frac{s+1}{s} v_{p(2)} \right) \right.$$

$$\left. - v_{1(1)} v_{2(1)} v_{p(1)}^{s-1} \left(1 - \frac{s+1}{s} v_{p(1)} \right) \right\} \Bigg]$$

$$+ \chi_{13}^0 \Bigg[(v_{1(2)} - v_{1(1)}) - (v_{1(2)} v_{p(2)} - v_{1(1)} v_{p(1)})$$

$$- \sum_{q=1}^{n_q} p_{13,q} (v_{1(2)} v_{p(2)}^{q+1} - v_{1(1)} v_{p(1)}^{q+1})$$

$$+ \sum_{q=1}^{n_q} \frac{p_{13,q}}{q+1} \left\{ v_{1(2)} \left(q + 1 - \frac{v_{p(2)}}{1 - v_{1(2)}} \right) \left(\frac{v_{p(2)}}{1 - v_{1(2)}} \right)^q \frac{1 - (v_{2(2)} + v_{p(2)})^{q+1}}{v_{1(2)}} \right.$$

$$\left. - v_{1(1)} \left(q + 1 - \frac{v_{p(1)}}{1 - v_{1(1)}} \right) \left(\frac{v_{p(1)}}{1 - v_{1(1)}} \right)^q \frac{1 - (v_{2(1)} + v_{p(1)})^{q+1}}{v_{1(1)}} \right\} \Bigg]$$

$$+ \chi_{23}^0 \Bigg[(v_{2(2)} - v_{2(1)}) - (v_{2(2)} v_{p(2)} - v_{2(1)} v_{p(1)})$$

$$- \sum_{r=1}^{n_r} p_{23,r}(v_{2(2)}v_{p(2)}^{r+1} - v_{2(1)}v_{p(1)}^{r+1})$$

$$+ \sum_{r=1}^{n_r} \frac{p_{23,r}}{r+1}\left\{v_{2(2)}\left(q+1-\frac{v_{p(2)}}{1-v_{2(2)}}\right)\left(\frac{v_{p(2)}}{1-v_{2(2)}}\right)^r \frac{1-(v_{1(2)}+v_{p(2)})^{r+1}}{v_{2(2)}}\right.$$

$$\left.- v_{2(1)}\left(q+1-\frac{v_{p(1)}}{1-v_{2(1)}}\right)\left(\frac{v_{p(1)}}{1-v_{2(1)}}\right)^r \frac{1-(v_{1(1)}+v_{p(1)})^{r+1}}{v_{2(1)}}\right\}\Bigg] \tag{25}$$

Substitution of equation (21) into equation (18) yields equation (26) (the left-hand side of eq. 26 is represented by A):

$$A \equiv \ln\frac{v_{1(2)}}{v_{1(1)}} + (v_{p(2)} - v_{p(1)}) - \left(\frac{v_{p(2)}}{\bar{X}_{n(2)}} - \frac{v_{p(1)}}{\bar{X}_{n(1)}}\right)$$

$$+ \chi_{12}^0\Bigg[(v_{2(2)} - v_{2(1)}) - (v_{1(2)}v_{2(2)} - v_{1(1)}v_{2(1)})$$

$$+ \sum_{s=1}^{n_s} p_{12,s}\{(v_{2(2)}v_{p(2)}^s - v_{2(1)}v_{p(1)}^s)$$

$$- (s+1)(v_{1(2)}v_{2(2)}v_{p(2)}^s - v_{1(1)}v_{2(1)}v_{p(1)}^s)\}\Bigg]$$

$$+ \chi_{13}^0\Bigg[(v_{p(2)} - v_{p(1)}) - (v_{1(2)}v_{p(1)} - v_{1(1)}v_{p(1)})$$

$$+ \sum_{q=1}^{n_q} p_{13,q}\{v_{p(2)}^{q+1}(1-v_{1(2)}) - v_{p(1)}^{q+1}(1-v_{1(1)})\}\Bigg]$$

$$+ \chi_{23}^0\Bigg[(v_{2(2)}v_{p(2)} - v_{2(1)}v_{p(1)})$$

$$+ \sum_{r=1}^{n_r} p_{23,r}(v_{2(2)}v_{p(2)}^{r+1} - v_{2(1)}v_{p(1)}^{r+1})$$

$$+ \sum_{r=1}^{n_r} p_{23,r}\left(\frac{r}{r+1}\right)\left\{v_{2(2)}\left(\frac{v_{p(2)}}{1-v_{p(2)}}\right)^{r+1} \frac{1-(v_{1(2)}+v_{p(2)})^{r+1}}{v_{2(2)}}\right.$$

$$\left.- v_{2(1)}\left(\frac{v_{p(1)}}{1-v_{p(1)}}\right)^{r+1} \frac{1-(v_{1(1)}+v_{p(1)})^{r+1}}{v_{2(1)}}\right\}\Bigg] = 0 \tag{26}$$

Substitution of equation (22) into equation (19) gives equation (27) (the left-hand side of eq. 27 is represented by B):

$$
\begin{aligned}
B \equiv \ln \frac{v_{2(2)}}{v_{2(1)}} &+ (v_{p(2)} - v_{p(1)}) - \left(\frac{v_{p(2)}}{\bar{X}_{n(2)}} - \frac{v_{p(1)}}{\bar{X}_{n(1)}} \right) \\
&+ \chi_{12}^0 \Bigg[(v_{1(2)} - v_{1(1)}) - (v_{1(2)}v_{2(2)} - v_{1(1)}v_{2(1)}) \\
&+ \sum_{s=1}^{n_s} p_{12,s} \{ (v_{1(2)}v_{p(2)}^s - v_{1(1)}v_{p(1)}^s) \\
&- (s+1)(v_{1(2)}v_{2(2)}v_{p(2)}^s - v_{1(1)}v_{2(1)}v_{p(1)}^s) \} \Bigg] \\
&+ \chi_{23}^0 \Bigg[(v_{p(2)} - v_{p(1)}) - (v_{2(2)}v_{p(1)} - v_{2(1)}v_{p(1)}) \\
&+ \sum_{r=1}^{n_r} p_{23,r} \{ v_{p(2)}^{r+1}(1 - v_{2(2)}) - v_{p(1)}^{r+1}(1 - v_{2(1)}) \} \Bigg] \\
&+ \chi_{13}^0 \Bigg[(v_{1(2)}v_{p(2)} - v_{1(1)}v_{p(1)}) \\
&+ \sum_{q=1}^{n_q} p_{13,q} (v_{1(2)}v_{p(2)}^{q+1} - v_{1(1)}v_{p(1)}^{q+1}) \\
&+ \sum_{q=1}^{n_q} p_{13,q} \left(\frac{q}{q+1} \right) \Bigg\{ v_{1(2)} \left(\frac{v_{p(2)}}{1 - v_{p(2)}} \right)^{q+1} \frac{1 - (v_{2(2)} + v_{p(2)})^{q+1}}{v_{1(2)}} \\
&- v_{1(1)} \left(\frac{v_{p(1)}}{1 - v_{p(1)}} \right)^{q+1} \frac{1 - (v_{2(1)} + v_{p(1)})^{q+1}}{v_{1(1)}} \Bigg\} \Bigg] = 0
\end{aligned}
\tag{27}
$$

When the original polymer is dissolved into solvent (ordinarily, solvent 1) and nonsolvent (ordinarily, solvent 2) is added, two-phase separation finally occurs. Expressing the volumes of solvents 1 and 2 and the polymer as V_1^0, V_2, and V_3^0, respectively, and the volume of starting solution V_0 by $V_1^0 + V_3^0$, the starting concentration v_p^s is given by

$$
v_p^s = \frac{V_3^0}{V_0} = \frac{V_3^0}{V_1^0 + V_3^0} \tag{28}
$$

Under the two-phase equilibrium condition, the total volume of the system is $V_1^0 + V_2 + V_3^0 (\equiv V)$ and initial concentration v_p^0 is expressed by

$$
v_p^0 = \frac{V_3^0}{V} = \frac{V_3^0}{V_1^0 + V_2 + V_3^0} \tag{29}
$$

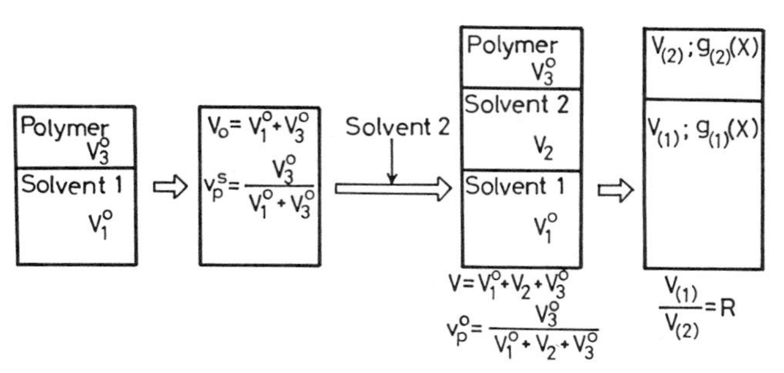

Figure 5. Phase equilibrium experiment of a quasi-ternary system: V_1^0, volume of solvent 1, V_2, volume of solvent 2; V_3^0, volume of polymer; $V_0(= V_1^0 + V_3^0)$, volume of starting solution; v_p^s, polymer volume fraction of the starting solution; V, total volume of the solution at phase equilibrium; $V_{(1)}$, volume of the polymer-lean phase; $V_{(2)}$, volume of the polymer-rich phase; R, phase volume ratio; $g_{(1)}(X)$; molecular weight distribution (MWD) of the polymer in the polymer-lean phase; $g_{(2)}(X)$, MWD of the polymer in the polymer-rich phase.

Figure 5 is schematic representation of the computer experiment on the phase equilibrium of a quasi-ternary solution (54).

If ρ_p is set in advance as an initial condition ρ_p^g, σ, A, and B become the functions of four variables $v_{2(1)}$, $v_{2(2)}$, R^a, and $\bar{X}_{n(2)}^a$ (54, 55).

$$\sigma = \sigma(v_{2(1)}, v_{2(2)}, R^a, \bar{X}_{n(2)}^a) \tag{30}$$

$$A = A(v_{2(1)}, v_{2(2)}, R^a, \bar{X}_{n(2)}^a) \tag{31}$$

$$B = B(v_{2(1)}, v_{2(2)}, R^a, \bar{X}_{n(2)}^a) \tag{32}$$

where R^a and $X_{n(2)}^a$ are the assumed values of R and $X_{n(2)}$ and $X_{n(2)}$ is

$$\bar{X}_{n(2)} = \frac{\sum_{i=1}^{m} X_i g_{(2)}(X_i)}{\sum_{i=1}^{m} g_{(2)}(X_i)} \tag{33}$$

Combining equations (13b), (14b), and (33), we see that ρ_p and $\bar{X}_{n(2)}$ are finally functions of σ and R^a. We define C and D by (54, 55)

$$C \equiv \rho_p(\sigma(v_{2(1)}, v_{2(2)}, R^a, \bar{X}_{n(2)}^a), R^a) - \rho_p^g \tag{34}$$

$$D \equiv \bar{X}_{n(2)}(\sigma(v_{2(1)}, v_{2(2)}, R^a, \bar{X}_{n(2)}^a), R^a) - \bar{X}_{n(2)}^a \tag{35}$$

By solving the nonlinear simultaneous equations (31), (32), (34), and (35), we

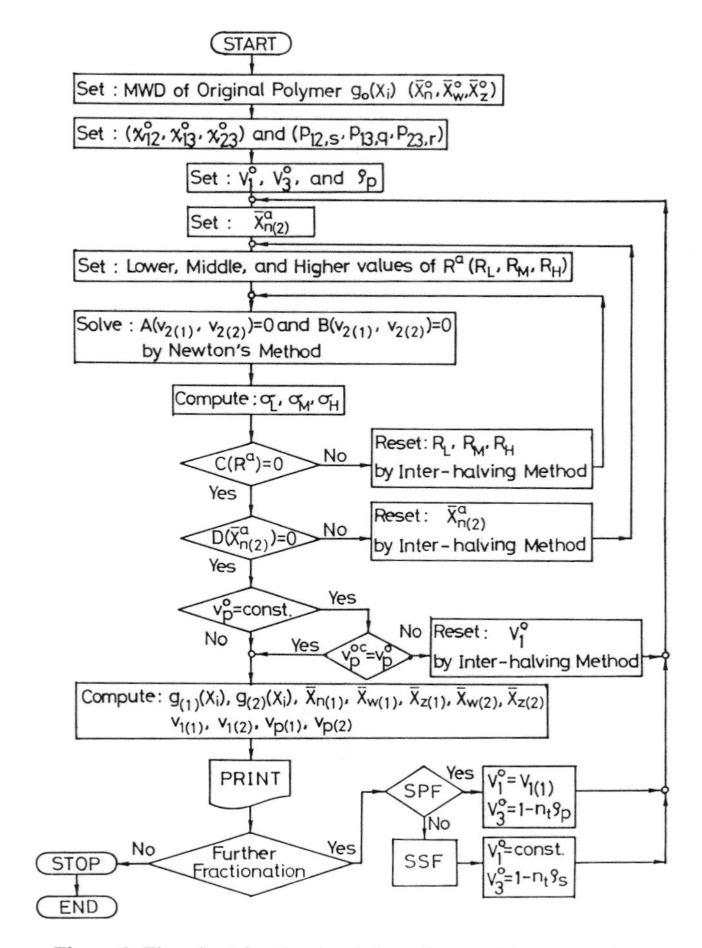

Figure 6. Flowchart for the simulation of a quasi-ternary system.

determine $v_{2(1)}$, $v_{2(2)}$, R, and $\bar{X}_{n(2)}$. Substituting these four values into equation (30), we can calculate σ and other phase separation characteristics. Figure 6 shows the main flowchart of simulation.

Figure 7 shows the plots of $\bar{M}_w/\bar{M}_n$ versus $\bar{M}_w$ for successive solution fractionation runs (56). The data points represent a series of PS fractions ($n_t = 22$), separated successively using methyl ethyl ketone as solvent and methanol as nonsolvent ($v_p^s = 6.15 \times 10^{-3}$). The line is a theoretical curve calculated assuming $\chi_{12} = 0.2$, $\chi_{13} = 0.05$, $\chi_{23} = 0.8$, and $\rho_p = 1/22$. The coincidence of the experimental data with theoretical calculation is satis-

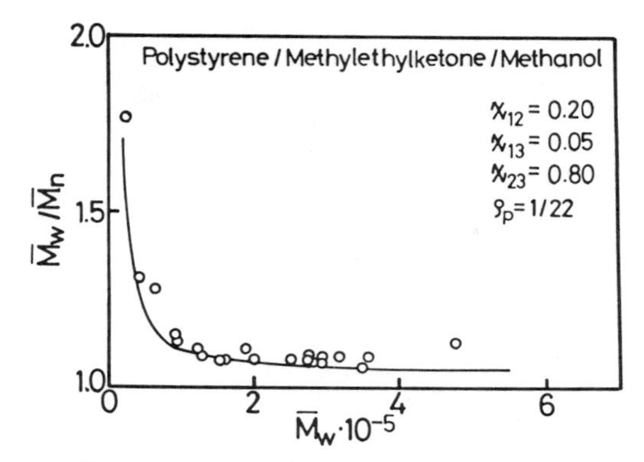

Figure 7. The ratio $\bar{M}_w/\bar{M}_n$ plotted against $\bar{M}_w$ of the fractions separated by successive solution fractionation from a quasi-ternary system; experimental data points on polystyrene–methyl ethyl ketone–methanol are shown; curve is theoretical, and $v_p^0 = 6.15 \times 10^{-3}$.

factory. However, note that the three χ parameters are not experimentally determined, and Figure 7 does not imply a direct comparison between experimental evidence and the theory of phase separation of quasi-ternary solutions. This problem is still open for research.

In this manner, Kamide et al. concluded that the accuracy of fractionation theory and the associated simulation technique must be accepted with reservations and such data regarded as quantitative. On this basis analytical SPF has been studied in a very systematic manner and definitive conclusions have been reached. In contrast, although analytical SSF is expected to be much superior to analytical SPF, the theoretical studies of analytical SSF were qualitative and not comprehensive until 1977. Therefore, Miyazaki and Kamide (5) attempted to derive conclusions covering more general cases.

EXPERIMENTAL PROCEDURE

Successive Precipitation and Solution Fractionation

An SSF–SPF apparatus was constructed for specially designed use as outlined in Figure 8 (42), where A is a solvent measuring vessel, C is a five-necked, round bottom fractionation vessel with mechanical sealed motor attached for agitation; D is a bath, thermostated to $\pm 0.1°C$ in the temperature range 5–150°C; E is a five-necked, round bottom storage vessel for the polymer-lean phase; F is a bath controlled to within $\pm 0.5°C$ and ranging from 20–140°C; G is an evaporator made by Tokyo Rika Ltd. (Tokyo); H is a solvent or

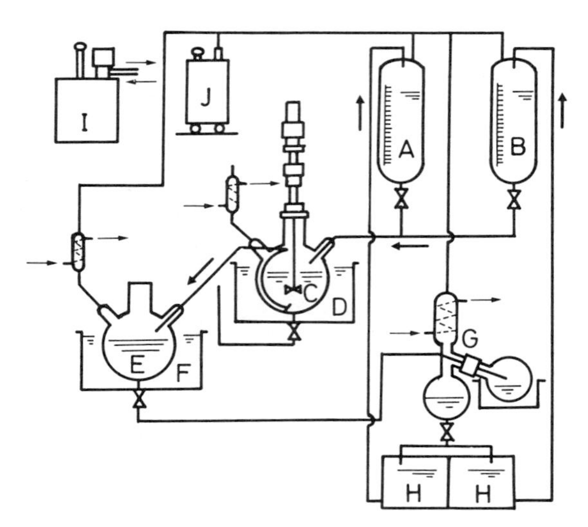

Figure 8. Outline of large-scale apparatus for SSF and SPF. A and B, solvent and nonsolvent measuring vessel; *D*, thermostated bath; E, storage vessel; F, bath; G, evaporator; H, solvent and nonsolvent recovery vessel; I, thermostated bath; J, vacuum pump.

nonsolvent recovery vessel; I is a low temperature thermostat bath [Haake T-33 made by Haake-Tecknik (Berlin)], supplying cooled water (5–10°C); and J is a vacuum pump. These measuring, fractionating, and storage vessels and the pipes connecting them are all made of glass manufactured by Shibata Chemical Appliance Manufacturing Company Ltd. (Tokyo). The leakage of solvent and nonsolvent vapor from the apparatus was completely prevented by allowing water at 5°C to flow through a 1 m^2 heat exchanger installed in the fractionation and storage vessels. All liquids, such as solvent, nonsolvent (if necessary), and polymer-lean phase, were transferred by the pressure due to the difference in liquid levels or by a vacuum.

Table 2 shows some typical operating conditions of SSF and $\bar{M}_w/\bar{M}_n$ of fractions isolated by SSF (38, 42, 57–60). The operating conditions of SSF are extensively listed in the *Polymer Handbook* (61) and are not repeated here.

Refractionation

Relatively low order fractions of SPF have a large tail in the low degrees of the polymerization region and consequently MWDs become broad. This long tail is the main reason for low accuracy of SPF as an analytical fractionation method and for the frequent occurence of reversed-order fractionation (ROF) in SPF (see later). To improve this shortcoming, relatively low order fractions of SPF are often refractionated into several fractions.

Table 2. Operating Conditions $\bar{M}_w/\bar{M}_n$ of Fractions Isolated by SSF

Sample	Solvent–Nonsolvent	Temperature (°C)	Polymer weight	v_p^0	n_t	$\bar{M}_w/\bar{M}_n$	Ref.
HDPE[a]	CH	98–133	—	0.0108	20	—	42
PS	MCH	20–55	—	0.0094	14	—	42
	MCH	15–55	—	0.0094	23	1.05–1.2	42
	MCH	7–55	—	0.0094	20	1.05–1.2	42
	MCH	10–58	—	0.0094	30	1.05–1.2	42
	MEK[b]–methyl alcohol	35	—	0.0068	22	1.05–1.2	42
Poly(α-MS)[c]	CH	—	—	0.0081	12	1.0–1.07	38
CA(2.92)[d]	Epichlorohydrin/*n*-hexane	35	60	0.0108	13	1.3 –1.5	57
CA(2.92)	Epichlorohydrin/*n*-hexane	35	—	0.0050	14	1.3 –1.5	42
CA(2.46)[e]	Acetone–ethyl alcohol	30	—	0.0210	21	1.2 –1.5	42, 58
CA(2.46)	Acetone–ethyl alcohol	30	—	0.0210	16	1.2 –1.5	42
CA(1.75)[f]	Acetone–water(7–3 vol)–methyl alcohol	—	—	—	10	1.28 –1.36	59
CA(0.49)[g]	Water–methyl alcohol	25	—	0.0056	15	1.3	60

[a] High density polyethylene.
[b] Methyl ethyl ketone.
[c] Methylstyrene.
[d] Cellulose acetate degree of substitution (DS) = 2.92.
[e] DS = 2.46.
[f] DS = 1.75.
[g] DS = 0.49.

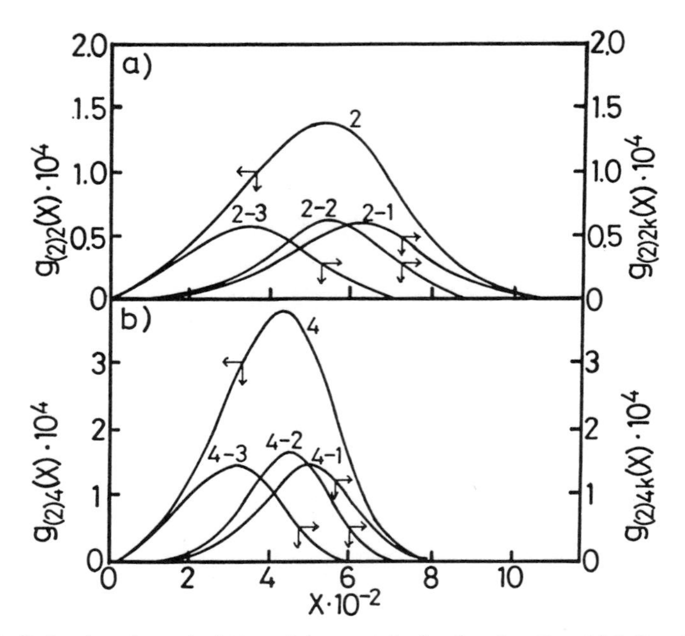

Figure 9. Refractionation calculation of the precipitation fractionation: (*a*) 2–3 and (*b*) 4–3.

Figure 9 shows a few examples of the MWD of refractionated fractions (subfractions) (21). The first three fractions were refractionated into four subfractions: $v_p^0 = 1\%$, $\rho_p = 1/15$, SZ2 (Schulz–Zimm type distribution of $\bar{X}_w^0/\bar{X}_n^0 = 2$), and $\bar{X}_w^0 = 300$. Here, $(j-k)$ denotes the kth subfraction of refractionation of the jth fraction; for example, the kth fraction and lth subfraction are called the $(k-0)$ fraction and $(k-1)$ subfractions, respectively. Although the MWD of the $(j-1)$ and $(j-2)$, $(j = 1\text{–}4)$, subfractions approaches the normal distribution more closely than that of the jth fraction, components having a low degree of polymerization cannot be removed perfectly. The $\bar{X}_{w(2)}/\bar{X}_{n(2)}$ ratios of the $(j-1)$ and $(j-2)$ subfractions are smaller than that of the jth fraction, but $\bar{X}_{w(2)}/\bar{X}_{n(2)}$ of the $(j-3)$ subfraction is larger than that of the jth fraction; $\bar{X}_{w(2)}/\bar{X}_{n(2)}$ of the $(j-k)$th subfraction is not always smaller than that of jth fraction (21). Cumulative weight fraction distribution of refractionation, evaluated by Schulz's method with $v_p^0 = 1.0\%$ (see discussion of eqs. 42 and 43) approaches the true distribution for the high degree of polymerization region, as indicated in Figure 10. But the relative error of the refractionation method is larger than that of simple fractionation evaluated by Schulz's method with $v_p^0 = 0.1\%$. The refractionation method does not serve to evaluate the MWD of the original polymer more precisely.

According to the results mentioned above, a simple refractionation of the

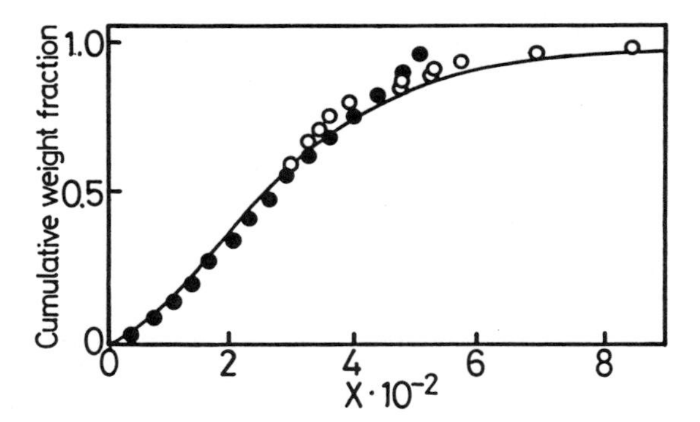

Figure 10. Integral distribution curve estimated from refractionation data by using Schulz's method. ○, plots of the results of a computer simulation in which the first four fractions were refractionated into 12 equal subfractions; ●, plots of data from single fractionation.

first few fractions is not sufficient to elucidate accurately the MWD of the original polymer. To obtain sharp fractions, however, the refractionation method is preferable to the use of simple SPF from a solution of a lower initial concentration. Taking into account this point, Kamide et al. (32) examined the three types of the successive precipitation refractionation (SPRF) shown in Figures 11–13: 15-1, 5-3, and 3-5. A $v_p^0 = 0.1\%$ solution of the original polymer, dissolved in a single solvent, was cooled in such a manner that an equal amount [i.e., $(j+1)/ij$ of the initial polymer in weight] was succesively separated by cooling the solution. The amount of polymer remaining in the polymer-lean phase after the last [i.e., the $(k-j)$th] refractionation process for

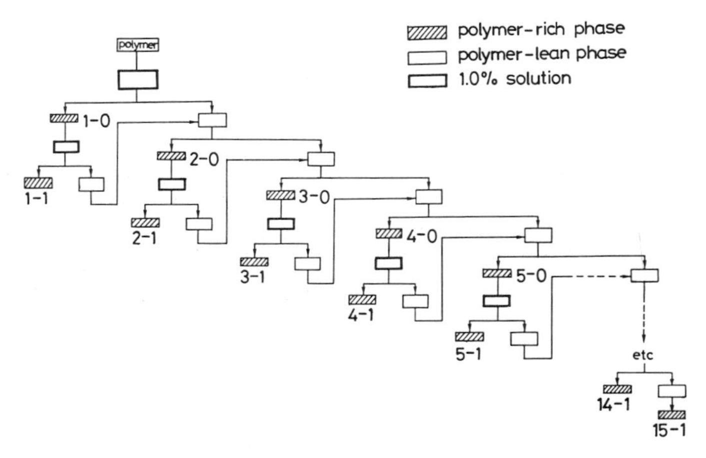

Figure 11. Schematic representation of the 15-1 type of successive precipitation refractionation.

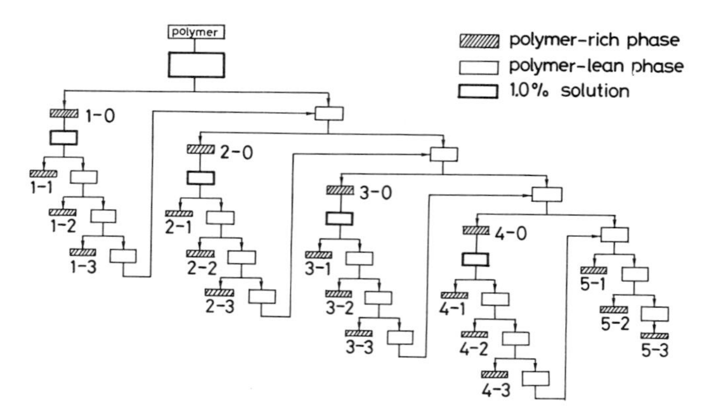

Figure 12. Schematic representation of the 5-3 type of successive precipitation refractionation.

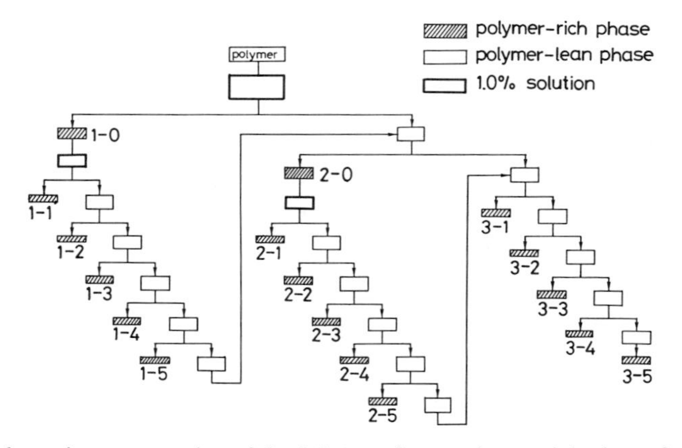

Figure 13. Schematic representation of the 3-5 type of successive precipitation refractionation.

a given fraction was always kept to be $1/ij$ of the total polymer. Figures 14–16 show the MWDs obtained by the 15-1, 5-3, and 3-5 types of refractionation, respectively, together with that of the original polymer (SZ2, $X_w^0 = 300$) shown as a broken curve. Other types of SPRF were investigated by Schulz and Dinglinger (62), Meffroy-Biget (63), Thurmond and Zimm (64), and Meyerhoff (65).

The 5-3 and 3-5 types of SPRF method (32) were found to be the most efficient in obtaining sharp fractions from the whole polymer without laborious and time-consuming operations. In general, there are many ways of selecting a pair i and j for a given value of the total subfractions. The results of the calculation given by computer indicate that the pair of i and j should be

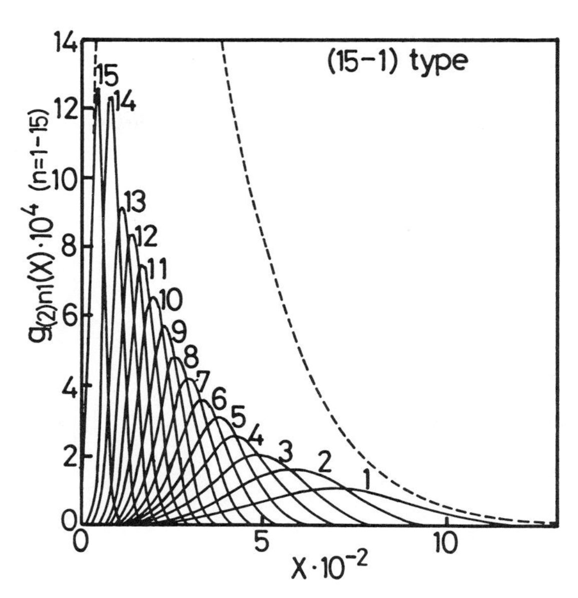

Figure 14. Hypothetical 15-1 type of refractionation from a 1.0 vol % solution of the polymer having the most probable distribution $\bar{X}_n = 150$. The 15 subfractions are represented by numbered curves; the broken curve is the original polymer.

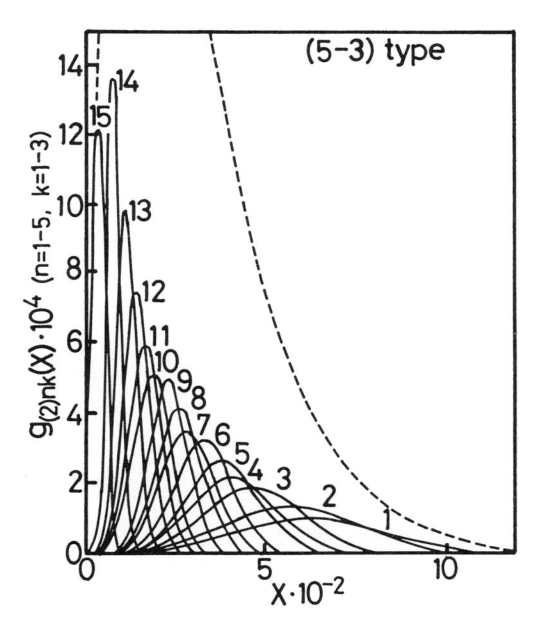

Figure 15. Hypothetical 5-3 type of refractionation from a 1.0 vol % solution of the polymer having the most probable distribution $\bar{X}_n = 150$; data arranged as in Figure 14.

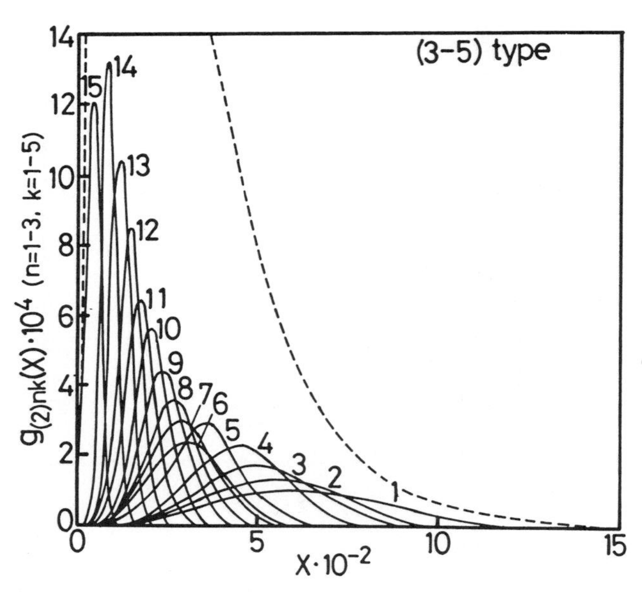

Figure 16. Hypothetical 3-5 type of refractionation from a 1.0 vol % solution of the polymer having the most probable distribution $\bar{X}_n = 150$; data arranged as in Figure 14.

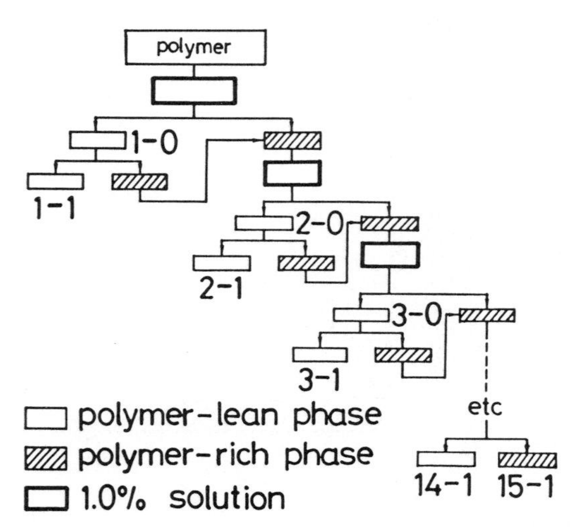

Figure 17. Schematic representation of the 15-1 type of successive solution refractionation.

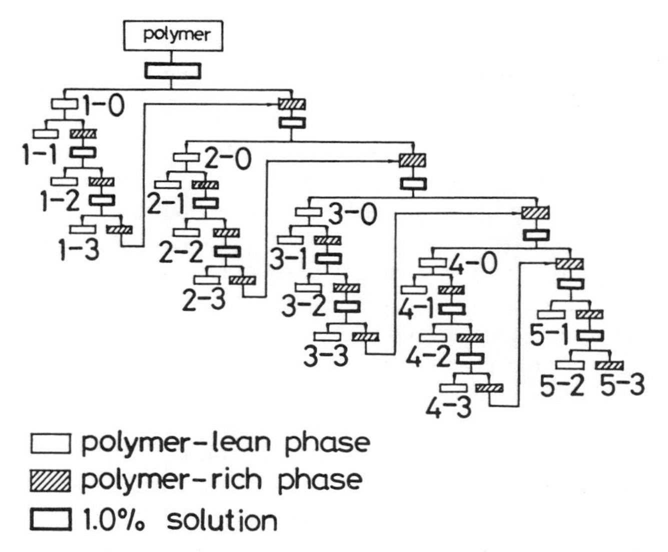

Figure 18. Schematic representation of the 5-3 type successive solution refractionation.

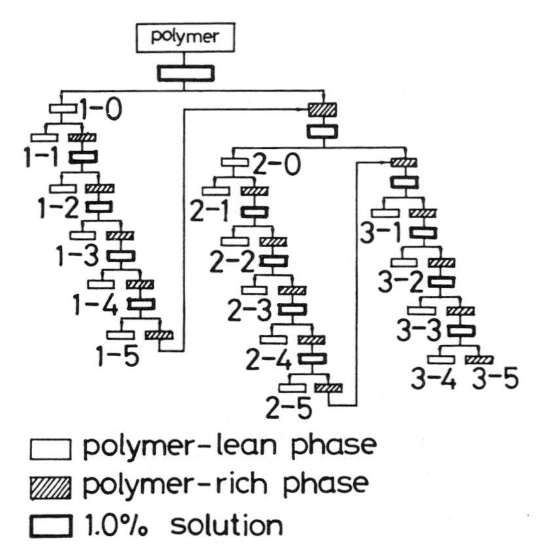

Figure 19. Schematic representation of the 3-5 type of successive solution refractionation.

chosen in such a way that the values of i and j do not differ very much in magnitude (32).

The successive solution refractionation (SSRF) approach was also examined by Kamide and his coworkers (66). Schematic representations of a typical SSRF procedure are shown in Figures 17–19 (66), where the total number of subfractions ij is kept constant (in this case $ij = 15$). The polymer concentration of the polymer-lean phase is controlled at 1% by evaporating an adequate amount of the solvent in that phase. The first subfraction of amount $1/ij$ of the initial polymer by weight is separated from the solution as the polymer-lean phase by lowering the temperature. Solvent is added to the polymer-rich phase to yield a 1.0% solution, which is used as the mother solution, for further fractionation. In total, j subfractions of equal amount are refractionated from the first fraction. The polymer-rich phase in the first separation step is employed for successive fractionations. The difference of the $\bar{X}_w/\bar{X}_n$ versus $\bar{X}_w$ relation in 15-1 and 0-15 fractions is not large, as compared in the case of SPRF (Fig. 20). This implies that the refractionation technique is not very effective for separating polymers into very narrow fractions (66).

SPRF and SSRF are in general not suitable for the purpose of analytical fractionation, but they are useful in preparative fractionation (32, 66).

Reversed-Order Fractionation

As the fractionation proceeds—that is, as the fractionation order i increases—the average molecular weight of the ith fraction obtained by the SPF

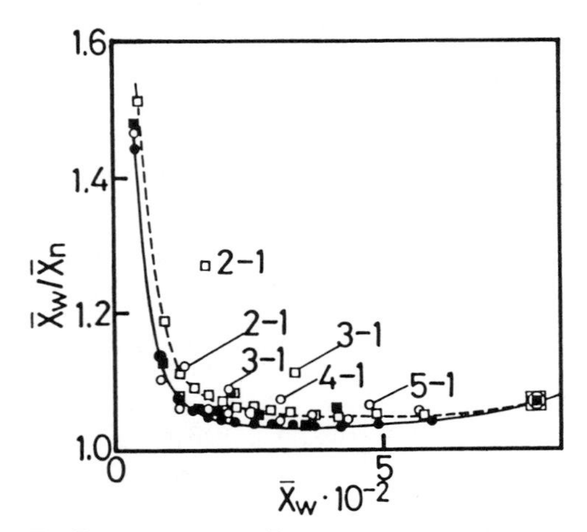

Figure 20. Values of $\bar{X}_w/\bar{X}_n$ plotted against $\bar{X}_w$ for the subfractions obtained under the different conditions. ●, 15-1 type, ○, 5-3 type; ■, 3-5 type; □, 0-15 type.

decreases, irrespective of the nature of the average molecular weight in the fractionation concerned. However, it is occasionally observed in practical SPF that the average molecular weight of the $(i+1)$th fraction is larger than that in the ith fraction (67). This phenomenon is commonly termed reversed-order fractionation (ROF). Until the mid-1970s, ROF was mainly attributed phenomenologically to an overly high v_p^0 value, although no explicit explanation has been given (68). In addition, according to SPF experiments on polyacrylonitrile by Fujisaki and Kobayashi (68–71), ROF occurs in SPF probably even from a solution of comparatively low v_p^0 in the case of the specific combination of a good solvent and a poor solvent. Matsumoto (72) predicted after some approximate hypothetical calculations that ROF would be observed if the fractionation were carried out under conditions in which the value of X_a, the value of X_i corresponding to $f_x = 0.5$, is adopted as large as possible for the first fraction and as small as possible for the second fraction.

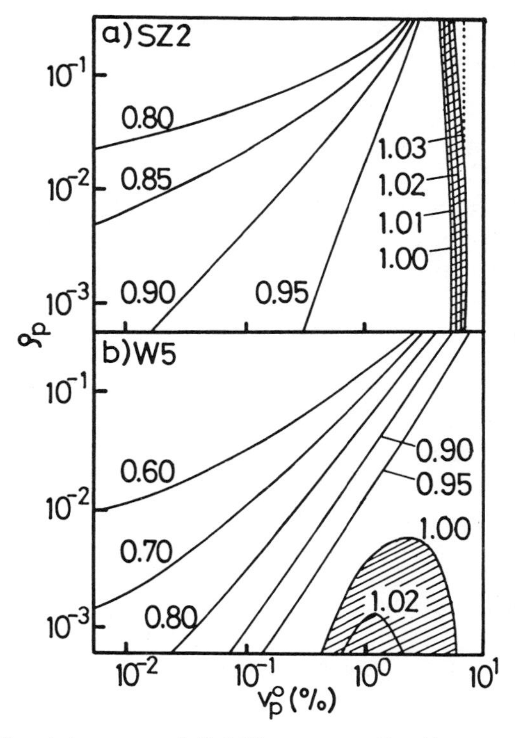

Figure 21. Correlations between ρ_p and v_p^0 yielding constant $\bar{X}_{n,2}/\bar{X}_{n,1}$ in SPF: original polymer, $\bar{X}_w^0 = 300$, $p_1 = 0$. Numbers on curves denote $\bar{X}_{n,2}/\bar{X}_{n,1}$; hatched region of $\bar{X}_{n,2}/\bar{X}_{n,1} > 1$. (*a*) Schulz–Zimm distribution, $\bar{X}_w^0/\bar{X}_n^0 = 2$ (SZ2). (*b*) Wesslau distribution, $\bar{X}_w^0/\bar{X}_n^0 = 5$ (W5).

Here, f_x denotes the probability of the X_i-mer partitioned in a polymer-rich phase. The hypothetical calculations performed later by Kamide et al. (20, 21) indicate that X_a increases eventually with an increase in v_p^0 and with a decrease in p_1 of χ (see eq. 2) and with a decrease in ρ_p. Accordingly ROF may occur if a very small amount of the first fraction is separated from a dilute solution. However, the computer simulation of Kamide et al. (21) indicated that Matsumoto's theoretical prediction lacks general character, such that the parameter X_a cannot be regarded as a reasonable measure of ROF, and that Matsumoto could not exemplify ROF theoretically because of the relatively narrow range of operating conditions adopted (73).

Kamide and Miyazaki (73) showed that ROF may occur theoretically in SPF from a dilute solution under operating conditions which are not unusual even when the phase equilibrium is fully achieved. For example, Figure 21 shows the operating conditions v_p^0 and ρ_p, yielding the fractions with a given value of $\bar{X}_{n,2}/\bar{X}_{n,1}$ (subscripts 1 and 2 show the order of fractions), which are fractionated by SPF from the two original polymers [SZ2 with $\bar{X}_w^0 = 300$ and the Wesslau-type distribution of $\bar{X}_w^0/\bar{X}_n^0 = 5$ (W5) with $\bar{X}_w^0 = 300$] (73). In Figure 21 the shaded area represents $\bar{X}_{n,2}/\bar{X}_{n,1} \geqslant 1$. For the SZ-type polymer, ROF occurs at a very limited range of v_p^0, almost independent of ρ_p (i.e., 4.2–5.2% at $\rho_p = 10^{-2}$ and 4.7–6.5% at $\rho_p = 10^{-3}$); but for the W-type polymer, ROF occurs at a relatively wide range of v_p^0 and small ρ_p (73). Following this, in the former, v_p^0 rather than ρ_p plays an essentially important role in ROF, and in the latter, both ρ_p and v_p^0 contribute effectively to ROF. In the latter, the v_p^0 range yielding ROF becomes wide as ρ_p is small. In this sense, the type of MWD in the original polymer controls predominantly the ease of occurrence of ROF, and clearly, ROF occurs even if a common value of ρ_p is employed through the first and second fractions. This is inconsistent with prediction of Matsumoto (72).

The possibility for the occurrence of ROF is large for solutions of comparatively high initial concentration in which the polymer having larger $\bar{X}_w^0/\bar{X}_n^0$ is dissolved. Among many experimental operating conditions, $\bar{X}_w^0$, $\bar{X}_w^0/\bar{X}_n^0$, v_p^0, and p_1 appear to play an essential role in the occurrence of ROF. Moreover the optimum external conditions for ROF are fundamentally dependent on the MWD type of the original polymer (73). In contrast to this, ROF does not occur in SSF under any operating conditions. ROF should be avoided for analytical fractionation.

Spencer's Method

In 1948 Spencer (74) proposed a method of fractionation into summative fractions which is occasionally termed Spencer's method. A represented schematically in Figure 22, Spencer's method is, in principle, an approximate

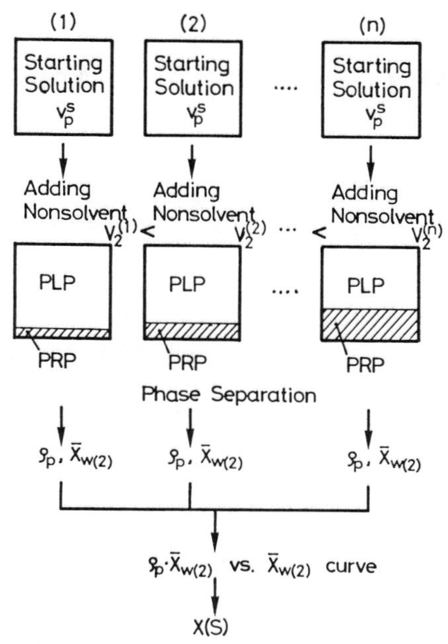

Figure 22. Schematic representation of Spencer's method.

method, based on the assumption that all polymer molecules above a certain degree of polymerization $X(S)$ remain in the polymer-rich phase (PRP) all molecules below $X(S)$ remain in the polymer-lean phase (PLP), where (S) denotes a given condition of fractionation. The procedure used in this method is to measure the weight fraction of the precipitate $\rho(S)$ and the weight average X_i, $\bar{X}_{w(2)}(S)$ of the precipitated polymer, from which one may obtain the value of $X(S)$ from the slope of the plot of $\rho(S)\bar{X}_{w(2)}(S)$ against $\rho(S)$. Finally, the plot of $1 - \rho(S)$ as a function of $X(S)$ affords directly the integral X_i distribution.

Until now Spencer's method has been examined by Billmeyer and Stockmayer (75), Matsumoto (76), Broda et al. (77, 78), Okamoto (79, 80), and Kamide et al. (31) in more detail. Kamide et al. (31) pointed out that a rigorous examination cannot be achieved without comparing the integral distribution elucidated from the fractionation data with the true distribution. The hypothetical fractionations carried out by Billmeyer and Stockmayer (75) and Matsumoto (76) involved approximations, which are impractical and not always correct, for the sake of simplifying the computation.

For polymers having several types of distribution, Billmeyer and Stockmayer (75) established the relationships between $\rho(S)$ and the parameter Z, representing the breadth in the X_i distribution in the precipitate. This parameter Z is given by

$$Z = \frac{\rho(S)(\bar{X}_{w(2)}(S) - \bar{X}_w^0)}{\bar{X}_w^0} \tag{36}$$

where $\bar{X}_w^0$ is $\bar{X}_w$ of the initial polymer. In this case, they used the assumption regarding $X(S)$ as described above. Using this parameter, they concluded that Spencer's method cannot be expected to disclose any fine details of distribution. However, it is theoretically apparent that the reliability of the integral distribution constructed by Spencer's method can be improved without limit by increasing the number of the fractions, if the assumption with respect to $X(S)$ is adopted. In addition, as Matsumoto (76) was the first to indicate, Z is less sensitive in expressing the X_i distribution than had been expected.

The hypothetical fractionation was attempted by Matsumoto (76) on the basis of the theory of Flory and Huggins for heterogeneous polymers of various types of distribution, assuming again a constant R. The fractionation data obtained were treated by Spencer's method and compared with the true distribution. Matsumoto concluded that Spencer's method would give a fairly reliable distribution if the experimental error were sufficiently small. His conclusion is thus contrary to that of Billmeyer and Stockmayer (75).

It seems very difficult to compare the results calculated by Billmeyer and Stockmayer and by Matsumoto with experimental data, since the fractionation at a constant R can be performed only by changing the initial concentration in a very complicated manner. Thus, it is more desirable to carry out the hypothetical fractionation with a constant initial concentration.

According to Okamoto (79, 80) $\rho(S)$ is given by the relation

$$\rho(S) = \int_{X(S)}^{\infty} g(X)\, dX \tag{37}$$

with

$$X(S) = \frac{d\{\rho(S)\bar{X}_{w(2)}(S)\}/d\rho(S)}{1 + \dfrac{\pi}{6}\left\{\dfrac{2}{(\ln R(S))^2} - \dfrac{4}{(\ln R(S))^3}\dfrac{d\ln R(S)}{d\ln[d\{\rho(S)\bar{X}_{w(2)}(S)\}/d\rho(S)]}\right\}} \tag{38}$$

where $g(X)$ represents the normalized differential weight X_i distribution of the initial polymer. Equation (38) was derived from the theory of Flory and Huggins, using some approximation, and it is valid when the order of the magnitude of $|X^n d^n g(X)/dX^n|$ is comparable to or smaller than $g(X)$. The value of $X(S)$ corresponding to a given value of $\rho(S)$ can be calculated precisely from equations (37) and (38). Hence, the integral distribution can easily be obtained from the plot of $(1-\rho(S))$ versus $X(S)$. In this revised Spencer's method, an exact value of $X(S)$ can be determined if the data on $\rho(S)$, $R(S)$, and $\bar{X}_{w(2)}(S)$ are available. The validity of this method has not been confirmed either by the hypothetical fractionations or by experiments.

An artificial assumption on the existence of the value of $X(S)$ made in Spencer's method cannot be justified by modern solution theory. Nevertheless, Spencer's method offers a convenient and simple way to determine integral distribution with fairly good accuracy. In some cases Spencer's method would be more advantageous than Schulz's method (see discussion of eqs. 42 and 43) applied to simple SPF data (31).

When the volume ratio $R(S)$ of the polymer-lean phase to the polymer-rich phase is measured together with $\bar{X}_{w(2)}(S)$ and $\rho(S)$, Okamoto's method can be applied, and it gives the most reliable integral X_i distribution (31).

Turbidimetric Titration

In 1945 Morey and Tambley (81) proposed a method of determining MWD by use of turbidimetric titration. They assumed that the precipitation particles were the same size and that the particles did not agglomerate. Figure 23 shows the apparatus for following turbidity changes (82). Dilute solutions with concentration less than 0.005 g/cm^3 are prepared, and a mechanical pump

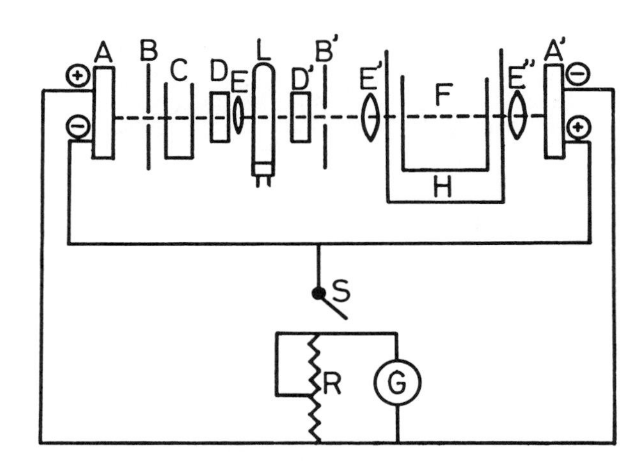

Figure 23. Apparatus for following turbidity changes. A and A′, photo cells; B and B′, adjustable diaphragms; C, water cell; D and D′, filters; E, E′, and E″, lenses; H, thermostat; F, turbidity cell; S, switch; R, resistance; G, galvanometer.

delivers additional precipitant at a slow rate to the solution, which is continuously and rapidly stirred during pumping.

Morey and Tambley (81) derived an empirical equation relating the critical volume of nonsolvent to cause precipitation v_2^{cp} and the polymer volume fraction at a precipitation point v_p^{cp}

$$v_2^{cp} = k \log v_p^{cp} + f(M) \tag{39}$$

where k is a constant and $f(M)$ is a function of the molecular weight. Then v_p^s is related to v_p^{cp} by (81)

$$v_p^{cp} = v_p^s(1 - v_2^{cp}) \tag{40}$$

Determination of v_2^{cp} as the point at which initial turbidity becomes detectable enables us to plot $\log v_p^{cp}$ versus v_2^{cp} for a fraction of known molecular weight; the slope of the plot gives a k value and the intercept on the v_2^{cp} axis at $\log v_p^{cp} = 0$ a numerical value of $f(M)$ for a given M (Fig. 24) (81, 82). By repeating this procedure for series of fractions of different molecular weights, it is possible to obtain $f(M)$ values corresponding to each molecular weight (81, 82). An MWD is obtained by determining the turbidity T as a

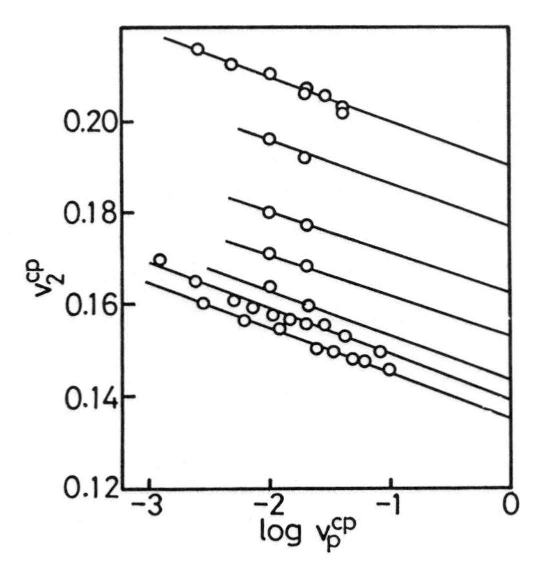

Figure 24. Relation between addition of nonsolvent v_2^{cp} to cause initial turbidity and polymer concentration at the point of precipitation.

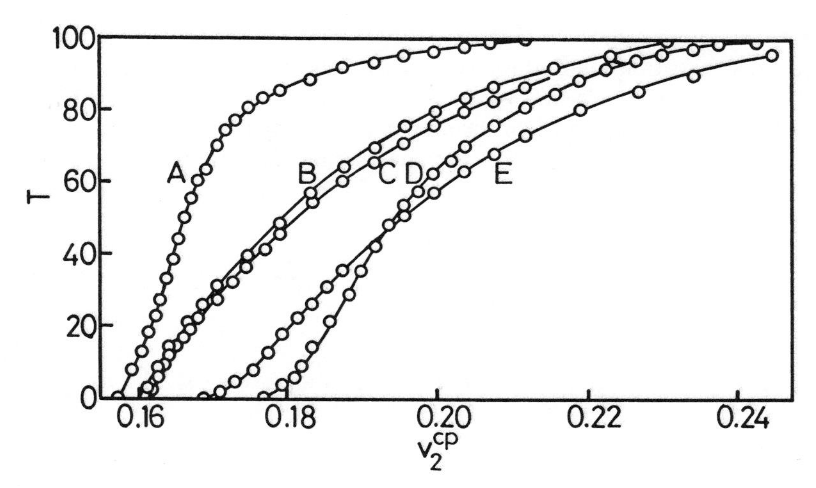

Figure 25. Turbidity curves for different polymers. A, low conversion polymer; B, "Diakon" F ($v_p^0 = 6 \times 10^{-5}$); C, "Diakon" F ($v_p^0 = 3 \times 10^{-5}$); D, Plexiglas V100; E, experimental photopolymer.

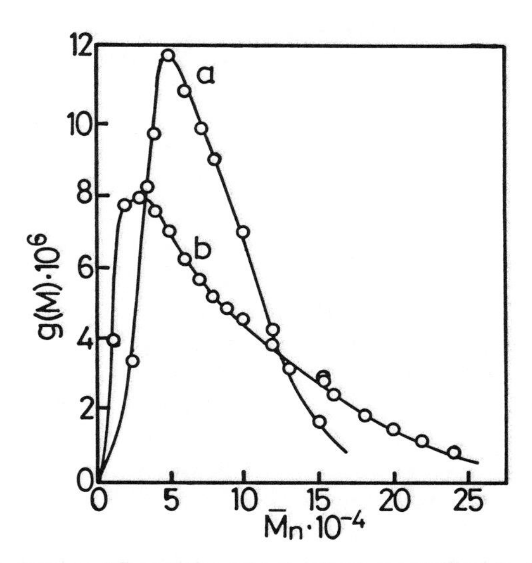

Figure 26. Curves showing differential weight distributions: (a) Plexiglas V100 and (b) Lucite H.M. 122.

function of v_2^{cp}, the amount of precipitant added, the former being defined as the ratio of the amount of light absorption at complete precipitation (Fig. 25) (81, 82). Complete precipitation therefore corresponds to 100% turbidity. If ΔT is the change in turbidity for the addition of Δv_2 of precipitant at the point given on the turbidity curve by v_2^{cp}, then Morey and Tamblyn (81) obtained for the weight fraction $g(X_i)$ of the X_i-mer:

$$g(X_i) = \frac{\Delta T}{1 - 10^{\Delta v_2/k}\left[\dfrac{1 - v_2^{cp}}{100 - v_2^{cp} - \Delta v_2}\right]} \quad (41)$$

Figure 26 shows the $g(X_i)$ calculated by use of equation (41) (81, 82).

ANALYTICAL TREATMENT OF FRACTIONATION DATA

Using the fractionation data [ρ_p (or ρ_s) and X_w], the MWD of the original polymer can be evaluated by means of the method of Schulz (62) and that of Kamide et al. (20,34), proposed on the basis of the modern fractionation theory. The principles of these methods are summarized next.

Method of Schulz

The cumulative weight fraction $I(X_{w,j})$ corresponding to $X_{w,j}$ of the jth fraction is given by

$$I(X_{w,j}) = \sum_{k=j+1}^{n_t} \sum_{x_i=0}^{\infty} g_k(X_i) + \tfrac{1}{2} \sum_{x_i=0}^{\infty} g_j(X_i) \qquad \text{(for SPF)} \quad (42)$$

$$I(X_{w,j}) = \sum_{k=1}^{j-1} \sum_{x_i=0}^{\infty} g_k(X_i) + \tfrac{1}{2} \sum_{x_i=0}^{\infty} g_j(X_i) \qquad \text{(for SSF)} \quad (43)$$

where $\sum_{x_i=0}^{\infty} g_k(X_i)$ denotes the weight fraction of the kth fraction. Figure 27 is a schematic representation of Schulz's method (72). Figure 28 shows that the MWD calculated by Schulz's method underestimates the high molecular weight region even if either $\bar{X}_w$ or $\bar{X}_n$ is used for the average X_i (20). If we use $\bar{X}_w$ for average X_i, the low molecular weight region is also underestimated. As is evident from equation (42) (or 43), the Schulz method assumes that the MWD of the jth fraction exists between $\bar{X}_{w,j-1}$ and $\bar{X}_{w,j+1}$ (20). But the true MWD of the jth fraction does not satisfy this assumption (see Fig. 4).

According to the results of the SPF simulation by Tung (18), the Schulz method gives fairly good coincidence with the true distribution (see Fig. 4 of

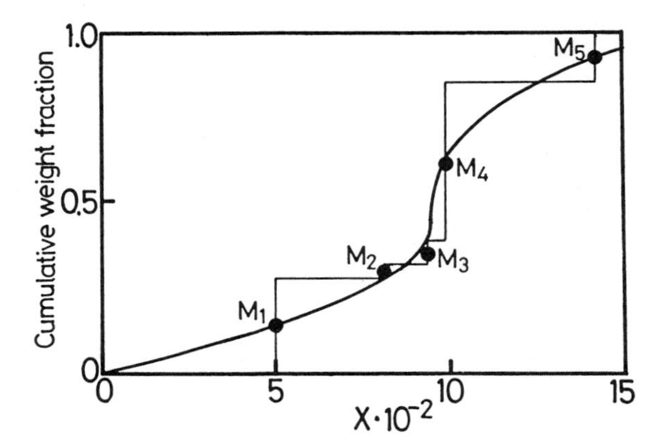

Figure 27. Method for determining integral distribution curves.

Ref. 18). Strictly speaking, as is shown by Tung (Fig. 4, Ref. 18), MWDs calculated by using the Schulz method underestimate the regions of low and high molecular weight (20).

Method of Kamide et al. for SPF and SSF

The MWD of the fraction separated by SPF can be expressed with fairly good accuracy by an equilateral triangle having its peak at $\bar{X}_w$ in the range $0 \leqslant \bar{X}_w \leqslant 2\bar{X}_w$ (Fig. 29*a*). The distribution function of the *j*th fraction $g_j(X_i)$

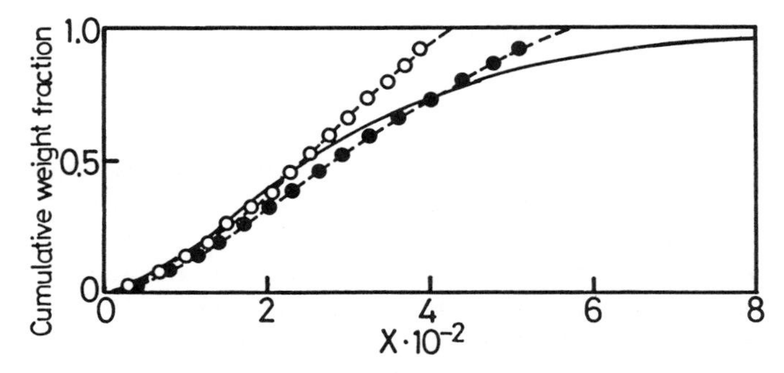

Figure 28. Integral distribution curve estimated from Schulz's method and comparison with the true distribution curve. ●, $\bar{X}_w$ is taken as the average degree of X of the fraction; ○, $\bar{X}_n$ is taken as the average degree of the fraction.

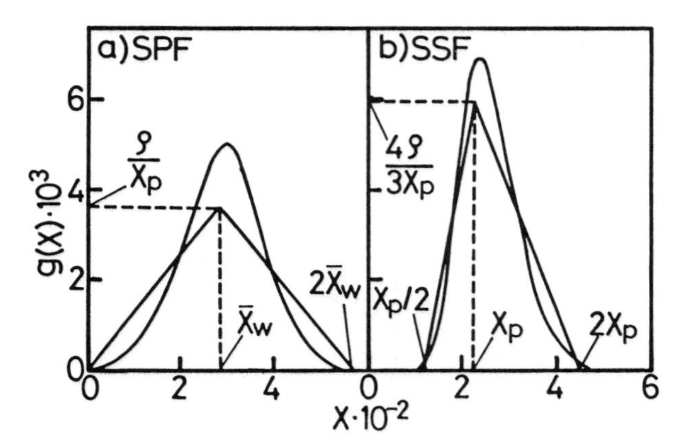

Figure 29. Schematic representation of a triangular molecular weight distribution (MWD). Solid curves are the true MEDs of the respective fractions: (*a*) successive precipitation fractionation and (*b*) successive solution fractionation.

can be expressed as

$$g_j(X_i) = \frac{\left[\sum_{x_i=0}^{\infty} g_j(X_i)\right] X_i}{\bar{X}_{w,j}^2} \qquad \text{for} \quad 0 \leqslant X_i \leqslant \bar{X}_{w,2} \tag{44}$$

and

$$g_j(X_i) = \frac{\left[\sum_{x_i=0}^{\infty} g_j(X_i)\right] (2\bar{X}_{w,j} - X_i)}{\bar{X}_{w,j}^2} \qquad \text{for} \quad \bar{X}_{w,j} \leqslant X_i \leqslant 2\bar{X}_{w,j} \tag{45}$$

and

$$g_j(X_i) = 0 \qquad \text{for} \quad X_i \geqslant 2\bar{X}_{w,j} \tag{46}$$

where $\sum_{x_i=0}^{\infty} g_j(X_i)$ denotes the weight fraction of the jth fraction to the total weight of the polymer. The summation of $g_j(X_i)$ [i.e., $\sum_{j=1}^{n} g_j(X_i)$] given by equations (44)–(46) with respect to X_i provides the normalized differential MWD $g_e(X_i)$ of the original polymer estimated from the fractionation data, and its integration yields the cumulative distribution curve $I_e(X_i)$ (5).

An improvement of the original method of Kamide et al. for the case of SSF can be made by introducing a modified triangle distribution, ranging from $\frac{1}{2}X_p$ to $2X_p$, where X_p is the value of X_i at the peak and the triangle's $\bar{X}_w$ value is identical to that of the fraction separated by SSF. After a simple calculation,

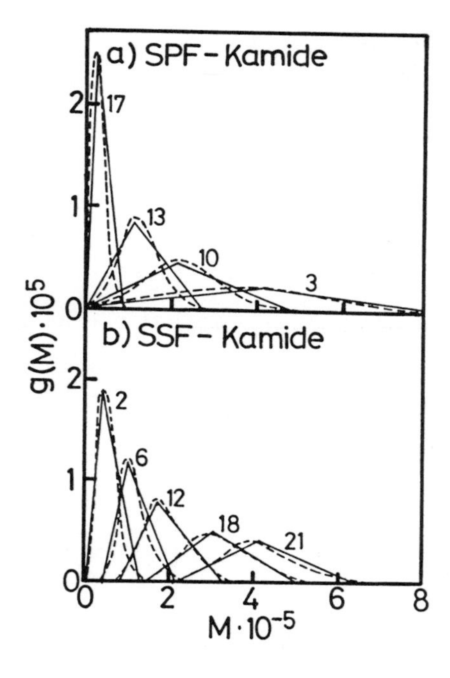

Figure 30. Comparison of the MWD determined by gel permeation chromatography with triangular distribution employed in Kamide's method for SPF (*a*) and for SSF (*b*).

not given here, it can readily be shown that X_p is related to $\bar{X}_w$ (34)

$$X_p = \tfrac{6}{7}\bar{X}_w \tag{47}$$

The modified triangle distribution is illustrated in Figure 29*b*. Figure 30 shows the examples of triangle distribution of SPF and SSF.

A parameter E, defined by equation (48), is used as a measure of the accuracy of the method for estimating the MWD of the original polymer:

$$E = \int |g_0(X) - g_e(X)|\,dX \tag{48}$$

where $g_0(X)$ is the normalized true differential MWD of the original polymer.

For convenience, combinations of fractionation procedure (SPF or SSF) and analytical method (Schulz or Kamide) are referred to hereafter as follows: SPF/Schulz (or SSF/Kamide) means that the fractionation data were obtained by SPF (or SSF) and the data were treated by the Schulz (or Kamide) methods.

Figure 31 compares the MWD curves of the original polymer, constructed

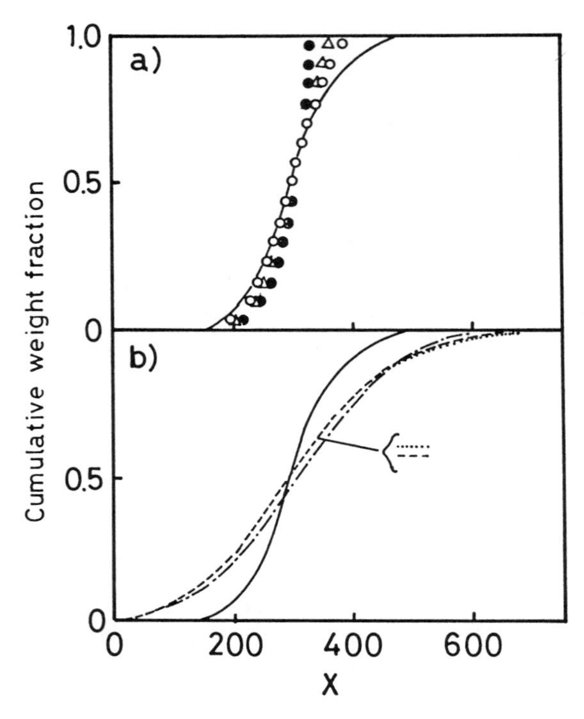

Figure 31. Cumulative weight distribution constructed by the Schulz method (*a*) and the Kamide method (*b*): original polymer; Schulz–Zimm distribution ($\bar{X}_w^0/\bar{X}_n^0 = 1.05$, $\bar{X}_w^0 = 300$), $p_1 = 0$, $\rho = 1/15$, $v_p^0 = 1\%$ (●, –·–·–), 0.1% (△, ––––), 0.01% (○, ····), true distribution curve (——).

by using the Schulz method and the Kamide method from the fractionation data thus obtained, with the true MWD. Inspection of Figure 31 shows that the Schulz method yields a narrower distribution as compared with the true distribution and that is grossly underestimates the higher region of degree of polymerization X (83): $I(X) = 0.99$ is estimated to be 340 at $v_p^0 = 1\%$ and 410 at $v_p^0 = 0.01\%$.

These values are much smaller than the true value ($X = 500$). The accuracy improves appreciably by lowering v_p^0 to 0.01%, but it is still far from satisfactory. In the Kamide method the estimated MWD curves are broader than the true distribution and are not significantly affected if $v_p^0 \leqslant 0.1\%$. We concluded from Figure 31 that the analytical fractionation method, based on solubility behavior, is unfortunately ineffective for very narrow MWD polymers obtained by an anionic polymerization mechanism, irrespective of the conditions of fractionation (83).

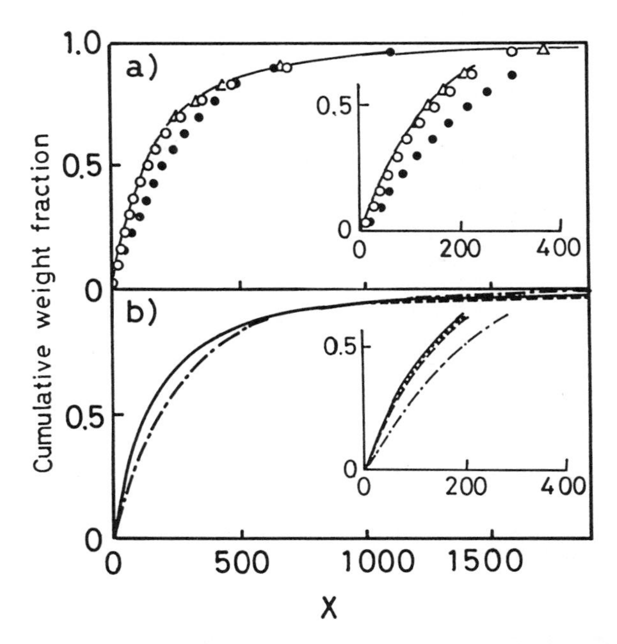

Figure 32. Cumulative weight distribution constructed by the Schulz method (*a*) and the Kamide method (*b*): original polymer, Wesslau distribution ($\bar{X}_w^0/\bar{X}_n^0 = 5$, $\bar{X}_w^0 = 300$), $p_1 = 0$, $\rho = 1/15$, $v_p^0 = 5\%$ (●, —·—·—), 1% (△, ----), 0.1% (○, ····), true distribution curve (——).

It may be advantageous from a practical point of view to employ higher v_p^0 for a broader original polymer. To examine this hypothesis, the polymer having the Wesslau type of distribution with $\bar{X}_w^0/\bar{X}_n^0 = 5.0$ and $\bar{X}_w^0 = 300$ was fractionated by an SPF method into 15 equal fractions from 5, 1, and 0.1% solutions, assuming $p_1 = 0$. The Schulz plot and the Kamide plot are shown in Figure 32, together with the true MWD of the original polymer (83). Both methods significantly underestimate in the lower molecular weight region (i.e., $X_i < 500$) for $v_p^0 = 5\%$, indicating that an increase in v_p^0 up to 5% is not acceptable even for a broad original polymer (83).

Effect of Average Molecular Weight and Polydispersity of the Original Polymer

Figure 33 plots the relative error E versus $\bar{X}_w^0$ (original polymer, SZ2; $p_1 = 0$, $n_t = 15$, $v_p^0 = 1\%$) (5). From inspection of Figure 33, the accuracy of an analytical fractionation appears to depend much more on the analytical method than on the operating conditions. The average molecular weight of the original polymer has a negligible effect on the peculiar features of both methods for

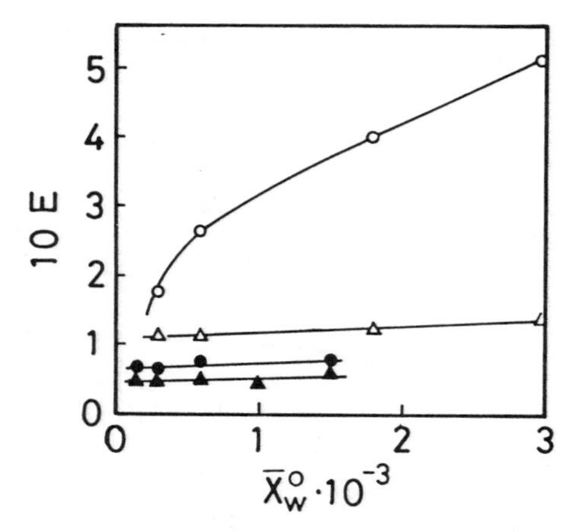

Figure 33. Effect of the weight average degree of polymerization of the original polymer $\bar{X}_w^0$ on the relative error E. In this case $g_e(X)$ in equation (48) was estimated from the fractionation data by SPF (open symbols) or SSF (closed symbols) by use of the Kamide method (triangle) or the Schulz method (circle): original polymer, Schulz–Zimm distribution ($\bar{X}_w^0/\bar{X}_n^0 = 2$), $p_1 = 0$, $n_t = 15$, $v_p^0 = 1\%$.

estimating the MWD of the original polymer from fractionation data, except for the case of the Schulz SPF method. There is a marked difference in E between SSF and SPF.

Figure 34 plots the relative error E as a function of the ratio $\bar{X}_w^0/\bar{X}_n^0$ of an original polymer having SZ or W distributions with $\bar{X}_w^0 = 300$, fractionated

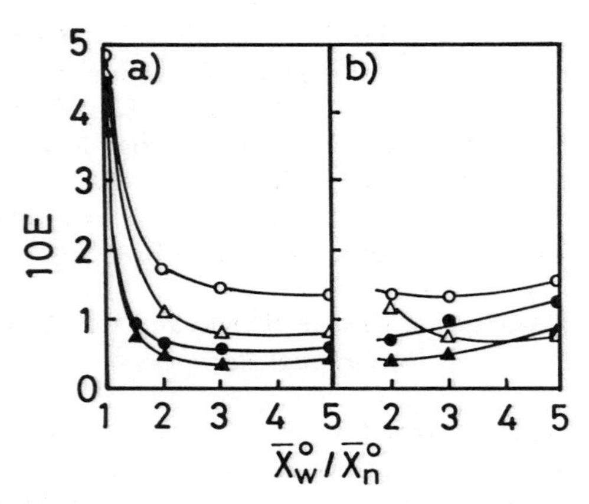

Figure 34. Effect of the polydispersity of the original polymer $\bar{X}_w^0/\bar{X}_n^0$ on the relative error E; symbols and parameters as in Figure 33. (*a*) Schulz–Zimm polymer ($\bar{X}_w^0 = 300$). (*b*) Wesslau polymer ($\bar{X}_w^0 = 300$).

into 15 equal fractions from a 1.0% solution (5). With progressively increasing $\bar{X}_w^0/\bar{X}_n^0$, the accuracy of the methods for estimating the original MWD improves considerably, irrespective of the type of MWD (SZ or W distribution). This appears to be mainly a consequence of the fact that the approximations adopted in both treatments become more accurate as $\bar{X}_w^0/\bar{X}_n^0$ increases. At least as far as a 1.0% polymer solution is concerned, the Kamide method is preferable to the Schulz method, regardless of the type and breadth of the MWD in the original polymer. For a very narrow MWD polymer (e.g., a polymer having $\bar{X}_w^0/\bar{X}_n^0 < 1.10$), $g_e(X_i)$ or $I_e(X_i)$ evaluated by using the methods above deviated significantly from the true original MWD (5). This has been observed, qualitatively, in the case of SPF.

Effect of Initial Polymer Volume Fraction

The dependence of the accuracy of the two evaluation methods on v_p^0 is shown in Figure 35. In both SSF and SPF, E obtained by the Kamide method exhibits an intense (in SPF) or a poorly defined (in SSF) minimum at a specific value of v_p^0. On the other hand, E found by the Schulz method decreases as v_p^0 increases. For example, for the polymer with SZ-type distribution ($\bar{X}_w^0 = 300$ and $\bar{X}_w^0/\bar{X}_n^0 = 2$), the Kamide method is more accurate at $v_p^0 \geqslant 1\%$, but the Schulz method becomes adequate at $v_p^0 \leqslant 1\%$. However, for the W type of

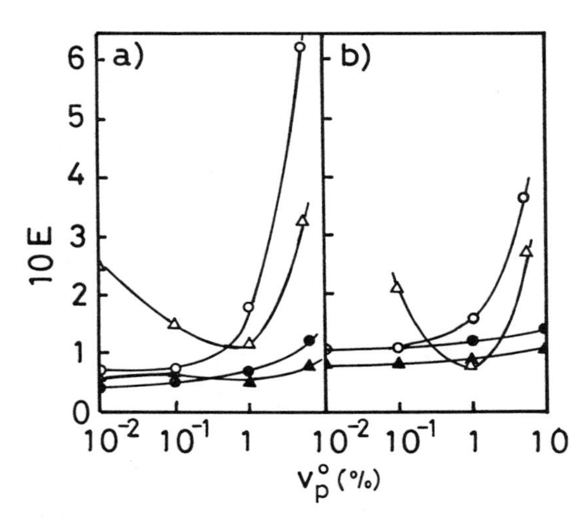

Figure 35. Effect of the initial polymer volume fraction v_p^0 on the relative error E; symbols as in Figure 33 ($n_t = 15$, $p_1 = 0$). (*a*) Schulz–Zimm polymer ($\bar{X}_w^0 = 300$, $\bar{X}_w^0/\bar{X}_n^0 = 2$). (*b*) Wesslau polymer ($\bar{X}_w^0 = 300$, $\bar{X}_w^0/\bar{X}_n^0 = 5$).

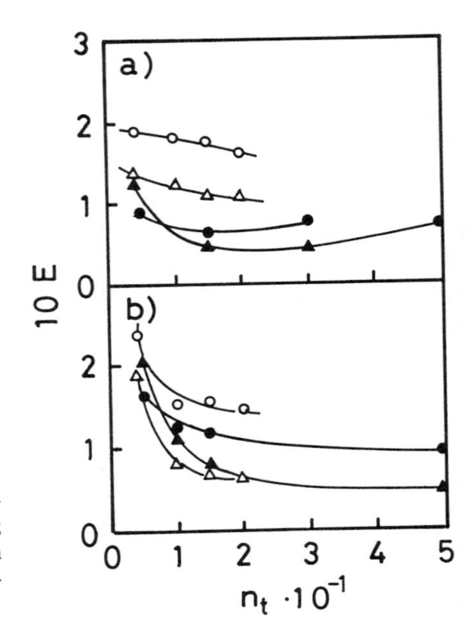

Figure 36. Effect of the total number of fractions in a given run, n_t, on the relative error E; symbols as in Figure 33. (*a*) Schulz–Zimm polymer ($\bar{X}_w^0 = 300$, $\bar{X}_n^0 = 2$). (*b*) Wesslau polymer ($\bar{X}_w^0 = 300$, $\bar{X}_w^0/\bar{X}_n^0 = 5$).

distribution the Kamide method is more accurate than the Schulz method over a whole range of v_p^0 from 10^{-2} to 10%, irrespective of the analytical procedures employed (5).

Effect of Total Number of Fractions

The effect of n_t on E is demonstrated in Figure 36 (5). In both methods, E decreases sharply at first and then approaches a limiting values as n_t increases

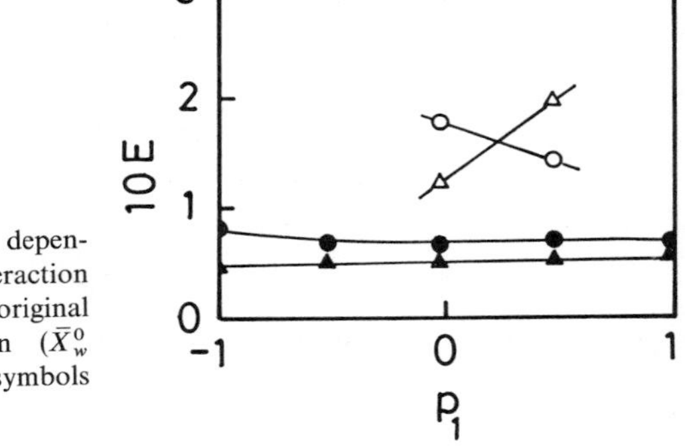

Figure 37. Effect of the concentration dependence p_1, of the polymer–solvent interaction parameter χ, on the relative error E: original polymer, Schulz–Zimm distribution ($\bar{X}_w^0 = 300$, $\bar{X}_w^0/\bar{X}_n^0 = 2$), $v_\mathrm{p}^0 = 1\%$, $n_t = 15$; symbols as in Figure 33.

progressively. It is immediately evident from Figure 36 that the Kamide method is preferable to the Schulz method at a given n_t, at least in the range $n_t > 10$, as far as the polymer with W-type distribution ($X_w^0 = 300$ and $X_w^0/X_n^0 = 5$) is concerned. Miyazaki and Kamide (5) concluded from inspection of the cumulative weight distribution $I_e(X_i)$ of the original polymer, constructed from the data by SPF with SZ, that in contrast to the Schulz plot, the Kamide plot approaches the true distribution of the original polymer with increasing n_t. Quantitative analysis of fractionation data, obtained by SSF with SZ2, as shown in Figure 36, indicates that there is no difference in the effect of n_t between the Schulz method and the Kamide method (5).

Effect of p_1

The effect of p_1 on E is demonstrated in Figure 37 (5). With an increase in p_1, E decreases in SPF/Schulz, but E increases in SPF/Kamide. The effect of p_1 (i.e., the effect of the type of solvent employed in the fractionation run) is almost insignificant in the case of SSF. Over a wide range of p_1, SSF/Kamide is the more accurate (5).

Optimum Operating Conditions for Analytical SPF and SSF

Figure 38 shows the relationships between v_p^0, n_t, and E for a polymer with SZ-type distribution ($\bar{X}_w^0 = 300$ and $\bar{X}_w^0/\bar{X}_n^0 = 2$) (5). The curves denote contour lines, relating v_p^0 to n_t, for a fixed value of E.

The relative error E obtained by Schulz's method approaches closely to zero if $v_p^0 \rightarrow 0$ and $n_t \rightarrow \infty$ are concurrently realized. In contrast, E exhibits a remarkable minimum E_{min} if the Kamide method is employed [in this case, E_{min} is 10% for SPF/Kamide and 4.8% for SSF/Kamide, respectively] (5). By use of a plot like Figure 38, we can optimize the fractionation procedure of any polymer sample for analytical purposes.

Effect of Average Molecular Weight of the Fractions

The average molecular weights besides $\bar{X}_w$ commonly used in analytical fractionation are $\bar{X}_n$ and the viscosity average degree of polymerization $\bar{X}_v$. If we define $X(r)$ by

$$X(r) = \left[\frac{\int X^r g_0(X)\,dX}{\int g_0(X)\,dX} \right]^{1/r} \tag{49}$$

we can readily express $\bar{X}_n$, $\bar{X}_v$, and $\bar{X}_w$ in terms of $X(r)$:

$$\bar{X}_n = X(-1), \qquad \bar{X}_v = X(\mathbf{a}), \qquad \text{and} \qquad \bar{X}_w = X(1) \tag{50}$$

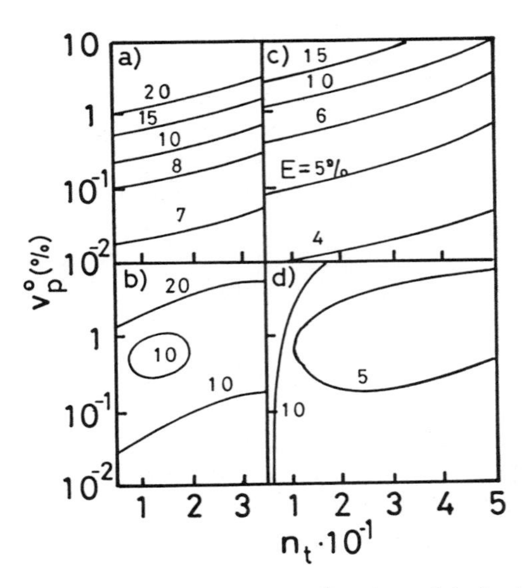

Figure 38. Correlation between the relative error E, v_p^0, and n_t: original polymer, Schulz–Zimm distribution ($\bar{X}_w^0 = 300$, $\bar{X}_w^0/\bar{X}_n^0 = 2$), $p_1 = 0$. (*a*) SPF (Schulz). (*b*) SPF (Kamide). (*c*) SSF (Schulz). (*d*) SSF (Kamide).

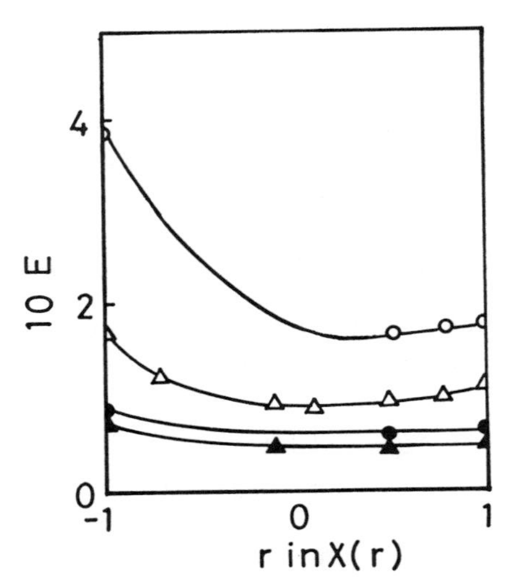

Figure 39. Effect of the moment r in $X(r)$ in equation (49) on the relative error E: original polymer, Schulz–Zimm distribution ($\bar{X}_w^0 = 300$, $\bar{X}_w^0/\bar{X}_n^0 = 2$), $v_p^0 = 1\%$, $n_t = 15$; symbols as in Figure 33.

where **a** is the exponent in the Mark–Houwink–Sakurada equation

$$[\eta] = K_m M^{a} \tag{51}$$

Here $[\eta]$ is the limiting viscosity number, M is molecular weight, and K_m and **a** are parameters characteristic of the polymer–solvent combination and temperature.

Figure 39 shows the effect of r in $X(r)$ on E for analytical SPF and SSF, where the original polymer with SZ distribution ($\bar{X}_w^0 = 300$ and $\bar{X}_w^0/\bar{X}_n^0 = 2$) was fractionated from a 1.0% solution into 15 equal fractions (5). The effect of the nature of the average (in this case, r in eq. 49) on the molecular weight of the fractions becomes pronounced in SPF. For example, in the case of SPF/Schulz the value of E doubles if $\bar{X}_n$ is employed in place of $\bar{X}_w$. The change in E of SPF if $\bar{X}_v$ is used instead of $\bar{X}_w$ is, however, negligible. Contrary to this, the dependence of E in SSF on r in $X(r)$ is small. It is very interesting to note that using $X(r)$ $(0 \leqslant r \leqslant 1)$ gives the minimum relative error, irrespective of the fractionation method and the analytical procedure (5).

Correlations Between Estimated Values of $\bar{X}_w/\bar{X}_n$ and E

The correlations between the value of $\bar{X}_w/\bar{X}_n$, estimated from the fractionation data, and the relative error E are shown in Figure 40, where the polymer with

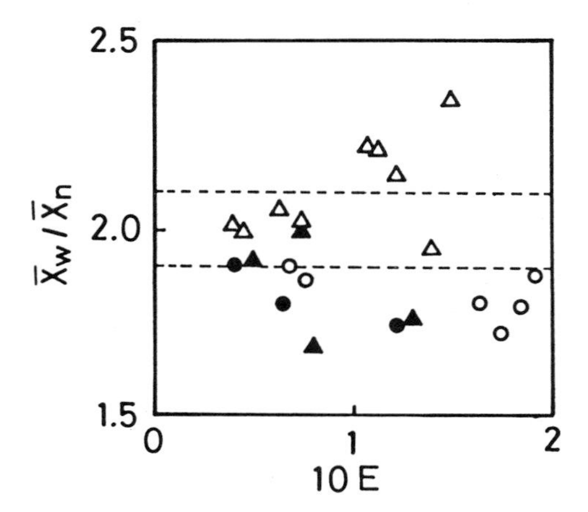

Figure 40. Correlations between the value of $\bar{X}_w/\bar{X}_n$ of the original polymer, estimated from the fractionation data, and the relative error E: original polymer, Schulz–Zimm distribution ($\bar{X}_w^0/\bar{X}_n^0 = 2$); open symbols, SPF; closed symbols, SSF; circles, Schulz method; triangle, Kamide method.

SZ distribution ($\bar{X}_w^0 = 300$ and $\bar{X}_w^0/\bar{X}_n^0 = 2$) was hypothetically fractionated by means of SPF and SSF and the data [X_w and ρ_p(or ρ_s)] obtained were analyzed according to the Schulz or Kamide procedure (5). Evidently, as E increases, the estimated value of $\bar{X}_w/\bar{X}_n$ tends to scatter to a greater extent, and with a gradual decrease in E to zero, the value of $\bar{X}_w/\bar{X}_n$ tends to the limiting value, which is equal to the true value (in this case, $\bar{X}_w/\bar{X}_n = 2$). Both SPF and SSF treated by the Schulz method have a clear tendency to give smaller values of $\bar{X}_w/\bar{X}_n$ than the true values (i.e., $\bar{X}_w^0/\bar{X}_n^0$). The value of $\bar{X}_w/\bar{X}_n$ obtained by SPF/Kamide is usually larger than the true value, but that estimated by using SSF/Kamide is always lower than $\bar{X}_w^0/\bar{X}_n^0$. As E becomes larger, the degree of deviation of $\bar{X}_w/\bar{X}_n$ from the true value becomes larger (5).

If one wishes to estimate $\bar{X}_w/\bar{X}_n$ of the original polymer using the fractionation method with an error of $\pm 5\%$, the conditions should be chosen so that E is, at least, less than 5%. Note, however, that SPF with a precision of $E \simeq 5\%$ is realizable only under comparatively restricted conditions (see Fig. 38). In this connection, SSF/Kamide is recommended for general use (5).

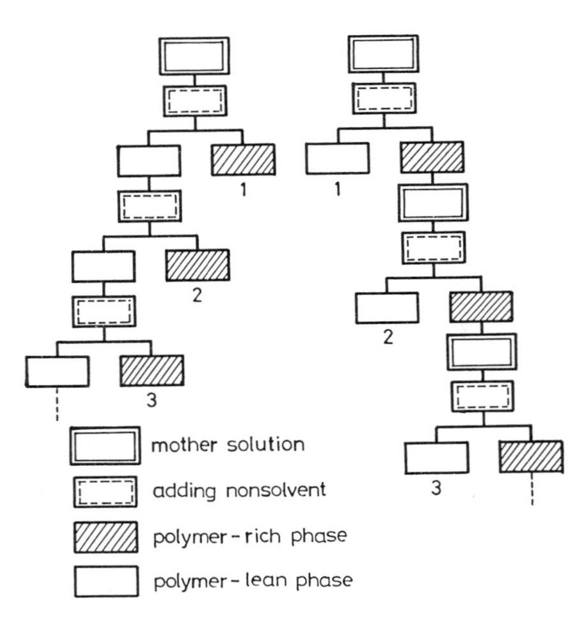

Figure 41. Schematic representation of SPF and SSF of quasi-ternary systems consisting of multicomponent polymers in a binary solvent mixture. Numbers 1–3 denote fraction numbers.

Effect of χ_{12}, χ_{13}, and χ_{23}

Figure 41 schematically represents SPF and SSF for quasi-ternary systems consisting of multicomponent polymers plus solvent plus nonsolvent. Nonsolvent was added to cause phase separation, instead of changing the temperature of the solution.

Figure 42 illustrates the effect of χ_{12}, χ_{13}, and χ_{23} on the relation between ρ_p and total volume V (54). In this case $V_1^0 = 100$ and $V_3^0 = 1$ (accordingly, $v_p^s = 0.01$) are assumed. As χ_{12} increases, the amount of solvent 2 needed for a given ρ_p increases. But all V–ρ_p relations are superposable by shifting them along the horizontal axis. The effect of χ_{13} on this relation is just the reverse to

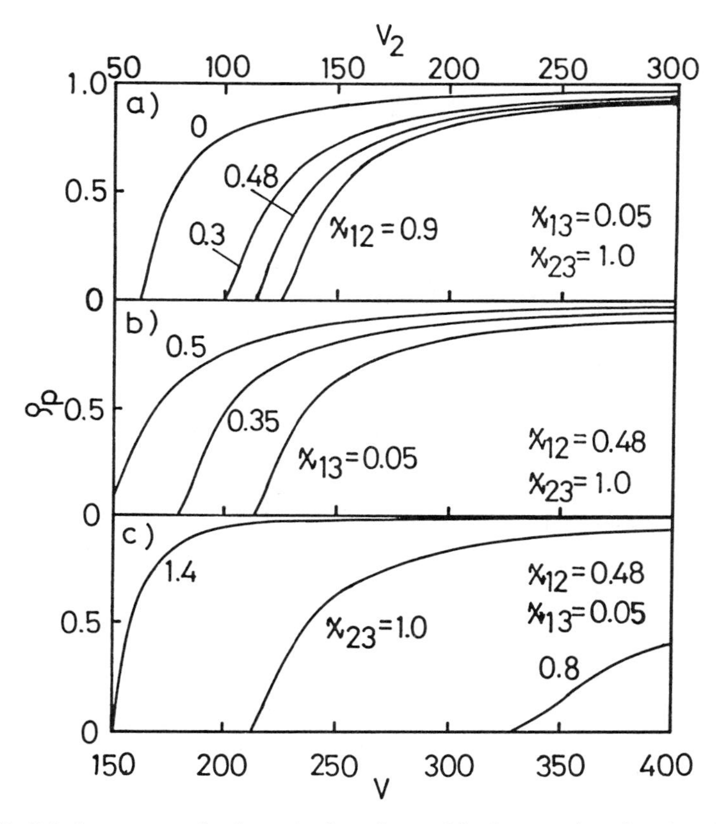

Figure 42. Relative amount of polymer in the polymer-rich phase ρ_p plotted against the total volume V and the volume of solvent 2 V_2: original polymer, Schulz–Zimm distribution ($\bar{X}_w^0 = 300$, $\bar{X}_w^0/\bar{X}_n^0 = 2$), $v_p^s = 0.01$.

that of χ_{12}. As χ_{23} decreases, the maximum ρ_p value decreases and V becomes large. To precipitate a given relative amount of polymer using the smallest possible amount of solvent 2, it is necessary to use a combination of solvents 1 and 2 having small χ_{12} and large χ_{13} and χ_{23}. When the mutual miscibility of solvents 1 and 2 is bad, the solubility power of solvent 1 is large, and the precipitation capacity of the solvent is weak, we should add a large amount of solvent 2 to the quasi-binary solution of the polymer and solvent 1 to bring about phase separation (54). Among the three χ parameters, χ_{23} plays an important role in controlling V (54). In a quasi-ternary system, ρ_p is controlled by adding solvent 2 to the quasi-binary system (see Fig. 5). As a result, the total volume increases unavoidably for larger ρ_p, approaching the experimental

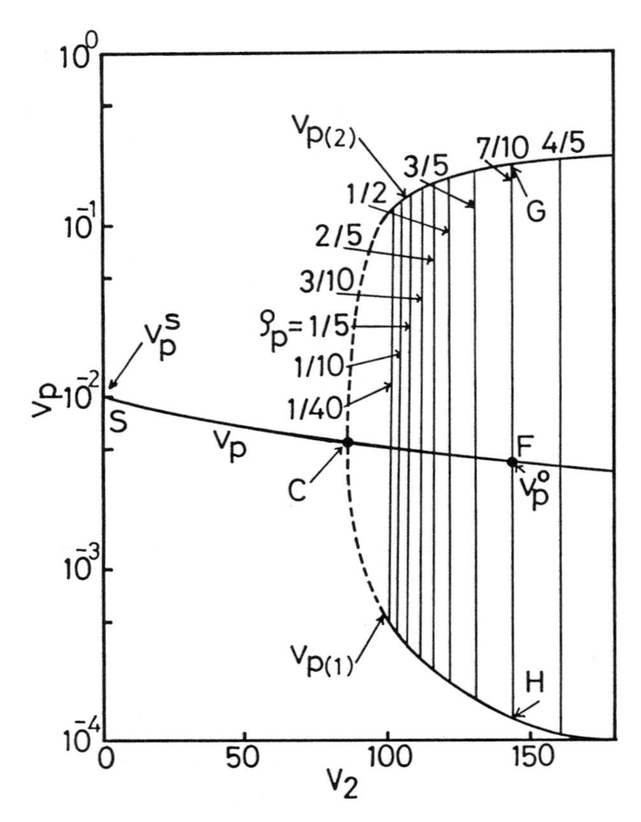

Figure 43. Change in the polymer volume fraction v_p with the addition of solvent 2 (volume of solvent 2, V_2) in the phase equilibrium experiment: S, polymer volume fraction of starting solution v_p^s; F, see text; C, cloud point; original polymer, Schulz–Zimm distribution ($\bar{X}_w^0 = 300$, $\bar{X}_w^0/\bar{X}_n^0 = 2$); $v_p^s = 0.01$, $\chi_{12} = 0.48$, $\chi_{13} = 0.2$, $\chi_{23} = 1.0$.

limit that is accessible, when the same solvent 1–solvent 2 pair is employed. For a quasi-ternary system, there are numerous combinations of solvents 1 and 2, enabling us to separate a polymer-rich phase effectively over a wide range of ρ_p. Figure 42 demonstrates that when we try to carry out a molecular weight fractionation based on the phase equilibrium phenomena of the quasi-ternary system, solvent 2 should be carefully chosen. For SPF, a small χ_{23} (and, if possible, large χ_{13}) is favorable; for SSF, a large χ_{23} is desirable for keeping total volume as small as possible (54). It is very rare that when a drop of solvent 2 is added to the polymer–solvent 1 system, the solution becomes instantly turbid, indicating the occurrence of a phase separation. Usually, after a measurably large amount of solvent 2 has been added, the cloud point of the solution (point C in Fig. 43) is observed (54). Further addition of solvent 2 rapidly increase ρ_p, as shown in Figure 43, where point S is the composition of a starting solution whose polymer volume fraction is v_p^s. When solvent 2 is added, the polymer volume fraction v_p changes along the full line leading to the cloud point C. Further addition of solvent 2 after the cloud point makes ρ_p large and, for example, at point F a phase equilibrium is attained. A fine line passing through F is a tie line connecting the polymer-rich phase G and

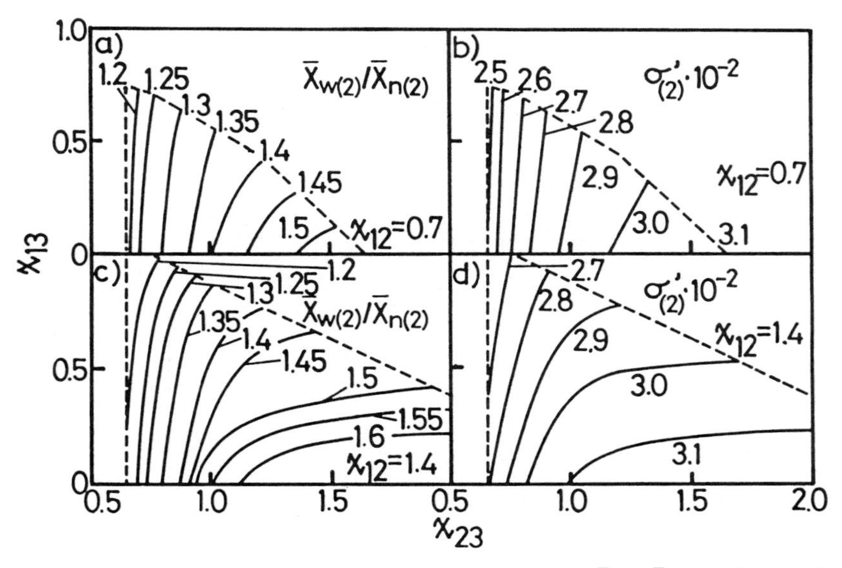

Figure 44. Correlation between χ_{13} and χ_{23} yielding constant $\bar{X}_{w(2)}/\bar{X}_{n(2)}$ or the standard deviation $\sigma'_{(2)}$ in the polymer-rich phase at a given χ_{12} [$=0.7$ for (*a*) and (*b*), 1.4 for (*c*) and (*d*)]. Original polymer: Schulz–Zimm distribution ($\bar{X}_w^0 = 300$, $\bar{X}_w^0/\bar{X}_n^0 = 2$, and $\sigma'_0 = 212.2$), $\rho_p = 1/15$, $v_p^s = 0.01$.

polymer-lean phase H. The polymer volume fraction at F is defined as an initial polymer volume fraction (i.e., initial "concentration") and denoted by v_p^0. In the small ρ_p region, an approximation of $v_{p(1)} = 0$ should not be employed when detailed phase separation characteristics are to be evaluated (54).

Figure 44 shows the effects of χ_{13} and χ_{23} on the ratio $\bar{X}_{w(2)}/\bar{X}_{n(2)}$ and the standard deviation of $g_{(2)}(X)$, $\sigma'_{(2)}$ (54). Both these parameters represents the breadth of the molecular weight distribution of the polymer in a polymer-rich phase. In the small χ_{23} region, the polydispersity of the polymer remaining in the polymer-rich phase is χ_{23}-dependent, being smaller as χ_{23} decreases. In preparing polymer fractions as a polymer-rich phase to be separated from a quasi-ternary system, a less poor solvent should be used as solvent 2 (54).

Figure 45 shows the effects of χ_{13} and χ_{23} on the polydispersity of the polymer-lean phase when $\chi_{12} = 1.30$ and $\rho_s = 1/15$ (i.e., $\rho_p = 14/15$) (54). As χ_{23} decreases, $\bar{X}_{w(1)}/\bar{X}_{n(1)}$ and $\sigma'_{(1)}$ decrease very gradually, approaching limiting values (i.e., $\bar{X}_{w(1)}/\bar{X}_{n(1)} = 1.48$ and $\sigma'_{(1)} = 22.8$). The effect of χ_{23} is the most predominant of the three χ parameters, but its magnitude is small compared to the effects of the χ parameters on $\bar{X}_{w(2)}/\bar{X}_{n(2)}$ and $\sigma'_{(2)}$ (54).

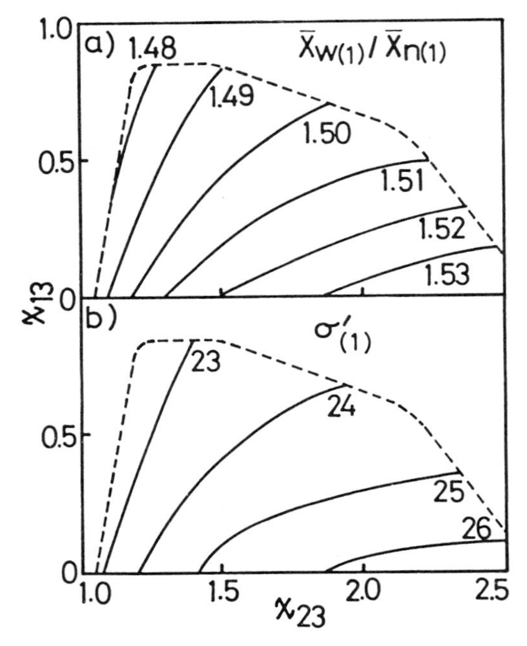

Figure 45. Correlation between χ_{13} and χ_{23} yielding constant $\bar{X}_{w(1)}/\bar{X}_{n(1)}$ or the standard deviation $\sigma'_{(1)}$ in the polymer-lean phase at a given χ_{12} (= 1.3): original polymer: Schulz–Zimm distribution ($\bar{X}_w^0 = 300$, $\bar{X}_w^0/\bar{X}_n^0 = 2$, and $\sigma'_0 = 212.2$), $\rho_p = 14/15$ ($\rho_s = 1/15$), $v_p^s = 0.01$.

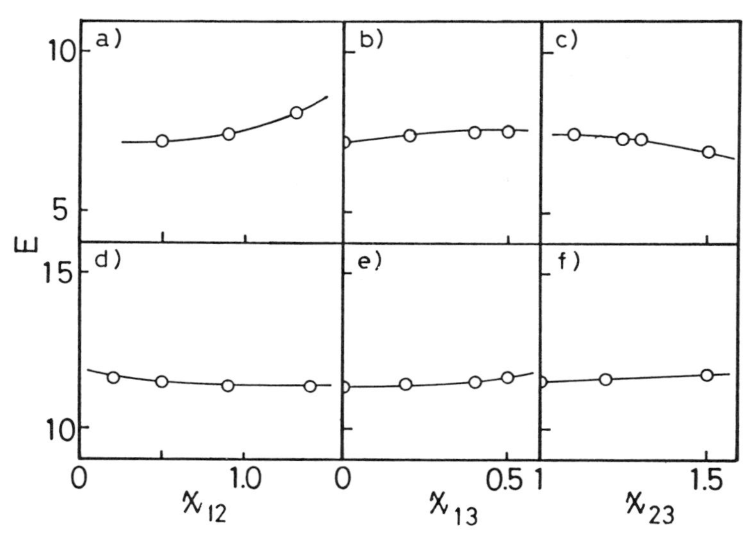

Figure 46. Effect of the three χ parameters on the relative error E: original polymer, Schulz–Zimm distribution ($\bar{X}_w^0 = 300$, $\bar{X}_w^0/\bar{X}_n^0 = 2$), $v_p^0 = 1\%$, $n_t = 1/15$. (*a*)–(*c*) SSF (Kamide). (*d*)–(*f*) SPF (Kamide). (*a*) $\chi_{13} = 0.2$, $\chi_{23} = 1.3$. (*b*) $\chi_{12} = 0.9$, $\chi_{23} = 1.3$. (*c*) $\chi_{12} = 0.9$, $\chi_{13} = 0.2$. (*d*) $\chi_{13} = 0.2$, $\chi_{23} = 1.0$. (*e*) $\chi_{12} = 0.5$, $\chi_{23} = 1.0$. (*f*) $\chi_{12} = 0.5$, $\chi_{13} = 0.2$.

Figure 46 shows the effect of χ_{12}, χ_{13}, and χ_{23} on E. With a decrease in χ_{12} and χ_{13} and an increase in χ_{23}, E decreases for SSF. With an increase in χ_{12} and a decrease in χ_{13} and χ_{23}, E decreases for SPF. Of the Kamide procedures, SSF is more accurate than SPF.

Experimental Verification of Modern Theory of Analytical Fractionation

Molecular weight distribution curves of the PS fraction isolated by SPF and SSF from a 0.94% solution in methylcyclohexane were determined by GPC. These were compared with the triangle distribution, which was approximated by Kamide et al., as exemplified in Figure 47 (5). Except for very low order fractions, the triangle distribution employed by Kamide et al. proved successful. Therefore, Figure 47 directly supports the validity of the approximations in Kamide's method (5).

The maximum and minimum values of the molecular weight, M_{max} and M_{min}, in the distribution for the PS fractions obtained by SPF and SSF were determined conventionally as the values at which the magnitude of $g_e(M)$ of a fraction is equal to 4×10^{-8}. Figure 48 plots $M_{min}/\bar{M}_w$ and $M_{max}/\bar{M}_w$ and shows the results of the computer simulation for this polymer–solvent system.

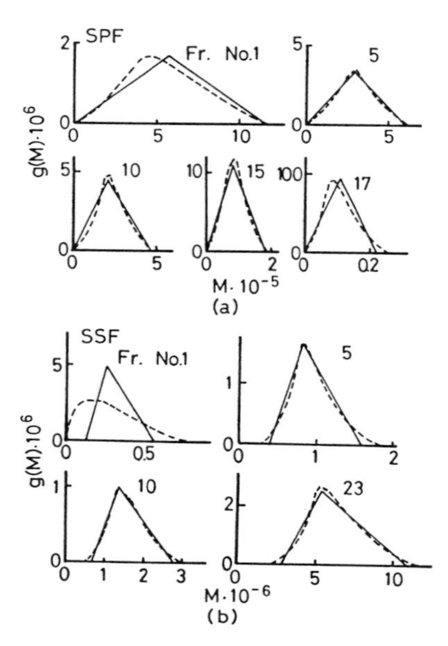

Figure 47. Comparison of the molecular weight distribution determined by gel permeation chromatography with the triangular distribution employed in Kamide's method for atactic polystyrene fractions. (*a*) SPF: $v_p^0 = 0.94\%$, $n_t = 18$. (*b*) SSF: $v_p^0 = 0.94\%$, $n_t = 23$.

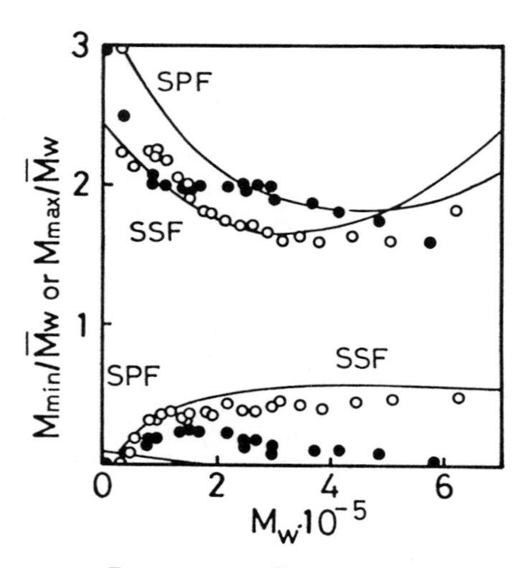

Figure 48. The ratio of $M_{max}/\bar{M}_w$ (and $M_{min}/\bar{M}_w$) plotted as a function of $\bar{M}_w$ for atactic polystyrene fractions separated by SSF (○) and SPF (●) experiments. Original polymer $\bar{M}_w = 23.9 \times 10^4$ and $\bar{M}_w/\bar{M}_n = 2.77$; solvent, methylcyclohexane; $v_p^0 = 0.94\%$; $n_t = 23$ in SSF and 18 in SPF. The curves are theoretical ($p_1 = 0.7$):

The agreement shown by $M_{min}/\bar{M}_w$ and $M_{max}/\bar{M}_w$ between experiment and theory is fairly satisfactory (5).

The simplest way of determining $\bar{X}_w$ and $\bar{X}_n$ is given by

$$\bar{X}_w = \sum_j \rho_j \bar{X}_{w,j} \tag{52}$$

or

$$\bar{X}_w = \sum_j \rho_j \bar{X}_{n,j} \tag{52'}$$

and

$$\bar{X}_n = \frac{1}{\sum_j (\rho_j / \bar{X}_{w,j})} \tag{53}$$

or

$$\bar{X}_n = \frac{1}{\sum_j (\rho_j / \bar{X}_{n,j})} \tag{53'}$$

where ρ_j, $\bar{X}_{w,j}$, and $\bar{X}_{n,j}$ are the relative amounts $\bar{X}_w$ and $\bar{X}_n$ of the jth fraction, respectively. In deriving equations (52) and (53), the polydispersity of the fractions is neglected. In consequence, by use of a kind of average degree of polymerization ($\bar{X}_{w,j}$ or $\bar{X}_{n,j}$) one cannot calculate both $\bar{X}_w$ and $\bar{X}_n$ accurately (5).

In the strict sense, $\bar{X}_w$ and $\bar{X}_n$ of the whole polymer should be calculated from

$$\bar{X}_w = \int X g_e(X)\, dX \tag{54}$$

$$\bar{X}_n = \frac{1}{\int \frac{g_e(X)}{X} dX} \tag{55}$$

where $g_e(X)$ is the differential MWD curve estimated by the Schulz method or the Kamide method.

Values of $\bar{M}_n$, $\bar{M}_w$, and E calculated from actual and hypothetical fractionation data by SPF and SSF on atactic PS in MCH, are compiled in Table 3. In this table, E' of the experiment is conveniently defined by:

$$E' = \int |g_G(X) - g_e(X)|\, dX \tag{56}$$

where $g_G(X)$ is the differential MWD obtained by GPC.

It should be stressed that in the case of computer simulation, the relative error E in equation (48) is the irreducible error inherent in the fractionation procedure or analytical method, and in experiments a second type of error accompanying the practical determination of $\bar{X}_w$ and $\bar{X}_n$ should be added to the above-mentioned intrinsic error. Morevoer, in the hypothetical fractionations it was assumed that equal amounts of the fractions were separated. Then, a variation in fraction size brings about another kind of error to the value of $\bar{X}_w/\bar{X}_n$. Consequently, the large experimental error makes a detailed comparison with theory impossible in practice.

The value of $\bar{M}_n$ obtained from the fractionation data by means of equation (53) using the Schulz method and the Kamide method has, even from the theoretical point of view, a large uncertainty compared with the value of M_w and is always appreciably larger than the true value (in this case, 8.6×10^4). The Kamide method is in principle superior to the Schulz's method with respect to the determination of $\bar{M}_n$, because the former takes better account of "the tailing effect" (5). The hypothetical fractionation data indicates that the Kamide method without exception affords a more accurate value of $\bar{M}_w/\bar{M}_n$ than the Schulz method does (5).

The operating conditions employed here, are appropriate for preparative

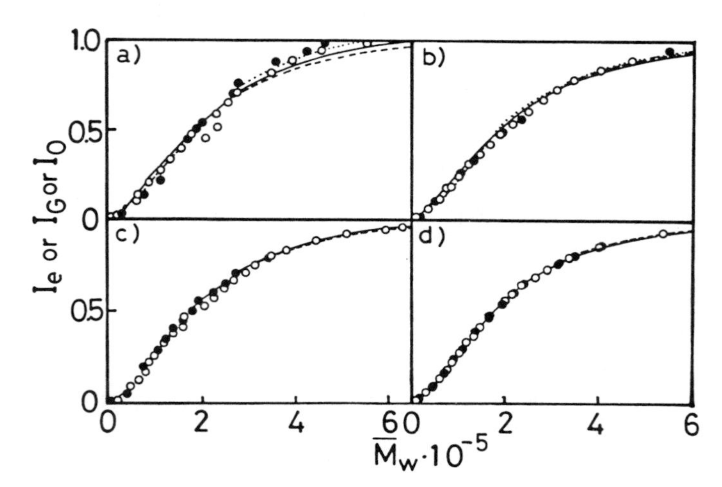

Figure 49. Cumulative weight fraction constructed from the analytical fractionation data I_e compared with the true cumulative weight fraction I_0 and that estimated by GPC I_G. (*a*) SPF experiment: ----, I_G; ····, Kamide method ($n_t = 18$); ——, Kamide method ($n_t = 13$); ●, Schulz method ($n_t = 13$); ○, Schulz method ($n_t = 18$). (*b*) SPF theory ($p_1 = 0.7$): ----, I_0; other symbols as in (*a*). (*c*) SSF experiment: ----, I_G; ····, Kamide method ($n_t = 14$); ——, Kamide method ($n_t = 23$); ●, Schulz method ($n_t = 14$); ○, Schulz method ($n_t = 23$). (*d*) SSF theory ($p_1 = 0.7$): ----, I_0; other symbols as in (*c*).

fractionation, since the majority of the fraction isolated by SSF has a polydispersity of $\bar{M}_w/\bar{M}_n \leqslant 1.10$ (for example, 9(8) and 19(20) fractions have $\bar{M}_w/\bar{M}_n \leqslant 1.10$ at $n_t = 14$ and 23 theoretically (experimentally)). As is shown from Figure 49a, b, c, and d, where I_e and I_0 or I_G are the cumulative MWD evaluated from the hypothetical fractionation data and that of the original polymer or the cumulative MWD as determined by GPC, the experiment (Figure 49a and c) is in satisfactory agreement with the theory (Figure 49b and d) (5). SPF/Schulz (Figure 49a and b) is shown to be rather less successful even for estimating the MWD curve of the whole polymer. Contrast to this, analytical SSF (Figure 49c and d) is very suitable for evaluating the total MWD profile (5). The theoretical calculation indicates that the operating conditions used in Table III do not fulfill the requirement of $E < 5\%$. Therefore, in order to estimate accurately the $\bar{M}_w/\bar{M}_n$ of the whole polymer, much more appropriate operating conditions should be chosen carefully.

This prediction was accurately examined by computer simulation. Plots of $(\bar{M}_w/\bar{M}_n)_t/(\bar{M}_w/\bar{M}_n)_0$ versus v_p^0, thus obtained, are shown in Figure 50, where $(\bar{M}_w/\bar{M}_n)_t$ is the $\bar{M}_w/\bar{M}_n$ value for the original polymer, evaluated from the SSF data by use of the Schulz or the Kamide method and $(\bar{M}_w/\bar{M}_n)_0$ is the $\bar{M}_w/\bar{M}_n$ value for the original polymer, assumed in advance for the simulation (5). Obviously, as v_p^0 diminishes $(\bar{M}_w/\bar{M}_n)_t/(\bar{M}_w/\bar{M}_n)_0$ approaches unity (5). The Schulz method has a tendency to give smaller values of $(\bar{M}_w/\bar{M}_n)_t/(\bar{M}_w/\bar{M}_n)_0$ than the Kamide method, as was found in Table 3. If one wishes to estimate $\bar{M}_w/\bar{M}_n$ of the original polymer with an error of less than $\pm 5\%$, $v_p^0 \leqslant 5\%$ at $n_t = 20$ or $v_p^0 \leqslant 2\%$ at $n_t = 50$ should be chosen as suitable operating conditions in the case of the Kamide method (5). In contrast, if the Schulz method is employed, the operating conditions of $v_p^0 \geqslant 0.01\%$ and $n_t \leqslant 50$ do not fulfil the foregoing requirement of $0.95 < (\bar{M}_w/\bar{M}_n)_t/(\bar{M}_w/\bar{M}_n)_0 < 1.05$.

Unfortunately, the value of $\bar{X}_w/\bar{X}_n$ (or $\bar{M}_w/\bar{M}_n$) obtained from analytical

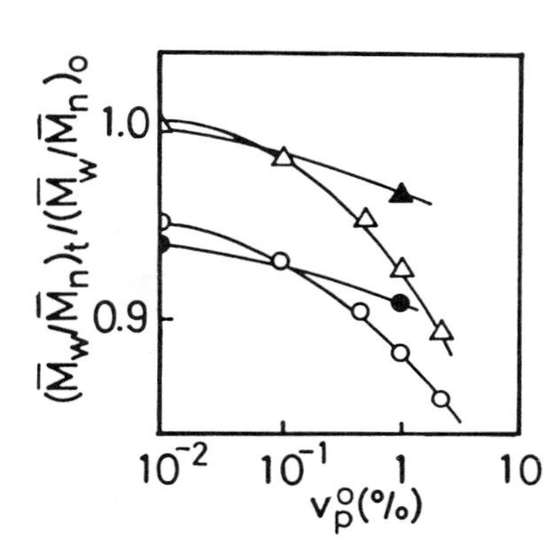

Figure 50. Effect of the initial polymer volume fraction v_p^0 on the ratio $(\bar{M}_w/\bar{M}_n)_t/(\bar{M}_w/\bar{M}_n)_0$: circles, Schulz method; triangles, Kamide method; open symbols, $n_t = 20$; solid symbols, $n_t = 50$; original polymer, Schulz–Zimm distribution ($\bar{M}_w = 23.9 \times 10^4$, $\bar{M}_w/\bar{M}_n = 2.77$).

Table 3. Values of $\bar{M}_n$, $\bar{M}_w$, and E Found by Actual Experiment on a Polystyrene–Methylcyclohexane System (experiment) and Computer Simulation with $p_1 = 0.7$ (theory)

Fractionation Procedure	Operating Conditions		Analytical Method (equations)	$\bar{M}_n \times 10^{-4}$		$\bar{M}_w \times 10^{-4}$		$\bar{M}_w/\bar{M}_n$		E'^a:	E^b:
	v_p^0 (%)	n_t		Expt	Theory	Expt	Theory	Expt	Theory	Expt	Theory
SPF	0.94	13	(51) and (52)	15.8	11.8	24.7	23.1	1.61	1.96	—	—
	0.94	13	(51′) and (52′)	12.1	8.8	17.4	17.4	1.44	1.98	—	—
	0.94	13	Schulz	12.8	11.4	22.6	23.4	1.76	2.03	0.252	0.078
	0.94	13	Kamide	10.9	8.4	24.6	23.2	2.26	2.76	0.181	0.078
	0.94	18	(51) and (52)	12.5	11.3	23.1	23.1	1.85	2.04	—	—
	0.94	18	(51′) and (52′)	10.0	8.8	18.0	17.9	1.80	2.02	—	—
	0.94	18	Schulz	13.1	10.8	23.8	23.7	1.81	2.20	0.222	0.156
	0.94	18	Kamide	9.5	8.7	23.1	24.1	2.43	2.77	0.123	0.103
SSF	0.94	14	(51) and (52)	10.4	11.5	21.7	23.1	2.10	2.00	—	—
	0.94	14	(51′) and (52′)	8.1	8.8	19.9	21.5	2.44	2.44	—	—
	0.94	14	Schulz	10.6	10.6	24.8	23.1	2.34	2.16	0.091	0.068
	0.94	14	Kamide	8.9	9.5	21.9	23.0	2.46	2.41	0.095	0.068
	0.94	23	(51) and (52)	11.7	10.9	22.1	23.0	1.88	2.11	—	—
	0.94	23	(51′) and (52′)	8.9	8.8	20.0	21.9	2.24	2.49	—	—
	0.94	23	Schulz	11.6	10.9	23.8	23.8	2.05	2.17	0.163	0.051
	0.94	23	Kamide	9.7	10.2	22.1	23.1	2.28	2.26	0.102	0.069
GPC	—	—	—	8.6 (8.9)[c]	—	23.9 (23.2)[d]	—	2.27	—	—	—

[a] E' was defined by equation (56).
[b] E was defined by equation (48).
[c] Membrane osmometry.
[d] Light scattering.

fractionation data by means of equations (52) and (53) or by the Schulz method has so far been compared directly only with the theoretical value, which has been derived by polymerization theory and has been correlated with some mechanical properties. It is obvious from Table 3 that much attention should be paid, especially when $\bar{X}_w/\bar{X}_n$ of the original polymer is to be determined from fractionation data.

CONCLUDING REMARKS

A combination of successive solution fractionation and the Kamide method (SSF/Kamide) is strongly recommended for estimating the molecular weight distribution of original polymers.

REFERENCES

1. G. V. Schulz, *Z. Phys. Chem.*, **B32**, 27 (1936).
2. J. Porath and P. Flodin, *Nature*, **183**, 1657 (1957).
3. M. F. Vaughan, *Nature*, **188**, 55 (1960).
4. J. C. Moore, *J. Polym. Sci.*, **A2**, 835 (1964).
5. Y. Miyazaki and K. Kamide, *Polym. J.*, **9**, 61 (1977).
6. D. Berek, D. Bakos, T. Bleha, and L. Soltes, *Makromol. Chem.*, **176**, 391 (1975).
7. J. V. Dawkins and M. Hemming, *Makromol. Chem.*, **176**, 1777 (1975).
8. J. V. Dawkins and M. Hemming, *Makromol. Chem.*, **176**, 1795 (1975).
9. J. V. Dawkins and M. Hemming, *Makromol. Chem.*, **176**, 1815 (1975).
10. M. R. Ambler and D. McIntyre, *J. Polym. Sci., Polym. Lett.*, **13**, 589 (1975).
11. G. V. Schulz, *Z. Phy. Chem.*, **B46**, 137 (1940).
12. G. V. Schulz, *Z. Phy. Chem.*, **B47**, 155 (1940).
13. R. L. Scott, *J. Chem. Phys.*, **13**, 178 (1945).
14. M. Matsumoto and S. Ohyanagi, *Kobunshi Kagaku*, **11**, 7 (1954).
15. T. Kawai, *Kobunshi Kagaku*, **12**, 63 (1955).
16. T. Kawai, *Kobunshi Kagaku*, **12**, 71 (1955).
17. C. Booth and L. R. Beason, *J. Polym. Sci.*, **42**, 81 (1960).
18. L. H. Tung, *J. Polym. Sci.*, **61**, 449 (1962).
19. M. Ueda, *Makromol. Chem.*, **90**, 139 (1966).
20. K. Kamide, T. Ogawa, M. Sanada, and M. Matsumoto, *Kobunshi Kagaku*, **25**, 440 (1968).
21. K. Kamide, T. Ogawa, and M. Matsumoto, *Kobunshi Kagaku*, **25**, 788 (1968).
22. R. Koningsveld and A. J. Staverman, *Kolloid-Z. Z. Polym.*, **218**, 114 (1967).

23. R. Koningsveld and A. J. Staverman, *J. Polym. Sci., A2*, **6**, 305 (1968).
24. R. Koningsveld and A. J. Staverman, *J. Polym. Sci., A2*, **6**, 367 (1968).
25. R. Koningsveld and A. J. Staverman, *J. Polym. Sci., A2*, **6**, 383 (1968).
26. R. Koningsveld, *Adv. Polym. Sci.*, **7**, 1 (1970).
27. M. Gordon, H. A. G. Chermin, and R. Koningsveld, *Macromolecules*, **2**, 107 (1969).
28. R. Koningsveld, W. H. Stockmayer, J. W. Kennedy, and L. A. Kleintjens, *Macromolecules*, **7**, 73 (1974).
29. M. Bohdanecky, *J. Polym. Sci., C*, **23**, 257 (1968).
30. K. Kamide and C. Nakayama, *Makromol. Chem.*, **129**, 289 (1969).
31. K. Kamide, T. Ogawa, and C. Nakayama, *Makromol. Chem.*, **132**, 65 (1970).
32. K. Kamide, T. Ogawa, and C. Nakayama, *Makromol. Chem.*, **135**, 9 (1970).
33. K. Kamide and K. Sugamiya, *Makromol. Chem.*, **139**, 197 (1970).
34. K. Kamide and K. Sugamiya, *Makromol. Chem.*, **156**, 259 (1972).
35. K. Kamide, *Pure Appl. Chem., Macromol. Chem.*, **8**, 147 (1972).
36. See, for example, K. Kamide, in *Fractionation of Synthetic Polymers*, L. H. Tung, Ed., Dekker, New York, 1977, Chap. 2.
37. K. Kamide, Y. Miyazaki, and T. Abe, *Polym. J.*, **9**, 395 (1977).
38. I. Noda, H. Ishizawa, Y. Miyazaki, and K. Kamide, *Polym. J.*, **12**, 87 (1980).
39. K. Kamide, K. Sugamiya, T. Kawai, and Y. Miyazaki, *Polym. J.*, **12**, 67 (1980).
40. K. Kamide and Y. Miyazaki, *Polym. J.*, **12**, 205 (1980).
41. K. Kamide and Y. Miyazaki, *Polym. J.*, **13**, 325 (1981).
42. K. Kamide, Y. Miyazaki, and T. Abe, *Br. Polym. J.*, **13**, 168 (1981).
43. K. Kamide, T. Abe, and Y. Miyazaki, *Polym. J.*, **14**, 355 (1982).
44. See, for example, K. Kamide, *High Polymer Summer Symposia*, 1972, p. 117.
45. K. Kamide, S. Matsuda, T. Dobashi, and M. Kaneko, *Polym. J.*, **16**, 839 (1984).
46. K. Kamide, Y. Miyazaki, and T. Abe, *Makromol. Chem.*, **177**, 485 (1976).
47. P. J. Flory, *J. Chem. Phys.*, **12**, 425 (1944).
48. R. L. Scott, *J. Chem. Phys.*, **17**, 268 (1949).
49. H. Tompa, *Trans. Faraday Soc.*, **45**, 1142 (1949).
50. M. Nakagaki and H. Sunada, *Yakugaku Zasshi*, **83**, 1147 (1963).
51. W. R. Krigbaum and D. K. Carpenter, *J. Polym. Sci.*, **14**, 241 (1954).
52. K. W. Suh and D. W. Liou, *J. Polym. Sci., A2*, **6**, 813 (1968).
53. P. J. Flory, *Principles of Polymer Chemistry*, Cornell University Press, Ithaca, NY, 1953.
54. K. Kamide, S. Matsuda, and Y. Miyazaki, *Polym. J.*, **16**, 479 (1984).
55. S. Matsuda, *Polym. J.* **18**, 993 (1986).
56. K. Kamide and S. Matsuda, *Polym. J.*, **16**, 515 (1984).
57. K. Kamide, Y. Miyazaki, and T. Abe, *Polym. J.*, **11**, 523 (1979).
58. K. Kamide, T. Terakawa, and Y. Miyazaki, *Polym. J.*, **11**, 285 (1979).

59. M. Saito, *Polym. J.*, **15**, 249 (1983).
60. K. Kamide, M. Saito, and T. Abe, *Polym. J.*, **13**, 421 (1981).
61. J. Bandrup and E. H. Immergut, Eds., *Polymer Handbook*, 2nd ed., Wiley, New York, 1974.
62. G. V. Schulz and A. Dinglinger, *Z. Phy. Chem.*, **B43**, 47 (1939).
63. A. M. Meffroy-Biget, *Compt. Rend.*, **240**, 1707 (1955).
64. C. D. Thurmond and B. H. Zimm, *J. Polym. Sci.*, **8**, 477 (1952).
65. G. Meyerhoff, *Makromol. Chem.*, **12**, 45 (1945).
66. K. Kamide and K. Yamaguchi, *Makromol. Chem.*, **167**, 287 (1973).
67. A. Kotera, in *Polymer Fractionation*, M. J. R. Cantow, Ed., Academic Press, New York, 1967.
68. Y. Fujisaki and H. Kobayashi, *Kobunshi Kagaku*, **18**, 305 (1961).
69. Y. Fujisaki and H. Kobayashi, *Kobunshi Kagaku*, **18**, 312 (1961).
70. Y. Fujisaki and H. Kobayashi, *Kobunshi Kagaku*, **19**, 49 (1962).
71. Y. Fujisaki and H. Kobayashi, *Kobunshi Kagaku*, **19**, 69 (1962).
72. M. Matsumoto, in *Polymer Chemistry*, (*Experimental Chemistry*, Vol. 8), A. Kotera, Ed., Maruzen Publishing Co., Tokyo, 1957.
73. K. Kamide and Y. Miyazaki, *Makromol. Chem.*, **176**, 2393 (1975).
74. R. S. Spencer, *J. Polym. Sci.*, **4**, 606 (1948).
75. F. W. Billmeyer, Jr., and W. H. Stockmayer, *J. Polym. Sci.*, **5**, 121 (1949).
76. M. Matsumoto, *Kobunshi Kagaku*, **11**, 182 (1954).
77. A. Broda, T. Niwinska, and S. Polowinski, *J. Polym. Sci.*, **22**, 343 (1958).
78. A. Broda, B. Bawronska, T. Niwinska, and S. Polowinski, *J. Polym. Sci.*, **29**, 183 (1958).
79. H. Okamoto, *J. Polym. Sci.*, **41**, 535 (1959).
80. H. Okamoto, *J. Phys. Soc. Jpn.*, **14**, 1388 (1959).
81. D. R. Morey and J. W. Tamblyn, *J. Appl. Phys.*, **16**, 419 (1945).
82. I. Harris and R. G. J. Miller, *J. Polym. Sci.*, **7**, 377 (1951).
83. K. Kamide, Y. Miyazaki, and K. Yamaguchi, *Makromol. Chem.*, **173**, 175 (1973).

LIST OF SYMBOLS

E, E'	relative error of analytical fractionation
$g_0(X_i)$	normalized molecular weight distribution of original polymer
$g_{(1)}(X_i)$	molecular weight distribution of the polymer partitioned in the polymer-lean phase
$g_{(2)}(X_i)$	molecular weight distribution of the polymer partitioned in the polymer-rich phase
$g_k(X_i)$	molecular weight distribution of the kth fraction

$I(X_{w,j})$	cumulative weight fraction corresponding to the $X_{w,j}$
K_m	parameter characteristic of the polymer–solvent combination and temperature in the Mark–Houwink–Sakurada equation
k'	molecular weight-dependent parameter of χ_i
M	molecular weight
n_t	total number of fractions
p_j	concentration-dependent parameter of χ_i
$p_{12,s}$	concentration-dependent parameter of χ_{12}
$p_{13,q}$	concentration-dependent parameter of χ_{13}
$p_{23,r}$	concentration-dependent parameter of χ_{23}
R	volume ratio of polymer-lean to polymer-rich phase
$\tilde{R}$	gas constant
T	Kelvin temperature
v_0	volume fraction of solvent
v_1	volume fraction of solvent 1
v_2	volume fraction of solvent 2
v_p	total volume fraction of polymer
v_{X_i}	volume fraction of X_i-mer
v_p^s	starting concentration of polymer (by volume)
v_p^0	initial concentration of polymer (by volume)
v_2^{cp}	v_2 at cloud point
v_p^{cp}	v_p at cloud point
V_1^0	volume of solvent 1
V_2	volume of solvent 2
V_3^0	volume of polymer
V_0	$= V_1^0 + V_3^0$
V	$= V_1^0 + V_2 + V_3^0$
X, X_i	degree of polymerization
$\bar{X}_n$	number average of X_i
$\bar{X}_w$	weight average of X_i
$\bar{X}_v$	viscosity average of X_i
X_p	X_i giving the peak of a modified triangle distribution
$\bar{X}_{n,j}$	$\bar{X}_n$ of the jth fraction
$\bar{X}_{w,j}$	$\bar{X}_w$ of the jth fraction
Z	parameter representing the breadth in the X_i distribution in the precipitate in Spencer's method
a	exponent in the Mark–Houwink–Sakurada equation (characteristic of polymer–solvent combination and temperature)
α	probability of polymerization
ΔT	change in turbidity
$\Delta\mu_0$	chemical potential of solvent
$\Delta\mu_1$	chemical potential of solvent 1

$\Delta\mu_2$	chemical potential of solvent 2
$\Delta\mu_{X_i}$	chemical potential of X_i-mer
$[\eta]$	limiting viscosity number
ρ	fraction size
ρ_j	relative amount of the *j*th fraction
ρ_s	weight fraction of polymer in the polymer-lean phase
ρ_p	weight fraction of polymer in the polymer-rich phase
σ, σ_i	partition coefficient
σ_0	molecular-weight-independent term of σ_i
σ_{01}	molecular-weight-dependent parameter of σ_i
σ'	standard deviation of MWD
χ, χ_i	thermodynamic interaction parameter between solvent and polymer
χ_{00}	temperature-dependent factor of χ_i (independent of the concentration and molecular weight of polymer)
χ_{12}	thermodynamic interaction parameter between solvent 1 and solvent 2
χ_{13}	thermodynamic interaction parameter between solvent 1 and polymer
χ_{23}	thermodynamic interaction parameter between solvent 2 and polymer
χ_{12}^0	temperature-dependent factor of χ_{12}
χ_{13}^0	temperature-dependent factor of χ_{13}
χ_{23}^0	temperature-dependent factor of χ_{23}

CHAPTER

10

DETERMINATION OF MOLECULAR WEIGHT DISTRIBUTION BY GEL PERMEATION CHROMATOGRAPHY

MARK G. STYRING

ICI Petrochemicals and Plastics Division, Wilton, Middlesbrough, Cleveland, England

and

ARCHIE E. HAMIELEC

McMaster Institute for Polymer Production Technology, McMaster University, Hamilton, Ontario, Canada

INTRODUCTION

GPC (gel permeation chromatography) is an acronym in wide usage for the chromatographic separation of macromolecules according to their size. The term size exclusion chromatography (SEC) is the general name for this process, emphasizing the mechanism of separation rather than directing attention to a particular class of column-packing materials (the gels) or macromolecules. Gel filtration chromatography (GFC) is the term most frequently found in biochemical applications, where the method is primarily applied to aqueous solutions of biopolymers. Hydrodynamic chromatography (HDC) has become the accepted name for the separation of colloidal species, mainly polymer latices. It has been argued by Giddings that HDC does not involve partition of solute between two phases (i.e., mobile solvent in the interstices and stationary solvent trapped in the pores). In this light, HDC is regarded as a subset of field-flow fractionation (FFF). The end effect, namely separation according to size, is the same however.

GPC is the name most commonly found in the polymer literature, referring to an analytical technique for the determination of polymer molecular weight (MW) averages and distributions (MWD). This chapter examines primarily GPC, although much of what is said applies to the other methods. GPC is a liquid column chromatographic technique in which a sample solution is introduced onto a column filled with a rigid porous gel and is carried through

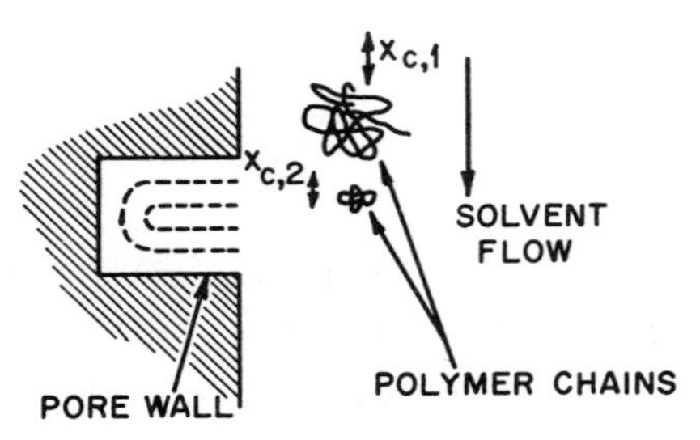

Figure 1. Schematic representation of mechanism of size separation in GPC: $x_{c,i}$ = characteristic solute dimension.

the column by solvent. Ideally, size separation is achieved by differential pore permeation (see Fig. 1). All molecules experience a solute-to-wall exclusion effect inside the pore. Owing to greater steric interference, larger molecules are kept away from the wall of the pore, demarcated by the inner dashed line in Figure 1. Smaller molecules can approach the pore wall more closely (outer dashed line). The volume of the pore which is effectively accessible is thus greater for a small molecule than for a large one. Under the influence of the solvent stream passing down the column, larger molecules are eluted from the column earlier than smaller ones, and are detected by means of some suitable instrument. Figure 2 shows a typical chromatogram, which is effectively a retention volume distribution. Clearly, if V, the retention volume, can be directly related to MW by means of an appropriate calibration, then in principle a chromatogram can be made easily to yield MW averages and distributions. The key words here are "in principle". In practice, there are frequently great problems associated with the establishment of instrumental calibrations for both MW and spreading, that is, instrumental effects which cause the MWD calculated for the chromatogram to be significantly broader than the true one. This aspect is illuminated below.

The first report of chromatographic separation of macromolecules according to size was made by Porath and Flodin (1) in 1959. Cross-linked, semirigid polydextran gels, swollen in aqueous media, were used to examine various

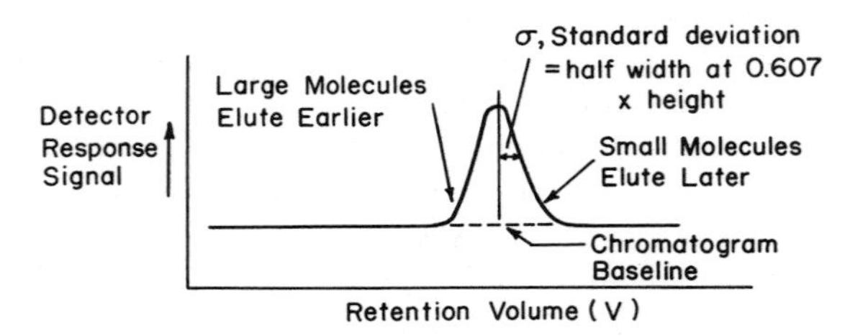

Figure 2. Typical GPC chromatogram.

water-soluble polymers. Such systems are still in wide use in GFC applications. The introduction of GPC is generally attributed to Moore (2), who showed in 1964 that rigid cross-linked porous polystyrene beads could be used to separate synthetic polymers in organic solvents. It was immediately recognized that at last a relatively simple, rapid, powerful, and inexpensive tool for polymer characterization had been discovered. The current importance of the technique is reflected by the volume of literature which has since been published and the time and money invested by the instrument makers in systems development. Numerous review articles are published each year, some containing more than 500 references (see, e.g., Ref. 3). Several texts have appeared (4), one of the most authoritative and readable being that of Yau et al. (4a), published in 1979. Balke's recent text (4c) is particularly thorough in regard to statistical and error analysis.

Over the two decades succeeding Moore's disclosure, much effort has been expended in the design and manufacture of a wide range of gels compatible with a variety of polymer and solvent types. In parallel, much has been done to improve the chromatographic hardware (solvent delivery systems, ancillaries capable of withstanding high pressures, and ever more sensitive and powerful detectors). Early, low pressure GPC systems could yield accurate values of whole polymer MW and MWD in a matter of hours, a significant advance over classical techniques such as osmometry, light scattering, and ultracentrifugation. However, the advent of microparticulate packings, capable of withstanding high eluant flow rates and pressures, together with rapid on-line data acquisition and processing techniques, have culminated in the design of systems capable of yielding similar information within 10 minutes.

Basic Instrumentation

Figure 3 is a schematic diagram of a typical modern analytical high performance (HP) GPC apparatus for use in separations by gel phase chromatography.

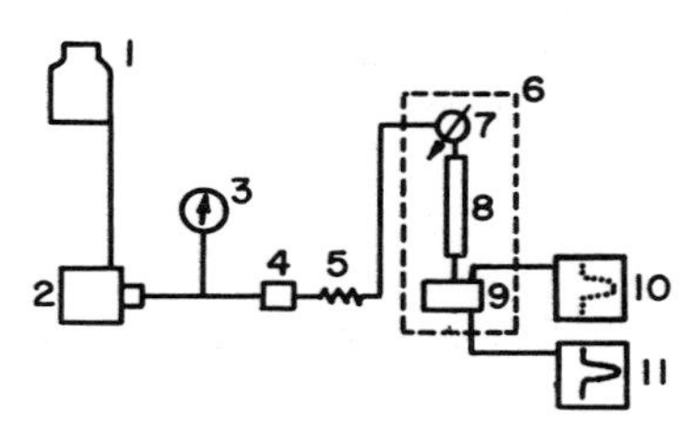

Figure 3. Schematic diagram of typical HPGPC apparatus: 1, solvent reservoir and degasser; 2, pump; 3, pressure guage; 4, in-line filter; 5, pulse dampener; 6, thermostated oven; 7, sample injector; 8, packed column(s); 9, detector(s); 10, on-line data station and processor; 11, strip-chart recorder.

Solvent Reservoirs

Reservoir capacities vary according to the particular application. These units should be fairly robust and inert to solvent and are typically made of stainless steel or glass. Solvents frequently need to be degassed to eliminate the formation of bubbles in the solvent stream. This may be achieved in situ, either by evacuating the tank, by purging with some inert gas, or by incorporating some heating and stirring device.

Pumps

A constant, reproducible solvent flow rate is the single most important parameter in accurate quantitative MW analysis. Variations in flow times of just a few seconds are sufficient to produce significant errors in MW and MWD estimates, particularly when the total throughput time of the system may be as short as 10 minutes. Modern pumps have to be able to produce high pressures (>1000 psi) to overcome the flow resistance due to the small particles packed in the columns and also due to the pulse dampener (typically, a coil of stainless steel tubing having a narrow internal diameter). Most of them are of the reciprocating-piston design, frequently with two and sometimes three pistons to reduce pressure fluctuations from each pulse. The parts in contact with solvent are usually made of stainless steel or Teflon to resist chemical attack by polymer, solvent, and cleansing materials, such as concentrated nitric acid, which is often recommended for flushing if it is suspected that deposits have formed inside the pump.

Filters

Particulate matter can cause severe problems with regard to pump reliability and longevity, blocking of the column end fittings (typically, stainless steel frits having small pores, 0.5–2 μm), and spurious detector signals. To avoid such difficulties, it is common practice to insert a filter in the solvent feed line from the reservoir and often a second one just downstream from the pump and to filter the sample solution before injection.

Injectors

An ideal injector should introduce a sharp plug of sample into the column with minimal peak spreading. Reproducibility of the amount of solution injected is an important consideration also, since in some instances the total amount of polymer injected needs to be known before one can proceed with MW calculations. The most commonly used device is the microsampling injector valve. In the "load" position, filtered solution is injected by means of a syringe into an external loop of predetermined volume, while the solvent flows through the columns. In the "inject" position, the valve rotor diverts the solvent through the filled loop, with minimal disturbance to flow rate, thereby introducing the sample onto the column.

Columns and Packings

GPC columns are typically stainless steel tubes from 20 cm to 1 m long, having a constant and well-polished bore of between 2 and 10 mm. The smaller dimension columns are those used in HPGPC. Columns may be packed by a variety of techniques (dry or slurry packings) and numerous types of column-packing equipment of various levels of sophistication are available. A perfectly homogeneously packed gel bed is required in the column to minimize peak-spreading effects, hence the need for very careful packing procedures.

Concerning the gels themselves, the polydextrans first used in Porath and Flodin's pioneering work are currently available under the brand name Sephadex. The materials developed from Moore's work are known as Styragels. In the late 1960s, technology was developed for the production from both silica and borosilicates of glass granules having reproducible pore diameters. Such packings have the two notable advantages of perfect rigidity (i.e., the pore diameter is invariant with solvent type) and compatibility with both organic and acqueous eluants. In the case of organic eluants, packings are almost exclusively Styragel or silica based. The peculiar nature of polymer–solvent packing interactions in aqueous GPC (see section entitled "Nonidealities," below) has led to the employment of considerable ingenuity in the design of a wide range of materials. In fact, no single, universally applicable packing type exists; the final choice is dictated by a user's specific requirements. Currently available for aqueous GPC/GFC are Sepharoses (agarose based), polyacrylamides, polyacrylomorpholines, and agarose–polyacrylamide composites, which are semirigid. Recently, a range of organic-based, tightly cross-linked hydrophilic gels suitable for aqueous HPGPC have become available (e.g., polyhydroxymethacrylates, sulfonated Styragels, and a range of hydroxylated gels of unknown structure from the Toyo Soda Company in Japan). Silica or borosilicate glasses can be used either untreated or coated with a variety of compounds which modify the surface characteristics to minimize the effects peculiar to individual solute–solvent systems. References 4 and, particularly, 5 give good surveys of the various packing types.

Detectors

There are several types of detector available for GPC which continuously monitor the column effluent through measurement of some physical property of what is essentially a very dilute polymer solution.

Differential Refractometry. Detectors which measure the differential refractive index (DRI) increment are the most commonly used. The principle of operation is to split a beam of monochromatic light into two beams, diverting one into a transparent cell through which is passing column effluent (the "sample" cell) and the other into the "reference" cell, containing the pure GPC eluant of interest. When the sample and reference cell contents are the same (i.e., when only pure solvent is eluting from the columns), the refractive indices are identical and the light beams fall at a certain location on the surface of a position-sensitive photodetector. The photodetector produces an electric signal proportional to the position of the light, which, for the time being, remains constant, thus establishing the baseline of the chromatogram (see Fig. 2). When polymer beings to elute from the column and enters the sample cell, the RI changes, causing a deflection in the location of the light beam on the photodetector (directly proportional to the mass concentration of polymer in solution), giving rise to the chromatographic peak due to the polymer.

DRI detectors have the advantage of being able to respond to all solutes by proper solvent choice. Subject to solubility criteria, maximum sensitivity is obtained when solute and solvent differ in RI as much as possible. However, one major weakness of DRI detectors is their relative low sensitivity, rendering them unsuitable for measuring solutes at low concentrations. In addition, temperature instability can be a problem.

Spectrophotometry. Detectors which measure the ultraviolet (UV) and, less commonly, the infrared (IR) absorption of solutes are in use, although their application in GPC is frequently hampered by the requirements that the polymer of interest be completely soluble in a solvent which is transparent to the incident light and contains a chromophore. UV photometers in particular are very sensitive, however, and in some cases, a few nanograms of solute is sufficient to give a signal. The UV instruments are also much less temperature sensitive than DRI detectors. Again, the signal amplitude is proportional to mass concentration.

In the case of mass concentration detectors (DRI/UV/IR), correct interpretation of the chromatogram requires the presence of the same number of chromophores per gram of polymer or, in the case of DRI detection, that refractive index be independent of MW. Furthermore, when dealing with multicomponent polymers as opposed to homopolymers, the composition of the sample should be homogeneous.

Molecular Weight Detectors. Despite considerable experimental drawbacks and high costs, the idea of continuously monitoring the MW of the effluent as well as solute concentration has proven attractive, and powerful instrumentation has appeared over the past decade. Use of an MW detector and a concentration detector in tandem provides data that can be used to make absolute MW calibrations (more on generalized calibration in the section entitled "Chromatogram Interpretation for the Ideal Case").

The four main presently used classical methods of MW determination—osmometry, viscometry, light scattering, and ultracentrifugation—yield different MW averages; these are, respectively, the number, viscosity, weight, and z averages. Each method, if it could possibly be coupled to a GPC apparatus, ought to yield the same MW values, since if solute spreading effects are not too important at each sampling interval (which can be made infinitesimally small in the case of continuous monitoring), the solute should be of sufficiently narrow MWD that $\bar{M}_n \simeq \bar{M}_v \simeq \bar{M}_w \simeq \bar{M}_z$. A suitable detector, however, must have a response that is fast enough to follow rapid changes in MW in the effluent; it must also be sensitive to low solute concentration, and it must have a small mixing volume (to minimize peak spreading). These requirements have given rise to the development of light-scattering and viscometric detectors, these being the most amenable to application in GPC.

Low Angle Laser Light-Scattering Photometry (LALLSP). The earliest flow-through, light-scattering sample cell using a laser light source was described in 1966 by Cantow et al. (6). Pioneering work on a more usable system was done in the mid 1970s by Ouano et al. (see, e.g., Ref. 7), which led to the commercialization in the late 1970s of a reliable instrument, the KMX-6, by Chromatix. Essentially, the intensity of light scattered by the polymer solution in the detector cell is proportional to both MW and concentration, just as in classical "static" light scattering. The MW can be evaluated across the whole chromatogram if a concentration-sensitive detector (usually a DRI), is connected in line with the LALLSP, eliminating the need for an MW calibration for the instrument. For this reason, the LALLSP-DRI combination is referred to as an absolute detector system.

Viscometry. The earliest viscometric detectors (8–10) in use to monitor MW changes in the effluent had a major drawback in that sampling was discontinuous. The method consisted of collecting and charging effluent fractions into a series of Ubbelohde-type viscometers and measuring the flow times of the fractions through the viscometers. The earliest continuous viscometer, which measured the pressure drop across a capillary to monitor the intrinsic viscosity $[\eta]$, of the effluent was described by Ouano (11) in 1972. The main problem with this early design was a noisy response signal due to extreme sensitivity of the pressure transducer in the flow cell to variations in solvent flow rate. Only recently has this problem been overcome, and a capillary bridge viscometer (12–14), having a reportedly good signal to-noise ratio, is now marketed by Viscotek. When combined with a concentration detector, the dual viscometer–concentration system also becomes capable of evaluating MW across the entire chromatogram.

Since each type of detector mentioned in this section measures a different physical property of the column effluent, a different method of data manipulation is necessary in each case, as extensively discussed in connection with equations (19)–(45), below.

Data Acquisition and Processing

A basic GPC chromatogram is, as we have seen, a two-dimensional plot of retention volume versus detector response, which is proportional to concentration or MW. The first of these parameters, V, can be monitored in a number of ways. The first is simply to assume constancy of pump flow rate, in which case, retention time may be taken as the elution parameter. It is good practice to make periodic checks on the pump flow rates if this is in fact the preferred method. More commonly, direct measurements of volumetric flow rates are used. Flow meters which directly measure laminar flow are available commercially. Siphon counters are more widely used, however, owing to their simplicity, reliability (with reservations mentioned below and, in the case of aqueous eluants, in the Section entitled "Nonidealities") and low cost. Column effluent, after passing through the detector(s), is fed into a siphon of known volume. When full, the siphon "dumps" into a suitable solvent container, an event which actuates a photoelectric switch feeding a signal to the recording device. Care must be taken, however, to eliminate solvent evaporation, which has been a problem at elevated temperatures with low boiling solvents such as tetrahydrofuran (THF) (15). Alternatively, a drop counter may be used, which, as the name implies, counts the number of drops issuing from the tube exiting the detector: V may be calculated from knowledge of the volume per drop, which varies according to the solvent used.

The second parameter, detector response, can also be monitored in a number of ways. Until the relatively recent revolution in the electronics industry, which gave birth to inexpensive desktop computers, simple chart recorders were almost exclusively used in both investigative and analytical GPC modes. Chart recorders are still widely used, yielding at a glance such valuable information as number, size, shape and retention volumes of peaks, spacing between peaks, detector response, noise levels, and baseline drift. Such observations may be used to judge the overall quality of an experiment in the investigative mode. Use of such chromatograms to calculate values of MW and MWD on a routine analytical basis is quite tedious, however. The chromatogram has to be converted manually to digital form, thus inviting human error, for off-line processing by a computer, traditionally using in-house software. Recently, real-time data acquisition followed by processing immediately after the end of the experiment has become de rigueur in many analytical laboratories following from the remarkable reductions in cost and increase in computing power afforded by developments in microcomputers. Detector responses are measured at time intervals determined by the operator while the experiment is running, and data processing begins as soon as the end point (again determined by the operator) is reached.

Calibration in GPC

The advantages of GPC over other techniques of MW determination, especially in terms of simplicity of operation and the ability to yield MWDs as well as MW averages, have already been stressed. One great disadvantage, however, is that unless a molecular weight detector is coupled to the instrument, GPC cannot be used as an absolute method. Rather, a calibration procedure must be used, which in practice can present considerable difficulties. The simplest type of calibration is a peak position calibration using polymer standards of narrow MWD, which is best explained with reference to Figure 4. In this instance, four standards of quite different average MW, determined by some absolute method, are injected simultaneously onto the column and a calibration curve of log MW against V is constructed (the solute retention mechanism which gives rise to such calibrations is discussed in the next section). However, such calibrations are valid only for the same polymer–solvent–temperature combination. One cannot employ a calibration established using linear polystyrenes to calculate values of MWs for, say, poly(ethylene oxides) (PEOs). Different polymers of the same MW usually have different chain dimensions, and separation in GPC occurs according to size, not according to MW. Another problem is that standards in a wide range of accurately known MWs and of narrow MWD are available for only a very few polymer types, most notably linear polystyrenes and PEOs.

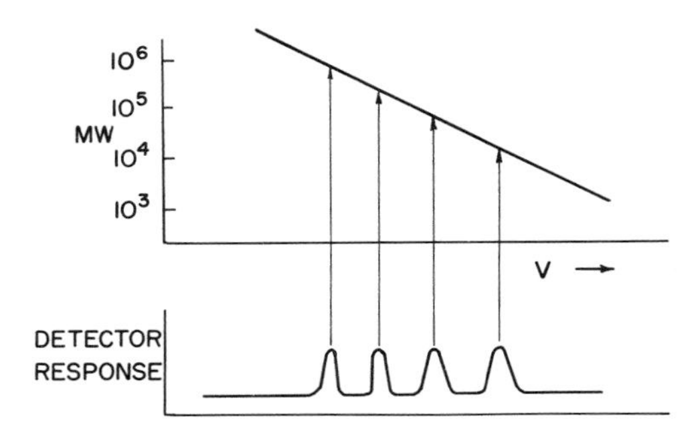

Figure 4. Schematic of peak position calibration.

It is noteworthy that at high MWs in particular, chromatograms from standards of narrow MWD are often asymmetrical, usually with a skew to the low MW (high retention volume) end. In such cases, it is appropriate to use the mean elution volume, rather than that at the peak in constructing the calibration.

It is convenient at this stage to define two types of solute system according to the ease of establishing a quantitative calibration. In the "ideal" case, we have separation due purely to size exclusion of linear homopolymers and of linear copolymers having uniform composition and compositional distribution (e.g., a calibration established for an A–B block copolymer will not be valid for an alternating A–B–A–B copolymer). The second or "nonideal" case covers a wide range of instances. In many systems, particularly aqueous ones, phenomena giving rise to retention mechanisms other than pure size exclusion occur. The gel permeation chromatography of "complex" polymers [i.e., homopolymers and uniform copolymers with long chain branching (LCB) and nonuniform copolymers] comes under this category. With these definitions in mind, it is now possible to proceed with a more thorough discussion of the solute retention and peak-spreading mechanisms in GPC.

CHROMATOGRAM INTERPRETATION FOR THE IDEAL CASE

Mechanism of Solute Retention and Instrumental (Peak) Spreading

Solute Retention and MW Calibration

In treating GPC on a quantitative basis, we first split the total volume within the packed column into three distinct regions (see Fig. 5) such that

$$V_c = V_d + V_i + V_p \tag{1}$$

where V_c is the total internal column volume, V_d is the "dead" volume occupied

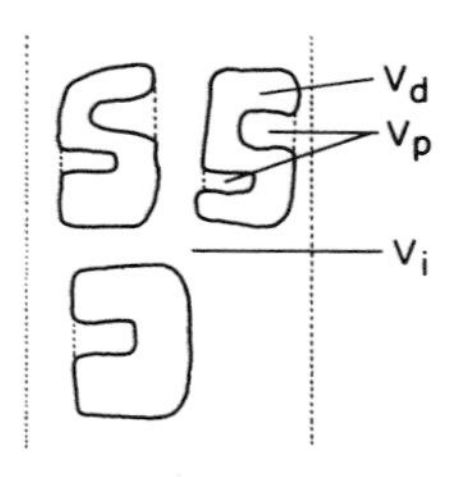

Figure 5. The three volume regions in a packed GPC column.

by the bulk of the packing material, V_p is the porous volume containing stationary solvent, and V_i is the interstitial volume occupied by moving eluant (sometimes referred to as the void volume V_0). Separation in GPC is a partitioning of solute between mobile (interstitial) and stationary (in the pores) solvent; V_d thus takes no part in the chromatographic process. Defining separation in terms of a GPC distribution coefficient K_{GPC}, we write

$$V = V_0 + K_{GPC} V_p \tag{2}$$

where V is the retention volume of a particular solute. For very small solutes, which can freely access the entire porous volume for the column, as well as the interstitial volume, $K_{GPC} = 1$; that is,

$$V = V_0 + V_p \tag{3}$$

Solutes too large to fit in any of the pores can access only the interstitial volume; thus $K_{GPC} = 0$ and

$$V = V_0 \tag{3'}$$

For solutes of intermediate size, there will be some fraction of the porous volume that is effectively accessible, depending on the size and shape (i.e., $0 < K_{GPC} < 1$).

Many workers (16–19) and, in particular, Casassa (20–22), have attempted to model K_{GPC} in terms of the size and shape of both solute and pore. With reference to Figure 6, we shall briefly consider the simplest model: that of a spherical solute of radius R, a pore of length l, and radius a. An excluded-volume effect prevents the center of the solute molecule from approaching the wall any closer than a distance R. This reduces the volume accessible to the solute to a smaller cylinder of radius $(a - R)$. In the smaller cylinder, the solute concentration is the same as in the interstices, while the shell of thickness R contains no solute. Thus, the average solute concentration in the pore as a whole is less than that outside. The fraction of the external concentration in the

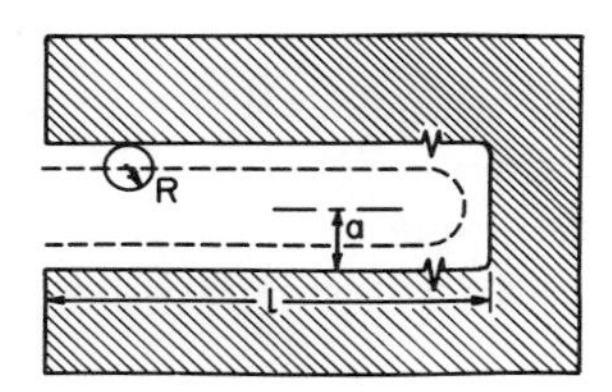

Figure 6. Schematic illustration of size exclusion of a spherical solute in a cylindrical pore.

pore is given by the ratio of the two volumes, that is,

$$K_{\mathrm{GPC}} = \prod (a-R)^2 l \Big/ \prod_a^2 l = (a-R)^2/a^2$$

$$= 1 - 2R/a + R^2/a^2 \tag{4}$$

If the solute dimensions are much smaller than the pore, the squared term vanishes and we obtain

$$K_{\mathrm{GPC}} = 1 - 2R/a \tag{5}$$

This argument can be extended to random coils by visualizing the coil domain as a sphere with r_g, the radius of gyration, taking the place of R. Equation (5), now expressed as an exponential series, becomes

$$K_{\mathrm{GPC}} = \exp(-cr_g) \tag{6}$$

Since r_g is proportional to MW for a given polymer, we have

$$K_{\mathrm{GPC}} = \exp(-kM) \tag{7}$$

Substitution into equation (2) gives

$$V = V_0 + V_p \exp(-kM) \tag{8}$$

Figure 7 is semilogarithmic plot according to equation (8), where we have arbitrarily chosen $V_0 = V_p = 50\,\mathrm{cm}^3$ and $k = 10^{-4}\,\mathrm{mol/g}$. The plot extends from $M = 10^3$ to $M = 10^6$ and is clearly appreciably linear over the range $3 \times 10^4 < M < 2 \times 10^5$.

It is commonplace in reality to find GPC systems for which, through judicious choice of column packings (i.e. as wide a range of packing porosities as possible to accommodate solutes of widely varying sizes), the calibration curve is approximately linear over four decades of M.

The form of the V versus M calibration thus reduces to

$$V = a - b \log M \tag{9}$$

where a and b are constants depending on instrumental parameters. The slope b defines the resolution of the instrument. A shallow slope (low b) is favorable, since there is a relatively large change in V for a relatively small change in M; that is, two species of nearly equivalent M are better separated from each other when b is small.

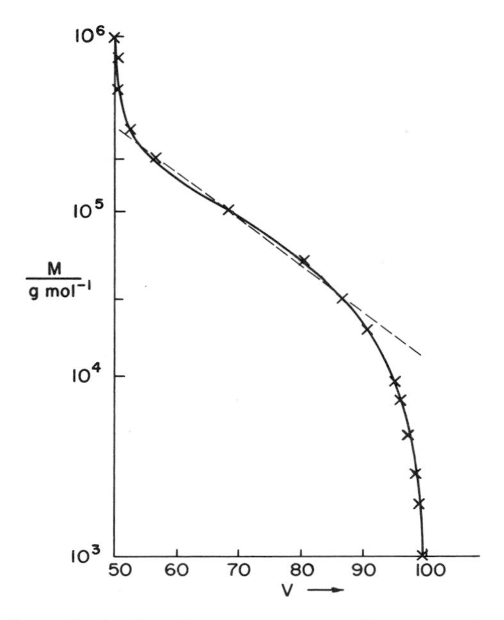

Figure 7. Semilogarithmic plot of V versus M according to equation (8) (see text).

If polymer standards of known MW (from absolute methods such as light scattering or osmometry) and narrow MWD are available for a given polymer type, then calibration of a GPC instrument is a simple matter, as schematized in Figure 4. The standards are injected into the instrument to find their peak retention volumes and plot of V versus $\log M$ is constructed.

MW calibrations of the foregoing type suffer the considerable drawback of being valid for one particular polymer–solvent–temperature combination. A calibration valid for a wide range of polymer and solvent types chromatographed on the same instrument would be most useful, and this is the topic of the next section.

Universal Calibration

Looking back to equation (6), we see that r_g is the size parameter determining the extent of pore permeation in GPC rather than MW. The flory–Fox (23) equation states

$$[\eta] = \Phi \langle \bar{r}_g^2 \rangle^{3/2} / M \tag{10}$$

where $[\eta]$ is the intrinsic viscosity of the polymer in the solvent of interest, Φ is a constant which has values ranging from $2.86 \times 10^{23}\,\text{mol}^{-1}$ for a theta solvent to $1.7 \times 10^{23}\,\text{mol}^{-1}$ for a good solvent, $\langle \bar{r}_g^2 \rangle$ is the mean-square radius of gyration for the whole-polymer sample. From equation (10), the product $[\eta]M$—that is, the hydrodynamic volume—is thus proportional to r_g. It was Benoit et al. (24) who first proposed the use of $[\eta]M$ as the universal calibration parameter. Their original investigation (24), using THF as eluant from their GPC instrument, showed that a plot of log $[\eta]M$ against V gave a common curve for the elution of solutes of widely varying molecular architecture (e.g., linear, "comb," and "star" polystyrenes; linear poly(methyl methacrylates), polybutadienes, polyphenylsiloxanes, and various copolymers). Numerous investigations have since borne out the soundness of the concept.

A second approach due to Coll and Prusinowsky (25) is based on the use of r_g directly as the universal parameter. According to Ptitsyn and Eizner (26), r_g is defined by

$$[\eta] = \Phi(\bar{r}_g^3/M) \tag{11}$$

which has the same form as equation (10). In equation (11), the parameter Φ is defined as

$$\Phi = \Phi_0(1 - 2.63f(\epsilon) + 2.86f^2(\epsilon)) \tag{12}$$

where Φ and Φ_0 are constants and

$$f(\epsilon) = \frac{2a - 1}{3} \tag{13}$$

where a is the exponent in the Mark–Houwink equation. The universal parameter arising from this approach is thus $[\eta]M$ divided by $f(\epsilon)$.

A third approach first suggested by Dawkins (27) recommends the use of the root-mean-square (RMS), end-to-end distance of the unperturbed molecule as the universal parameter $\langle \bar{r}_\theta^2 \rangle^{1/2}$. The basic equation is

$$[\eta] = \Phi(\langle \bar{r}_\theta^2 \rangle/M)^{3/2} M^{1/2} \alpha^3 \tag{14}$$

analogous to equations (10) and (11), where α is a parameter defining the degree of expansion of the molecule relative to its dimensions in a theta solvent. Subsequent studies by Dawkins et al. (28–34) compared various proposed calibration parameters. It was concluded that the parameters $[\eta]M$ and $\langle \bar{r}_\theta^2 \rangle^{1/2}$ were the most satisfactory in the range $10^3 < M < 10^6$. Above this range, the latter seemed preferable. Coll (35), however, in another survey, gave his preference to the use of $[\eta]M$, which today is the most widely used of the so-called universal parameters.

Other Types of Calibration

In addition to the types of calibration mentioned above, which are generally peak position calibrations relying on the availability of well-characterized standards of narrow MWD, there are several other procedures for which only a single standard of broad MWD is required. However, a major drawback of such procedures is that unlike the peak position methods, instrumental peak broadening (see below) must be correctly accounted for if accurate results are to be obtained.

The "integral MWD" method requires the use of a standard for which the MWD is accurately known. The GPC calibration curve is obtained by matching those MW and V values which correspond to the same value of sample weight fraction on the MWD and GPC elution curves separately.

The first of the so-called linear broad MWD standard calibration methods was developed by Balke et al. (36). All that is required is a single broad standard of accurately known $\bar{M}_w$ and $\bar{M}_n$. A linear calibration of the form

$$M(V) = D_1 \exp(-D_2 V) \tag{15}$$

where V represents the GPC retention volume, is assumed. Values of $\bar{M}_w$ and $\bar{M}_n$ are computed from the normalized experimental chromatogram $F_{\mathrm{n}}(V)$ according to

$$\bar{M}_w = \sum_{v} F_{\mathrm{n}}(V) M(V) \tag{16}$$

and

$$\bar{M}_{\mathrm{n}} = \frac{1}{\sum_v F_{\mathrm{n}}(V)/M(V)} \tag{17}$$

At the heart of the method is a computer program which iteratively adjusts D_1 and D_2 to satisfy equations (16) and (17) for the known values of $\bar{M}_w$ and $\bar{M}_n$. Since peak spreading is not accounted for in this method, an "effective" rather than the true calibration is obtained. Improved versions of this procedure which do account for peak spreading, called GPCV2 and GPCV3, have been proposed by Yau et al. (37, 38).

Instrumental Peak Spreading

In a GPC experiment, a small volume of sample solution is injected as a very narrow band at the top of the column. As the band moves down the column, its

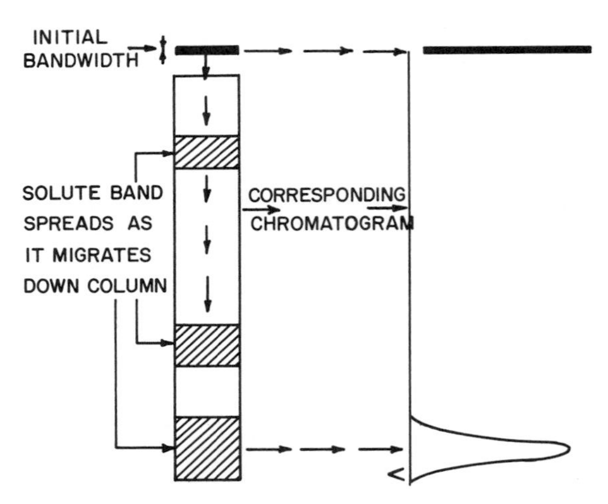

Figure 8. Effect of column dispersion processes on the peak width of a monodisperse solute.

width increases, partly because of size exclusion (smaller molecules being retained in the pores longer than larger ones), but also as a result of mechanical dispersion processes. These latter phenomena are responsible for the finite widths of peaks for monodisperse species which should otherwise elute with the same bandwidth as the injected "slug". This is illustrated in Figure 8. All such effects are detrimental to the resolution of the instrument; any amount of spreading interferes with the MWD information in the chromatogram. A small distortion in the shape of the chromatogram can cause large errors in calculating MWs and MWDs.

The chromatographer can minimize spreading but never completely eliminate it. Excessive spreading occurs, for example, if the gel bed is not homogeneously packed in the column, or if any flow channeling arises, necessitating careful packing procedures. In addition, the volume of all connecting tubing, detector cells, and any other extracolumnar elements should be minimized to reduce the possibility of solute mixing. Such features as solvent filters and pulse dampeners should, for example, be installed before the sample injection valve.

Eddy diffusion, in which diffusing solute molecules take different paths through the gel bed, is an important contributory factor to spreading. Another is mobile phase mass transfer, which arises because of the velocity profile of the solvent through the interstices between the packing granules. As in pure laminar flow, solvent velocity reaches a maximum at the center of the flow stream. A third process is stationary phase mass transfer, which arises from the time taken for solute to diffuse into and out of the pores in the gel. Large

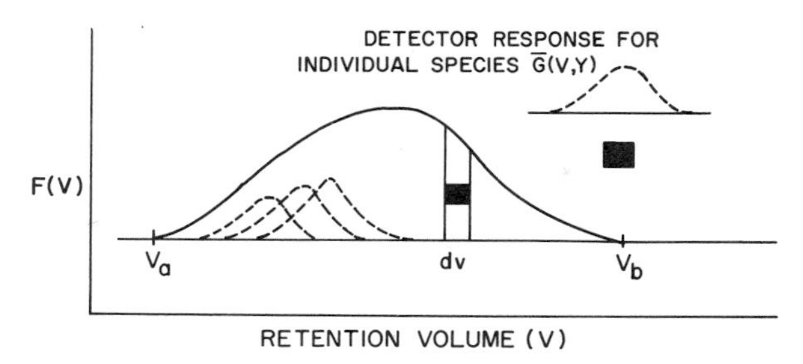

Figure 9. Typical detector response as measured by a strip-chart recorder.

solutes with low diffusion coefficients (e.g., latex particles) are particularly badly affected by this phenomenon.

A polymer sample injected into a GPC instrument typically contains 10^3–10^5 individual species differing in MW by the monomer repeat unit. Owing to instrumental spreading, it is highly doubtful that resolution in GPC will ever reach the point of achieving baseline separation of the individual species. In Figure 9, we show how a chromatogram having limits defined by V_a and V_b (solid curve) is composed of a series of overlapping single-species detector responses (dashed curves). With reference to this figure, we can also introduce some mathamatical concepts in use for quantification of instrumental spreading. We let the function $G(V, Y)$ represent the normalized detector response for an individual species whose mean retention volume is Y, henceforth referred to simply as species Y; V is also a retention volume variable. The distinction between V and Y is that owing to instrumental spreading, some fraction of polymer which would elute at Y in its absence actually elutes at over a range of V. This fraction at V is $W(Y)G(V, Y)$. We use $G(V, Y)$ to refer to the spreading function for species Y and $W(Y)$ to represent the detector response for Y corrected for spreading. The chromatogram for the whole sample is then expressed as the integral

$$F(V) = \int_{-\infty}^{+\infty} W(Y)G(V, Y)\,dY \tag{18}$$

This is known as *Tung's integral equation* after the author's pioneering GPC work. Note that, for the sake of generality, the integration limits V_a and V_b have been replaced by $-\infty$ and $+\infty$. The inverse problem is to solve for $W(Y)$ knowing both $F(V)$ and the spreading function $G(V, Y)$.

Calibration of GPC instruments for spreading [i.e., experimental determination of $G(V, Y)$] is not straightforward. Two methods of note are the reversed-flow (39) and recycle (40) techniques, both of which are rather demanding experimentally. In the former, more commonly used method, a sample is injected into the instrument in the normal way, but when the sample peak is half-way through the column, the flow is reversed. On the return to the top of the column, separation due to size exclusion becomes completely canceled, but instrumental spreading effects continue. When the peak reaches the detector, now situated at the top of the column, its width is assumed to be entirely due to instrumental spreading. Tung and Runyon (39) performed reversed-flow experiments for a series of narrow MWD polystyrene standards to obtain a spreading calibration for their instrument, which they chose to formalize as a σ versus V function, where σ is the standard deviation of the flow-reversed Gaussian elution curve (see Fig. 2). The method does not work for skewed $G(V, Y)$, since asymmetry in spreading does not cancel out upon flow reversal.

Since 1970, a wide variety of solutions to the integral equation (eq. 18) have been proposed. Some account only for Gaussian spreading; that is, $G(V, Y)$ has a purely Gaussian form. Others allow for non-Gaussian effects by admitting, for example, a skewing factor. The numerical methods have been reviewed by Friis and Hamielec (41) and evaluated by Silebi and McHugh (42). Friis and Hamielec concluded that the method of Ishige et al. (43), known as "method 2," performed better than any other available at the time. Three analytical solutions have been presented by Hamielec et al. (44–46) and their relative merits discussed by Penlidis et al. (47).

Detector Types and Equations Governing Their Use

Generalities

Very simply, the function of a detector is to continuously monitor the column effluent and to produce a signal somehow proportional to the amount or MW of the polymer in the solvent.

We may represent the normalized spreading function for species Y as $G(V, Y)$ where

$$G(V, Y) = \bar{G}(V, Y)/A(Y) \tag{19}$$

Here, $A(Y)$ is the area under the detector response for species Y [and equals $W(Y)dY$]; $G(V, Y)$ is independent of detector type, but this is not so for $A(Y)$. The detector response for any species depends on the physical quantity being monitored. For example, both DRI and UV detectors produce a signal proportional to the mass concentration of polymer in effluent. For a signal to be produced in the case of DRI, however, there must be a difference in RI between solvent and solution. Likewise in the case of the UV, the polymer must contain one or more chromophones in each individual species.

The detector of choice should be checked for linearity of response over the range of solute concentrations likely to be encountered under typical experimental conditions. This may be carried out by injecting polymer solutions at varying concentrations directly into the detector. The detector signal should vary linearly with concentration. An alternative is to use the whole apparatus and once again inject solutions of different concentrations. In this case, the chromatograms obtained should be normalized.

$$F_{\mathrm{n}}(V)=\frac{F(V)}{\int_{-\infty}^{+\infty} F(V)\,dV} \tag{20}$$

and $F_{\mathrm{n}}(V)$ should be invariant with concentration. In fact, if $F_{\mathrm{n}}(V)$ is concentration invariant at all V, it can also be argued that the separation obeys the principle of linear superposition and that polymer–polymer interactions are negligible. This principle is not obeyed when polymer coil size changes with changing polymer–solute concentration in the columns.

Mass Concentration Detectors (MCD)

Under ideal operating conditions, an MCD response is proportional to the total solute concentration in the detector cell. We can illustrate the steps involved in MWD calculations with reference to Figure 10. Our aim is to convert the chromatogram $F(V)$ into an MWD, $H(M)$, by using the appropriate calibration. If $F(V)$ is to be corrected for spreading, it is customary to do this before performing MWD calculations.

Now, the weight fraction dH of polymer eluting between V and $V-dV$ (Fig. 10*a*), is given by

$$dH=\frac{-F(V)\,dV}{\int_{V_{\mathrm{a}}}^{V_{\mathrm{b}}} F(V)\,dV} \tag{21}$$

that is,

$$\frac{dH}{dV}=\frac{-F(V)}{\int_{V_{\mathrm{a}}}^{V_{\mathrm{b}}} F(V)\,dV} \tag{22}$$

The same fraction dH contains species lying in the molecular weight range

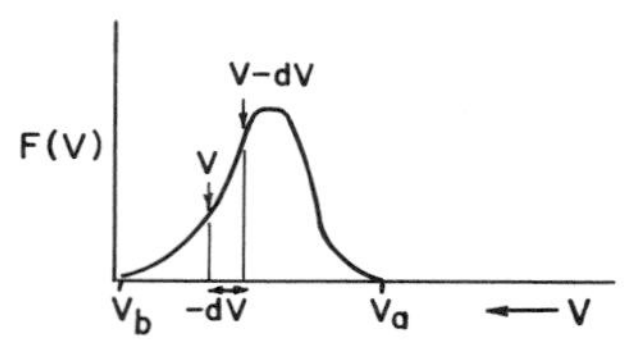

b) CORRESPONDING MWD

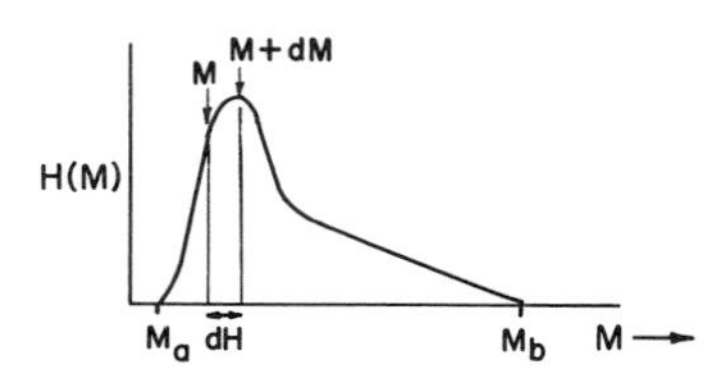

c) APPROPRIATE CALIBRATION CURVE

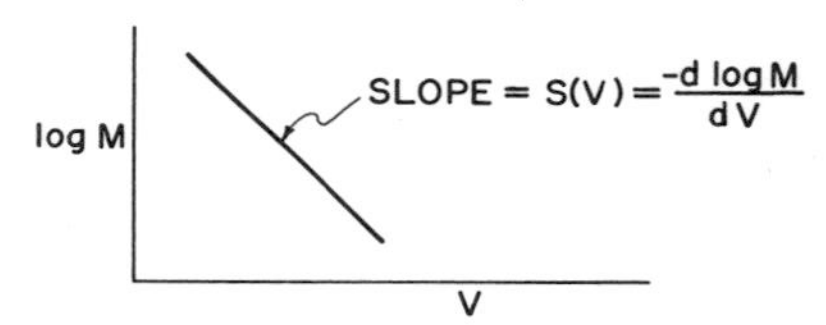

Figure 10. Relationship between the chromatogram and the molecular weight distribution.

M to $M + dM$ (Fig. 10*b*), so we write

$$dH = \frac{H(M)\,dM}{\int_{M_b}^{M_a} H(M)\,dM} \tag{23}$$

that is,

$$\frac{dH}{dM} = \frac{H(M)}{\int_{M_b}^{M_a} H(M)\,dM} \tag{24}$$

Now, dH/dM, the actual MWD, may be obtained using the chain rule of differentiation:

$$\frac{dH}{dM} = \frac{dH}{dV}\frac{dV}{d\log M}\frac{d\log M}{dM} \tag{25}$$

The first term in equation (25) is the normalized chromatogram height at V

(eq. 22), the second is the inverse slope of the calibration curve at V (see Fig. 10c), and the third reduces to the inverse molecular weight at V from the calibration curve. Equation (25) thus reduces to

$$H(M) = F_{\mathrm{n}}(V)\frac{1}{S(V)}\frac{1}{M(V)} \tag{26}$$

If a universal calibration is being used, we are assuming that at any retention volume, hydrodynamic volumes of calibrant (subscript 1) and unknown (subscript 2) are equal according to

$$[\eta]_1 M_1 = [\eta]_2 M_2 \tag{27}$$

Provided the Mark–Houwink constants for both calibrant (K_1 and a_1) and unknown (K_2 and a_2) are available, we compute M_2 from

$$\log M_2 = \frac{1}{1 + a_2}\log\left(\frac{K_1}{K_2}\right) + \frac{1 + a_1}{1 + a_2}\log M_1 \tag{28}$$

The calculations summarized in equations (20)–(28) may be conveniently performed using a computer, together with the molecular weight averages from

$$\bar{M}_x = \frac{\int_0^\infty H(M)M^\beta dM}{\int_0^\infty H(M)M^{\beta-1}dM} \tag{29}$$

where $\beta = 0$ corresponds to $\bar{M}_n$, $\beta = 1$ to $\bar{M}_w$, $\beta = 2$ to $\bar{M}_z$, and so on.

Note: Values of the Mark–Houwink parameters K and a have been determined for a very wide range of polymers and solvents at various temperatures (see, e.g., Ref. 48).

The LALLSP Detector

To determine molecular weights from light-scattering experiments, we compare the intensity of light scattered at some angle θ or range of angles, with that of the incident beam impinging on the scattering volume. The ratio of these intensities may be written

$$R_\theta = \frac{I' r^2}{I_\theta(1 + \cos^2\theta)} \tag{30}$$

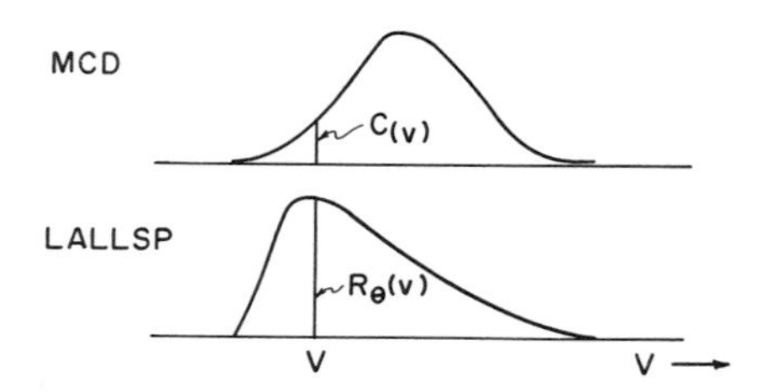

Figure 11. Chromatograms obtained from a dual MCD-LALLSP detector system.

where R_θ is known as the Rayleigh ratio, I_θ and I' are, respectively, the intensity of incident light and of that scattered at angle θ, and r is the distance of the observer from the center of scattering.

In GPC applications, it is customary to pass column effluent sequentially through a LALLSP and an MCD detector. One obtains two chromatograms, as shown in Figure 11. Notice the difference in peak profiles. The LALLSP detector is more sensitive to the relatively few solute molecules of high molecular weight, which explains the apparent skewness toward the low retention volume side. Combination of both detector responses leads to the absolute molecular weight at any chosen value of V across the entire chromatogram, as will be made plain below.

The LALLSP detector gives directly the Rayleigh ratio as a function of V, $R_\theta(V)$. The quantity used to calculate molecular weight at V, however, is the excess Rayleigh ratio $\bar{R}_\theta(V)$:

$$\bar{R}_\theta(V) = R_\theta(V)(\text{sample}) - R_\theta(V)(\text{solvent}) \tag{31}$$

In the conditions of the GPC experiment, light scattered at very low forward angles (typically $\theta = 5\text{–}7°$) is examined; thus the intramolecular interference correction (extrapolation to zero angle customarily applied in classical "static" light-scattering experiments) is considered to be unnecessary (49). Then $\bar{R}_\theta(V)$ is related to $M(V)$ through

$$\frac{Kc(V)}{\bar{R}_\theta(V)} = \frac{1}{M(V)} + 2A_2(V)c(V) \tag{32}$$

or

$$M(V) = \left[\frac{Kc(V)}{\bar{R}_\theta(V)} - 2A_2(V)c(V)\right]^{-1} \tag{33}$$

where $c(V)$ is the polymer concentration at V, K is an optical constant, and $A_2(V)$ is the second virial coefficient; $c(V)$ can be calculated from the MCD response if the chromatogram is divided into volume increments of width dV. Then, from the injected mass of polymer m and the fraction $F(V)\,dV/\int_0^\infty F(V)\,dV$ of the total peak area represented by the increment, we have

$$c(V)\,dV = \frac{mF(V)\,dV}{\int_0^\infty F(V)\,dV} \tag{34}$$

The optical constant K varies from instrument to instrument, according to the light source (plane polarized or not) and instrumental geometry. A typical form (50), however, is

$$K = \frac{2\Pi^2 n^2}{\lambda_0^4 N_A}\left(\frac{dn}{dc}\right)^2 (1 + \cos^2\theta) \tag{35}$$

where n is the solution refractive index, dn/dc is the specific refractive index increment, N_A is Avogadro's number, and λ_0 is the wavelength of the light *in vacuo*.

Finally, the second virial coefficient (A_2) is a quantity which is a complex function of temperature, pressure, MW, size, and the thermodynamic parameters of the particular polymer–solvent system (51). The pertinent value of A_2 may be found in the literature; alternatively, it may have to be determined from a static MW analysis of the polymer in the solvent used for chromatography. In many cases A_2 may be assumed to be independent of MW (hence V) without introducing significant error. However, when samples of wide MWD are considered this is often not valid (52). From static analysis of narrow MWD fractions of the polymer of interest, we obtain the constants α and β in the equation

$$A_2(M) = \beta M^{-\alpha} \tag{36}$$

where $A_2(M)$ may be used interchangeably with $A_2(V)$. The appropriate value of $A_2(V)$ is used to obtain $M(V)$ from equation (33). In absolute terms, however, the magnitude of the contribution of $A_2(V)c(V)$ to $M(V)$ in equation (33) is relatively small, owing to the low polymer concentrations in GPC effluents. It becomes most important around the peak maximum, where $c(V)$ is at its highest.

It is clear that the LALLSP-MCD system requires no calibration. Once K, $c(V)$, and $A_2(V)$ are known, equation (33) is used to calculate the absolute value of M across the entire chromatogram, hence leading to $H(M)$, the

MWD. Sample molecular weight averages may then be calculated from equation (29).

The Viscometric Detector (VISC)

As mentioned in the first section, only very recently has a continuous VISC detector with a reportedly high signal-to-noise ratio become commercially available. Its incorporation into GPC instrumentation is very much akin to that of the LALLSP in that it is used simultaneously with an MCD.

The VISC detector continuously monitors the pressure drop P of the column effuent as it flow through a capillary bridge in the instrument. Assuming laminar flow obeying Poiseuille's law, P is proportional to the viscosity η of the medium flowing through the capillary:

$$P = \frac{8Ql\eta}{\Pi r^4} \tag{37}$$

where Q is the flow rate and l and r are, respectively, the length and radius of the capillary tube. Since the concentration of polymer in the column effluent is exceedingly low, the intrinsic viscosity is in fact continuously monitored by the VISC:

$$[\eta](V) = \frac{1}{c(V)} \ln\left(\frac{P(V)}{P_0}\right) \tag{38}$$

where $c(V)$ and $P(V)$ are, respectively, the concentration and pressure drop of solution eluting from the column at retention volume V, and P_0 is the pressure drop of the pure solvent. Chromatograms obtained from a dual VISC-MCD system are very similar in outline to those obtained from the LALLSP-MCD combination schematized in Figure 11. The difference is that the VISC chromatogram height is $P(V)$, rather than $R_\theta(V)$. Once again, the VISC detector is skewed to the high MW (low V) side owing to its greater sensitivity to high MW solutes. The value of $c(V)$ is determined from the MCD detector in the customary manner by use of equation (34).

If the Mark–Houwink constants of the eluting polymer are known, the VISC becomes an absolute detector in the same sense as the LALLSP, since

$$[\eta](V) = KM(V)^a \tag{39}$$

According to our definitions at the close of the first section, polymers with long chain branching (LCB) fall strictly into the class of "complex" polymers, giving rise to nonideal GPC behavior, which is discussed separately, below. It is pertinent to mention polymers with LCB at this juncture, however, since one of the more powerful properties of the MCD-VISC system is its potential for estimating degrees of LCB from a single experiment. Lecacheux et al. have demonstrated the utility of such a system in measuring LCB frequencies λ in polyethylenes (53) and in copolymers of ethylene and vinyl acetate (54).

Polymers with LCB have, by virtue of their more compact conformation, smaller hydrodynamic volumes than their linear counterparts. In a GPC experiment, a branched polymer will therefore elute at higher V than a linear one having the same MW. Qualitative information on whether a polymer is branched could thus be obtained by simple inspection of an MCD chromatogram. Quantitative evaluation of λ, however, requires measurement of intrinsic viscosities and some model relating $[\eta]$ to λ. A number of models exist, but in each case, the starting point is the structure parameter g':

$$g' = [\eta]_b/[\eta]_l \tag{40}$$

where the subscripts b and l refer to branched and linear species having the same MW respectively, and $[\eta]_l$ is given by

$$[\eta]_l(V) = K_l M_l(V)^{a_l} \tag{41}$$

The $[\eta]$ is continuously monitored across the chromatogram of a whole branched polymer sample, and we can evaluate g' over very narrow increments of V or simply obtain an average value for the whole sample if desired.

It is necessary to assume a relationship between g' and g, the ratio of mean square radii of gyration of branched and linear samples of the same MW (55)

$$g' = \langle \bar{r}_g^2 \rangle_b / \langle \bar{r}_g^2 \rangle_l \tag{42}$$

A frequently used relationship (53) is

$$g' = g^{1.2} \tag{43}$$

Finally, the number of branches per macromolecule n is obtained through one of the Zimm–Stockmayer relationships (56), for example,

$$g = \frac{3}{2}\left(\frac{\Pi}{n}\right)^{1/2} - \frac{5}{2n} \tag{44}$$

Then λ is calculated from

$$\lambda(V) = \frac{n(V)}{M(V)} \tag{45}$$

Values for λ in polyethylene reported in Reference 53 compare favorably with those obtained from a longer established technique, ^{13}C NMR spectroscopy, (57), and with those obtained using GPC with a discontinuous-aliquot viscometric detector (58). One significant advantage of the continuous VISC-MCD over the latter system is, however, a fourfold decrease in analysis time. The technique is in general limited in the sense that only in a few cases are certified linear analogues of branched polymers available. Another theoretical objection is that the Zimm–Stockmayer models (56) strictly apply only when comparing branched and linear polymers of the same molecular weight. In GPC, however, at any one time the detector cell contains solutes having the same hydrodynamic volumes (if instrumental spreading is negligible) but possibly very different molecular weights. (This applies unless the whole polymer sample is known to have exactly the same degree of LCB across the entire range of MWs, which in reality is never the case.) It is not clear how the Zimm–Stockmayer branching models could be applied rigorously to such a mixture.

As pointed out by Hamielec and Meyer (59), the best possible detector system to employ for polymers with LCB would be a combined MCD-LALLSP-VISC, which would provide complete information on the polymer MWD and MW dependence of λ.

Data Acquisition and Analysis

Strip-Chart Raw Chromatogram (MCD Response)

The simplest and most common form of data acquisition makes use of a strip chart. The American Society for Testing and Materials has produced a standard method, the ASTM Standard Method for Polystyrene (ANSI/ASTM D3536-76), which gives the recommended calculational procedure for this case. Details are well presented by Yau et al. (4a). We consider here a few salient features.

Correct identification of the baseline and integration limits is imperative if accurate MW averages are to be obtained. At the high MW end of the chromatogram (V_a in Fig. 9), there is normally little difficulty. However, many chromatograms are skewed to the high V side and sometimes have a pronounced low MW "tail" due, for example, to impurities. This can be troublesome with regard to the accurate determination of $\bar{M}_n$, which is influenced strongly by any low MW component in the sample.

Having defined a baseline and integration limits, the recommended procedure is then to use equal intervals in V (constant ΔV) and take at least 50 chromatogram heights $F(V)$ across the whole detector response—more in the case of samples having a broad MWD. The integration of the chromatogram to obtain the desired MW averages and distributions is then performed numerically (e.g., using a lower order Simpson's rule). If at least three significant figures are employed in measuring individual $F(V)$'s, errors in MW averages due to numerical calculation procedures should the negligible ($< 2\%$).

Fully Automated Systems

Particularly when two or more detectors are incorporated in series in a GPC instrument, the amount of information accumulated on strip-chart recorders would make subsequent manual chromatogram height measurements particularly laborious. The form of a GPC chromatogram lends itself well to automated data acquisition and analysis, and a wide range of automated systems with single as well as dual detectors are available from the instrument manufacturers. A real-time, multichannel A–D converter is employed to convert the detector signals to digital (chromatogram height) form. Such systems are more accurate than manual data handling, since noise in detector responses can be averaged out and a larger number of heights per chromatogram can be collected without difficulty. The interested reader can consult References 60 and 61 for details of data handling systems employed with MCD-LALLSP detectors and Reference 62, which describes a system applicable to an MCD-VISC combination.

NONIDEALITIES

We shall now discuss factors which give rise to non-size-exclusion effects in the basic mechanism of GPC as well as other phenomena which detract from proper quantitative analysis of GPC elution curves. Because gel permeation chromatography of water-soluble polymers is plagued with such problems, it is appropriate to devote a separate section to this field.

GPC of Water-Soluble Polymers

Although the first successful reported GPC experiments were performed in an aqueous system (1), the development of aqueous GPC to the point of being able to yield rapid, accurate estimates of MW and MWD, as is feasible for a multitude of organic systems, has been very slow. The main reason is that superimposed on the basic size exclusion mechanism are numerous nonsteric effects that arise because for column-packing materials to be compatible with aqueous eluants, charged and/or polar groups are required at the packing surface. Such polar groups also confer water solubility on macromolecules.

Interactions between these groups give rise to the nonsteric effects, which fall into five categories. Specifically affecting ionic solutes are ion exchange, ion exclusion, ion inclusion (sometimes referred to as the Donnan effect), and the ionic strength dependence of polyelectrolyte chain dimensions. Adsorption to the packing material is a phenomenon which also occurs for nonionic solutes. Problems arising herefrom are briefly discussed below. The effects are well explained and documented in several reviews articles covering the field of aqueous GPC (5, 63–67).

(i) *Ion Exchange.* Sephadex (68), agarose and polyacrylamide (69), and particularly chemically unmodified silica packings (70, 71) have all been shown to display ion-exchange properties in which ionic groups at the packing surface exchange with those along the backbone of the polyelectrolyte chain. The siliceous packings behave as weak cation exchangers at $pH \geqslant 4$. Various devices which eradicate the problem have been reported—for example, using eluants of $pH < 4$ (72) or adding electrolyte to the eluant to compete for anionic sites (73). Cationic modifiers have been shown to solve simultaneously this problem and that of adsorption of certain types of polymer to siliceous packing (74).

(ii) *Ion Exclusion.* If the surface of a packing has net charge, solutes of similar charge will be repelled from the surface because of electrostatic repulsion. Rochas et al. (75) performed GPC experiments using simple electrolytes on columns packed with borosilicate gels and observed that when distilled water was used as eluant, the salt peak always eluted at V_0, the void volume of the column, instead of at V_t, which would be the expected elution volume from a purely size exclusion viewpoint. This condition suggested complete exclusion from the pores and was overcome by addition of electrolyte to the eluant to screen the electrostatic repulsions. An ionic strength of $50\,\text{mmol/dm}^3$ was required to completely overcome this effect.

(iii) *Salt Sensitivity of Polyelectrolyte Chain Dimensions.* It is well established (76–78) that the introduction of ionizing groups to a polymer chain causes changes in molecular size to an extent that depends on the degree of ionization and on the ionic strength of the solvent. A number of theories have been proposed, which quantify the effect of ionic strength on such physically measurable quantities as intrinsic viscosity. The theories were reviewed by Noda et al. (79) in 1970. The most successful ones predict proportionality between $[\eta]$ and $I^{-1/2}$(80–84); that is, for a polyelectrolyte having a fixed degree of ionization, the following relationship holds:

$$[\eta] = [\eta]_{\infty} + kI^{-1/2} \tag{46}$$

where $[\eta]_{\infty}$ is the intrinsic viscosity at the limit of infinite ionic strength (electrostatic repulsions between charged groups on the backbone fully suppressed) and I is the solvent ionic strength. This relationship has been shown to hold in numerous instances (see, e.g., Refs. 76–78 and 85). This salt sensitivity of chain dimensions has important ramifications in the GPC of polyelectrolytes. For a given polymer, V will increase with I as a result of the decrease in coil dimensions. The problem of calibration becomes hereby further complicated in that both polymer chain composition and degree of ionization determine the overall chain dimensions, as discussed further below.

(iv) *Ion Inclusion.* It has been observed that when a polyelectrolyte is eluted through a column in an eluant containing salt, two peaks occur (75, 86): one at low V due to the polyelectrolyte and one at high V due to electrolyte in the eluant. As eloquently explained by Tanford (87), the packing material acts as a semipermeable membrance (permeable to small ions; impermeable to macro-ions), permitting the establishment of a so-called Donnan equilibrium. As the band of eluant containing polyelectrolyte solute travels down the column, the activity of small ions outside the pores becomes higher because the overall ionic concentration is higher. Some of the small ions are thus driven into the pores by the difference in activity. They become eluted later than the solute, giving rise to the "Donnan peak." Great care must be taken in such instances not to confuse this peak with that which may be due to polyelectrolyte of low MW.

If the eluant contains no salt, and provided ion exclusion effects are absent, it is the smaller, more permeating polyelectrolyte molecules from within the sample which become retarded on the column in this manner, producing a spurious tail on the low end of the MWD.

(v) *Adsorption.* Adsorptive effects occur in the cases of both ionic and nonionic solutes. They may be either physical or chemical, and a variety of methods have attempted to circumvent the accompanying difficulties. Adsorption can arise from hydrogen bonding and from hydrophobic (88, 89) and ionic interactions. The latter render GPC work with cationic polyelectrolytes particularly difficult, since the majority of packings for aqueous GPC acquire a net negative charge under chromatographic conditions. Polycations frequently become irreversibly adsorbed on the packings. For complex macromolecules such as proteins, several mechanisms of adsorption may be present. For solutes of high MW, the problem can be compounded by the possibility of multiple contacts as well as cooperative effects.

Two basic approaches are used to combat the problem. The first is to include in the eluant some species which preempts adsorption of the solute. This can be either an oligomer of the same or similar chemical nature to the solute under investigation, or some surface-active compound. For example, by including in the aqueous eluant system polyvinylpyrrolidone

(MW $\simeq$ 10,000 g/mol), Letot et al. (90) observed that PEOs and polydextrans could be chromatographed on porous silica glasses without adsorption. It is important, however, that the additive in the eluant not interact in any significant way with the solute species. For example, Tanford et al. (91, 92) showed that the guanidine hydrochloride added to their eluant caused denaturation of the globular proteins under investigation to give linear, random coils.

The second method of dealing with adsorption, widely used in the case of porous glasses, is to chemically bond some adsorption-reducing species to the packing surface. This generally results in a more stable coating than when physically bonding additives are incorporated into the eluant. Regnier et al. (93, 94) where the first to introduce a chemically bonded glyceryl propylsilyl silica packing for aqueous GPC. Such packings, called glycophase-bonded supports, were first produced by reacting glycidoxypropyl trimethoxysilane with a borosilicate glass (93). Silica supports for GPC subsequently became available (94). Use of such packings permitted chromatography of a number of proteins and polynucleotides with high sample recoveries in the absence of modifiers in the eluant.

Each of the above-mentioned effects can vary quite markedly in relative magnitude from system to system. The basic tactics employed in achieving successful separation of species on a "size-only" basis involve the tailoring of chromatographic conditions in terms of choice of packing materials, eluants and modifiers to suit the particular case.

Calibration in Aqueous GPC

Given the foregoing remarks, it may be surprising that elution can be achieved at all in aqueous systems. It is true in many cases that chromatograms are distorted to the point of virtual uselessness or, at best, yield only qualitative information about MW and MWD (e.g., showing that a distribution is bimodal). However, a number of systems have been examined in which proper calibration has proven possible. The simplest approach, as in organic GPC, is to optimize conditions for a given packing–solvent–solute combination to obtain a straightforward peak position calibration. This has been done by Letot et al. (90) for PEOs, by Klein and Westerkamp (95) for polyacrylamides (PAM), and by Malawer et al. (96) for polyvinylpyrrolidones (PVP). A similar methodology has been adopted by Hamielec et al., who have chromatographed a wide range of both nonionic and anionic water-soluble polymers on porous borosilicates glasses—for example, PAMs (97), polydextrans (98), and sodium poly(styrene sulfonates) (NaPSS) (99). Stickler et al. (100, 101) have recently reported the efficacy of GPC in obtaining accurate values of MW and MWD for fairly broad fractions of the cationic polyelectrolyte poly(2-trimethylammonium ethyl methacrylate chloride) (PTMAC). Absolute determinations of MW and MWD were made by light-scattering, viscometry, and ultracentrifugation analyses. GPC yielded comparable results using a specially modified silica packing and an eluant of high ionic strength to screen the greater part of the electrostatic repulsions.

Many investigators have attempted to investigate the applicability of the "universal" ($[\eta]$M) calibration procedure to aqueous systems. Kato et al. (102) chromatographed PEO, dextran, and pullulan standards on TSK Gel PW and found the universal procedure to be entirely appropriate for these nonionic polymers. Studies involving ionic polymers have also been undertaken, many of which report once again that $[\eta]M$ is an appropriate calibration parameter. Most such investigations have involved the elution of both nonionics (chain dimensions insensitive to salt) and polyelectrolytes (chain dimensions very sensitive to salt) from the same column set. The fundamental assumption behind such experiments is that for each packing–solvent–solute system there exists a certain ionic strength above which both polyelectrolyte expansion and electrostatic repulsions between column packings and solute molecules become insignificant. Once such conditions have been achieved, it becomes permissible to compare directly the elution behavior of nonionics and of polyelectrolytes. Several groups of workers have used this methodology. Spatorico and Beyer (103) examined the behavior of samples of NaPSS, dextran, and copolymers of acrylic acid and ethyl acrylate. Columns were packed with borosilicate glasses and eluants containing Na_2SO_4 at $0.2M$ and $0.8M$ were chosen. Intrinsic viscosities were measured in the same solvents. In both cases, single linear plots of $\log[\eta]M$ against V were obtained for the various polymer types; thus the universal calibration procedure was deemed to be applicable.

Rochas et al. (104) have used unmodified silica gels to examine the elution of NaPSS, dextrans, and sodium polyglutamates. When aqueous $0.1M$ $NaNO_3$ was used as eluant, universal calibration was reportedly successful. In a more recent report from the same laboratory (105), silica gels suitably modified to render them compatible with cationics (quaternary methyl ammonium derivatized) were used to study a very wide variety of solute types, including dextrans, PEOs, poly(trimethylacrylic acetate), polyvinylpyridine, and quaternized amylopectin. Once again, $[\eta]M$ was deemed to be an appropriate calibration parameter. However, for almost every report of a successful use of $[\eta]M$ in calibrating for water-soluble polymers, the literature contains a failure. Talley and Bowman (106), for example, using nominally identical gels to those mentioned in Reference 105, could not get polyvinylpyridines and dextrans to fall on a common plot. Bose et al. (107) used a support of Sepharose CL-6 to separate NaPSS and dextran samples over a wide range of ionic strengths, using both NaCl and NaOH, separately, as the simple electrolytes. The universal calibration procedure was invalid for all experimental conditions reported here. One of the authors of this chapter has made an extensive study of the hydrodynamic properties of NaPSSs, including GPC behavior (85). Although universal calibration was valid for comparing the elution of polystyrene standards in THF solvent from the same column set (85, 108), the data for NaPSSs could not be made to fit on the same line. Such conflicing reports are indicative of the high degree of uncertainty pervading the field of aqueous GPC.

Measurement of Retention Volume in Aqueous GPC

In aqueous GPC, use of the siphon counter must sometimes be avoided or approached with some reservations. Problems arise from differences between the surface tension of pure eluant and that of eluant containing solute, as it issues from the detector into the siphon counter. As the surface tension varies, so does the volume of liquid comparising each count of the siphon. Such effects are much less marked with organic solvents. It has also been observed that surface contamination of the siphon by sample occurs with certain water-soluble polymers. Wolkoff and Larose (109) have reported the use of a simple device to overcome such problems. As the eluant leaves the detector, it is fed straight into a sealed container filled with an organic liquid such as inhibisol. The inhibisol is displaced as the eluant trickles in and is routed to the siphon counter, which now comes in contact only with pure liquid.

Concentration and Related Effects

Dependence of Refractive Index on MW

All MCDs are in some way sensitive to monomer units in the polymer backbone, and to chain ends which are chemically different from the main backbone units (e.g., in polyethylene backbone units are $—CH_2—$, whereas chain ends are $—CH_3$) and thus give a different detector response. Chain end effects are generally negligible for high MW polymers (since chain ends are "dilute" in such cases) but can cause considerable problems when one is examining oligomers. One study (110) has shown that the detector response alters significantly for polystyrenes when $M < 5000$ g/mol using a differential refractometer as MCD. In fact, several authors have demonstrated the MW dependence of the refractive index of various polymer solutions, including polystyrenes (111), polyethylenes (112), and PEOs (113).

Overloading Effects

The overloading phenomenon manifests itself as a substantial loss in column efficiency and resolution, due to excessively high sample concentration and/or

MW. Several individual causes have been isolated. Viscous fingering (114) is the name given to the effect arising when localized sample viscosity is so high that proper pore penetration is prevented. Solute hydrodynamic volume has been shown to be a function of concentration (115). Coil dimensions tend to decrease with increased concentration, hence crowding. Further more, at high concentration, partition of solute between the gel and mobile solvent may not be at equilibrium (116). To prevent problems of this kind, a rule of thumb is that an injected sample solution should have a viscosity no greater than twice that of the mobile phase (72).

One major practical problem arising from overloading effects is the difficulty in obtaining an accurate MW calibration for solutes of high MW. Coil size for such species is far more sensitive to concentration than for low MW species. The chromatographic peak profile is thus more highly dependent on the sample concentration. Usually, high MW solutes show skewed $G(V, Y)$ as a result of this effect, which remains to date an important unsolved problem in the GPC of high MW polymers.

Degradation effects

GPC experiments have shown that polystyrenes of high MW degrade significantly at high solvent flow rates (116, 117). This is most probably caused by hydrodynamic forces (elongation and shear) acting on the polymer chains. Reference 118 shows that for a 7.1×10^6 MW polystyrene, V increases with flow rate on a column of porous silica (diameter $\simeq 8\ \mu$m), whereas standards of lower MW have a fairly constant V under the same conditions. When the effluent from the 7.1×10^6 material was collected, then reinjected, the chromatogram was indicative of a material of lower MW and broader MWD than the original, showing that the starting material had indeed undergone degradation during the first separation. It is noteworthy that the sample may undergo shear degradation during filtration, as well as during its passage through the column.

Complex Polymers

In the context of GPC, a complex polymer is one for which a unique relationship between hydrodynamic volume and MW does not exist. Examples include heterogeneous copolymers, homo- and copolymers with LCB, and mixtures of polymers of different types.

Since GPC separates macromolecules according to size, under conditions of negligible peak spreading the contents of the detector cell at any retention volume will be uniform in size. However in the case of complex polymers, uniformity of size does not mean uniformity of MW. There is thus an infinity of calibration curves, one for each species of polymer of fixed molecular architecture. It is thus impossible, given the current state of the art in GPC, to obtain a true MWD for complex polymers. Despite this, it may be feasible using multidetector systems to obtain some useful information about the deviation of the "complex" polymer from "simple" behavior. We can define the polydispersity index at retention volume V of the detector cell contents as

$$P(V) = \bar{M}_w(V)/\bar{M}_n(V) \tag{47}$$

If $P(V)$ is unity, the solute in the detector cell is a simple polymer. Values higher than unity would be indicative of complex behavior. Taking the specific example of polymers with LCB synthesized by free-radical polymerization, one would expect a higher $P(V)$ at low V in a given sample, since the degree of branching increases normally with conversion and MW.

The LALLSP-VISC-MCD system can provide measurements of $[\eta](V)$ and $\bar{M}_w(V)$ directly. As was shown by Hamielec and Ouano in 1978 (119), the correct MW average to use in generalized universal calibration is $\bar{M}_n$. Hence,

$$[\eta](V)\bar{M}_n(V) = [\eta]_{\text{ref}}(V)M_{\text{ref}}(V) \tag{48}$$

where the subscript ref denotes the reference polymer used in establishing the universal calibration. Equation (48) could then be used to find $\bar{M}_n(V)$. In this manner, we might obtain a value of $P(V)$ across any portion of the chromatogram we desire.

CONCLUDING REMARKS

We hope to have conveyed in this chapter an idea of the importance of gel permeation chromatography as the preferred routine analytical tool for polymer characterization in general and for the determination of molecular weight distributions in particular. We have tried not to obscure the essential simplicity of the technique with too many equations. In addition, we hope to have conveyed to the reader the power of GPC for analyzing what we have defined as "simple" polymers (the "ideal" case), as well as the limitations regarding "complex" polymers (the "nonideal" case).

ACKNOWLEDGMENTS

The authors are grateful to the Natural Sciences and Engineering Research Council of Canada for financial support. MGS is particularly indebted to the

Science and Engineering Research Council of the United Kingdom for the award of an Overseas Postdoctoral Fellowship.

REFERENCES

1. J. Porath and P. Flodin, *Nature*, **183**, 1657 (1959).
2. J. C. Moore, *J. Polym. Sci.*, A, **2**, 835 (1964).
3. G. L. Hagnauer, *Anal. Chem.*, **54**, 265R–276R (1982).

4a. W. W. Yau, J. J. Kirkland, and D. D. Bly, *Modern Size-Exclusion Liquid Chromatography: Practice of Gel Permeation and Gel Filtration Chromatography*, Wiley, New York, 1979.

4b. J. Janca, Ed., *Steric Exclusion Liquid Chromatography of Polymers*, Dekker, New York, 1984.

4c. S. T. Balke, *Quantitative Column Liquid Chromatography: A Survey of Chemometric Methods*, Elsevier, Amsterdam, 1984.

5. H. G. Barth, *J. Chromatogr. Sci.*, **18**, 409 (1980).
6. H.-J. Cantow, E. Siefert, and R. Kuhn, *Chem.-Eng.-Tech.*, **38**, 1032 (1966).
7. A. C. Ouano and W. Kaye, *J. Polym. Sci., A-1*, **12**, 1151 (1974).
8. G. Meyerhoff; *Makromol. Chem.*, **118**, 265 (1968).
9. D. Goedhart and A. Opschoor, *J. Polym. Sci., A-2*, **8**, 1227 (1970).
10. G. Meyerhoff, *Sep. Sci.*, **6**, 239 (1971).
11. A. C. Ouano, *J. Polym. Sci., A-1*, **10**(7), 2169 (1972).
12. M. A. Haney, U.S. Patent No. 4, 463, 598 (1984).
13. M. A. Haney, *Am. Lab.*, **17**, 41 (1985).
14. M. A. Haney, *Am. Lab.*, **17**, 116 (1985).
15. W. W. Yau, H. L. Suchan, and C. P. Malone, *J. Polym. Sci., A-2*, **6**, 1349 (1968).
16. J. C. Giddings, E. Kucera, C. P. Russell, and M. N. Myers, *J. Phys. Chem.*, **72**, 4397 (1968).
17. T. C. Laurent and J. Killander, *J. Chromatogr.*, **14**, 317 (1964).
18. A. G. Ogston, *Trans. Faraday Soc.*, **54**, 1754 (1958).
19. M. E. van Kreveld and N. van den Hoed, *J. Chromatogr.*, **83**, 111 (1973).
20. E. F. Casassa, *J. Phys. Chem.*, **75**, 3929 (1971).
21. E. F. Casassa, *Sep. Sci.*, **6**, 305 (1971).
22. E. F. Casassa, *Macromolecules*, **9**, 182 (1976).
23. P. J. Flory and T. G. Fox, *J. Am. Chem. Soc.*, **73**, 1904 (1951).
24. H. Benoit, Z. Grubisic, and R. Rempp, *J. Polym. Sci., B*, **5**, 753 (1967).
25. H. Coll and P. Prusinowsky, *J. Polym. Sci., B*, **5**, 1153 (1967).
26. O. B. Ptitsyn and Y. E. Eizner, *Sov. Phys. Tech. Phys.* (*Eng. trans.*), **4**, 1020 (1960).
27. J. V. Dawkins, *J. Macromol. Sci., B*, **2**, 263 (1968).

28. J. V. Dawkins, R. Denyer, and J. W. Maddock, *Polymer*, **10**, 154 (1969).
29. J. V. Dawkins, *Eur. Polym. J.*, **6**, 831 (1970).
30. J. V. Dawkins, J. W. Maddock, and D. Coupe, *J. Polym. Sci., A-2*, **8**, 1803 (1970).
31. J. V. Dawkins and J. W. Maddock, *Eur. Polym. J.*, **7**, 1537 (1971).
32. J. V. Dawkins and M. Hemming, *Makromol. Chem.*, **155**, 75 (1972).
33. J. V. Dawkins, J. W. Maddock, and A. Nevin, *Eur. Polym. J.*, **9**, 327 (1973).
34. J. V. Dawkins, *Eur. Polym. J.*, **13**, 837 (1977).
35. H. Coll, *Sep. Sci.*, **5**, 273 (1970).
36. S. T. Balke, A. E. Hamielec, B. P. LeClair, and S. L. Pearce, *Ind. Eng. Chem., Proc. Res. Dev.*, **8**, 54 (1969).
37. W. W. Yau, H. J. Stoklosa, and D. D. Bly, *J. Appl. Polym. Sci.*, **21**, 1911 (1977).
38. W. W. Yau, H. J. Stoklosa, C. R. Ginnard, and D. D. Bly, paper P013, presented at the 12th Middle Atlantic Regional Meeting of the American Chemical Society, April 1978.
39. L. H. Tung and J. R. Runyon, *J. Appl. Polym. Sci.*, **13**, 2397 (1969).
40. J. L. Waters, *J. Polym. Sci., A-2*, **8**, 411 (1970).
41. N. Friis and A. E. Hamielec, *Adv. Chromatogr.*, **13**, 41 (1975).
42. C. A. Silebi and A. J. McHugh, *J. Appl. Polym. Sci.*, **23**, 1699 (1979).
43. T. Ishige, S. I. Lee, and A. E. Hamielec, *J. Appl. Polym. Sci.*, **15**, 1607 (1971).
44. A. E. Hamielec and S. Singh, *J. Liq. Chromatogs.*, **1**(2), 187 (1978).
45. A. Husain, A. E. Hamielec, and J. Valchopoulos, *J. Liq. Chromatogr.*, **4**(3), 425 (1981).
46. A. Husain, J. Valchopoulos, and A. E. Hamielec, *J. Liq. Chromatogr.*, **2**(2), 193 (1979).
47. A. Penlidis, A. E. Hamielec, and J. F. MacGregor, *J. Liq. Chromatogr.*, **6**(S-2), 179 (1983).
48. J. Brandrup and E. H. Immergut, Eds., *Polymer Handbook*, 2nd ed., Wiley, New York, 1975.
49. Application Note LS-1, Chromatix Corporation (suppliers of LALLSP instruments), 1978.
50. R. C. Jordan, *J. Liq. Chromatogr.*, **3**(3), 439 (1980).
51. P. J. Flory, *Principles of Polymer Chemistry*, Cornell University Press, Ithaca, NY, 1953.
52. A. C. Ouano, *J. Chromatogr.*, **118**, 303 (1976).
53. D. Lecacheux, J. Lesec, and C. Quivoron, *J. Appl. Polym. Sci.*, **27**, 4867 (1982).
54. D. Lecacheux, J. Lesec, L. Quivoron, R. Prechner, R. Panaras, and H. Benoit, *J. Appl. Polym. Sci.*, **29**, 1569 (1984).
55. B. H. Zimm and R. W. Kilb, *J. Polym. Sci.*, **37**, 19 (1959).
56. B. H. Zimm and W. H. Stockmayer, *J. Chem. Phys.*, **17**, 1301 (1949).
57. F. A. Bovey, F. C. Schilling, F. L. McCrackin, and H. L. Wagner, *Macromolecules*, **9**, 76 (1976).

58. Société Nationale Elf Aquitaine (Production), 64170 Lacq, France.
59. A. E. Hamielec and H. Meyer, "On-line Molecular Weight and Long-Chain Branching Measurement Using SEC and Low-Angle Laser Light Scattering," in *Developments in Polymer Characterization*, Vol. 5, Applied Science Publishers, Banking, 1985.
60. M. L. McConnell, *Am. Lab.*, **10**(5), 63 (1978).
61. Application Notes LS-2, LS-3 and LS-5; Chromatix Corporation, 19–1979.
62. M. E. Koehler, A. F. Kah, T. F. Niemann, C. Kuo, and T. Provder, *Prepr. Org. Coat. Plast. Chem., Div. Am. Chem. Soc.*, **48** (1983).
63. A. E. Hamielec and M. G. Styring, *Pure Appl. Chem.*, **57**, 955 (1985).
64. A. R. Cooper and D. S. VanDerveer, *J. Liq. Chromatogr.*, **1**(5), 693 (1978).
65. P. L. Dubin, Sep. *Purif. Methods.*, **10**(2), 287(1981).
66. P. L. Dubin, *Am. Lab.*, **15**(1), 62, 64–67, 70–73 (1983).
67. J. E. Rollings, A. Bose, J. M. Caruthers, G. T. Tsao, and M. R. Okos, *Adv. Chem. Ser.*, **203**, 345–360 (1983).
68. P. A. Neddermeyer and L. B. Rogers, *Anal. Chem.*, **40**, 755 (1968).
69. A. R. Cooper and D. P. Matzinger, *J. Appl. Polym. Sci.*, **23**, 419 (1979).
70. D. N. Strazhesko, V. B. Strelko, V. N. Belzakov, and S. C. Rubanik, *J. Chromatogr.*, **102**, 191 (1974).
71. K. K. Unger, *Porous Silica*, Elsevier, Amsterdam, 1979.
72. L. R. Snyder and J. J. Kirkland, *Introduction to Modern Liquid Chromatography*, 2nd ed., Wiley, New York, 1979.
73. T. Mizutani and A. Mizutani, *Anal. Biochem.*, **83**, 216 (1977).
74. F. A. Buytenhuijs and F. P. B. van der Maeden, *J. Chromatogr.*, **149**, 489 (1978).
75. C. Rochas, A. Domard, and M. Rinaudo, *Eur. Polym. J.*, **16**, 135 (1980).
76. S. A. Rice and M. Nagasawa, *Polyelectrolyte Solutions*, Academic Press, New York, 1961.
77. F. Oosawa, *Polyelectrolytes*, Dekker, New York, 1971.
78. A. Eisenberg and M. King, *Ion Containing Polymers*, Vol. 2, *Physical Properties and Structure*, Academic Press, New York, 1977.
79. I. Noda, T. Tsuge, and M. Nagasawa, *J. Phys. Chem.*, **74**, 710 (1970).
80. H. W. Chien and A. Isihara, *J. Polym. Sci., Polym. Phys. Ed.*, **14**, 1015 (1976).
81. H. W. Chien, C. H. Isihara, and A. Isihara, *Polym. J.*, **8**(3), 288 (1976).
82. N. Imai, *Rep. Prog. Polym. Phys. Jpn.*, **23**, 95 (1980).
83. M. Fixman, *J. Chem. Phys.*, **41**, 3772 (1964).
84. F. Oosawa, *Biopolymers*, **6**, 145 (1968).
85. M. G. Styring, Ph.D. thesis, University of Manchester, 1984.
86. B. Stenlund, *Adv. Chromatogr.*, **14**, 37 (1976).
87. C. Tanford, *Adv. Protein Chem.*, **23**, 121 (1968).
88. C. Horvath, W. Melander, and I. Molnar, *J. Chromatogr.*, **125**, 129 (1976).

89. C. Horvath and W. Melander, *Am. Lab.*, **10**(10), 17–18, 21–24, 29–32, 35–36 (1978).
90. L. Letot, J. Lesec, and C. Quivoron, *J. Liq. Chromatogr.*, **4**(8), 1311 (1981).
91. W. W. Fish, K. G. Mann, and C. Tanford, *J. Biol. Chem.*, **244**, 4989 (1969).
92. C. Tanford, *Adv. Protein Chem.*, **23**, 121 (1968).
93. F. E. Regnier and R. Noel, *J. Chromatogr. Sci.*, **14**, 316 (1976).
94. S. H. Chang, K. M. Gooding, and F. E. Regnier, *J. Chromatogr.*, **125**, 108 (1976).
95. J. Klein and A. Westerkamp, *J. Polym. Sci., Polym. Chem. Ed.*, **19**, 707 (1981).
96. E. G. Malawer, J. K. De Vasto, S. P. Frankoski, and A. J. Montana, *J. Liq. Chromatogr.*, **7**(3), 441 (1984).
97. C. J. Kim, A. E. Hamielec, and A. Benedek, *J. Liq. Chromatogr.*, **5**, 1277 (1982).
98. C. J. Kim, A. E. Hamielec, and A. Benedek, *J. Liq. Chromatogr.*, **5**, 425 (1982).
99. S. N. E. Omorodion, A. E. Hamielec, and J. L. Brash, *J. Liq. Chromatogr.*, **4**, 1903 (1981).
100. M. Stickler and F. Eisenbeiss, *Eur. Polym. J.*, **20**(9), 849 (1984).
101. M. Stickler, *Angew. Makromol. Chem.*, **124**, 85 (1984).
102. T. Kato, T. Tokuya, and A. Takahashi, *J. Chromatogr.*, **256**, 61 (1983).
103. A. L. Spatorico and G. L. Bayer, *J. Appl. Polym. Sci.*, **19**(11), 2933 (1975).
104. C. Rochas, A. Domard, and M. Rinaudo, *Eur. Polym. J.*, **16**, 135 (1980).
105. A. Domard and M. Rinaudo, *Polym. Commun.*, **25**, 55 (1984).
106. C. P. Talley and L. M. Bowman, *Anal. Chem.*, **51**(13). 2239 (1979).
107. A. Bose, J. E. Rollings, J. M. Caruthers, M. R. Okos, and G. T. Tsao, *J. Appl. Polym. Sci.*, **27**(3), 795 (1982).
108. M. G. Styring, C. Price, and C. Booth, *J. Chromatogr.*, **319**, 115 (1985).
109. A. W. Wolkoff and R. H. Larose, *J. Chrom. Sci.*, **14**, 51 (1976).
110. American Society for Testing and Materials, ASTM Standard Method for Polystyrene, ANSI/ASTM D3536-37, ASTM, Philadelphia.
111. E. M. Barrall, M. J. Cantow, and J. Johnson, *J. Appl. Polym. Sci.*, **12**, 1373 (1968).
112. H. L. Wagner and C. A. J. Hoeue, *J. Polym. Sci. A-2*, **9**, 1763 (1971).
113. R. A. Rheim and D. Lawson, *Chem. Tech.*, 122 (1971).
114. J. C. Moore, *Sep. Sci.*, **5**, 723 (1970).
115. A. Rudin, *J. Polym. Sci., A-1*, **9**, 2587 (1971).
116. A. C. Ouano, *J. Polym. Sci., A-1*, **9**, 2179 (1971).
117. E. L. Slagowski, L. J. Fetters, and D. McIntyre, *Macromolecules*, **7**, 394 (1974).
118. J. J. Kirkland, *J. Chromatogr.*, **125**, 231 (1976).
119. A. E. Hamielec and A. C. Ouano, *J. Liq. Chromatogr.*, **1**(1), 111 (1978).

CHAPTER

11

PHASE DISTRIBUTION CHROMATOGRAPHY

GEORG S. GRESCHNER

Mainz, Federal Republic of Germany

INTRODUCTION

There are two main domains in which exact knowledge of the molecular weight distribution (MWD) of a given polymer is essential, namely in the characterization of the substance as well as in the judgment of its application range, and in the investigation of the kinetics of polymerization as applied to polymer preparation. Three examples may illustrate this: (1) It is well known that the mechanical properties of synthetic polymers depend not only on the chemical composition of their monomeric units, but also on their MWD, being often characterized by the averages $\bar{M}_n$, $\bar{M}_w$, and $\bar{M}_z$ of their molecular weight. (2) The knowledge of the (usually very narrow) MWD of biological macromolecules is often very helpful in the investigation of their biological activity and efficiency. (3) An accurate determination of the likewise very narrow MWD of, for example, polystyrene samples polymerized anionically, is indispensable in checking a proposed kinetic scheme and in gaining new, otherwise unobtainable, insights into the mechanism of the reactions taking place.

Three methods are available today for the determination of the molecular weight distribution of polymers: gel permeation chromatography (GPC), Baker–Williams fractionation (BWF), and phase distribution chromatography (PDC).

Gel permeation chromatography is a very useful, rapid, and comfortable routine method for the determination of molecular weight distributions that are not extremely narrow. Narrow MWDs cannot be obtained directly from GPC elution curves because the influence of spreading (axial dispersion) on the chromatogram becomes comparable with that of the column resolution, and complicated mathematical methods are necessary to obtain the correct MWD. The direct application to these calculations of the existing general theory of spreading phenomena in chromatographic columns is difficult because the spreading alone cannot be measured in GPC—unlike the PDC case, where column resolution can completely be eliminated at temperatures not far below the theta point of the system (consisting of an injected polymer, its high molecular weight linear gel, and a theta solvent) on which a particular PDC column is based.

The same difficulties arise when broad MWDs are calculated from GPC measurements performed on samples having a very high weight average molecular weight. In this case, the high fraction of larger pores in the GPC gel leads not only to stronger spreading, but also to lower mechanical stability of the gel, such that the calibration curve becomes less accurate. Moreover, no generally valid mathematical expression is available for GPC calibration curves because the GPC resolution mechanism has not yet been described mathematically to a sufficient extent. Hence, semiempirical or purely empirical statements must be applied in GPC.

Baker-Williams fractionation usually serves well for the determination of narrow molecular weight distributions. However, BWF has proved to be a laborious method, and wide experience is required to obtain correct MWDs. It is well known that the resolution of the BW column is based on an interaction of the transported polymer with a solvent–precipitant mixture having a suitable gradient along the column axis. The column is filled with fractionated small glass beads (ballotines, diameter 50–100 μm), usually treated with $(CH_3)_2SiCl_2$ to avoid polymer sorption on their surface; the column is usually held at a suitable temperature gradient along its axis to intensify its resolution power. The eluate is collected into about 40 fractions, in which the average molecular weight of the polymer is determined analytically, usually by viscometry.

Unfortunately, the standard numerical method based on the inversion of integral operators as mentioned above cannot be applied to BWF because the transport equation proposed was solved numerically only (1). Hence, only the laborious viscometric–gravimetric method remains. In addition, much experience is needed to adjust the correct solvent–precipitant and temperature gradients in the BW column; any wrong adjustment of either of these gradients inevitably leads to a false MWD. The correct operating conditions of the BW column for the fractionation of polystyrene were investigated by Böhm et al. (2: especially eq. 3 and Table II).

Today there is available a third routine column method for the determination of molecular weight distributions: phase distribution chromatography (PDC), proposed by Casper and Schulz (3, 4) in the early 1970s. It does not have the serious disadvantages of GPC in the case of very narrow MWDs and also does not require the great experimental effort typically needed for BWF.

To some extent, PDC represents a modification of BWF and an inversion of GPC. When the precipitant as well as the cross-linking of the gel are omitted, the high molecular weight species "dissolve" in the gel phase of the column to a greater extent than do those of lower molecular weight, and therefore leave the column later. But unlike BWF and GPC, PDC is based on an interaction of the injected dilute solution of the analyzed polymer (mobile phase) with a non-cross-linked polymer gel of the same kind (abbreviated as "linear" gel) immobilized, as a thin film, on the surface of small glass beads. This interaction takes place below the theta temperature of the system (i.e., the injected polymer, its high molecular weight linear gel, and the theta solvent) at constant temperature, and strongly depends on the molecular weight of the injected polymer species, if the column temperature is chosen to be low enough.

The operating conditions of the column immediately show that PDC does not represent a universal method because the injected sample must be of the same chemical type as the linear gel used in the column. This is a serious disadvantage of PDC compared with GPC. On the other hand, no problems arise in the calculation of the very narrow MWDs of polystyrene samples obtained from anionic polymerization because the PDC column, based on the polystyrene (PS)–HMW linear gel–cyclohexane (CHX) system investigated throughout this chapter, has the following essential properties: (1) it shows a powerful resolution (PDC effect) in the dynamic region, where the calibration curves also can be measured with the same high accuracy, and (2) its resolution practically vanishes near the theta point (but below it), where the spreading phenomena alone can be measured exactly. In addition, the elution curves measured in these very dissimilar regimes of the PDC column can easily be fitted to the corresponding mathematical expressions, both representing results of a closed theory. Under such circumstances, the mathematical method mentioned above leads to no difficulties even when extremely narrow MWDs of polystyrenes are calculated from PDC measurements. Often it can even be omitted if the MWD is not very narrow. Moreover, PDC proved to be an absolute method because the same MWD can be calculated from the different PDC chromatograms of the same sample obtained at two column temperatures and showing a very different shape, as discussed in the section entitled "Determination of Narrow Molecular Weight Distributions." All these facts represent a great advantage of PS–cyclohexane PDC in comparison to GPC and BWF.

Although only one system has been studied so far, namely polystyrene–cyclohexane, the theory of PDC resolution deduced from these measurements predicts the existence of a greater family of systems showing the PDC effect. In fact, there is no reason for this effect to be limited to the system investigated (i.e., PS–HMW gel–CHX). The question is, however, whether the design of the column is practicable for any given system. In addition, it must be ascertained whether the measurements can be performed with sufficiently high accuracy.

DESIGN OF AN AUTOMATIC PRECISION PD CHROMATOGRAPH

The first PD chromatograph based on the polystyrene–cyclohexane system was designed in 1969 by Casper (see Ref. 3). Already this rather primitive apparatus, operating manually with an average reproducibility of only a few percent, showed strong dependence of the column resolution on the column temperature (3–5). To explain the unexpected high resolution of the PDC column at suitably low column temperatures on the one hand, and the spreading phenomena taking place only slightly below the theta temperature where the column resolution practically vanishes on the other, a new automated precision PD chromatograph was designed in 1974 by Greschner (6). This apparatus, also based on the reliable system polystyrene–cyclohexane (PS–CHX) with the very favorable theta temperature of 34°C, and operating with a reproducibility of 0.1% in the temperature range 10–29°C, is shown in Figure 1. The sample column is composed of four V4A-steel tubes, filled with sharply fractionated glass beads of an average diameter of 76 μm, up to a nearly closest-packed spherical structure. The glass beads are covered with a homogeneous film (mean thickness, 305 nm at 27°C, swollen) of the same polymer as used previously by Casper (3, 4). The column temperature is kept constant with an accuracy of ± 0.01°C by the system of thermostats shown in Figure 1 (*a* and *c*). The column operates under a pressure of 3–6 bars with a flow rate of 10 to 40 cm^3/h cyclohexane, the constancy of the flow rate being better than $\pm 0.1\%$ in a room thermostated to 25 ± 0.5°C. No reference column was applied. The wiring diagram of the electronics governing the equipment is shown in Figure 1*b*.

Packing of the Column Units

The four identical V4A-steel tubes shown schematically in Figure 1*a*, each of diameter $d = 1.0$ cm and length $l = 146$ cm measured between the metallic frits at both ends connected by the conventional GPC metallic capillaries as short as possible (Fig. 1*a*), have a total volume $\sum V = \pi d^2 l = 458.7\ cm^3$, which is filled with sharply fractionated and gel-coated glass beads as densely and as homogeneously as possible.

First 2 L of rough glass beads (ballotines) having a declared diameter of 70–100 μm are fractionated between two cylindrical sieves, once between 70–100 μm, and again between 70 and 80 μm, taking measures to ensure that electrostatic charging does not hinder the fractionation. The middle fraction thus obtained is poured into 2 L of a concentrated H_2SO_4, 98% p.a. (= purity grade "pro analysi"), in which 110 g of solid $NaNO_3$ p.a. has been dissolved. The mixture is cautiously stirred from time to time, and the ballotines are decanted after 24 hours and alternately washed and boiled with bidistilled water until the flush water is practically neutral. After pouring the clean mass into 1 L of chromatographically pure cyclohexane, removing the solvent in a vacuum desiccator at 20°C, and drying the mass for 24 hours at 120°C and homogenizing it carefully, approximately one liter of sharply fractionated and chromatographically pure ballotines of a diameter $76 \pm 3\ \mu$m (measured in an X-ray photosedimentometer) is obtained. The ballotines must immediately be coated with a polystyrene film. Since the surface of the glass beads is sufficiently clean, neutral, and rough, no special surface treatment is necessary, as it usually is in BW columns, to ensure that the system obtained after the packing of the column is well defined.

To obtain stable, homogeneous, elastic, and linear gel films on the surface of these glass beads, 790 cm^3 of ballotines, obtained above and measured in a graduated cylinder after obtaining a "meniscus" by tapping, is poured very showly and under careful stirring into the viscous solution of 0.55 g of unfractionated high molecular weight polystyrene (average polymerization degree $\bar{P}_w = 65{,}000$) and of 1.95 g of fractionated high molecular weight polystyrene (average polymerization degree $\bar{P}_w = 80{,}000$) in 280 cm^3 of benzene (p.a., bidistilled) and homogenized for nearly 1 hour in a closed system to avoid the evaporation of the solvent. The volume ratio of solution to ballotines must be chosen such that no superfluous PS solution may occur above the homogenized mass. The uniform mash then is dried for 10 hours in a vacuum desiccator at 20°C. To avoid the formation of a hard "cake"—which would indicate the failure of the experiment—a few cubic centimeters of methanol (p.a., bidistilled) are added to the mass at the end of the drying procedure but just before hardening of the mass sets in.

When the solvent has evaporated, the mass is sprinkled with methanol and a loose mash is obtained from which the precipitant is also removed in vacuo. The roughly coated ballotines are now cautiously suspended in 280 cm^3 of chromatographically pure cyclohexane and left at rest in the closed desiccator for 24 hours to homogenize and stabilize the polystyrene films by swelling. Then the cyclohexane is removed in vacuo, the mass is sprinkled with methanol to make it loose, and the precipitant is also removed in vacuo at 20°C. This procedure yields nearly 750 cm^3 of correctly coated ballotines as a dazzling white and very movable powder, which must immediately be used for filling the column.

The coating procedure is rather tricky, and both experience and experimental skill are necessary to obtain good results. If, for example, the glass surface is not clean enough and not neutral, stable, homogeneous, elastic films are not obtained. The procedure also fails if there occurs a superfluous polymer solution above the homogenized mass or if the timely hardening of the mass is prevented because a loose high molecular weight polymer is set free in the mass after the precipitant (methanol for polystyrene) has been added. An uncoated glass surface can then be found among the beads, leading to poorly reproducible and concentration-dependent elution curves due to polymer sorption on the glass, and to a diminution of the PDC resolution. The same, of course, takes place if there is not enough coating mass in the mixture or if its degree of homogenization is poor. On the other hand, it is not easy to measure

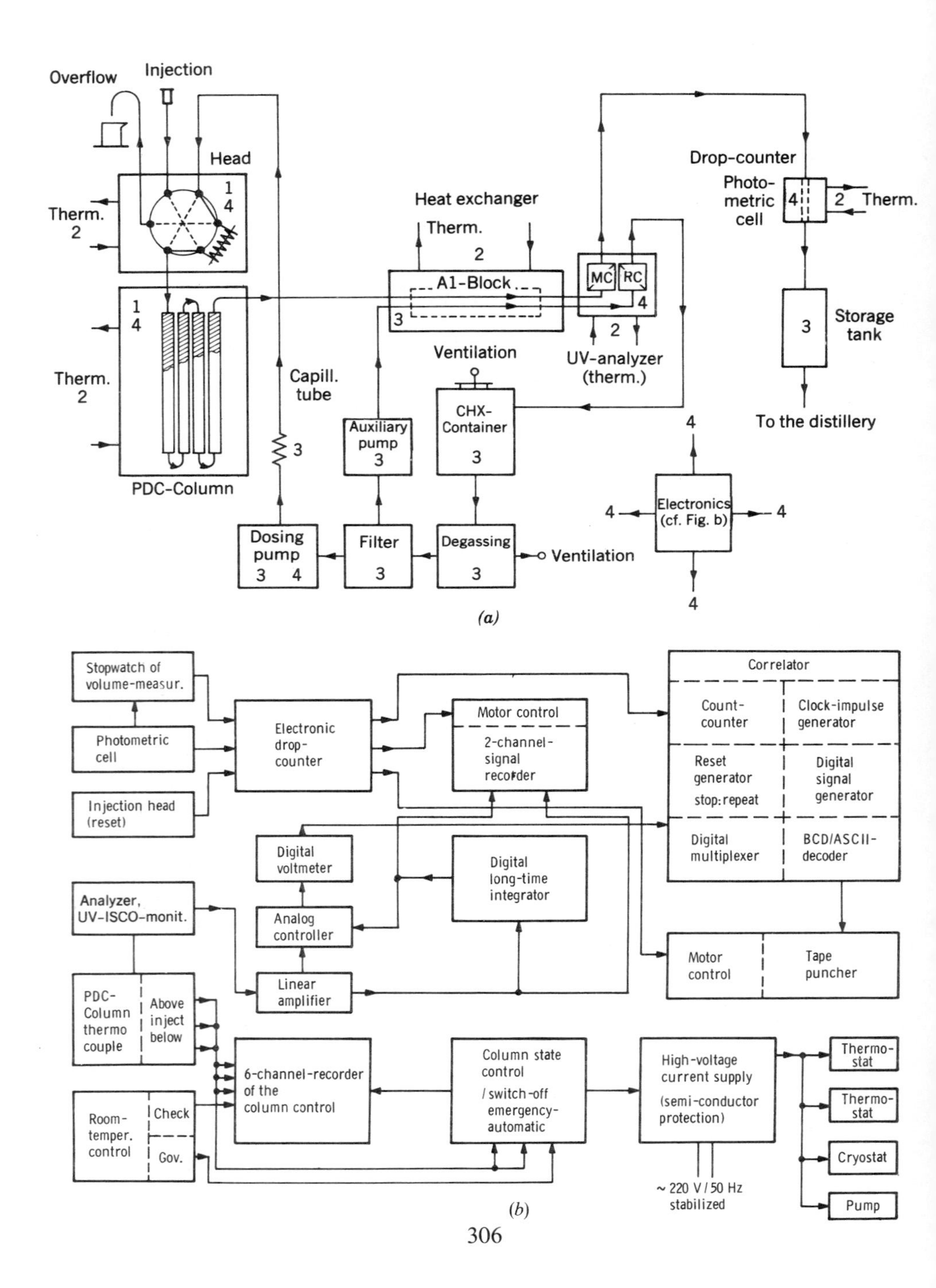

Overflow
Injection
Head
Therm.
2
1
4
Therm.
2
PDC-Column
Capill. tube
3
Heat exchanger
Therm.
2
A1-Block
3
MC
RC
4
2
UV-analyzer (therm.)
Ventilation
CHX-Container
3
Auxiliary pump
3
Dosing pump
3 4
Filter
3
Degassing
3
Ventilation
Drop-counter
Photo-metric cell
4
2
Therm.
Storage tank
3
To the distillery
Electronics (cf. Fig. b)
4
(a)
Stopwatch of volume-measur.
Photometric cell
Injection head (reset)
Electronic drop-counter
Motor control
2-channel-signal recorder
Correlator
Count-counter
Clock-impulse generator
Reset generator stop: repeat
Digital signal generator
Digital multiplexer
BCD/ASCII-decoder
Digital voltmeter
Digital long-time integrator
Analyzer, UV-ISCO-monit.
Analog controller
Linear amplifier
Motor control
Tape puncher
PDC-Column thermo couple
Above inject below
6-channel-recorder of the column control
Column state control / switch-off emergency-automatic
High-voltage current supply (semi-conductor protection)
Thermo-stat
Thermo-stat
Cryostat
Pump
Room-temper. control
Check
Gov.
~ 220 V / 50 Hz stabilized
(b)

Figure 1. The automatic precision phase distribution chromatograph. (*a*) Block diagram of the apparatus. (*b*) Wiring diagram. (*c*) Physical installation.

the volume of a loose glassy mass consisting of uniform small spheres, and to homogenize the mass without damaging the films or even the beads.

The second great difficulty arises from the demands on the structure of the gel film: it must consist of a linear gel, that is, of a non-cross-linked structure of polymer chains as long as possible. The cross-linking must in any case be avoided in PDC because PDC resolution is based on an effect opposite to that of GPC: Contrary to GPC, the smaller macromolecules always pass through the column earlier than the larger ones, and the more strongly the PDC effect sets in, the more pronounced the effect. Hence, any attempt to connect the long polymer chains with the glass surface chemically leads inevitably to a strong diminution of PDC resolution caused by the opposite GPC effect on a cross-linked gel formed in the grafting procedure. Only elastic gel films on the beads are available. To obtain them, the high molecular weight polymer used for coating the beads must contain some component of relatively short polymer chains representing a sort of plasticizer for the long polymer chains in the film formed. Therefore, both unfractionated and fractionated high molecular weight polystyrene were applied in the coating procedure described above.

The filling of the column with sharply fractionated, coated ballotines can be performed by the conventional method. The four V4A-steel tubes mentioned above, each closed by the lower metallic frit, attached to a cylindrical glass funnel (600 ml) filled with the chromatographically pure theta solvent (cyclohexane for polystyrene) and connected by its metallic capillary to the water-jet vacuum pump, are vertically and stiffly attached to a metallic support being periodically hit once a minute by means of a motor-driven device. The low frequency vibration of the column segments achieved in this way must not exhibit any knots along the column axis, which would cause inhomogeneities in the filling of the segments.

After having stopped the supply of the theta solvent in each funnel until the meniscus has just reached half of the funnel height, $1.25 \times 458.7/4 \simeq 145\ \mathrm{cm}^3$ of the coated ballotines suspended in $280\ \mathrm{cm}^3$ of theta solvent are poured at once into each funnel at a flow rate of $150\ \mathrm{cm}^3$ of theta solvent per hour through each tube. The volume of the exhausted liquid is continually replaced in all funnels. The filling procedure is finished in nearly 12 hours and yields four identical column segments packed uniformly up to 88% of the closest packing of equivalent spheres.

The column segments are now closed by the upper metallic frits, connected by its GPC metallic capillaries as short as possible, and installed in the thermostated column jacket as shown schematically in Figure 1*a*. The PDC column is then flushed with the chromatographically pure theta solvent at a constant temperature chosen below the theta temperature but as near the theta point as possible without damaging the gel films (this must be found out experimentally: the temperature $29.0 \pm 0.1°\mathrm{C}$ was found for the PS–CHX system with the theta point 34°C). The flow rate of the theta solvent is held nearly constant at $15\ \mathrm{cm}^3/\mathrm{h}$. The procedure is continued until no light scattering on the elute is directly visible and until less than 1 mg of polymer per liter of eluate is found gravimetrically. In this first flushing period (which took nearly 2 weeks in the PS–CHX system), the frits of the column must be tightened from time to time; no detector is inserted into the measuring channel (see Fig. 1*a*) in this first period to avoid formation of a thin gel film on the windows of the measuring cell. In the 2 weeks that follow, the column temperature found above is held constant with the accuracy of the measurements (29.0 ± 0.01°C for PS–CHX) at a theta solvent flow rate of $15.0 \pm 0.015\ cm^3/h$ in a room thermostated to 25 ± 0.5°C; a sufficiently sensitive detector (UV-ISCO-spectrometer in case of PS–CHX, see Fig. 1*a*) is now inserted into the measuring channel and the zero line is recorded up to the highest sensitivity of the detector marking the balance of the PDC column (extinction up to $E = 0.01$ in case of PS–CHX). Elution curves related to this temperature can then be measured; the changes of the column temperature and the flow rate, however, must be made slowly to avoid a damage of the films changing its swelling degree. In the PS–CHX system at first elution curves at 25.0 ± 0.01°C were measured at the volume rates 5–30 cm^3 of CHX per hour and at concentrations of 0.5–5 mg/ml of PS in CHX in the injected sample. Reproducibilities of ± 0.1–0.2% were obtained as measured at the maximum of the calibration chromatograms; in the PDC theory, the mathematical expectation $\bar{V}$ of the elution volume in the calibration chromatogram $D(V)$ is taken instead of V_{max}: the value $\bar{V}$ is calculated in analogy to $\bar{P}_w$,

$$\bar{V} = \int_{-\infty}^{+\infty} V D_{cal}(V)\, dV \tag{1}$$

and lies near V_{max}. Additional details and a summary of all important control and column data can be found in Reference 6, where also the purification of cyclohexane p.a. up to the chromatographic purity is described in detail.

To be able to control the stability of the gel film on the surface of the glass beads, Casper (3) prepared a radioactive polystyrene (tritium-marked chain having the activity 220 Bq/mg). Because the injected samples are not radioactive, the origin of the polymer in the eluate could be well identified. The same radioactive high molecular weight polystyrene as prepared by Casper, both fractionated and unfractionated, was used in the coating procedure described above. Both the experiments published previously by Casper and Schulz (4), summarized in Reference 5, and the experiments with the precision chromatograph described in this chapter, showed that the gel films are very stable at all column temperatures below 30°C—that is, up to 4°C below the theta point. In fact, the high reproducibility of the precision chromatograph remained unchanged for nearly 2 years until the project was stopped. During this time, measurements between 10 and 29°C were performed on narrowly distributed polystyrene samples within the broad concentration range mentioned above; the most important results are shown in the sections that follow. The cyclohexane transport through the column was stopped only when necessary, and then for a short time, and the column temperature was changed very slowly. Damage to the column was avoided by the automatic monitoring and controlling equipment shown in Figure 1*b*; a very careful one-point grounding of all equipment as well as the use of zero-current switches in the high voltage supply are necessary (thermostats, pump, and above all a strong motor for the air-conditioning plant). To save money, the theta solvent is continually removed from the eluate by distillation, purified, and recycled.

Some Experimental Results

The high resolution of the precision PD chromatograph described above can be seen by comparing the GPC-chromatogram in Figure 2 with the PDC chromatogram of the same sample in Figure 3 (1:1 mixture by weight of the anionic standard polystyrenes PCC-K 390,000 and PCC-K 110,000). Whereas only partial separation of this mixture is achieved by the GPC in Figure 2, complete separation is observed in case of PDC, even at 25°C, as shown in Figure 3. A further improvement in column resolution could easily be achieved in PDC by measuring at 23°C instead of 25°C, which would lead to a considerably longer part of the zero line separating the peaks in Figure 3, whereas the resolution of the GPC column remains constant.

The possibility of obtaining accurate calibration curves of carefully characterized polystyrene calibration samples, representing BW middle fractions of anionic PCC polystyrene standards (see Ref. 6), is of particular importance. These calibration curves (Figs. 4 and 5) exhibit a sigmoidal shape, completely contradicting the reversible thermodynamical treatment of the chromatographic process proposed by Schulz and Casper (4) for PDC. For

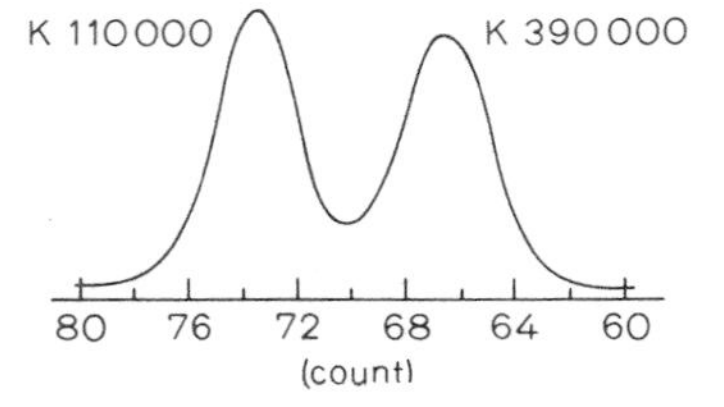

Figure 2. Separation of a 1:1 mixture by weight of the anionic standard polystyrenes PCC K-390,000 and PCC K-110,000; high performance GPC (1 count $= 0.025\ cm^3$ of tetrahydrofuran).

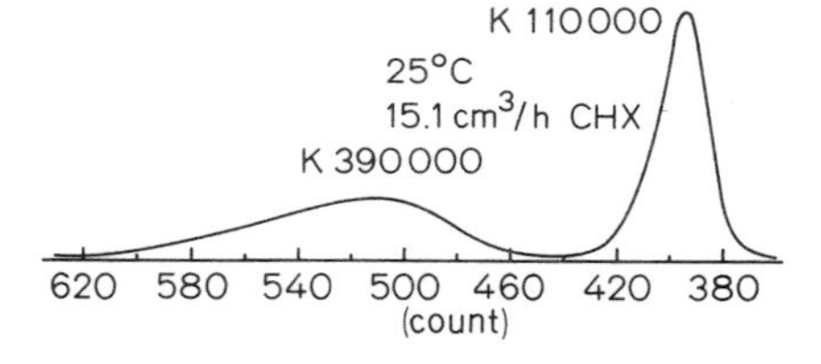

Figure 3. The same system as in Figure 2, but separated with PDC (25°C, 1 count $= 0.514\ cm^3$ of cyclohexane at a flow rate of $15.1\ cm^3/h$).

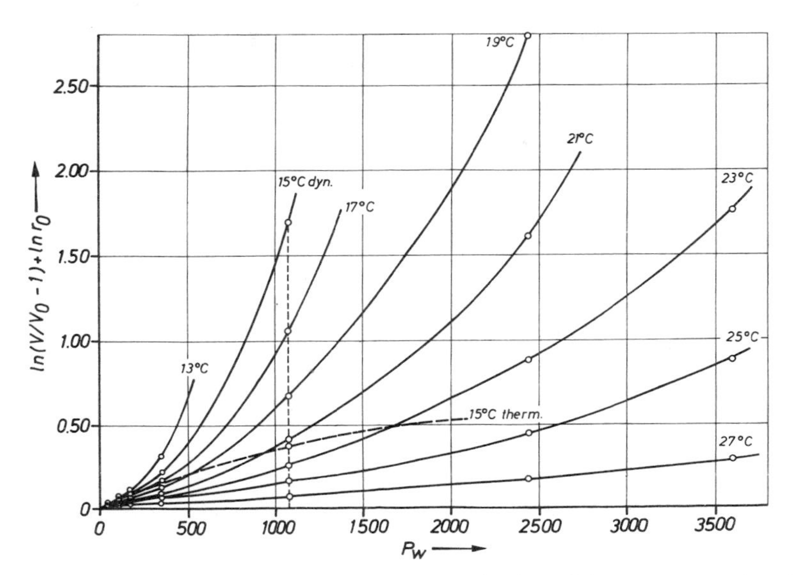

Figure 4. PDC calibration curves for polystyrene samples in cyclohexane measured at eight temperatures as indicated, and an overall cyclohexane flow rate of $15\,\mathrm{cm}^3/\mathrm{h}$ (ordinate is normalized as indicated). The 15°C calibration curve "dyn" is measured, whereas the dashed 15°C curve "therm" is extrapolated from the measured part of the "dyn" curve (cf. Fig. 5) and corresponds to reversible-thermodynamic equilibrium of the PDC column. The difference between the curves shows a pronounced PDC effect at 15°C for $P = 1082$. Elution volume $V \equiv V_e(P)$ and zero volume $V_0 \equiv V_e^0$ are expressed in counts (1 count $= 0.51423\,\mathrm{cm}^3$). For the definition of r_0 see equation (9).

the calibration curves considered, this concept predicts the straight lines

$$\ln[V_e(P)/V_e^0 - 1] = -\ln r_v + \epsilon P, \qquad \epsilon > 0 \tag{2}$$

$V_e(P)$ being the measured elution volume associated to the degree of polymerization P of the sample, V_e^0 the void volume of the column (i.e., the volume of the space between the coated beads),

$$r_v = V_{sol}/V_{gel} \tag{3}$$

the overall volume ratio of sol and gel in the column, and ε the coefficient in the partition function $K(P) = \exp(\epsilon P)$, known from static separation experiments.

The calibration curves from the detail (Fig. 5) show clearly that there are two different temperature regions for the PDC column: a region in which the calibration curves run below their tangent with a decreasing slope for an

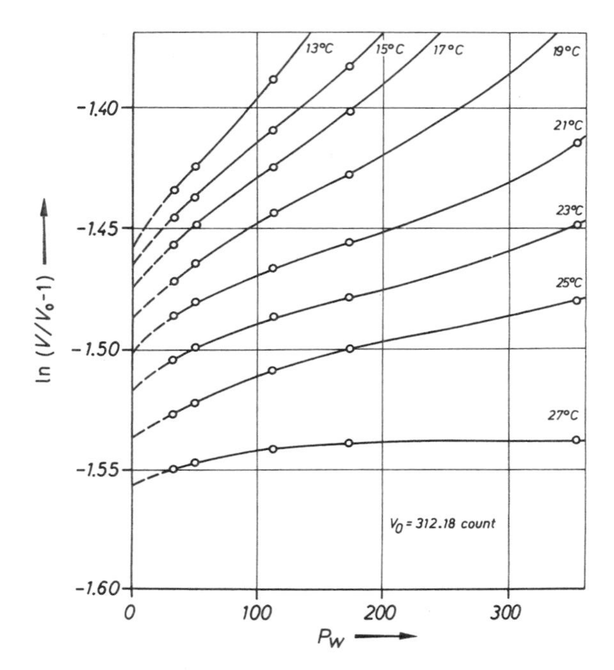

Figure 5. Reversible-thermodynamic regions of the calibration curves from Figure 4 drawn in detail up to the point of inflection; ordinate is not normalized here; the intersections with the ordinate are equal to $-\ln r_0(T)$, obtained by spline extrapolation.

increasing degree of polymerization, and a region in which they run above their tangent with an increasing slope. The first region (e.g., $\bar{P}_w < 300$ at 27°C, or $\bar{P}_w < 100$ above 20°C up to the corresponding points of inflection in Fig. 5), characterized by low resolution of the chromatographic process, corresponds to equation (2) with P^v instead of P for $0 < v < 1$ depending on the temperature; it is called the reversible-thermodynamical region of the column and is indicated by "therm" or "rev." The second region, characterized by high column resolution, is in complete contradiction to equation (2) because P^v with $v > 1$ cannot be assumed instead of P in this equation; this range, called dynamic region of the column, is denoted by "dyn" to indicate that irreversible thermodynamics must be applied here to the chromatographic process.

The difference between the column resolution in these two regions is clearly demonstrated in Figure 4, where the measured 15°C calibration curve "dyn" is compared with its extrapolated reversible part "therm" (dashed) for $P = 1082$. This pronounced effect is called the PDC effect. But how can it be explained?

OUTLINE OF THE THEORY OF PDC RESOLUTION

Let $\bar{m}_{P,s}$ and $\bar{m}_{P,g}$, respectively, be the polymer mass of a transported P-mer in the sol and gel of a PDC column in a reversible-thermodynamic equilibrium, $\bar{c}_{P,s}$ and $\bar{c}_{P,g}$ the corresponding concentrations, $u(P) = (dz/dt)_P$ the constant elution rate of the P-mer transported along the Z axis of the vertical PDC column, and $\bar{v}$ the average overall linear rate of the column liquid. Then trivial integration of the chromatographic transport equation (i.e., the thermodynamically and hydrodynamically defined retention coefficient R of the P-mer in the column)

$$\bar{m}_{P,s}/(\bar{m}_{P,s} + \bar{m}_{P,g}) = R(P) = \frac{(dz/dt)_P}{\bar{v}} \qquad \text{with } z = 0 \qquad \text{for} \quad t = 0 \tag{4}$$

yields the equation of PDC in a reversible-thermodynamic equilibrium

$$V_e(P) = V_e^0[1 + K(P)/r_v] \qquad r_v = V_s/V_g \qquad T = \text{const} \tag{5}$$

where $V_e(P) = q\bar{v}t_e(P)$ is the elution volume of the P-mer measured at the end $z = L$ of the column with the free cross section q at the elution time $t_e(P)$, and $V_e^0 = qL$ is the void column volume. The ratio r_v in the transported zone is assumed to be equal to the overall value from equation (3); $K(P)$ is the conventional partition function in which, as usual, the volume fractions of the P-mer in gel and sol are approximated by the corresponding equilibrium concentrations defining the ratio of the kinetic constant of the reversible polymer transport from the sol into the gel (k_s) and from the gel into the sol (k_g):

$$\bar{c}_g/\bar{c}_s = K(P) = k_s/k_g \tag{6}$$

Figure 5 immediately shows that equation (5) can serve well for the mathematical description of the calibration curves measured in the whole reversible-thermodynamic region of the column (i.e., up to their points of inflection in Fig. 5), if the dependence of the partition function on the degree of polymerization

$$K(P) = \exp\{\epsilon_0(T) + \epsilon_1(T)P^{\nu(T)}\} \tag{7}$$

with $0 < \nu(T) < 1$ is stated. Such a statement agrees well with the results published by Schulz and Jirgensons (7), Wolf et al. (8–9), and Kleintjens, Koningsveld, and Stockmayer (10). After introducing equation (7) into equation (5), we obtain equation for PDC in a reversible-thermodynamic

Table 1. Summary of the Important Thermodynamic Data of the PDC Column Described ($V_s = 160.54\,cm^3$)[a]

$t/(°C)$	$r_v = (V_s/V_G)_\infty$	$r_0 = (\bar{m}_s/\bar{m}_g)_0$	$K(0) = (\bar{c}_g/\bar{c}_s)_0$	$\varepsilon_0 = \ln K(0)$	$\varepsilon_1 \times 10^4$	ν
13	44.51	4.300	10.35	2.3371	4.82	0.995
15	40.92	4.330	9.45	2.2461	3.61	0.993
17	37.33	4.372	8.54	2.1446	2.73	0.990
19	33.75	4.426	7.62	2.0308	2.07	0.985
21	30.67	4.490	6.83	1.9214	1.45	0.98
23	27.59	4.564	6.04	1.7984	1.02	0.96
25	24.71	4.650	5.31	1.6696	0.68	0.90
27	22.09	4.744	4.66	1.5390	0.44	0.68

[a] Cf. Greschner (11, 5).

equilibrium (11)

$$\ln [V_e(P)/V_e^0 - 1] = -\ln r_0(T) + \epsilon_1(T) P^{\nu(T)} \tag{8}$$

which well describes the PDC calibration curves in the whole reversible-thermodynamic region of the PDC column (i.e., up to their points of inflection). It contains the new quantity

$$r_0 = r_v/K(0) = (V_s/V_g)(\bar{c}_{P,s}/\bar{c}_{P,g})_{P=0} = (\bar{m}_{P,s}/\bar{m}_{P,g})_{P=0} \tag{9}$$

which is closely related to the solvation of polymer chains near the gel front (see Refs. 11 and 12, summarized in Ref. 5.

A mathematical analysis published in the appendix of Reference 11 and applied to the whole shape of all the calibration curves from Figure 4 yields, besides the phenomenological function $\alpha(P, T)$ defined below, the temperature dependencies $r_v(T)$, $r_0(T)$, $\epsilon_0(T)$, $\epsilon_1(T)$, and $\nu(T)$ shown in Table 1 and Figure 6. They explain well the phenomena taking place in a PDC column operating at low resolution in its reversible-thermodynamic region, but they cannot explain the most important property of the column, namely the PDC effect mentioned above and shown in Figure 4 at 15°C for $P = 1082$: the real measured concentration $c_{P,g}$ of this P-mer in the gel, as indicated on the "15°C-dyn" chromatogram, is considerably higher than the corresponding reversible concentration $\bar{c}_{P,g}$ predicted by the partition function $K(1082)$ from equation (7), as indicated by the dashed extrapolated reversible part "therm" of the measured 15°C curve in Figure 4.

It was mentioned above that the PDC effect cannot be explained by any sorption of the transported polymer on eventually uncovered glass beads or

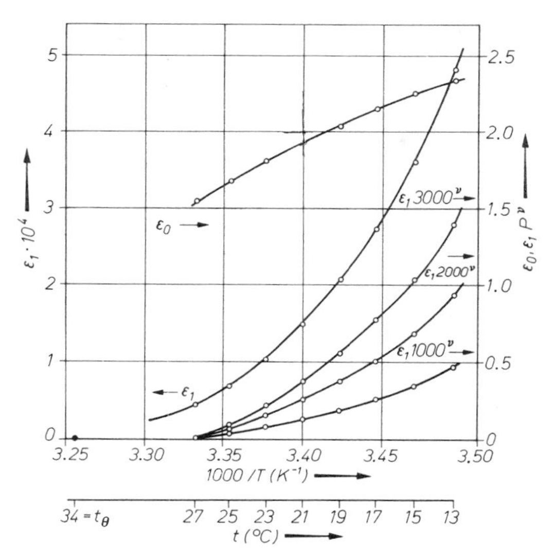

Figure 6. Arrhenius plot of the quantities ε_1, $\varepsilon_1 1000^\nu$, $\varepsilon_1 2000^\nu$, and $\varepsilon_1 3000^\nu$, with the function $\varepsilon_0(T)$ for comparison.

on metallic parts of the column. All elution curves are reproducible with the high accuracy of 0.1–0.2%, practically independent of the sample concentration from 0.5 to 5 mg/ml PS in CHX in the whole temperature range of the measurements (viz., 10–29°C). A concentration-dependent bend ("knick") occurred in the chromatograms after the temporary replacement of 1 cm^3 of the covered beads by uncovered ones, and high reproducibility was lost until the previous conditions were restored. This is typical for sorption of polymers. The gravimetrically determined amount of polymer washed out from the column during the whole flushing period described in the "Design" section was small compared to the polymer mass entered the coating procedure. The tendency of the eluate to form thin gel films along its way during the first 2 hours of the flushing procedure shows that also the metallic parts of the column were coated in this way with the superfluous polymer; however, this metallic surface is negligible when compared with the enormous surface of the glass beads in the column (235,000 cm^2).

Hence, the PDC effect must originate from interactions of the column liquid with the noncross-linked polymer gel layer during flow (i.e., within the transported zone), at a constant temperature chosen sufficiently below the theta point for the *P*-mer considered. Since the temperature-dependent thickness of the gel film on the glass beads (150–300 nm, swollen) can be neglected when compared with the radius of the beads (38 μm), the mobile phase of the column is pumped at a pressure of 3–6 bars and at constant temperature through a system of channels with tetrahedral symmetry related to the closest packing of equivalent spheres with the radius of 38 μm. It can be assumed that high and steep velocity gradients arise in the neighborhood of the gel front in the stream of the column liquid passing these channels and permanently changing its flow direction, although its average volume rate is low (typically 15 cm^3/h). Indeed, rough calculations confirm this (see Ref. 13). There is no reason not to use this approach for stating a closed theory, which then must be proved carefully by experiment. To this end, the following model was proposed (11, 12) for radial polymer transport between sol and gel within a zone transported axially through the PDC column at constant temperature.

Each macromolecule of the P-mer transported axially in the channel far enough from the gel front (e.g., near the channel axis, where the velocity gradient affecting the polymer chain is negligible) remains undeformed; it moves freely in the stream of the column liquid, rotates around its center of gravity, and shows a configuration of maximal thermodynamic probability W and consequently of maximal entropy $S = k \ln W$. Hence, any deformation of the polymer chain caused by the stress related to the velocity gradient on the way of the macromolecule toward the gel front (which can be assumed to be a plane perpendicular to the radial chain movement), must lead to some decrease of the coil entropy and therefore to an increase of the free energy $-T\Delta S_{\text{def}}$, of the coil that is being considered. Since the velocity gradient is steep in the neighborhood of the gel front, this function of state becomes significant just near the gel front where the radial transport of the P-mer from the sol into the gel begins; if the chain deformation can be considered to be purely elastic, the entropic term stated above represents simply the free enthalpy of deformation $\Delta G_{\text{def}}(P, T) = -T\Delta S_{\text{def}}(P, T) > 0$ mainly governing the polymer transport from the sol into the gel in the dynamic region of the PDC column.

According to this model, the polymer transport from the sol into the gel represents a spontaneous process only near the theta point, where it is governed solely by the partition function $K(P)$. In the dynamic region of the column where the PDC effect is pronounced, however, this polymer transport is no longer spontaneous, but is forced, in the way described above, at the cost of the kinetic and the thermal energy of the column liquid. It is governed not only by the partition function $K(P)$, but above all by the free enthalpy $\Delta G_{\text{def}}(P)$, which is connected with an irreversible process: after having been deformed by the stress related to the velocity gradient at a given position near the gel front, the polymer chain diffuses, forced from the sol into the gel up to some distance comparable with the mean depth of penetration Λ of the coil diffusion, and relaxes after reaching it (see Ref. 12). As a consequence, the coil entropy increases until a spontaneous rediffusion of the coil from the gel into the sol sets in after some retardation time, during which the coil receives the enthalpy of activation needed to loose the polymer–polymer contacts with the stationary gel phase, in order to repeat the whole procedure (forced diffusion, spontaneous rediffusion after a retardation time) on some other position of the gel front, down stream of the column liquid. This process must be very fast because of the powerful resolution of the PDC column in its dynamic region.

To be able to prove this concept by experiment and to avoid the formulation of an integrodifferential equation even more complicated than that equations (24a, b), below, and not solvable analytically, no differential approach was applied. Instead, an integral approach was proposed in Reference 11 (see also Ref. 5) assuming a steady state—that is, a flow equilibrium for polymer transport between sol and gel, governed by three flows: mass flux of the P-mer from the sol into the gel, mass flux of the P-mer from the gel into the sol, and flux of free energy transported through the gel front as explained above and maintaining the flow equilibrium energetically. To state a theory, the partition function $K(P)$ from equations (6) and (7) is replaced by the partition function $Q(P)$ of the flow equilibrium in equation (5),

$$V_e(P) = V_e^0[1 + Q(P)/r_v] \tag{10}$$

with $Q(P)$ defined in analogy to equation (6), but for the real measured concentration $c_{P,g} > \bar{c}_{P,g}$ of the P-mer in the gel as indicated in Figure 4 for $P = 1082$ at 15°C:

$$(c_g/c_s)_{flow} = Q(P) = k_s/k'_g \tag{11}$$

Here, $k'_g < k_g$ is the kinetic constant of the retarded redifussion of the P-mer from the gel into the sol in the flow equilibrium related to $Q(P)$, and k_s is the same as in equation (6). Both partition functions are related by the statement

$$Q(P) = K(P)\exp\left[\frac{\Delta G_{def}(\text{sol} \rightarrow \text{gel})}{RT}\right] \tag{12}$$

describing both the PDC effect $[\Delta G_{def} > 0,\ Q(P) > K(P)]$ and the reversible-thermodynamic equilibrium near the theta point $[\Delta G_{def} = 0,\ Q(P) = K(P)]$.

To avoid serious difficulties that would necessarily accompany any statistical description of the third flux mentioned above, only the nonclosed subsystem of the remaining two fluxes was considered, to which a perturbation calculus was applied according to the measured PDC effect (see Ref. 11):

$$Q(P) = K(P) + \delta Q(P) \tag{13}$$

Table 2. Calibration Data and Phenomenological Functions in Terms of Degree of Polymerization $\bar{P}_w$ at Eight Column Temperatures[a]

***a.* Summary of Measured Calibration Data**

	27°C		25°C		23°C		21°C		19°C		17°C		15°C		13°C	
$\bar{P}_w$	V_e	$\zeta(V_e)$	V_e	$\zeta(V_e)$	V_e	$\zeta(V_e)$	V_e	$\zeta(V_e)$	V_e	$\zeta(V_e)$	V_e	$\zeta(V_e)$	V_e	$\zeta(V_e)$	V_e	$\zeta(V_e)$
3594	399.49	−1.274	474.44	−0.654	712.60	+0.249	—	—	—	—	—	—	—	—	—	—
2444	389.71	−1.393	416.86	−1.093	476.71	−0.640	658.30	+0.103	1494.1	+1.331	—	—	—	—	—	—
1082	382.46	−1.491	390.67	−1.381	400.32	−1.265	417.00	−1.091	450.28	−0.816	517.10	−0.421	703.50	+0.226	—	—
353	379.23	−1.538	383.25	−1.480	385.51	−1.449	388.00	−1.415	392.05	−1.363	396.63	−1.307	402.00	−1.246	411.22	−1.148
172	379.14	−1.539	381.84	−1.500	383.28	−1.479	385.00	−1.456	387.00	−1.428	389.00	−1.402	390.49	−1.383	393.27	−1.348
111	379.00	−1.542	381.18	−1.509	382.74	−1.487	384.20	−1.467	385.83	−1.444	387.22	−1.426	388.39	−1.410	390.00	−1.389
49	378.65	−1.547	380.25	−1.523	381.90	−1.499	383.15	−1.481	384.26	−1.466	385.52	−1.448	386.29	−1.438	387.23	−1.425
32	378.46	−1.550	379.98	−1.527	381.56	−1.504	382.82	−1.486	383.82	−1.472	384.90	−1.457	385.70	−1.446	386.52	−1.435
Extr. $P=0$	−1.557		−1.537		−1.518		−1.502		−1.487		−1.475		−1.466		−1.459	

[a]The values corresponding to the dynamic regions of the calibration curves are listed above the lines in both tables (V_e in counts; 1 count = 0.51423 cm^3; $V_e^0 = 312.18$ count; $\zeta(V_e) = \ln(V_e/V_e^0 - 1)$; $w = 15.0\,cm^3/h$; $P_w/P_n - 1 = U < 0.05$; cf. Refs. 11, 5).

***b.* Shifted Phenomenological Functions $\alpha(P)$ of the Flow Equilibrium Corresponding to Calibration Data Above**

$\bar{P}_w$	$\alpha_{27}(P)$	$\alpha_{25}(P)$	$\alpha_{23}(P)$	$\alpha_{21}(P)$	$\alpha_{19}(P)$	$\alpha_{17}(P)$	$\alpha_{15}(P)$	$\alpha_{13}(P)$
3594	1.05476	1.22601	1.77690	—	—	—	—	—
2444	1.02942	1.08375	1.20954	1.62089	3.53581	—	—	—
1082	1.01094	1.02317	1.03583	1.06433	1.13025	1.27356	1.68921	—
353	1.00288	1.00797	1.00777	1.00799	1.01156	1.01617	1.02139	1.03394
172	1.00280	1.00540	1.00452	1.00449	1.00503	1.00567	1.00464	1.00622
111	1.00249	1.00407	1.00399	1.00385	1.00416	1.00407	1.00334	1.00350
49	1.00165	1.00206	1.00270	1.00258	1.00227	1.00268	1.00203	1.00200
32	1.00117	1.00148	1.00205	1.00190	1.00180	1.00200	1.00150	1.00165
$P=0$	1.0	1.0	1.0	1.0	1.0	1.0	1.0	1.0

In additional, an analytical expression for the kinetic constant k_s of the spontaneous (reversible) diffusion of the P-mer from the sol into the gel was found in Reference 12. As the main result, the general equation of the PDC calibration curves was found (11, 12)

$$\ln\left[\frac{V_e(P)}{\alpha(P;T)V_e^0} - 1\right] = -\ln r_0(T) + \epsilon_1(T)P^{\nu(T)} \tag{14}$$

which can easily be fitted in the whole range of the calibration chromatograms by the method given in Reference 11. The phenomenological function $\alpha(P;T)$ of the flow-equilibrium can be calculated directly from the measured PDC chromatograms and, together with the expression for $k_s(P)$ mentioned above, applied to a complete phenomenological description of the PDC resolution in any system showing the PDC effect. The phenomenological functions for the polystyrene–cyclohexane system investigated can be found in Table 2*b*, related to Figures 4 and 5. The elution volume V_e in Table 2*a* is identical to the mathematical expectation of the elution volume in the calibration chromatogram as given by equation 1.

Having obtained the dependencies $r_0(T)$, $\epsilon_1(T)$, $\nu(T)$, $\epsilon_0(T)$, and $\alpha(P;T) \geqslant 1$ from the fit leading to equation (14), we can find the partition function $K(P)$ from equation (7) and subsequently the kinetic constant $k_g(P)$ from equation (6), taking the expression for $k_s(P)$ from Reference 12. The relative perturbation of the reversible-thermodynamic equilibrium by the transport $\delta Q/K$ can be calculated from the relation (11, 12)

$$\frac{\delta Q(P)}{K(P)} = \frac{\alpha(P;T) - 1}{1 - \alpha(P;T)V_e^0/V_e(P)} \geqslant 0 \tag{15}$$

giving also the kinetic constant k_g' of the retarded rediffusion of the P-mer from the gel into the sol in the flow equilibrium related to $Q(P)$:

$$\frac{k_g}{k_g'} = 1 + \frac{\delta Q(P)}{K(P)} \geqslant 1 \tag{16}$$

These relations, together with the temperature dependence of the corresponding activation enthalpy (see Ref. 12) can now be used to prove the whole PDC concept by experiment. The results obtained above for the case of the reversible-thermodynamic equilibrium in a PDC column near the theta point remain unchanged because $\alpha(P) = 1$ is valid in this region for any P-mer. Equation (14) is then identical with equation (8), and $\delta Q(P)/K(P) = 0$ follows from equation (15) for any P, giving $k_g'(P) = k_g(P)$ in equation (16) and

$Q(P) = K(P)$ in equation (13) related to $\Delta G_{def}(P) = 0$ in equation (12) for any P, indicating that the resolution of the PDC column is low.

The column resolution vanishes completely at the theta point where $K_{chain}(P) = 1$ is reached (cf. Fig. 6 for $t_\theta = 34°C$ in the PS–CHX system). At the theta temperature, however, the PDC column is damaged.

To check the theory, we first compare Figures 7 and 8. Figure 7 shows the temperature dependence of the standard deviation σ calculated from four typical PDC chromatograms as indicated, without making use of any theory, while Figure 8 shows the temperature dependence of the relative perturbation $\delta Q/K$ calculated from the same chromatograms by making use of equation (15) and Table 2. All the corresponding curves are very similar in shape. Indeed, the same shape can be seen in Figure 9, which gives an Arrhenius plot of the corresponding kinetic constants k_g' of the retarded rediffusion from gel into sol in the flow equilibrium (solid curves), as calculated

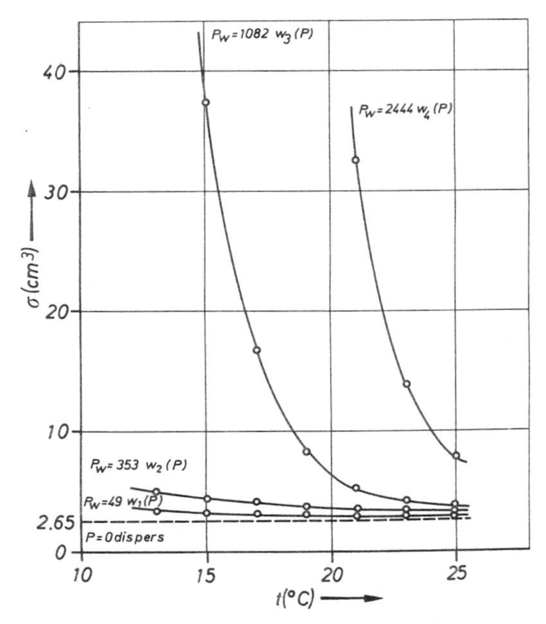

Figure 7. Temperature dependence of the standard deviation σ of the PDC elution curves for four typical polystyrene samples. The curves are parametrized by the molecular weight distribution $w(P)$ as indicated and explained in the next section 4 ("Determination of Narrow MWDs"). The asymptote $\sigma(0) = 2.65\ \text{cm}^3$ corresponds to a section $P = 1$ through the spreading surface of the PDC column (kernel, cf. next section).

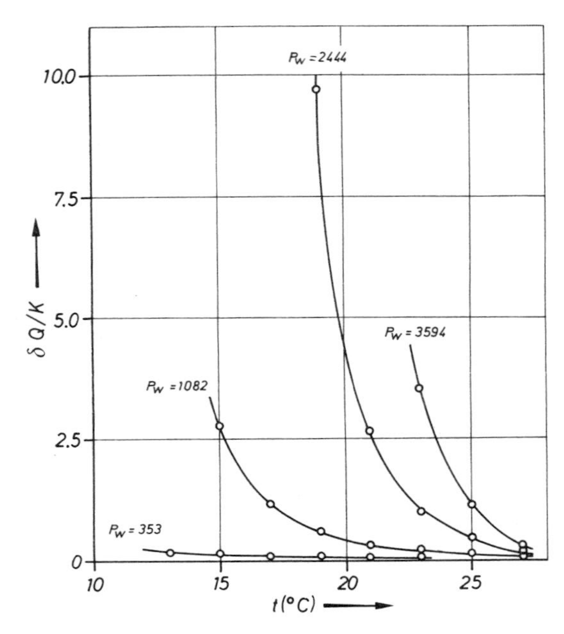

Figure 8. The temperature dependence of the perturbation function $\delta Q(P)/K(P)$ of the flow equilibrium proposed, calculated from PDC measurements performed on the samples from Figure 7. The curves are parametrized with the corresponding weight averages of the degree of polymerization.

from equation (16); the asymptotes (dashed curves) show in each case the limit k_g of k'_g related to the reversible-thermodynamic equilibrium with $\alpha = 1$ in Table 2b, $\delta Q/K = 0$ in Figure 8, and $\sigma = \sigma(0)$ in Figure 7 (axial dispersion). Finally, the same shape can also be seen from the Arrhenius plot of corresponding activation enthalpies in Figure 10 (solid curves), describing the flow equilibrium energetically; the corresponding asymptotes (dashed) show the reversible parts of these functions of state related to the low resolution of the PDC column in the reversible-thermodynamic equilibrium ($\alpha = 1$, $\delta Q/K = 0$, $k'_g = k_g$, and $Q = K$, related to $\Delta G_{def} = 0$). Figure 9 also shows that the column resolution is governed by fast kinetics, as was expected (typically $k'_g = 10^3$–$10^4\,s^{-1}$).

All these results well support the PDC concept explained above. The main evidence, however, is given in Table 3, which compares directly the normalized entropy of deformation $\Delta S_{def}/R$ calculated from the PDC measurements according to the theory of PDC resolution stated above (left-hand part of the table) with the same function of state $\Delta S_{def}/R$ calculated from the well-known theory of rubber elasticity under the operating conditions of the PDC column

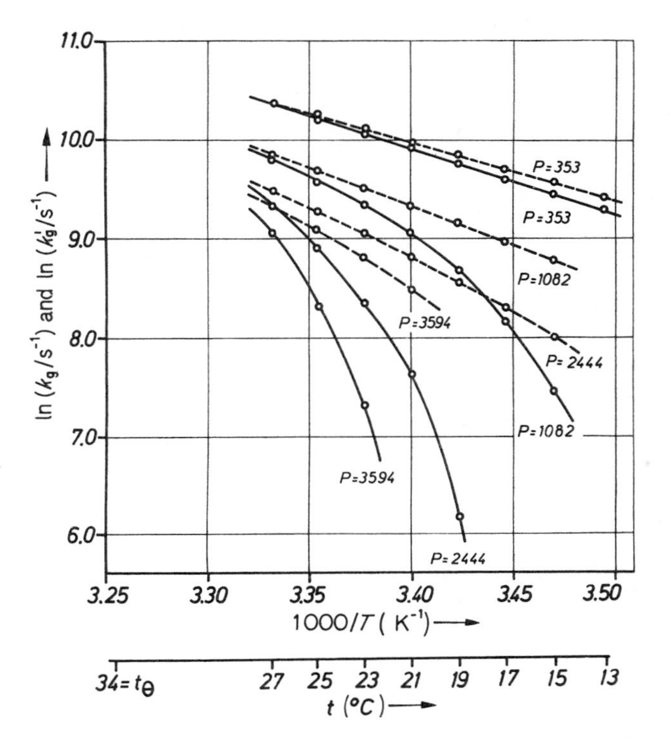

Figure 9. Arrhenius plot of the rate constant k'_g for the retarded polymer transfer from gel into the sol (solid lines), and k_g for the corresponding reversible-thermodynamic equilibrium in that transport (dashed lines).

(right-hand part of the table) in Reference 13. The entropies show excellent agreement, although the two methods are absolutely independent. Table 3 shows that only a few contacts are formed between the transported polymer and the stationary gel phase in the gel front, where the deformation mainly takes place. A detailed analysis even shows how a given macromolecule moves along the gel front within the transported zone (see Ref. 13). If a sorption can be assumed at all to explain the measured PDC effect shown in Figure 4, only a sorption of the transported polymer on the polymer gel layer of the stationary phase as indicated can be meant.

After having packed a PDC column based on a system other than polystyrene–cyclohexane, in which exact measurements are feasible, the calibration chromatograms must be fitted to obtain equation (14) by the method described in Reference 11. If the PDC effect is found, the phenomenological theory stated above can directly be adopted. If the measured PDC

Table 3. Comparison of Normalized Molar Entropies of Deformation[a]

Measurements at 20°C				Theory of Deformation Concept, 20°C				
Quantity	$P = 353$	$P = 1082$	$P = 2444$	Quantity	n	$P = 353$	$P = 1082$	$P = 2444$
k_s/s^{-1}	1.5×10^5	0.88×10^5	0.61×10^5	$\sigma\,(N/m^2)$		3×10^4	3×10^4	3×10^4
k_g/s^{-1}	1.96×10^4	1.025×10^4	5.80×10^3	$G\,(N/m^2)$	1	0.623×10^5	0.207×10^5	0.899×10^4
$k'_g\,s^{-1}$	1.87×10^4	7.08×10^3	9.92×10^2		2	1.246×10^5	0.413×10^5	1.80×10^4
$Q(P)$	8.021	12.43	61.49		3	—	—	2.70×10^4
$K(P)$	7.650	8.585	10.52	σ/G	1	0.482	1.449	3.333
$Q(P)/K(P)$	1.048	1.448	5.846		2	0.241	1.726	1.667
$\delta Q(P)/K(P)$	0.048	0.448	4.846		3	—	—	1.111
$-\Delta S_{def}/R$	0.047	0.37	1.766	$\lambda - 1$	1	0.190	0.769	2.420
					2	0.087	0.310	0.935
					3	—	—	0.536
				$-\Delta S_{def}/R$	1	0.052	0.675	4.97
					2	—	0.260	1.90
					3	—	—	1.06

[a] Calculated from the measurements (6) for polystyrene samples of degrees of polymerization P as indicated at 20°C (left-hand half of table), and for the same P at 20°C as calculated from the deformation concept with tensile stress $\sigma = 0.3 \times 10^5\ N/m^2$ in the PDC column (right-hand half of table) as indicated (n is the number of contacts between the sorbed molecule and the stationary surface of the gel in the column; cf. Refs. 13, 5).

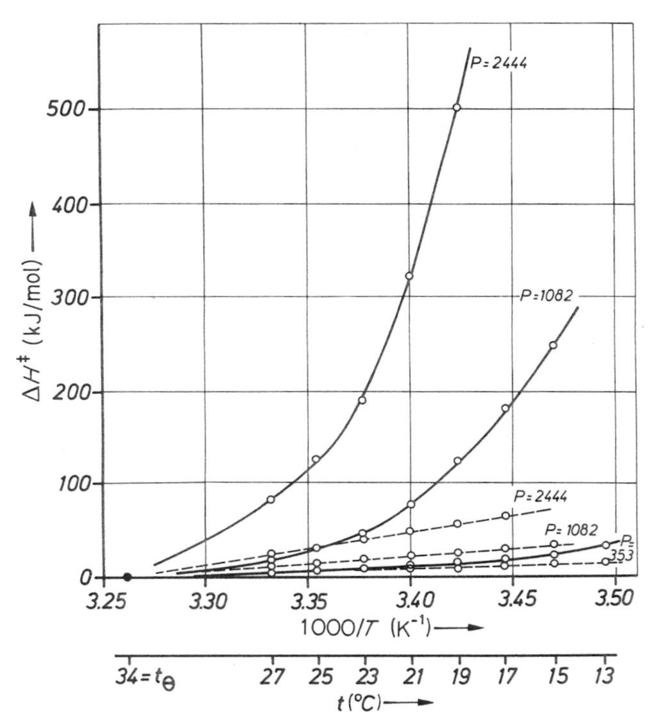

Figure 10. Arrhenius plot of the activation enthalpy in the flow equilibrium (solid lines) and the corresponding reversible contributions (dashed) for three typical degrees of polymerization of polystyrene, taken per mole of polymer.

effect is pronounced enough, this theory can be applied to the calculation of narrow molecular weight distributions, up to extremely small nonuniformities $U = (\bar{P}_w - \bar{P}_n)/\bar{P}_n$.

DETERMINATION OF NARROW MOLECULAR WEIGHT DISTRIBUTIONS

The definition of the molecular weight distribution $w(M) = dm_{red}/dM$ immediately shows how the MWD is related to an unspread chromatogram $D(V) = dm_{red}/dV$ normalized to 1 (i.e., enclosing a unit area). Taking the degree of polymerization $P = M/M_0$ instead of the molecular weight M, the well-known chain rule of differential calculus $w(P) = dm_{red}/dP = dm_{red}/dV(dV/dP) = D(V)/(dP/dV)$ together with the calibration curve $P = \psi(\bar{V})$ and equation (32), yield the relations

$$P = \psi(\bar{V}) \qquad \text{and} \qquad w(P) = \frac{D(V)}{|\psi'(\bar{V})|} \tag{17}$$

where $\psi'(\bar{V}) = dP/d\bar{V}$ represents the slope of the calibration curve at position P related to the average elution volume $\bar{V}$ defined in equation (1). As $D(V)dV = dm_{\text{red}} = w(P)dP$ follows from equations (17) and (32) in which $D(V)$ encloses a unit area, the MWD $w(P)$ is also normalized to 1,

$$\int_{-\infty}^{+\infty} D(V)\,dV = 1 = \int_{0}^{\infty} w(P)\,dP \tag{18}$$

so that the averages

$$\bar{P}_w = \int_0^{\infty} Pw(P)\,dP \qquad \text{and} \qquad \bar{P}_n^{-1} = \int_0^{\infty} P^{-1}w(P)\,dP \tag{19}$$

follow as usual.

The direct recalculation, equation (17), widely used in GPC, is known as *strip method* and is denoted by "S" in this section. In the case of PDC, the calibration curves are described by equation (14) with $V_e(P) \equiv \bar{V}$ from equation (1), giving the relations

$$\psi(\bar{V}) = \left[\frac{\ln\{\bar{V}/(\alpha V_e^0) - 1\} + \ln r_0}{\varepsilon_1}\right]^{1/\nu} \equiv P \tag{20}$$

and

$$\psi'(\bar{V}) = \frac{\alpha - \bar{V}\,d\alpha/d\bar{V}}{\alpha\varepsilon_1 \nu(\bar{V} - \alpha V_e^0)} P^{1-\nu} \equiv \frac{dP}{d\bar{V}} \tag{21}$$

which contain the quantities obtained in the preceding section; the differential quotient $d\alpha/d\bar{V}$ must be calculated numerically from the dependence $\alpha(\bar{V})$ shown in Table 2 with $V_e(P) \equiv \bar{V}$ calculated from equation (1) from the calibration chromatograms.

Before proceeding with the application of the theory to the measurements, we must define carefully the term "very narrow MWD." Conventionally, all MWDs with the nonuniformity $U = (\bar{P}_w - \bar{P}_n)/\bar{P}_n < 0.005$ are designated as "extremely narrow." In reality, however, the width of any MWD is defined only by its variance σ^2 calculated from the well-known statistical relation

$$\sigma^2 = \int_0^\infty (P - \bar{P}_w)^2 w(P)\, dP \qquad (22)$$

which easily yields the dependence of σ^2 on the z nonuniformity $U_z = (\bar{P}_z - \bar{P}_w)/\bar{P}_w$ related to the MWD:

$$\sigma^2 = \bar{P}_w^2 U_z \qquad (23)$$

This important relation is generally valid. In the case of very narrow MWDs, U_z is nearly equal to U, whereupon $\sigma^2 = \bar{P}_w^2 U$ is nearly valid. It is well known that the width of a Gaussian, measured between the sections of both inflection tangents with the P axis, is 4σ. This is nearly valid also for any narrow MWD $w(P)$ because all these MWDs tend with $U \to 0$ toward the Dirac delta function $\delta(P - \bar{P}_w)$, which can be represented by a Gaussian with vanishing variance. Hence, the width $4\sigma = 4\bar{P}_w\sqrt{U}$ can nearly be taken for any sufficiently narrow MWD $w(P)$. This quantity, however, depends not only on U, but also on $\bar{P}_w$. Therefore, any MWD with even $U = 0.001$ is in fact not very narrow if $\bar{P}_w$ is large enough. For $\bar{P}_w = 1000$, for example, the width $4000\sqrt{0.001} = 126$ follows, so that the MWD is significant between $P = 936$ and $P = 1064$. This can be seen well also from Figure 13 (below) and from Table 4, taking the normalized width $4\sigma/\bar{P}_w = 0.2$, which is not as low as could be anticipated from the corresponding nonuniformity $U = 0.0029$ alone: for $\bar{P}_w = 1498$, it follows that $\bar{P}_n = \bar{P}_w/(U + 1) = 1498/1.0029 = 1494$.

This brief statistical analysis yields two useful pieces of information. First, such "polymolecular indices" as $\bar{P}_w/\bar{P}_n$ or $\bar{P}_z/\bar{P}_w$, widely used in practice, are not of any statistical significance; only the corresponding nonuniformities $U = \bar{P}_w/\bar{P}_n - 1$ *and* $U_z = \bar{P}_z/\bar{P}_w - 1$ are significant. They represent, namely, a normalized variance of the mol fraction distribution $h(P)$ and of the MWD $w(P)$, respectively, as can be seen from equation (23): $U_z = \sigma_w^2/\bar{P}_w^2$ and $U = \sigma_n^2/\bar{P}_n^2$, both vanishing with $\sigma_w^2 \to \sigma_n^2 \to \sigma_{\text{Gauss}}^2 \to 0$ [with Dirac's delta function as limit for all $h(P)$ and $w(P)$]. Second, $\bar{P}_w$ and $\bar{P}_n$ cannot directly be measured in the case of very narrow MWDs because it would be impossible to reach the high accuracy required. For $\bar{P}_n = 1000$ and $U = 0.001$, for example, $\bar{P}_w = 1001$ follows; a relative error of nearly 1% in $\bar{P}_w$ would yield the nonuniformity $U = (1010 - 1000)/1000 = 0.01$, differing by 900% from the correct value 0.001. Hence $\bar{P}_w$ and $\bar{P}_n$ must be calculated according to equation (19) from $w(P)$ in the case of very narrow MWDs. It was shown above that such MWDs are not as narrow as they seem to be from a mere interpretation of, for example, $U = \bar{P}_w/\bar{P}_n - 1 = 0.001$. It is clear that $\bar{P}_w$ and $\bar{P}_n$ need not be integers because they represent *average* values, which must be accurate enough.

Unlike GPC, PDC is especially suited for the determination of narrow MWDs up to extremely low nonuniformities $U = (\bar{P}_w - \bar{P}_n)/\bar{P}_n$ if the measured PDC effect is pronounced enough. Namely, the contribution of the constant

Table 4. Characterization of Polystyrene Samples Investigated[a]

Sample	PDC		P_w						$U=(P_w-P_n)/P_n$					
				PDC					PDC					
	$t/(^{\circ}C)$	σ_D/σ	LS	S	G	K	BWF	GPC	S	G	K	BWF	GPC	δU_{GPC}[b]
K-110,000 PCC-PS-Std	15.0	0.08		1116	1116	1116			0.0091_5	0.0090_8	0.0090_8			0.039_7
(anionic)	21.0	0.59		1031	1106	—			0.0125_7	0.0093_2	—			
			1082				1065	1098				0.009_2	0.048_8	
	23.0	0.73		997	1105	—			0.0182_8	0.0100_1	—			
	25.0	0.81		898	1146	—			0.1572_9	0.0775_3	—			
K-233,000 PCC-PS-Std (anionic)	21.0	0.09	2444	2526	2509	2509	2360	2424	0.0109_2	0.0105_9	0.0105_7	0.025_3	0.071_6	0.061_0
9th BW fraction of sample K-110,000	15.0	0.11	—	—	1057	1057	1070	1065	—	0.0063_3	0.0063_2	< 0.005	0.045_7	0.039_4
0/2 Anionic THP/0°C	15.0	0.05	—	—	1225	1225	1090	1110	—	0.0102_5	0.0102_5	0.009_2	0.052_0	0.041_8
10th BW fraction of anionic sample Ti-Te-1	15.0	0.05	—	—	1498	1498	—	1480	—	0.0028_8	0.0028_8	< 0.005	0.049_7	0.046_8

[a] Here $\sigma_D = 2.65\ cm^3$ is the spreading variance obtained from the measured kernel sections and σ is the variance of the given polydisperse elution curve (cm^3); S indicates the strip method assuming a δ function as a kernal ($\sigma_D \rightarrow 0+0$), and G and K indicate inversion methods G39 and G40 from Ref. 17. In addition, LS, light scattering; PDC, phase distribution chromatography; BWF, Baker–Williams fractionation; GPC, gel permeation chromatography; cf. Ref. 20.

[b] Note that $\delta U_{GPC} = U_{GPC} - U_{PDC}$ is the correction of $U=(P_w-P_n)/P_n$ for spreading the GPC column used and specified as follows: 6 water-units: 5×10^5, 6.5×10^4, 5×10^4 to 1.5×10^4, 1.5×10^4–5000, 5000–2000, and 2000–700 Å, giving a slightly S-shaped calibration curve over the whole P region needed. Linear regression of the data in the last column against P_w from LS or PDC yields the straight line $\delta U_{GPC} = 0.023 + 0.000016\ P_w$ for $1000 \leqslant P_w \leqslant 3000$.

column resolution to the chromatogram now becomes comparable with that of the spreading in the case of GPC, whereas the strongly temperature-dependent resolution of the PDC column is always high in the dynamic region of the column (where the PDC effect is pronounced), vanishing asymptotically only in the vicinity of the theta point (where spreading phenomena alone can be measured with high accuracy).

If the MWD is not extremely narrow (e.g., for $U > 0.01$ at $\bar{P}_w > 500$ in the case of polystyrene), no use of complicated mathematics must be made, contrary to the situation for GPC; the fast strip method (17) (eqs. 20 and 21) can easily be applied to the PDC chromatogram of such a sample to obtain the MWD with sufficient accuracy. This can be seen well from Figure 11, showing the MWD of the standard polystyrene sample PCCK-110,000 (denoted by ∘, $P_w = 1116$, $U = 0.009$: see Table 4) calculated from the very broad PDC chromatogram of this sample at 15°C (Fig. 12). The breakdown of the strip method (17) applied to a GPC chromatogram of this sample is shown for comparison (triangles in Fig. 11). Other examples given in Table 4 show the weight average degree of polymerization $\bar{P}_w$ and the nonuniformities U

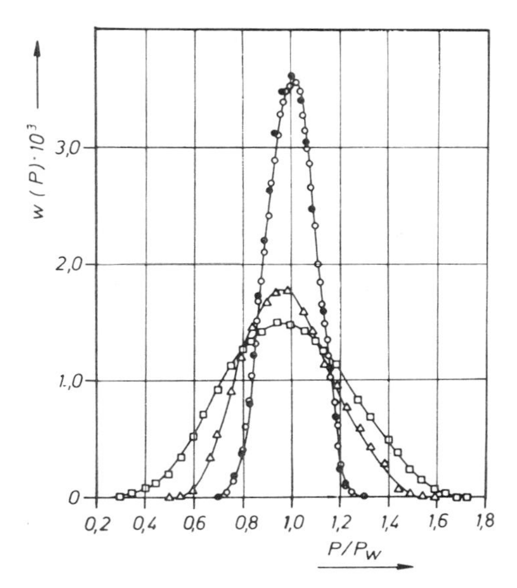

Figure 11. Distribution of the degree of polymerization of polystyrene sample PCC K-110,000 calculated from the elution curves of Figure 12 at 15°C (○), at 23°C (●), and at 25°C (□) (not shown in Fig. 12, as explained in text). For comparison: an incorrect distribution of the degree of polymerization of the same sample was calculated from GPC without correction for spreading (strip method, △).

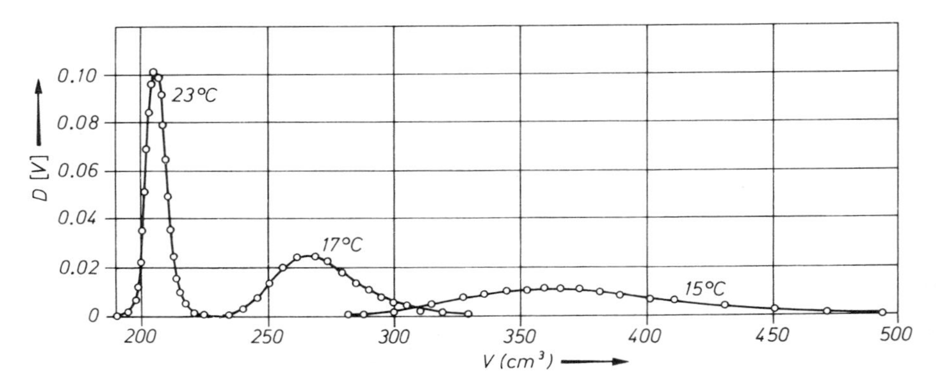

Figure 12. PDC elution curves $D(V)$ of the standard anionic polystyrene PCC K-110,000 ($\bar{P}_w = 1115$, $U = 0.009$) measured at three column temperatures. A comparison of data for 15 and 23°C shows a strong PDC effect at 15°C.

calculated according to equation (19) from MWDs obtained by means of the strip method (S) and by means of the exact inversion methods mentioned below (G and K) from PDC chromatograms of the indicated samples.

If the MWD is extremely narrow, or if it cannot be measured in a typical dynamic region of the PDC column, even the PDC chromatograms must be corrected for spreading. Contrary to GPC, however, the results of the general theory of spreading phenomena in chromatographic columns stated in Reference 14 can directly be applied to this procedure in PDC. It can be shown that the spreading of the concentration profile caused by axial dispersion in any chromatographic column consists of a symmetrical and an asymmetrical contribution. The symmetrical contribution is caused by hydrodynamic interactions and is described by the *elution rate* $u(P)$ and by the coefficient $D'(P)$ of axial dispersion, whereas the asymmetrical contribution is caused by kinetic interactions and is described by two first-order rate constants, namely by the coefficient of sorption $\lambda(P)$ and the coefficient of desorption $\lambda'(P)$.

Two very different sorts of kinetics must be distinguished in any chromatographic column: the fast kinetics of column resolution, and the slow kinetics of spreading, leading to a tailing of the concentration profile due to axial dispersion described mathematically by the last two terms on the left-hand-side of equation (24a).

The spreading itself is described by the integrodifferential equation (14)

$$\frac{\partial c(z,t)}{\partial t} - D'\frac{\partial^2 c(z,t)}{\partial z^2} + u\frac{\partial c(z,t)}{\partial z} + \lambda c(z,t) - \lambda\lambda' \int_0^t d\tau e^{\lambda'(\tau - t)} c(z,\tau) = 0 \quad \text{(24a)}$$

with the singular time boundary

$$c(z,0)=\delta(z) \tag{24b}$$

representing Dirac's delta function as a concentration impulse at the time $t=0$ of injection. This very complicated equation was solved analytically by Greschner (14), and the solution was recalculated to elution volumes. If the chromatogram exhibits a positive tailing, the spreading is completely described by the relatively simple spreading function (surface)

$$K(V,\bar{V}_{\mathrm{D}})=\frac{1}{\sigma_0\sqrt{2\pi}}\exp\left[\frac{-(V-\bar{V}_{\mathrm{D}})^2}{2\sigma_0^2}\right]$$
$$\times\left[1+\frac{\gamma_2}{2}\mathrm{He}_2\left(\frac{V-\bar{V}_{\mathrm{D}}}{\sigma_0}\right)+\frac{\gamma_3}{6}\mathrm{He}_3\left(\frac{V-\bar{V}_{\mathrm{D}}}{\sigma_0}\right)+\frac{\gamma_4}{24}\mathrm{He}\left(\frac{V-\bar{V}_{\mathrm{D}}}{\sigma_0}\right)\right] \tag{25}$$

representing the first four terms of a quickly convergent Fourier series related to the basis of modified Hermite polynomials

$$\mathrm{He}_n(\zeta)=2^{-n/2}\mathrm{H}_n(\zeta/\sqrt{2}) \qquad \text{with} \quad \mathrm{H}_n(\xi)=(-1)^n e^{\xi^2}\frac{\mathrm{d}^n}{\mathrm{d}\xi^n}e^{-\xi^2} \tag{26}$$

which can easily be calculated from the well-known recurrence relations

$$\begin{aligned} &\mathrm{He}_0(\zeta)=1 && \mathrm{He}_n(\zeta)=\zeta\mathrm{He}_{n-1}(\zeta)-(n-1)\mathrm{He}_{n-2}(\zeta)\\ &\mathrm{He}_1(\zeta)=\zeta && n\geqslant 2 \end{aligned} \tag{27a}$$

In equation (25), $\bar{V}_{\mathrm{D}}$ is the elution volume related to the calibration curve (eq. 20) for $P\equiv\bar{P}_{\mathrm{W}}$ of the sample measured near the theta point of the PDC column where the column resolution can be neglected; V is any elution volume of the corresponding chromatogram $D_{\mathrm{disp}}(V)\equiv K(V;\bar{V}_{\mathrm{D}})$ representing a section of the plane $\bar{V}_{\mathrm{D}}=\text{const}$ with the spreading surface (eq. 25). The Fourier coefficients γ_n are related to the set $(u(P), D'(P), \lambda(P), \lambda'(P))$ from equation (24a); they can easily be calculated from the statistical moments of the spreading chromatogram

$$\mu_n'=\int_{-\infty}^{+\infty}(V-\bar{V}_{\mathrm{D}})^n D_{\mathrm{disp}}(V)\,dV \qquad \text{for} \quad n=2,\ 3,\ \text{and}\ 4 \tag{27b}$$

with $\bar{V}_{\mathrm{D}}$ in analogy to equation (1):

$$\bar{V}_D = \int_{-\infty}^{+\infty} V D_{disp}(V)\, dV \qquad \sigma_D^2 = \mu_2', \; \gamma_D = \mu_3'/\sigma_D^3, \; \delta_D = \mu_4'/\sigma_D^4 - 3 \quad (27c)$$

and

$$\begin{aligned} \gamma_2 &= (\sigma_D/\sigma_0)^2 - 1 \\ \gamma_3 &= \gamma_D(\sigma_D/\sigma_0)^3 \\ \gamma_4 &= (\delta_D + 3)(\sigma_D/\sigma_0)^4 - 6(\sigma_D/\sigma_0)^2 + 3 \end{aligned} \quad (27d)$$

The statistical quantities σ_D^2, γ_D, and δ_D in these relations represent the variance, the skewness, and the kurtosis of the spreading eluogram $D_{disp}(V) = K(V; \bar{V}_D)$ designated as the kernel section for a given $P \equiv \bar{P}_w$ by the calculation of the MWD; the quantity σ_0^2 in equation (25) is the variance of the leading Gaussian in this tailed section. In most cases, even the simple relations

$$\sigma_0 = \sigma_D, \qquad \gamma_2 = 0, \qquad \gamma_3 = \gamma_D, \qquad \text{and} \qquad \gamma_4 = \delta_D \quad (28)$$

can well be applied in PDC, giving a very simple spreading surface (eq. 25) representing a kernel of the integral equation of the first kind

$$D(V) = \int_{-\infty}^{+\infty} K(V, \bar{V}) f(\bar{V})\, d\bar{V} \quad (29)$$

for the unspread chromatogram $f(\bar{V})$ related to the measured chromatogram $D(V)$ from which the MWD should be computed. It is a great advantage of PDC that the kernel sections $D_{disp}(V) = K(V; \bar{V}_D)$, and thus the whole kernel $K(V, \bar{V})$, can be measured directly in the PDC column near the theta point, whereas this cannot be achieved in GPC. The details of the theory and some results of the kernel measurements can be found in Reference 14. If the leading first term of equation (25) is inserted into the integral equation (29), the well-known semiempirical Tung equation is obtained (15).

To compute a very narrow MWD from a PDC chromatogram $D(V)$, the integral equation (29) with the measured kernel $K(V, \bar{V})$ must first be inverted. To do so, the integral operator

$$K = \int_{-\infty}^{+\infty} K(V, \bar{V}) \cdot d\bar{V} \quad (30a)$$

with the kernel (eq. 25) is introduced, and the compactly formulated integral equation (29)

$$D = Kf \tag{30b}$$

is inverted to

$$f = K^{-1}D \tag{30c}$$

by means of Hilbert space methods described in detail in Reference 16; for programming, one can use the inversion methods proposed by Golub (G) and by Köckler (K), both quoted in Reference 16 (cf. Ref. 17). Having once computed the unspread chromatogram (eq. 30c), the MWD $w(P)$ is calculated in analogy to equation (17):

$$P = \psi(\bar{V}) \quad \text{and} \quad w(P) = f(\bar{V})/|\psi'(\bar{V})| \tag{31}$$

Equations (20) and (21) remain unchanged. A comparison of equations (31)

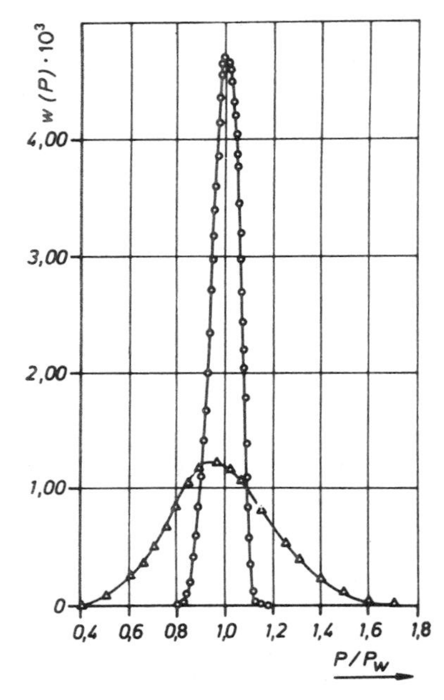

Figure 13. Distribution of the degree of polymerization of the 10th Baker–Williams fraction (middle fraction, $\bar{P}_w = 1500$, $U = 0.0029$) of polystyrene sample Ti-Te-1, polymerized anionically in 1, 2-dimethoxyethane at $-48°C$; PDC at 15°C (○, inversion methods K and G) and GPC, not corrected for spreading (strip method, Δ; see text).

and (17) immediately shows that the kernel (eq. 25) degenerates into Dirac's delta function $\delta(V-\bar{V}_D)$ in the case of the strip method, neglecting the spreading due to axial dispersion ($\gamma_n=0$ for $n \geqslant 2$ and $\sigma_0 \to 0$ in eq. 25). The integral equation (29) then degenerates into the identity

$$D(V)=\int_{-\infty}^{+\infty} \delta(V-\bar{V}) f(\bar{V})\, d\bar{V}=f(V) \tag{32}$$

and equations (31) and (17) are then identical.

Figure 13 compares a very narrow MWD of a BW middle fraction ($\bar{P}_w=1500$, $U=0.0029$; see Table 4), computed by these inversion methods, with the false MWD calculated from a GPC chromatogram of this sample obtained by means of the strip method (17) (denoted by triangles, uncorrected for spreading because of unknown kernel constants in eq. 25 applied to GPC). As another example, Figure 14 shows the integral MWD of the anionically polymerized and BW-prefractionated polystyrene sample 0/2 (see Table 4), computed by methods G and K (giving the same result, viz., circles, in Fig. 14) from PDC measurements at 15°C on the one hand and from Baker-Williams fractionation of this sample by Bohm (18) on the other hand (crosses in Fig. 14). The MWD obtained by Böhm is somewhat narrower because of somewhat different $\bar{P}_w$ values (cf. eq. 23).

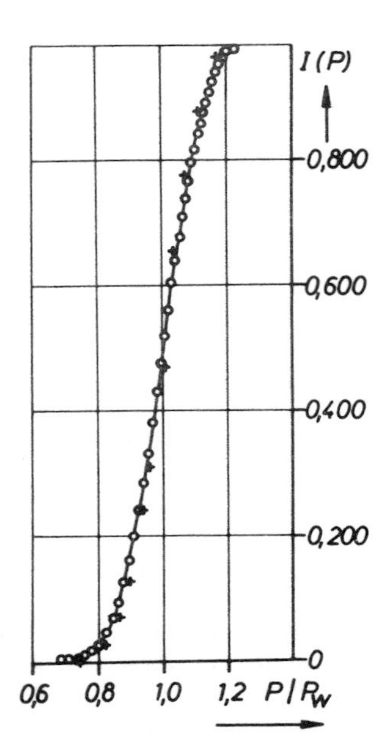

Figure 14. Integral distribution of the degree of polymerization for polystyrene sample 0/2 polymerized anionically in tetrahydropyran at 0°C ($\bar{P}_w=1225$, $U=0.010$) and Baker–Williams prefractionated: PDC at 15°C (○ by both inversion methods G and K), and Baker–Williams fraction of this sample by Böhm (18) (+; see Table 4 and text).

A comparison of Figures 11 and 12 and Table 4 is especially instructive. Figure 12 shows three measured PDC chromatograms of the same standard polystyrene sample PCC K-110,000 ($\bar{P}_w = 1115$ and $U = 0.009$; see Table 4) at 23°C (very narrow), 17°C (medium broad), and 15°C (very broad). Only the very broad elution curve allows a simple calculation of the MWD by means of the strip method (S), because only 8% of the curve width is caused by spreading, with 92% caused by resolution ($\sigma_D/\sigma = 0.08$ in Table 4), whereas an inversion method (G or K) must be applied when the MWD is calculated from the elution curve at 17°C or even 23°C, where 73% of the curve width is caused by spreading and only 27% by resolution ($\sigma_D/\sigma = 0.73$). The results of the MWD computations from Figure 12 are shown in Figure 11. The open circles give the MWDs computed from the very broad (15°C) chromatogram in Figure 12—here all three methods, S, G, and K, yield practically the same MWD—while the solid circles give the MWD computed by method G from the very narrow (23°C) chromatogram in Figure 12. It can be seen that these MWDs (solid and open circles) are almost identical, although the corresponding PDC chromatograms are extremely different. This important result again not only confirms the whole PDC concept stated in the preceding section, but also shows that PDC represents an absolute method for the determination of molecular weight distributions.

If the column temperature is chosen to be 25°C for the same sample (the corresponding chromatogram is not shown in Fig. 12 because it differs graphically only very slightly from that of 23°C), almost a kernel section is measured, showing that 81% of the curve width is caused by spreading and only 19% by resolution. Hence, the elution curve at 25°C contains so little information about the MWD that the mathematical inversion procedures G and K break down because of the condition number of the inverse operator K^{-1} in equation (30c) is too large (cf. Fig. 26 in Ref. 5 and Refs. 16 and 17). The false MWD obtained is also shown in Figure 11 (open squares); it is even broader than the likewise false MWD obtained from the strip method applied to an uncorrected GPC chromatogram of the same sample (open triangles in Fig. 11). This result shows well the kinds of problem arising if the contribution of the column resolution is too small compared with the spreading contribution. This is less important in the case of PDC, which shows a strongly temperature-dependent resolution, as it is with respect to GPC, where the column resolution is constant in any case.

In principle, broad MWDs (e.g., $U > 0.3$ at $\bar{P}_w > 100$ for polystyrenes) also could—up to some $\bar{P}_w$ limit depending on the $\bar{P}_w$ of the linear gel used in the column—be determined by PDC; GPC, however, is more convenient in this case. The strip method (17) can then easily be applied to such GPC chromatograms if the weight average of the degree of polymerization of the injected sample is not too high; the GPC calibration curve can then easily be fitted by means of spline polynomials. For samples with a high $\bar{P}_w$ value, electron microscopy [cf. Koszterszitz, Greschner, and Schulz (19)] or—in the future—light scattering would be the best method (cf. Ref. 16, Table 1, and the literature quoted).

ACKNOWLEDGMENTS

The author thanks Springer-Verlag and Hüthig & Wepf-Verlag for having granted permission to use the figures and tables in this chapter.

REFERENCES

1. G. V. Schulz, P. Deussen, and A. Scholz, *Makromol. Chem.*, **69**, 47 (1963).
2. L. L. Böhm, R. H. Casper, and G. V. Schulz, *J. Polym. Sci.*, **12**, 239 (1974).
3. R. H. Casper, *Thesis*, Mainz, 1970.
4. R. H. Casper and G. V. Schulz, *Sep. Sci.*, **6**(2), 321 (1971).
5. G. S. Greschner, *Adv. Polym. Sci.*, **73**, 1 (1986).
6. G. S. Greschner, *Makromol. Chem.*, **180**, 2551 (1979).
7. G. V. Schulz and B. Jirgensons, *Z. Phys. Chem. B*, **46**, 105 (1940).
8. B. A. Wolf and J. W. Breitenbach, *Makromol. Chem.*, **108**, 263 (1967).
9. B. A. Wolf, H. F. Bieringer, and J. W. Breitenbach, *Br. Polym. J.*, **10**, 156 (1978).
10. L. A. Kleintjens, R. Koningsveld, and W. H. Stockmayer, *Br. Polym. J.*, **8**, 144 (1976).
11. G. S. Greschner, *Makromol. Chem.*, **181**, 1435 (1980).
12. G. S. Greschner, *Makromol. Chem.*, **182**, 2845 (1981).
13. G. S. Greschner, *Makromol. Chem.*, **186**, 1047 (1985).
14. G. S. Greschner, *Eur. Polym. J.*, **19**, 881 (1983).
15. L. H. Tung, *J. Appl. Polym. Sci.*, **10**, 375 (1966); L. H. Tung, *ibid.*, **33**, 775 (1969).
16. G. S. Greschner, *Eur. Polym. J.*, **20**, 475 (1984).
17. G. S. Greschner, *Maxwellgleichungen*, Vol. 3, *Mathematische Hilfsmittel*, 1st ed., Hüthig & Wepf Verlag, Basel Heidelberg New York, 1981, Chap. 5, Sect. 21 and Chap. 7, Sect. 31, where also advice for programming is given.
18. L. L. Böhm, private communication.
19. G. Koszterszitz, G. S. Greschner, and G. V. Schulz, *Makromol. Chem.*, **178**, 1169 (1977).
20. G. S. Greschner, *Makromol. Chem.*, **183**, 2823 (1982).

CHAPTER

12

MOLECULAR WEIGHT DISTRIBUTION FROM FIELD-FLOW FRACTIONATION

J. CALVIN GIDDINGS, KARIN D. CALDWELL, and LAYA F. KESNER

Department of Chemistry University of Utah Salt Lake City, Utah

INTRODUCTION

Field-flow fractionation (FFF) is a rather broad-ranging family of techniques by means of which high molecular weight materials can be separated and characterized (1–8). The FFF methodology appears to be applicable to essentially all macromolecules and particles over a mass range covering 15 orders of magnitude. This includes species ranging in molecular weight from less than 1000 up to those whose effective molecular weight is 10^{18}, the latter corresponding to a 100-μm particle. The technique is applicable in both aqueous and nonaqueous systems; it has been applied to many particulate materials including virus particles, whole cells, environmental particles, latices, pigments, and emulsions. It has also been applied to a broad range of macromolecules including proteins, DNA, humic materials, and synthetic polymers, both lipophilic and water soluble. In this chapter we confine our attention to the macromolecular applications, particularly as related to synthetic polymers, although the principles are virtually the same as those required to deal with particulate material.

Field-flow fractionation resembles chromatography in experimental operation (1–8). With either approach, the heart of the system is a flow channel or column in which separation takes place. The separation occurs in a flow stream within the channel. A narrow sample pulse is introduced into this stream at the head of the channel. The flow rate is controlled by a pump. Fractionated macromolecules are eluted from the channel into a detector and, if desirable, a fraction collector. Figure 1 is a schematic diagram of an FFF system.

The major difference between FFF and chromatography lies in the mechanism of retention. In chromatography, differential retention is achieved by the equilibrium partitioning of macromolecules between a mobile phase and a stationary phase. In the case of size exclusion or gel permeation

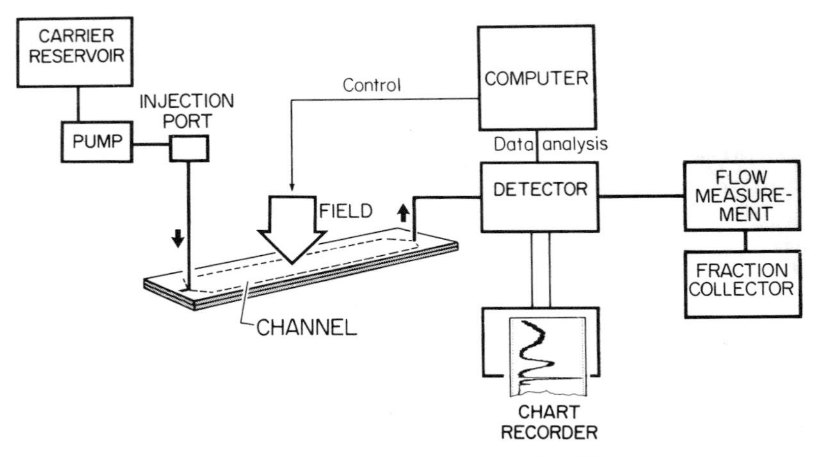

Figure 1. Schematic diagram of FFF channel and auxiliary equipment.

chromatography, the stationary phase is simply a porous matrix containing molecule-sized pores.

In FFF, retention is achieved by the action of an external field, which drives macromolecules into the low flow region near one of the walls of the separation channel. This mechanism is illustrated in Figure 2. We note that

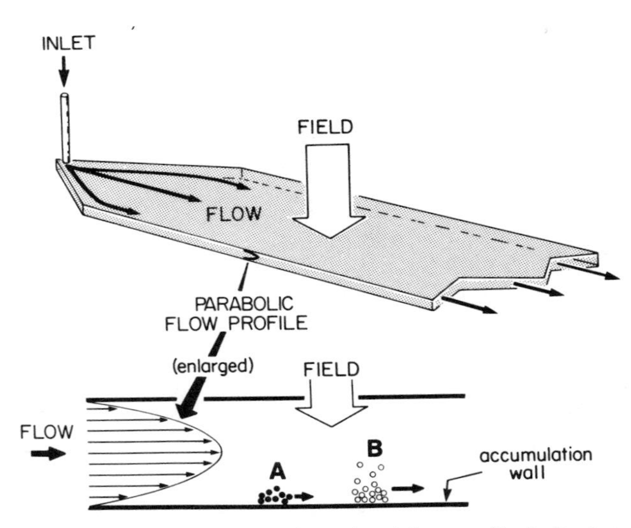

Figure 2. Illustration of structure of FFF channel and flow profile, indicating separation of components A and B.

the channels are normally of rectangular cross section with the external field applied across the face of the channel, driving macromolecules toward the accumulation wall, one of the two main walls of the channel. After the field has been applied and a short relaxation period has elapsed, the macromolecules approach a steady state condition in which the field-driven motion toward the wall is balanced by a diffusion-driven motion away from the wall. When the steady state condition has been approached, flow can be initiated; this marks the beginning of the separation process.

The flow profile in the FFF channel, also illustrated in Figure 2, is normally parabolic or very close to parabolic [we ignore the deparature from parabolic flow next to the channel edges (9)]. For such parabolic flow profiles, the flow velocity decreases as one approaches the accumulation wall; the velocity falls to zero upon reaching the wall. Thus the low flow region near the wall approaches a condition of stagnation, much like the stationary phase in chromatography. FFF is based on the principle that different kinds of macromolecules can be driven to different depths within this "stagnation" region next to the accumulation wall. Components driven close to the wall find themselves in a portion of the flow stream with a very low flow velocity. These components are consequently carried very slowly down the channel. Those species whose distributions extend into the faster streamlines near the channel center are carried more rapidly downstream. Consequently, separation is achieved based on how vigorously different macromolecular species are driven toward the wall by the externally applied field.

In most FFF systems, the force exerted on the macromolecule by the external field increases with molecular weight. Thus the highest molecular weight species tend to be driven closest to the accumulation wall, where they are carried downstream at the lowest velocity. Lower molecular weight species, having less force exerted on them, remain distributed in regions of more rapid stream flow and are thus displaced rapidly down the channel. Consequently, the normal elution sequence is one in which small macromolecules elute ahead of large macromolecules. This, of course, is the inverse of the elution order found with gel permeation chromatography.

A variety of fields or gradients can be used to induce retention in FFF (6, 8). Any externally applied influence capable of herding macromolecules toward the accumulation wall of the channel will yield FFF retention. Nearly all separations to date in FFF have been achieved using four different fields or gradients: a thermal gradient, a sedimentation field, an electrical field, and a cross-flow or hydraulic gradient. These four "fields" have led to the four distinct subtechniques of FFF called thermal FFF, sedimentation FFF, electrical FFF, and flow FFF, respectively. Each subtechnique has its own distinctive range of applicability, its characteristic selectivity, and other unique advantages and disadvantages as detailed later.

As a class, FFF methods have some unique advantages that merit their consideration for polymer molecular weight characterization. First of all, FFF is a high resolution separation method which breaks a polymer sample down into a detailed distribution curve. Fractions can be collected if necessary for further characterization, or the eluting stream can be continuously monitored by other techniques such as light scattering for further information on the sample. We note that the intrinsically high selectivity of FFF systems (greater than that of gel permeation chromatography) enhances the effectiveness of the separation process in characterizing macromolecules (10, 11).

The versatility of FFF is another advantage in applications to macromolecules. With the proper choice of applied field (or subtechnique), virtually any soluble macromolecular material is expected to be resolvable by FFF. Furthermore, because retention is induced by an external field, the field strength can be varied as desired to control retention levels and to influence resolution and separation speed. To shift to a sample with a different molecular weight range, it is necessary to change only the field strength, not the separation channel.

We note that the FFF system consists of an open channel without packing and therefore has minimal surface area. The planar surface of the accumulation wall with which the macromolecules come into contact can, if necessary, generally be coated to reduce adsorption and catalytic activity. We also note that the open structure of the FFF channel reduces shear forces, particularly the extensional shear characteristic of packed columns, and therefore leads to a reduced level of degradation of high molecular weight polymers and macromolecules.

The combination of the simple, open geometry of the FFF channel, the predictable (usually parabolic) flow, and the uniformly applied field leads to another advantage of FFF: a rigorous theory can be developed to relate FFF retention to macromolecular properties. Turned around, the theoretical equations yield the value of macromolecular properties in terms of observed retention levels. Consequently, one can often get distribution curves, including molecular weight distributions, without calibration standards or with a very limited use of calibrating samples. This theory-based approach to molecular weight distribution curves has a weak link, however, with thermal FFF, where one parameter controlling retention, the thermal diffusion coefficient, has no sound theoretical ties to polymer type and molecular weight (12). This case requires calibration but works out very simply, as will be shown.

We will show that the advantage of rigorous theory carries over to the characterization of the molecular weight distribution of narrow polymer standards. Because peak broadening in an FFF channel can be described exactly, this channel contribution can be subtracted from the total peak broadening to yield the band broadening due to polydispersity alone. By means of this procedure, the polydispersity of narrow samples can be readily obtained. Because of the high selectivity of FFF, polydispersities can be measured even for highly "monodisperse" ($\bar{M}_w/\bar{M}_n < 1.01$) samples (13).

The disadvantages of FFF are twofold. First of all, the technique does not readily adapt to the separation of large samples. For most analytical work, of course, the emphasis on small samples (< 1 mg) is a plus.

The second disadvantage of FFF is that it has not yet been widely applied to a large variety of macromolecular materials. Although more than two dozen research groups are now engaged in FFF studies, most of these have entered the field only recently. A large fraction of the work on macromolecules and colloidal systems has been done in our laboratory. [Recent work led by Dr. J. J. Kirkland at du Pont has also covered a broad base of applications (14–16).] Consequently, the experience base is not as broad as that found in more mature fields. This means that each new application requires the thoughtful choice of the best FFF subtechnique for the problem, along with the selection of effective operating conditions. The object of this chapter is to ease the path for those who see sufficient advantages in FFF methodology to apply it to their macromolecular problems.

THEORY

Role of the Driving Force

We have emphasized that the degree of retention depends on the applied force driving the macromolecules to the accumulation wall. The driving force in turn is proportional to the field strength S as shown by the following equation (8)

$$F = \phi S \tag{1}$$

in which F is the force per mole and ϕ is a constant of proportionality called the solute–field coupling constant. Parameter ϕ, which equals the force F at unit field strength, depends on both system and solute parameters, among the latter being the molecular weight.

Under the influence of applied force F the solute molecules form a Boltzmann-like distribution at the accumulation wall (17). This distribution is represented by

$$c/c_0 = e^{-(Fx/\Re T)} \tag{2}$$

where c is the concentration at distance x above the wall, c_0 is the concentration at the wall, and $\Re T$ is the thermal energy. If we write this distribution as

$$c/c_0 = e^{-x/l} \tag{3}$$

then the simple exponential constant l, representing the mean elevation of molecules above the wall, is found by comparison with equation (2) to be

$$l = \Re T/F \tag{4}$$

This equation shows that the mean distance of sample molecules from the wall (sometimes termed the mean layer thickness) is inversely proportional to the applied force.

The compression of the sample against the accumulation wall is best described in terms of a dimensionless form of l termed the retention parameter λ

$$\lambda = l/w = \Re T/Fw \tag{5}$$

where w is the channel thickness.

Whereas equations (4) and (5) show the inverse dependence of l and λ on F, it is sometimes useful to think of the formation of the steady state solute cloud against the accumulation wall in terms of the balance between mean velocity U induced by the field and the opposing diffusion coefficient D (see discussion in introductory section). This change in viewpoints can be expressed mathematically by writing

$$U = F/f \tag{6}$$

where f is the molar friction coefficient (defined as the force on one mole per unit velocity) of molecules being dragged through the solvent. If we substitute equation (6) into equation (5), we get

$$\lambda = \Re T/fUw \tag{7}$$

The term $\Re T/f$, according to the Einstein equation, is equal to the diffusion coefficient D. Therefore

$$\lambda = D/Uw \tag{8}$$

which shows that λ (along with l) increases with D but decreases as the induced velocity U toward the accumulation wall is made higher.

For completeness, we show the dependence of λ on field strength S by substituting equation (1) into equation (5). This yields

$$\lambda = \Re T/\phi S w \tag{9}$$

Since S, U, and F are all proportional to one another, it is not surprising to see the inverse dependence of λ on these parameters as found in equations (5), (8), and (9).

Up to this point we have focused on parameters controlling the steady state distribution of components against the accumulation wall. However, when the components first enter the channel they are generally distributed evenly over thickness w. The field must be applied for a finite time, approximated by the relaxation time τ, to drive the components into their steady state distributions. Time τ can be expressed as the time necessary to drive the components most remote from the accumulation wall, namely those near the opposite wall, fully across the channel, where their steady state distributions are to be formed. Thus the most remote components must be driven entirely across channel thickness w and, since they are driven by the field at velocity U, the time required for this transit is

$$\tau = w/U \tag{10}$$

To avoid undue zone distortion, the flow should be stopped for a time (the stop-flow time) equal to or greater than τ (18).

An alternate expression for τ can be obtained by substituting from equation (6) for the U in equation (10) and then replacing f by $\Re T/D$. With these substitutions we get

$$\tau = w\Re T/FD \tag{11}$$

Retention

We said earlier that the components compressed most tightly against the wall (usually the components of highest molecular weight) are caught up in relatively slow streamlines and thus migrate only slowly through the FFF channel. Consequently, their retention time t_r is relatively long and the volume of carrier liquid needed to sweep them through the channel, the retention volume V_r, is relatively large. These retention characteristics are generally measured in terms of the retention ratio R, which is related to V_r and t_r through

$$R = t^0/t_r = V^0/V_r \tag{12}$$

where t^0 is the channel void time (the elution time for a nonretained species) and V^0 is the void volume of the channel. Simply interpreted, R is the ratio of the downstream migration velocity of a given species relative to the mean velocity of the carrier liquid.

Clearly, R will be lowest for species compressed most firmly against the wall, that is, species with low l and λ values (17). When λ is much smaller than unity, it can be shown that there is a direct proportionality between R and λ

$$R = 6\lambda \tag{13}$$

However, for λ values in excess of 0.05 the error in equation (13) exceeds 10% and it is better to use the approximation (17)

$$R = 6\lambda - 12\lambda^2 \tag{14}$$

For still larger λ values, or for more accurate results, the following rigorous equation can be employed (1, 8)

$$R = 6\lambda[\coth(1/2\lambda) - 2\lambda] \tag{15}$$

If we replace λ by the expression of equation (5), each of the last three equations expresses a direct relationship between measured parameter R and the force F acting on one mole of the designated component. Thus equation (13) becomes

$$R = 6\mathfrak{R}T/Fw \tag{16}$$

and the more rigorous form, equation (15), becomes

$$R = 6\mathfrak{R}T/Fw[\coth(Fw/2\mathfrak{R}T) - (2\mathfrak{R}T/Fw)] \tag{17}$$

The forces exerted on macromolecules and particles by fields and gradients of most types are well known. Thus the force exerted by the sedimentation field is given by

$$F_s = M(\Delta\rho/\rho_s)G \tag{18}$$

where M is the molecular weight, ρ_s is the density of sample material, $\Delta\rho$ is the density difference between the sample material and the liquid carrier, and G is the sedimentation field strength expressed as acceleration. (More rigorously, ρ_s should be defined as the reciprocal of the partial specific volume of the sample component.) When this force expression is substituted into equation (5), we get the λ equation appropriate for sedimentation FFF (1)

$$\lambda_S = \frac{RT}{M(\Delta\rho/\rho_s)Gw} \tag{19}$$

This λ can be substituted into equation (13), (14), or (15) as another means of relating the experimental parameter R to component properties such as molecular weight M.

The λ expressions for other subtechniques of FFF can also be directly obtained using equations (5), (7), or (8). For flow FFF we obtain (19)

$$\lambda_F = DV^0/\dot{V}_c w^2 \tag{20}$$

in which $\dot{V}_c$ is the volumetric cross-flow rate. For polymeric materials, the diffusion coefficient D (which is the only term in this expression reflecting component properties) can be expressed by (20, 21)

$$D = A/M^b \tag{21}$$

in which A and b are constant for a given polymer type in a given solvent. (Exponent b generally lies in the 0.5–0.7 range.) This equation provides the tie-in with molecular weight M for this subtechnique.

In the case of thermal FFF the λ value is given by (22)

$$\lambda_T = \frac{D}{D_T(dT/dx)w} \simeq \frac{D}{D_T\Delta T} \tag{22}$$

where D_T is the thermal diffusion coefficient, dT/dx is the temperature gradient, and ΔT is the temperature drop between hot and cold walls. Unfortunately, a sound theoretical basis for D_T does not appear to be available. It is found empirically, however, that for a given polymer series D_T is a constant independent of molecular weight (12), and thus the ratio D/D_T can be described by the expression

$$D/D_T = A'/M^b \tag{23}$$

which again provides the desired molecular weight dependence in a simple form. For completeness we note that D_T, although independent of molecular weight, depends upon the composition of both polymer and solvent (23).

For electrical FFF the U of equation (8) can be written as μE, where μ is the mobility and E is the electrical field strength. Accordingly equation (8) assumes the form (24)

$$\lambda = D/\mu E w \tag{24}$$

In some cases the molecular weight dependence of λ will be determined solely by the variation of D with molecular weight, and in some cases μ will also depend on molecular weight; thus the overall dependence of the ratio D/μ on molecular weight requires consideration. Although in theory electrical FFF can be applied to charged particles and polymers of nearly all types, in practice the technique has not been widely applied because of experimental difficulties. Accordingly, we do not consider this subtechnique in any greater detail.

In all the FFF subtechniques above, λ depends on molecular weight M, and since retention ratio R depends on λ (see eq. 15), the level of retention invariably acquires an M dependence. Thus the components of a polymeric (or particulate) series are distributed across the experimental retention time or volume range according to molecular weight.

Selectivity

The degree to which an FFF process (or any other separation procedure) is able to pull apart close-lying molecular weights is determined by the mass selectivity S_M, which can be defined in two equivalent forms (10)

$$S_M = \left|\frac{d \ln V_r}{d \ln M}\right| = \left|\frac{d \ln R}{d \ln M}\right| \tag{25}$$

These expressions show that mass selectivity is equal to the percentage change in retention volume V_r corresponding to a 1% change in M.

When retention ratio R is adequately approximated by equation (13), the selectivity assumes the limiting form

$$S_M = \left|\frac{d \ln \lambda}{d \ln M}\right| = \left|\frac{d \ln \phi}{d \ln M}\right| \tag{26}$$

where the latter equality is based on equation (9). The expression for selectivity becomes somewhat more complicated when the more rigorous expression of equation (15) is used in place of equation (13).

Values of S_M can be immediately estimated using equation (26) and the various λ equations obtained above. Thus for sedimentation FFF the combination of equations (19) and (26) yields the limiting value $S_M = 1$. For flow FFF, in which λ is obtained by substituting equation (21) into equation (20), the selectivity becomes $S_M = b$, that is, about 0.5–0.7. (However, for populations of spherical particles $S_M = \frac{1}{3}$.) The λ required for selectivity in thermal FFF is obtained by substituting equation (23) into equation (22), which in conjunction with equation (26) shows again that $S_M = b$, the same result as found for flow FFF.

The results above show that for polymeric materials, values of selectivity S_M range between 0.5 and 1.0 as long as equation (13) is valid, which one generally strives for. By contrast, S_M values for size exclusion chromatography are considerably lower, in the vicinity of 0.1–0.2 (10).

While the high selectivity of FFF means that components of different molecular weight are separated by as large a relative increment as presently available by any method, the different molecular weight components provide information on the molecular weight distribution only if band broadening is sufficiently small that the components do not seriously overlap.

Band Broadening

As in chromatography, band broadening is expressed in terms of plate height H and plate number N. The plate height is defined simply as $H = \sigma^2/Z$, where σ^2 is the variance of the band and Z is the distance the component has traveled from the inlet of the channel (25). For a single monodisperse polymeric component, σ^2 can be obtained from the width at half-height $w_{1/2}$ by means of the following equation:

$$\sigma^2 = \frac{1}{8 \ln 2}\left(\frac{w_{1/2}}{Z}\right)^2 \tag{27}$$

Plate height H in chromatography or FFF is generally the sum of several contributing terms. For FFF we can write (1, 8)

$$H = H_n + H_p + \sum H_i \tag{28}$$

where H_n is a nonequilibrium term having origins very similar to the nonequilibrium term of chromatography and H_p is the polydispersity contribution to plate height. The summation term represents various end effects and channel nonuniformity disturbances that can be made very small or even negligible in well-designed systems. (The longitudinal diffusion term for macromolecules is so small that we have not bothered to show it in eq. 28.)

The two dominant terms of equation (28), H_n and H_p, have vastly different origins and significance. The term H_n is the major term for band broadening in the channel system. Quantity H_p is only an "apparent" band broadening; it arises by virtue of the migration at different velocities of components of different molecular weight in a sample, which therefore appear as part of a broadened band. However, this "broadening" actually reflects the separation of the individual components, which is essential to obtaining molecular weight distributions; the polydispersity broadening increases in magnitude (and in usefulness) for those systems with the highest selectivity (13).

It is useful to specify the nature of H_n and H_p in more detail. The first of these, the nonequilibrium term, is proportional to flow velocity as expressed by (17, 26)

$$H_n = C\langle v \rangle \tag{29}$$

As a consequence of the uniform channel geometry of FFF systems, the nonequilibrium coefficient C can be calculated from theory. Although the rigorous expression is rather cumbersome, it can be approximated by (26)

$$C = 24\lambda^3(1 - 8\lambda + 12\lambda^2)(w^2/D) \tag{30}$$

The polydispersity term H_p, by contrast, depends upon the polydispersity μ of the polymer sample, where μ is the ratio of the weight average to number average molecular weight. Specifically (13)

$$H_p = LS_M^2(\mu - 1) \tag{31}$$

where L is the channel length. We note that this term is independent of flow velocity. Accordingly, if we neglect the summation term in equation (28) and use equations (29) and (31) to emphasize the velocity dependence (or lack of it) of the two principal terms, the plate height becomes

$$H = C\langle v \rangle + LS_M^2(\mu - 1) \tag{32}$$

We observe that since C is available from theory or from the slope of an H versus $\langle v \rangle$ plot, the second or polydispersity term can be evaluated from experimental H measurements. In that S_M can be predicted from theory (see eq. 25), the polydispersity μ can be obtained as a consequence of plate height measurements. This approach is quite sensitive for rather small μ values in FFF because the polydispersity term is magnified by the large selectivity of FFF systems, as emphasized earlier. Consequently, narrow polymer samples with μ values down to about 1.003 have been characterized by FFF methods, detailed later.

Resolution and Fractionating Power

The fact that we can observe polydispersities down to $\mu = 1.003$ implies that the components of such extremely narrow fractions are adequately resolved for experimental observation. In general, the resolution R_s of two components differing in molecular weight by ΔM is given by (11)

$$R_s = \frac{(N)^{1/2}}{4} S_M \frac{\Delta M}{M} \tag{33}$$

where the number of plates N is given by

$$N = \frac{L}{H_c} = \frac{L}{H - H_p} \tag{34}$$

in which H_c is the channel contribution to the plate height, equal to the total plate height less the polydispersity contribution.

For continuous distributions of polymeric molecules, there is no specific ΔM on which to focus at the exclusion of other ΔM's. However, we can gage a system's intrinsic resolving power by using an index that expresses the resolution relative to ΔM; this is the fractionating power or, more specifically, the molecular-weight-based fractionating power, given by (27)

$$F_M = \frac{R_s}{\Delta M/M} \tag{35}$$

Using equation (33), we find that F_M acquires a form independent of ΔM, namely

$$F_M = \frac{(N)^{1/2}}{4} S_M \tag{36}$$

Equation (36) makes clear that the "specific" resolving or fractionating power F_M is determined by two parameters—N and S_M. While both terms are important, the second term S_M is the more influential because its power is unity instead of 1/2 as is the case for N. We note that S_M (or at least its maximum value) tends to be fixed by the mechanism of separation, while N is easily changed (e.g., by changes in column or channel length, flow velocity, flow uniformity, etc.).

In a recent study comparing the fractionating power of thermal FFF and size exclusion chromatography, it was observed that S_M was larger for FFF (as already pointed out) and that with current technology N can be made larger for size exclusion chromatography. Because of the greater influence of S_M in equation (36), it was found that thermal FFF generally has a higher fractionating power than size exclusion chromatography (11).

Programming

We note finally that when a very broad molecular weight distribution is subjected to FFF analysis, the high selectivity of FFF distributes the components over a very broad range of elution volumes. While this extensive "spreading out" along the retention volume scale provides maximum resolution, it has the disadvantage that some components do not emerge for a very long time, causing excessive delays in the completion of the experiment. In those cases a very simple technique, programmed field FFF, can be used to strike a compromise between fractionating power and analysis time (28). In programmed field FFF, the field strength is originally high but gradually declines in the course of the run. This gradual reduction in field strength forces the high molecular weight components that normally migrate most slowly to gradually increase their speed of migration to ensure that they emerge within a reasonable time (5).

Because of the changing field strength, the mathematical description of field-programmed FFF is somewhat more complicated than that of "isocratic" FFF. For the latter case the retention time t_r is simply calculated using equation (12) in combination with equation (15), where λ is chosen appropriately from the various equations we have given for the different subtechniques of FFF. In the field-programmed case, R is no longer constant (because λ varies with field strength; see eq. 9) and equation (12) cannot be used. In this instance retention time must be calculated from the integral equation (28)

$$L = \int_0^{t_r} R(t)\langle v\rangle(t)\,dt \tag{37}$$

where the time dependence of R is expressed as $R(t)$. We note that, in theory, the flow velocity v can also be made to vary with time as suggested by the equation. When flow velocity is changed in the course of the run, the procedure is referred to as flow programming (29). While flow programming is another promising approach for reducing the excessive analysis time of broad molecular weight distributions, only one experimental study has been carried out to investigate its efficacy (29).

INSTRUMENTATION

FFF Channel and Auxiliary Equipment

The distinguishing feature of FFF methodology is the FFF channel and column system. The column includes the separation channel as well as the apparatus for applying the external field. The ribbonlike channel, site of the fractionation process, may vary in dimensions, with length L typically 0.25–1.0 m, breadth b 1.0–3.0 cm, and thickness w 0.05–0.5 mm. The ends of the channel are tapered into trianglelike elements to provide a transition from the tubelike inlet flow to the ribbonlike channel flow (see Fig. 2).

The position of the channel in the total instrumental assembly was shown in Figure 1. The auxiliary equipment, consisting of the reservoir of carrier fluid, the pump to deliver the carrier, the injection port to introduce the sample, the detector to monitor the sample components, and the optional fraction collector, is similar for all subtechniques of FFF. Also similar for all subtechniques is the integration of a computer into the system, primarily for the control of operation and for data analysis. The auxiliary equipment generally resembles that used for liquid chromatography and is consequently widely available.

It is preferred that the pumping system for FFF provide a steady, controllable flow without pulses that may interfere with the detector response. Pressures need not be high; 1–10 atm is generally adequate, depending on the specific apparatus. The Kontron model LC1040 pump, the Chromatronix CMP-I and CMP-IV pumps from Laboratory Data Control, and the Gilson Minipuls 2 peristaltic pump have generally proved satisfactory. Simple gravity systems, which have the advantage of being totally pulseless, have also been used to deliver carrier liquid. In addition, a pneumatic pump has been designed and constructed in our laboratory (1). Recently a piston pump has also been designed for use in flow FFF. The single piston, driven by a stepping motor, simultaneously pumps and withdraws liquid at a preset rate. The controlled withdrawal of liquid is an important element of flow FFF because there are two simultaneous flow streams (channel flow and cross-flow) whose flows must be independently controlled.

In our laboratory, flow rates have generally been measured by collecting the channel effluent in a buret and observing the collected volume at fixed intervals of time. However, we have worked with other flow monitoring devices, including a special bubble meter (30).

The injection port has generally consisted of a septum device. More recently Rheodyne and Valco sampling valves have also been used.

The choice of detector is dictated by the sample property amenable to measurement. A highly sensitive detector is desirable (but not always available) because small FFF samples are advantageous to avoid overloading. We have most often used a refractive index detector for polymer characterization. However, for solvents that are transparent down to a sufficiently low wavelength, we have used ultraviolet detectors (such as the Laboratory Data Control model 1285 and Altex model 153). When the channel, in the case of thermal FFF, is maintained at an elevated pressure (to prevent boiling of the carrier liquid), a restriction (e.g., Nupro fine metering valve) at the channel exit reduces the pressure before the stream is introduced into the refractive index detector (31). The Altex UV detector, however, can tolerate high pressure (to 500 psi); the restriction can be located after the dectector with less extracolumn zone broadening.

Detector sensitivity may be enhanced by constructing FFF channels with a split outlet for the selective collection and sampling of the lower laminae of the flowstream in the FFF channel (32, 33). This works because, as pointed out in the section entitled "Theory," most of the solute is concentrated in the laminae near the accumulation wall.

The most effective splitting system is one in which a thin physical stream-splitting element extends across the breadth of the channel to evenly divide the channel flow (32). A somewhat simpler and less effective splitter consists of two concentric stainless steel tubes mounted with the inner tube protruding beyond the opening of the outer tube. The inlet of the inner tube is positioned very close to the channel accumulation wall; the entrance to the outer tube is flush with the upper (opposite) wall of the channel (33).

With flow splitting, the carrier flow rate of the sample-rich stream may be independently monitored by collecting the eluent in a buret. However, if the flow is well controlled and standardized, the eluent may be collected in a fraction collector so that separated portions of the sample can be examined by other techniques.

Thermal FFF

FFF channels of all types share important features. Therefore, the channel and column system for thermal FFF, the dominant subtechnique for polymer analysis, is discussed in detail as a reference system; the channels for other subtechniques are considered as they differ from the thermal case.

Figure 3 presents an exploded view of the thermal FFF channel. The channel shape is cut from a Mylar or Teflon spacer of known thickness, originally 0.25 mm but now usually 0.076 mm thick for improved efficiency (see eqs. 29 and 30 to observe the strong dependence of plate height on channel thickness w). The spacer is clamped between plates forming the channel walls. The thermal FFF plates are copper or copper alloy bars, generally 50 mm wide by 25 mm thick by 460–560 mm long. The copper bars are milled, machine ground to flatness, and polished to a mirror finish. In some cases the bars are electroplated with chromium before the final polishing to produce a harder, more inert surface, less susceptible to damage. In some earlier systems the lower bar was gold plated or plastic coated for inertness. Care in polishing is necessary because maximum resolution is achieved only with the most highly uniform channel walls (34).

The copper bars also serve as the heating and cooling plates to establish the

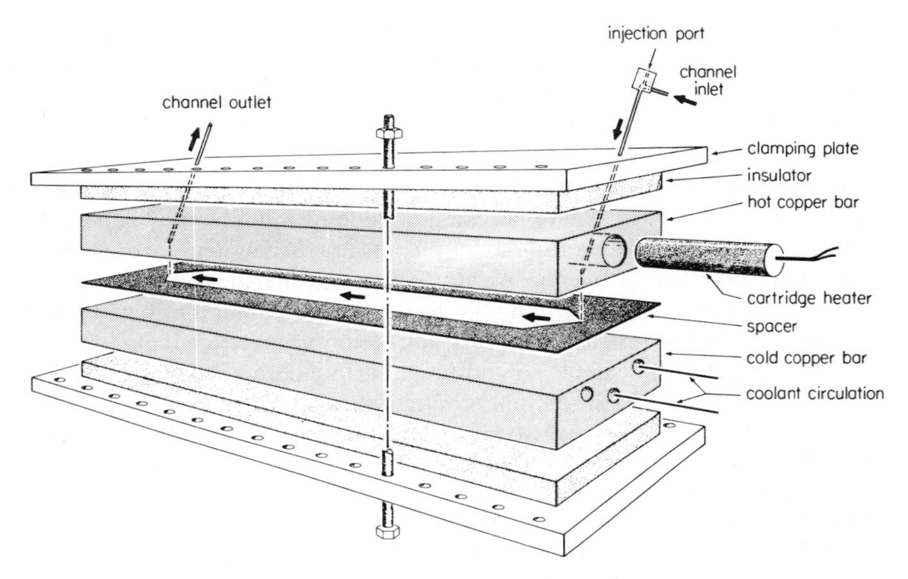

Figure 3. Sandwich assembly of thermal FFF system.

temperature gradient across the channel. The upper bar normally has a hole bored through its length to accommodate two cartridge heaters inserted from opposite ends. The heaters are controlled with either a variable transformer or a computer-controlled, solid state relay to regulate the upper plate temperature. To decrease the voltage for temperature programming, a motor-driven transformer can be used.

The lower plate has three holes bored through its length which are interconnected in a capped recessed area near the channel outlet. Tap water, most commonly used as coolant, enters the middle hole and returns along the two outer holes. Both bars have thermistor wells extending to within 0.76 mm of the channel surface. The temperature of the plates is measured by thermistors or thermocouples placed in the wells.

The copper bars are sandwiched between pieces of insulating board which, in turn, are sandwiched between aluminum or steel clamping plates. The clamping plates contain holes which accommodate bolts to hold the system together. The bolts are uniformly tightened with a torque wrench, to provide, as much as possible, a channel whose thickness is constant throughout.

Flow FFF

Flow FFF, in which the "field" is represented by a cross-flow of carrier liquid, requires channel walls permeable to solvent but not to the sample species (1).

Membranes fulfill the requirement but must be supported by a rigid layer of frit material to prevent them from flexing and distorting the channel geometry. Theoretically it should be possible to eliminate the membrance forming the upper channel wall, using only the frit. In practice, channels without an upper membrance wall sometimes show abnormally large void volumes and at low cross-flows (low field) show a loss of sample as well. In an alternative technology, the upper wall is made of a nonpermeable material such as glass (35); the operating requirements for such systems are substantially altered relative to operation with two permeable walls.

The lower membrane (forming the accumulation wall of the channel) must be very smooth and resistant to adsorption of sample material. It and the other membrane and frit elements should be uniformly permeable.

As in thermal FFF, the channel configuration is determined by the shape cut out of a spacer of Teflon. (Mylar and stainless steel spacers have also been used.) The channel thickness is typically 0.25–0.51 mm (somewhat greater than that generally used with other subtechniques), to offset the increased difficulty of obtaining precisely uniform channel walls.

The spacer is placed between the membrane walls, each supported by its backing material or frit. For the upper wall, a cellulose acetate membrane has been cast on a polyethylene frit of pore size 5 μm or a stainless steel frit of pore size 25 or 70 μm. The lower wall may be composed of a similar cellulose acetate membrane cast on a frit, but it may also be formed by a commercial membrane (e.g., Millipore PTGC membrane or Amicon type YM5) layered tightly over the frit.

The upper and lower frits are glued into recessed areas milled into plastic or metal clamping blocks to make the surface of the frit approximately even with the blocks. The clamping blocks are bolted together as in thermal FFF. Stainless steel inlet and outlet tubes lead through the clamping blocks into the frits to permit delivery of the cross-flow liquid. The cross-flow stream is delivered by means of the piston pump-unpump described above; another pump delivers the channel flow. The delivery of the piston pump may be varied to permit field programming of the cross-flow.

The split channel outlet consisting of concentric tubes as described above has been used in flow FFF (33). In the split-flow case, the two channel outlet flows are controlled by flow restrictors.

Sedimentation FFF

Stainless steel is commonly used to form both the spacer and the channel walls of the sedimentation FFF channel developed in this laboratory (1, 8). The stainless steel may be coated with Teflon or polyimide, as necessary for inertness, or replaced by titanium or Hastelloy C in experiments requiring physiological buffers that corrode stainless steel. The sample port is mounted at the inlet of the column, which is coiled to fit the inside circumference of a centrifuge basket. The centrifuge, which creates the field, can be controlled by using a feedback loop to increase the accuracy of speed regulation and to permit speed variation for field programming.

The greatest experimental challenge in sedimentation FFF has been the construction of the seal between the stationary and the spinning portions of the system. For aqueous systems, self-lubricating O-rings and highly polished drive shafts coated with aluminum oxide or chromium oxide ceramics have proved most dependable. A spring-loaded Teflon seal (Bal-Seal) used with a chromium oxide-coated shaft has been utilized for nonaqueous systems (36).

The alternative to the laboratory-built sedimentation FFF assembly discussed above is the du Pont sedimentation FFF (SF^3) instrument designed to be used with aqueous carriers. This unit is equipped with a specially designed face seal which permits operation at higher spin rates and thus increased force fields relative to those possible using the laboratory-built apparatus (37). This allows work to be done with species having molecular weights of 10^6 or lower, depending on density difference $\Delta\rho$ (see eq. 19). Thus for polymers, molecular weight determinations can be pursued at the high end of the molecular weight scale, which is widely recognized as difficult to work with. However, one must be cognizant of possible overloading effects (38) with such polymers because of their tendency to engage in chain entanglement.

The du Pont instrument also offers a special time-delayed exponential field decay programming system (14). This mode of programming is convenient for the interpretation of a spectrum of molecular weights but has less uniform resolving power than a recently developed "power-programmed" approach (39).

METHODOLOGY

Injection–Relaxation

For most analytical purposes, the sample is injected into the FFF channel as a small plug by means of a syringe or an injection valve. During the injection, and for a brief moment after this event, the axial flow is allowed to continue uninterrupted to ensure complete entrance of the sample into the channel. Once the zone has been fully entered, the axial flow is generally stopped to allow relaxation of the sample into its equilibrium distribution, which is specified by the nature and magnitude of the applied field as described above. Since the migration distance across the thin FFF channel is of the order of a few hundred micrometers or less, equilibrium is generally established in a matter of seconds or at most minutes. An approximate assessment of the time needed for full relaxation is given by the time τ required by the slowest moving component in the field (see eqs. 10 and 11). When τ is insignificant in comparison with the time required for elution of the sample, this stop-flow procedure is omitted.

Following relaxation of the sample, the axial flow is resumed, and the sample components are transported downstream with velocities determined by the thickness of their respective distributions. In general, a condition near equilibrium is maintained for each component in the direction of the field during the entire passage through the channel. A small departure from equilibrium, however, is caused by flow; this departure is responsible for nonequilibrium band broadening as described by equation (29). A significant departure is sometimes caused by field programming (see below), particularly where a rapid reduction in the field leads to a rapid change in the equilibrium distribution. Corrections for this "secondary relaxation" in a continually varying field have recently been established (40, 41); with the aid of these corrections, elution volumes can be translated into accurate information about sample molecular weight.

Retention Measurements

The experimental determination of the retention parameters R and λ is based on the determination of elution time t_r or elution volume V_r, and their relationship to the passage time t^0 or sweep volume V^0, associated with an inert compound. Either measurement requires accurate delivery of carrier by the pumping system, whether at a constant or systematically changing rate (29). For highest accuracy, the delivery of carrier is monitored on-line, using flow meters built in our laboratory (42, 30). Normally, however, spot checking of the flow by means of a buret and stop-watch gives adequate control.

The void volume V^0 is measured either as the elution volume recorded at zero field strength or as the elution volume of a compound which is unaffected by the applied field. Alternately, V^0 is assigned the value of the geometric volume of the channel.

For most purposes, elution volumes are measured at the point of maximum concentration in an eluting peak. An exception to this procedure is made in the analysis of polydisperse samples, where the shape of the peak relates to the molecular weight distribution in the sample. In this case, each volume element under the peak is assigned values for V_r and R, which in turn are related to the molecular weight at that point in the elution curve (see subsection entitled "Scale Correction," below).

Plate Height Determination

The most rigorous techniques for measuring the plate height of narrow distributions are based on moment analysis, whereby the variance σ^2 determined for an eluting peak is identical to its second moment. The plate height H is found by dividing σ^2 by the length of the channel as stated earlier. Often, peaks have a nearly Gaussian shape and their variances can be determined graphically from peak widths at half-height $w_{1/2}$ and the distances Z from the point of injection to the peak maximum (see eq. 27). In this procedure, both $w_{1/2}$ and Z can be measured directly from the fractogram.

The plate height associated with a given peak can be thought of as the sum of contributions from a variety of zone-broadening effects, as discussed in connection with equation (28). Of these, only the nonequilibrium contribution H_n shows a significant dependence on the mean velocity $\langle v \rangle$ of the carrier. By determining plate heights for a set of runs performed at identical fields but varying flows, and extrapolating to zero velocity, one finds the sum of the velocity independent H terms, of which the polydispersity contribution, the principal term, is of particular interest because it represents the molecular weight distribution.

Another approach to the determination of H_p, applicable also to GPC, is based on the principle of flow reversal (43). Here, the polydisperse sample is allowed to first pass through the channel in the normal manner, thereby generating a trace whose plate height H_s is determined. In a parallel experiment, the sample is allowed to move forward along the separation coordinate for only half the channel length. At this point the direction of flow is reversed, nullifying the partial separation realized for the sample's components. Most instrument- and procedure-based zone-broadening terms (certain end and edge effects excluded) remain the same as in the direct elution case; the difference in the plate height H_r observed for the reversed flow case and H_s therefore can be used to approximate H_p.

Scale Correction

Highly polydisperse samples will span a wide range of elution volumes. Although the recorded fractogram represents the distribution of molecular weights present in the sample, the amplitude of this distribution is normally distorted by departure from linearity between the molecular weight and elution volume scales. A correction for this distortion can be accomplished through a procedure in which the detector response at each elution volume is multiplied by a scale correction factor (44).

Thus, for a polydisperse sample whose molecular weight distribution is given by $m(M)$, the fractionation will result in a detector trace $c(V_r)$, where c is mass-based measure of concentration and V_r is the retention volume. Given $m(M)$, the mass of constituents of molecular weight lying between M and $M + dM$ can be expressed in terms of the concentration of sample in the corresponding volume interval dV_r

$$m(M)\,dM = c(V_r)\,dV_r \tag{38}$$

The molecular weight distribution $m(M)$ can therefore be obtained by multiplying the fractogram signal $c(V_r)$ by the scale correction function $|dV_r/dM|$, which is generally known from first principles in FFF or can be empirically established. Applications of this procedure are discussed later.

Overloading

For any FFF method to generate analytical information of high accuracy, it is necessary to work with small sample loads. This requirement is brought about by the very mechanism underlying the separation, in which the sample is forced to accumulate into a thin exponential distribution in the vicinity of a channel wall. Even at modest average inlet concentrations, the wall concentration c_0 may be significant if the sample is highly compressed and thus well retained (38). Therefore any sample prone to aggregation, chain entanglement, or other nonideal solute interaction should be injected at the lowest possible concentration yielding an adequate detector signal and should, if necessary, be studied at a reduced level of retention. For rigorous sample characterization, it is advisable to perform a systematic variation in load so that the desired analytical information can be obtained through extrapolation to zero concentration, or at least under conditions shown to be independent of load.

Dilute Samples

For dilute samples, large volumes must be injected to have enough material for detection. These large volumes cause excessive band broadening. Fortunately, techniques have been developed for the injection of samples whose volumes may exceed that of the column (normally 0.5–5 mL) by an order of magnitude or more (45). Such procedures are particularly useful for recycling or reinjection work, where moderately dilute cuts are collected and reintroduced into the channel. Here the sample introduction is performed in the presence of a field which is significantly stronger than that used for the actual analysis, and at a flow which is slow enough to permit relaxation of the sample immediately upon entering the column. To limit the detrimental zone-broadening effects incurred in this step, relaxation and subsequent downstream migration should be limited such that the sample is contained within the first 10% of the channel.

Upon completion of the feed, the channel flow may be temporarily interrupted to allow the sample zone to readjust itself to the reduced strength of the working field. Resumption of the flow at a faster working level leads to sample elution. The plate height H now has, in addition to the regular contributions summarized in equation (28), a term related to the increased width of the zone at the start of the analysis.

APPLICATIONS

Verification of Retention Equations

One great operational advantage of the field-flow fractionation methods is the ease with which experimental parameters can be changed to accommodate a particular sample. The need for system calibration is in principle obviated by the firm theoretical foundation which has been developed for these techniques over the past two decades. In practice, however, there are several reasons for seeking a verification of the relationships which have been established between retention and field strength on the one hand, and between retention and sample molecular weight on the other.

From the general treatment of FFF retention given above, we find the

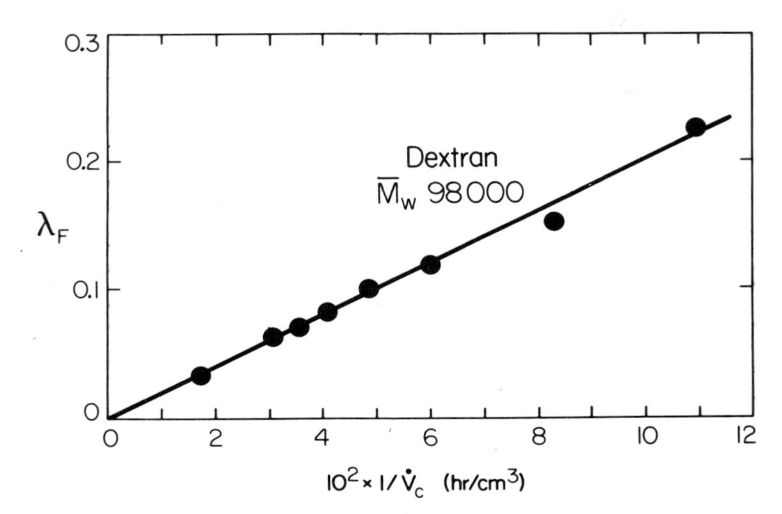

Figure 4. Relationship between retention-derived values for parameter λ for flow FFF (shown as λ_F) and the inverse of the applied cross-flow $\dot{V}_c$. Both the linearity of the plot and the fact that the line passes through the origin indicate that the sample (linear dextran with $\bar{M}_w = 98{,}000$) behaves ideally over a wide range of experimental conditions.

retention parameter λ to vary inversely with strength S of the applied field (see eq. 9), irrespective of type. Likewise, the ideally performing system shows an inverse relationship between λ and ϕ, the sample's field susceptibility, which for the sedimentation, flow, and thermal subtechniques is some function of molecular weight M (eqs. 19–23). These equations have all been established assuming the complete absence of interactions between sample molecules or between the sample and the accumulation wall. It is therefore prudent to verify their validity for each new class of samples and carrier solvents and for each new instrumental system.

Figure 4 shows a typical set of retention-derived λ values obtained in a series of flow FFF experiments at different cross-flow rates $\dot{V}_c$. The sample in this case is a linear dextran with a stated $\bar{M}_w$ of 98,000. The theoretically predicted linearity between λ and $1/\dot{V}_c$ (eq. 20) is seen to hold over a wide range of cross-flows in this system. A similarly well-behaved set of results from a thermal FFF system is illustrated in Figure 5, where a group of linear polystyrene samples were retained under different field strengths, here symbolized by ΔT (the temperature drop across the channel) (46). The linearity between the observed λ values and the inverse of the applied field strength is an indication that the system behaves ideally (i.e., is not affected by adsorption or sample aggregation) in the tested range of temperature drops across the channel.

Nonideal behavior is often seen in work with high molecular weight polymers, which are prone to aggregation or temporary chain entanglement at elevated concentration. Since the applied field concentrates sample molecules

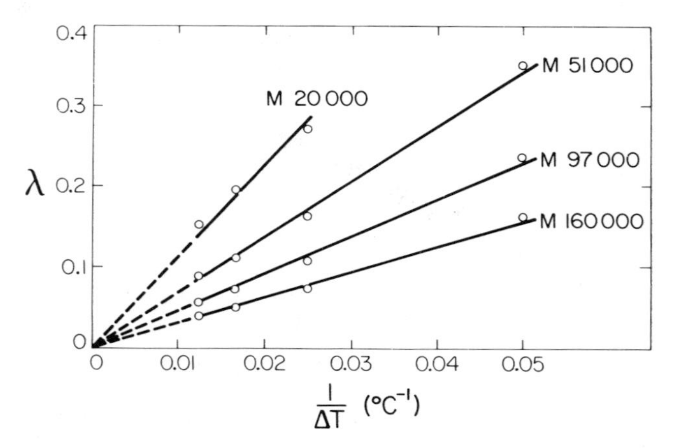

Figure 5. Plots of λ versus $1/\Delta T$ for thermal FFF runs of polystyrene solutes of different molecular weights in ethylbenzene. The cold wall temperature is 16°C.

into thin layers near the accumulation wall, and since the highest molecular weight samples form the most compact layers, there is a progressively stronger tendency toward concentration-induced nonideal behavior with increasing molecular weight. To retain normal behavior, therefore, the field-induced concentration must be offset by a reduction in the concentration of the injected sample, as discussed in a recent study on overloading effects (38).

Relationships between Molecular Weight and Retention

In most FFF subtechniques, the solute–field coupling constant ϕ is a function of sample molecular weight M as noted above. The limiting mass selectivity S_M of equation (26) expresses the power which relates ϕ and M for the different subtechniques. Thus for sedimentation FFF (SdFFF) with a limiting S_M of unity, ϕ is proportional to M, whereas for the flow and thermal subtechniques the weaker mass selectivity results from ϕ values proportional to $M^{0.5-0.6}$.

Although the mass selectivity favors the sedimentation FFF subtechnique, its applicable range is somewhat more restrictive [generally limited to masses $> 10^6$ (37)] than that offered by the flow and thermal techniques [$M > 10^3$ (47, 31)]. In addition, the present use of SdFFF is for all practical purposes limited to aqueous systems, although one low field study involving ethanol has been published (36). By contrast, equipment for thermal and flow FFF is generally designed to permit operation in either aqueous or nonaqueous solvents (33, 48–50), thus providing significant versatility in the selection of experimental conditions.

A sedimentation FFF system operated with constant field strength G should generate retention-derived λ values which vary linearly with $1/M$, in accordance with equation (19). For solid spherical test particles, M is proportional to particle diameter d raised to the third power, and a properly operating system should therefore generate λ values which are proportional to $1/d^3$. Good agreement between experiment and theory is demonstrated by the data set in Figure 6, collected for a series of standard polystyrene latex spheres retained under a constant field of 127.6 gravities (51). Such plots are helpful in assuring the integrity of a system's performance prior to practical use.

In flow FFF, the solute–field coupling constant is the sample's coefficient of friction f, which is inversely related to its diffusivity D. Measurements of flow FFF retention can therefore be directly translated into D values for the sample using a combination of equations (12), (15), and (20). The relationship between R and M, which reflects the dependence of D on M, generally begins with equation (21). The exponent b in equation (21), which equals the limiting mass selectivity S_M, will be 1/3 for solid spheres, 0.5 for linear polymers in θ solvents (20), and near 0.7 for polyelectrolytes in media of low ionic strength (52, 53). Figure 7 is a typical illustration of the relationship observed between the

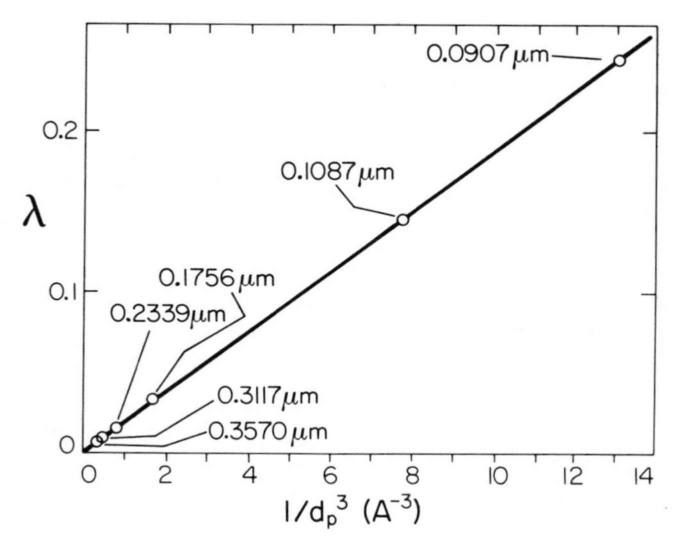

Figure 6. Plot of λ versus $1/(\text{particle diameter})^3$ at $G = 127.6$ gravities and 6 mL/h flow rate. The straight line is predicted from theory.

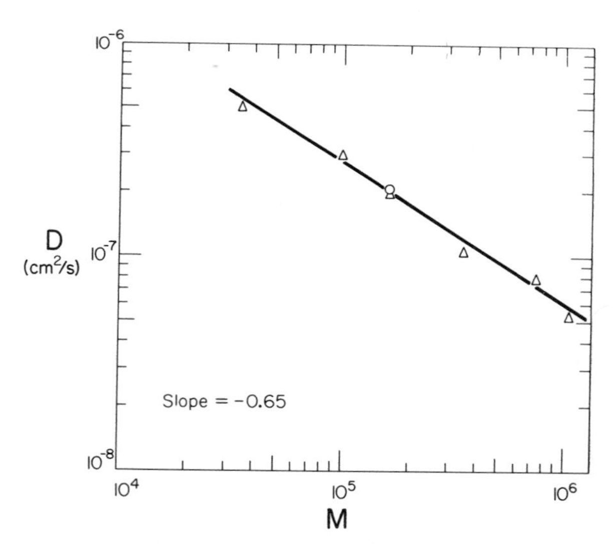

Figure 7. Relationship between retention-derived diffusion coefficients and molecular weight for a series of sulfonated polystyrene standards analyzed by flow FFF. The graph contains observations taken at two different cross-flows: Δ, 0.63 mL/min = 1.2 μm/s; O, 0.8 mL/min = 1.5 μm/s. The axial flow velocity $\langle v \rangle$ was held constant at 0.5 cm/s. A least-squares fit of the data gives a slope of −0.65 (this equals the negative of the exponent b in eq. 21).

retention-derived diffusion coefficient D and the corresponding molecular weight M of polystyrene sulfonate samples in an aqueous carrier (33). The slope of the line in this logarithmic plot is $b = -0.65$, which is reasonable for a linear polyelectrolyte in a medium of relatively low ionic strength (0.025 M).

From a strictly theoretical point of view, the relationship between retention and molecular weight in thermal FFF is more complex than in the case of flow FFF. Instead of having simply $\phi = f$, the solute–field coupling constant for this technique is equal to the product of friction coefficient f and thermal diffusion coefficient D_T. The link between f and M is well described by equation (21) in combination with Einstein's relationship between D and f. There is, however, no sound theoretical model available for relating D_T to sample molecular weight, temperature, or solvent–solute composition. Fortunately, a significant volume of experimental evidence suggests that D_T is independent of M (46, 49), although it is sensitive both to temperature (54) and to the chemical composition of solvent and polymer (23, 49, 55). Under fixed conditions relative to temperature and composition, it is therefore possible to establish a linear relationship between the logarithms of λ and M, as suggested by a combination of equations (22) and (23). The numerical value of the slope $-b$ of such a plot is then identical to the limiting mass selectivity S_M for the particular solvent–solute system under investigation.

The requirement of fixed experimental conditions sets a practical limit to the molecular weight range that can be covered in this type of plot, since the

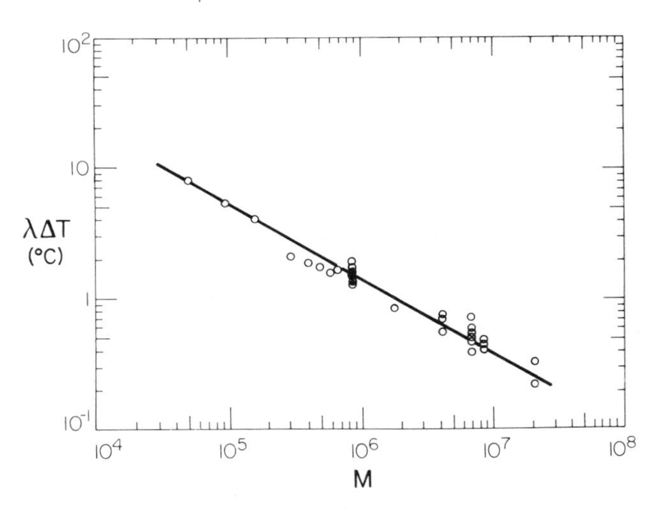

Figure 8. Thermal FFF retention data obtained for a broad range of molecular weights with temperature drops ΔT ranging from 8 to 81°C and a cold wall temperature of 15°C. A linear least-squares fit of the data gave a slope ($-b$) of -0.53.

high field (temperature drop ΔT) needed to retain low molecular weight polymers results in an unduly strong retention, and thus an excessively long retention time, for polymers of high M. The problem can be circumvented by allowing the temperature drop ΔT across the channel to vary in response to molecular weight, while maintaining the temperature T_c of the accumulation (cold) wall at a constant level. By a slight rearrangement of the combined equations (22) and (23), we find that a plot of the logarithms of $\lambda\Delta T$ and M should be linear, again with a slope of $-b$. Figure 8 is an example of the broad range of molecular weights accessible on a single calibration curve by this procedure. Here, linear polystyrene standards spanning the molecular weight interval of 51,000–25,000,000 are analyzed using temperature drops in the range of 8–81°C (22).

The diffusion coefficient D in equation (23) is sensitive to the branching structure of the polymeric sample, but branching does not appear to affect the thermal diffusivity D_T. Indeed, a recent thermal FFF study of a series of polystyrenes, differing both in molecular weight and in degree of branching, showed a strictly linear relationship between D and λ (12), implying a constant D_T throughout. However, D_T is sensitive to the chemical nature of the polymer; samples of different composition may show substantial differences in retention despite completely overlapping molecular dimensions, as discussed below (23).

Resolution of Monodisperse Components

To illustrate the resolving power of the FFF techniques for polymer characterization, it is instructive to make a comparison with size exclusion chromatography (SEC), whose performance is well understood and amply documented. Although flow and thermal FFF methods are both applicable to polymers and the two have comparable mass selectivity, the thermal subtechnique, which has been most widely developed and used for the characterization of polymers in nonaqueous solvents, is most suitable for comparison.

The ability of a separation process to resolve two components of a given relative difference in molecular weight is shown by equations (33) and (36) to depend on two independent factors: the efficiency or plate number N, which is subject to modification through experimental design, and the selectivity S_M, which is governed by the separation mechanism. The selectivity of FFF systems was described above. The efficiency of the FFF process has undergone a steady increase since the first reported separation of two decades ago. By decreasing the channel thickness and increasing its uniformity, it has been possible to achieve in a few minutes (56) the same resolution first observed in an 18-hour run (57). Despite these considerable improvements, the number of plates readily available in a thermal FFF channel is almost an order of

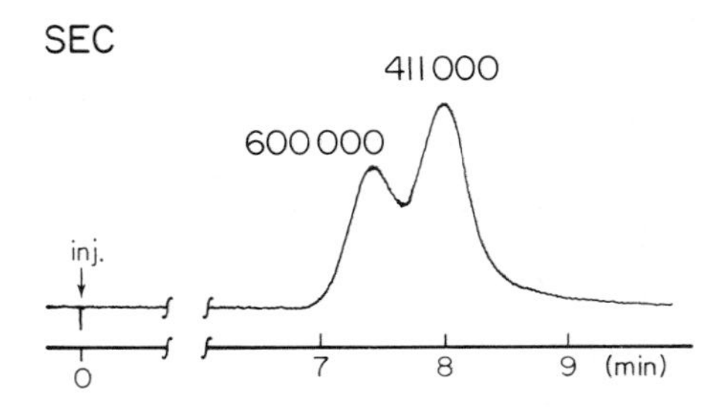

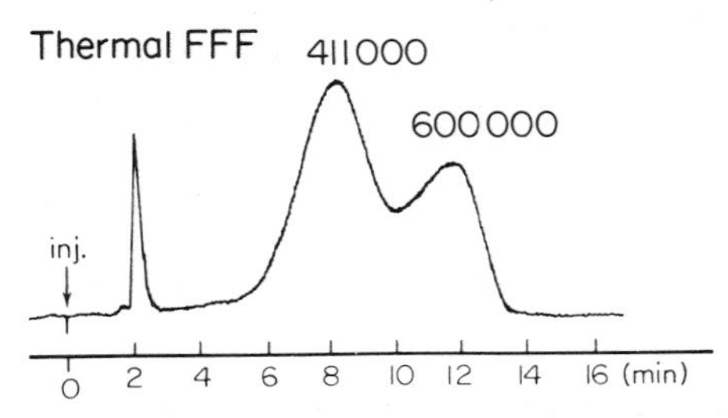

Figure 9. Separation of linear polystyrene samples with manufacturer-assigned molecular weights of 411,000 and 600,000. Upper trace: fractogram obtained on an Ultrastyragel column 30 cm long, from Waters Chromatography Division of Millipore Corporation, using a flow rate of 1.0 mL/min. Lower trace: fractogram produced by a thermal FFF unit operated at a ΔT of 50°C with the cold wall held at 22°C; the flow rate was 0.4 mL/min and the channel dimensions were 45 cm × 2.0 cm × 0.00762 cm. The solvent was tetrahydrofuran (THF) in both cases.

magnitude less than the number observed for present-day commercially available SEC columns. However, in comparing the two techniques one finds that the square root dependence of R_s on N (eq. 33) is more than offset by the linear dependence on S_M, which for linear polystyrene polymer in THF assumes maximum values of 0.62 for thermal FFF as compared to 0.12 for SEC (11). Figure 9 shows fractograms of a mixture of two polystyrene standards (600,000 and 411,000) generated by the two techniques. Conditions were selected to generate the fractograms in comparable times. The resolution was measured as 1.14 for thermal FFF as compared to 0.78 for the SEC fractionation (11).

Separation by SEC is based on differences in physical size existing between various components in a polymer mixture. Retention in thermal FFF is also related to molecular size, as just discussed. However, unlike SEC, retention in thermal FFF is not *exclusively* related to the physical size of the molecule, as reflected in D, but is also affected by the chemical nature of the solvent–solute pair. This dependency arises through D_T. In a recent study, the possibility of

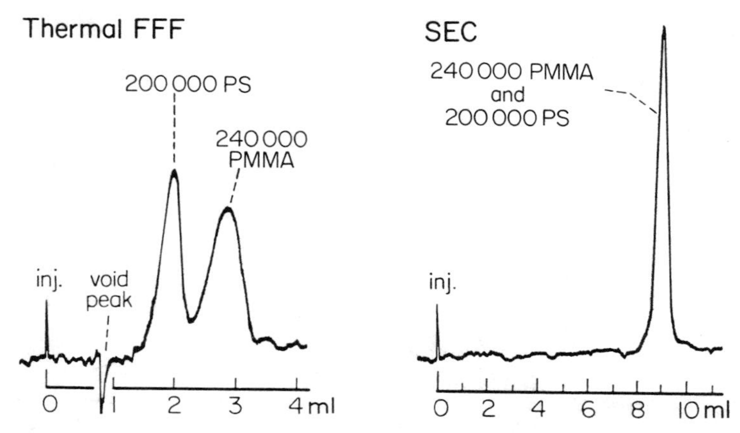

Figure 10. Runs with 2.4×10^5 MW poly(methyl methacrylate) (PMMA) and 2.0×10^5 MW polystyrene (PS) by thermal FFF and SEC. Experimental conditions: for SEC, flow rate = 1.0 mL/min; for thermal FFF, $T_c = 21°C$, $T_h = 62°C$, and flow rate = 0.13 mL/min.

resolving pairs of polymer samples of different composition but with roughly the same diffusivities was examined using SEC and thermal FFF (23). As predicted, the similarities in D, and thus in hydrodynamic size, precluded resolution of the components by SEC. By contrast, significant differences were noted in the thermal FFF retention patterns, as seen in Figure 10. Under the conditions used to develop the figure, the mixture of linear polystyrene and poly(methyl methacrylate), with molecular weights of 200,000 and 240,000, respectively, was virtually baseline resolved in thermal FFF, whereas no evidence of resolution was seen in SEC. By combining the two techniques in a systematic retention analysis of polymers with different composition, it was possible to determine values of both parameters, D and D_T, for different components (23). As expected, the latter parameter showed no dependence on M but was strongly influenced by the chemical nature of the sample.

Polydispersity of Narrow Samples

The effect of polydispersity on peak shape and width is illustrated by the flow FFF fractograms of narrow and broad polystyrene standards shown in Figure 11 (50), which gives the supplier's values for weight and number average molecular weights for the two samples. The ratio of these two numbers, $\mu = \bar{M}_w/\bar{M}_n$, for each sample correlates well with the observed peak width in accordance with equation (31). This suggests that the band broadening reflects primarily the molecular weight distribution of the polymer, with only a small contribution from nonequilibrium.

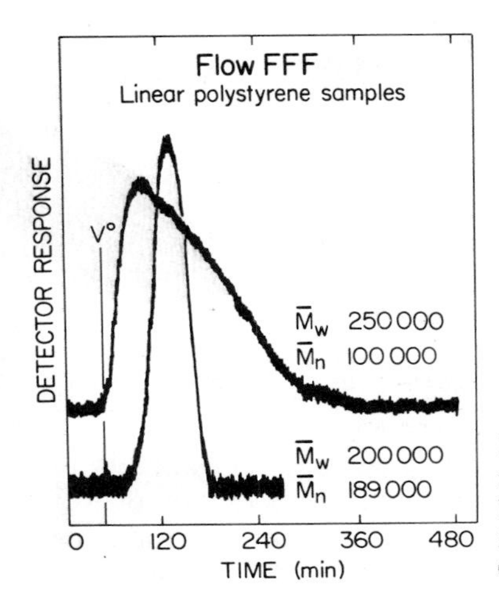

Figure 11. Flow FFF fractograms of broad and narrow polystyrene samples. Experimental conditions: longitudinal flow rate, 5 mL/h; cross-flow rate, 20 mL/h; channel volume, 3.90 mL; channel thickness, 0.0532 cm.

To subtract out the nonequilibrium effects for a more accurate evaluation of the polydispersity for narrow fractions, we plot experimental H values against carrier velocity $\langle v \rangle$ as suggested by equation (32). The polydispersity contribution H_p can then be accurately evaluated as the intercept of the plot. The procedure, outlined in connection with equations (28)–(31) is illustrated in Figure 12 using an ultranarrow linear polystyrene fraction with $\bar{M}_w =$

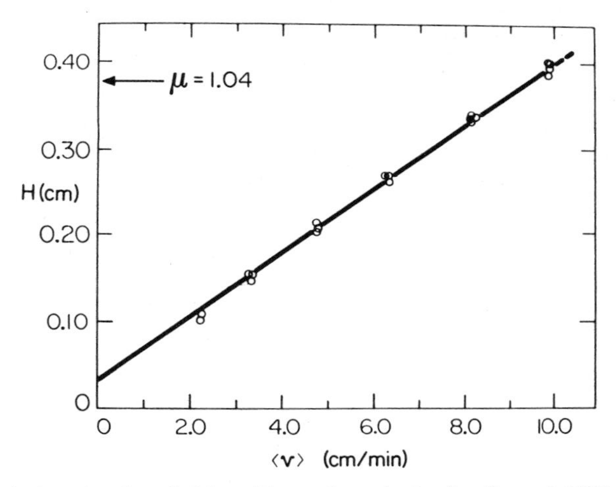

Figure 12. Variations in plate height with carrier velocity for thermal FFF runs of linear polystyrene, $\bar{M}_w = 170{,}000$, in ethylbenzene. Experimental conditions: $\Delta T = 30°C$ and cold wall temperature 21°C; the channel had the dimensions 34.4 cm × 2.3 cm × 0.00762 cm.

170,000 and with a polydispersity μ reported by the supplier only as "less than 1.06." The plot is seen to be linear, as predicted by equation (32), with an intercept of 0.0311 cm, which corresponds to a μ value of 1.0034 (13). For comparison, the figure indicates that a μ value of 1.04, although less than the upper limit given by the supplier, would give rise to an intercept more than an order of magnitude larger than the observed value. The high level of precision in this and similar determinations gives confidence in the ability of FFF to accurately assess the polydispersity of narrow polymer fractions.

Broad Distributions

Although the polydispersity index μ is a highly useful characteristic for samples of narrow to moderate distribution width, it does not provide sufficient information to characterize broad and sometimes multimodal distributions. In this case the sample becomes spread over a wide elution volume range by fractionation and all other broadening effects are negligible.

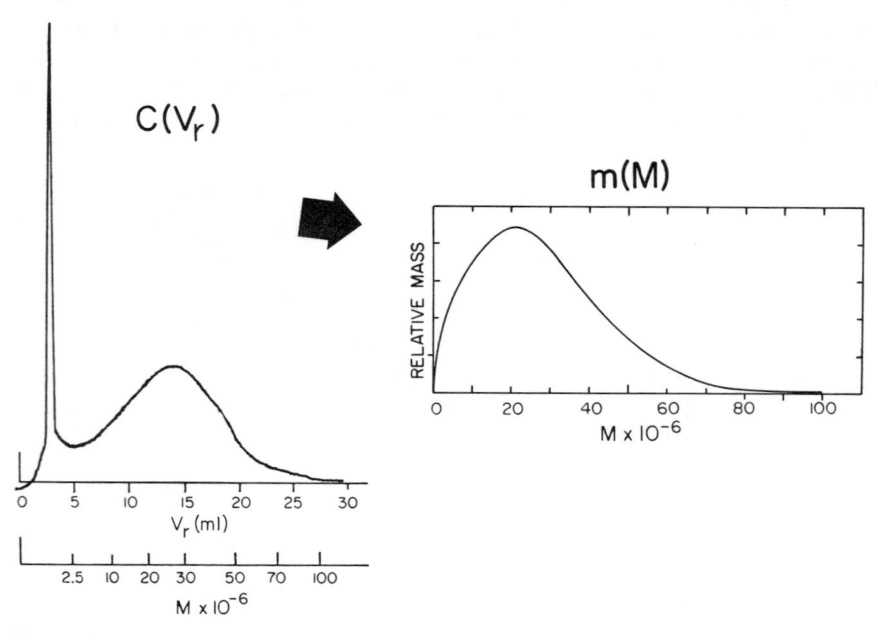

Figure 13. Transformation of a thermal FFF fractogram $c(V_r)$ into a molecular weight distribution curve $m(M)$. The fractogram on the left was recorded at a ΔT of 8°C and a cold wall temperature of 15°C using THF as the carrier (flow rate 7.2 mL/h) and a channel whose dimensions were 47.0 cm × 2.00 cm × 0.0254 cm.

The shape of the fractogram will then contain the information needed to give a rather complete picture of the molecular weight distribution. To transform the raw fractogram into an actual distribution curve, one must apply the necessary scale correction described in the preceding section (see eq. 38). This procedure is illustrated in Figure 13 for a polystyrene sample having a nominal $\bar{M}_w$ of 20,600,000. Here a scale correction has been used to transform the raw fractogram on the left into the distribution curve on the right. Similar procedures have been successfully applied to a wide variety of polydisperse samples, including latex particles (58) and emulsions (59) analyzed by sedimentation FFF, water-soluble polyacrylic acids analyzed by flow FFF (60), and polystyrenes (22) and polyethylenes (61) studied by thermal FFF.

Programming

The analysis of highly polydisperse materials poses a practical problem in the choice of field strength, since invariably some portion of the sample elutes poorly retained, and therefore lies in a region of low selectivity, whereas other portions may be excessively retained to the point where detectability becomes a problem. To avoid this problem and to reduce the run time, strategies have been developed for programming the field strength as noted earlier (see eq. 37). The fractogram in Figure 14 is a good illustration of the high resolution that can be achieved by this method over a very large mass range—in this case four decades (62). With the field strength under computer control, the form of the decay curve can be selected to optimize the resolution of a given sample (39).

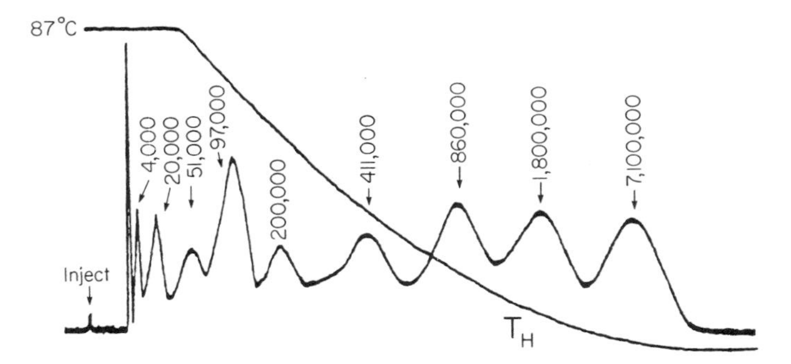

Figure 14. Thermal FFF fractogram of a polystyrene mixture using parabolic programming. Experimental conditions: Flow velocity $\langle v \rangle = 0.023$ cm/s. Initial $\Delta T = 70$°C with hot wall temperature $T_H = 87$°C. After initial time lag, T_H decays according to a parabolic program as shown.

ACKNOWLEDGMENT

This work was supported by Grant No. CHE-8218503 from the National Science Foundation.

REFERENCES

1. J. C. Giddings, M. N. Myers, K. D. Caldwell, and S. R. Fisher, "Analysis of biological macromolecules and particles by field-flow fractionation," *Methods of Biochemical Analysis*, Vol. 26, in D. Glick, Ed., Wiley, New York, 1980, p. 79.
2. J. C. Giddings, *Anal. Chem.*, **53**, 1170A (1981).
3. J. C. Giddings, M. N. Myers, and K. D. Caldwell, *Sep. Sci. Technol.*, **16**, 549 (1981).
4. J. C. Giddings, K. A. Graff, K. D. Caldwell, and M. N. Myers, "Field-Flow Fractionation: Promising approach for the Separation and Characterization of Macromolecules," in C. D. Craver (Ed.), *Polymer Characterization: Spectroscopic, Chromatographic, and Physical Instrumental Methods* (*ACS Advances in Chemistry Series*, No. 203), American Chemical Society, Washington, DC, 1983, p. 257.
5. J. C. Giddings and K. D. Caldwell, *Anal. Chem.*, **56**, 2093 (1984).
6. J. C. Giddings, *Sep. Sci. Technol.*, **19**, 831 (1984).
7. J. C. Giddings, "Separation in Thin Channels: Field-Flow Fractionation and Beyond," in *Chemical Separations*, Vol. 1, J. D. Navratil and C. J. King, Eds., Litarvan, Denver, 1986, p. 3.
8. J. C. Giddings and K. D. Caldwell, "Field-Flow Fractionation," in *Physical Methods of Chemistry*, Vol. IIIB, B. W. Rossiter, Ed., Wiley, in press.
9. J. C. Giddings and M. R. Schure, *Chem. Eng. Sci.*, **42**, 1471 (1987).
10. J. C. Giddings, *Pure Appl. Chem.*, **51**, 1459 (1979).
11. J. J. Gunderson and J. C. Giddings, *Anal. Chim. Acta*, **189**, 1 (1986).
12. M. E. Schimpf and J. C. Giddings, *Macromolecules*, **20**, 1561 (1987).
13. M. E. Schimpf, M. N. Myers, and J. C. Giddings, *J. Appl. Polym. Sci.*, **33**, 117 (1987).
14. W. W. Yau and J. J. Kirkland, *Sep. Sci. Technol.*, **16**, 577 (1981).
15. J. J. Kirkland and W. W. Yau, *Science*, **218**, 121 (1982).
16. J. J. Kirkland and W. W. Yau, *J. Chromatogr.*, **353**, 95 (1986).
17. J. C. Giddings, *J. Chem. Educ.*, **50**, 667 (1973).
18. F. J. Yang, M. N. Myers, and J. C. Giddings, *Anal. Chem.*, **49**, 659 (1977).
19. J. C. Giddings, F. J. Yang, and M. N. Myers, *Anal. Chem.*, **48**, 1126 (1976).
20. P. J. Flory, *Principles of Polymer Chemistry*, Cornell University Press, Ithaca, NY, 1953, Chap. 14.
21. C. Tanford, *Physical Chemistry of Macromolecules*, Wiley, New York, 1961, p. 362.

22. Y. S. Gao, K. D. Caldwell, M. N. Myers, and J. C. Giddings, *Macromolecules*, **18**, 1272 (1985).
23. J. J. Gunderson and J. C. Giddings, *Macromolecules*, **19**, 2618 (1986).
24. K. D. Caldwell, L. F. Kesner, M. N. Myers, and J. C. Giddings, *Science*, **176**, 296 (1972).
25. J. C. Giddings, *Dynamics of Chromatography*, Part I, Dekker, New York, 1965.
26. J. C. Giddings, Y. H. Yoon, K. D. Caldwell, M. N. Myers, and M. E. Hovingh, *Sep. Sci.*, **10**, 447 (1975).
27. J. C. Giddings, P. S. Williams, and R. Beckett, *Anal. Chem.*, **59**, 28 (1987).
28. F. J. F. Yang, M. N. Myers, and J. C. Giddings, *Anal. Chem.*, **46**, 1924 (1974).
29. J. C. Giddings, K. D. Caldwell, J. F. Moellmer, T. H. Dickinson, M. N. Myers, and M. Martin, *Anal. Chem.*, **51**, 30 (1979).
30. T. Koch and J. C. Giddings, *Anal. Chem.*, **58**, 994 (1986).
31. J. C. Giddings, L. K. Smith, and M. N. Myers, *Anal. Chem.*, **47**, 2389 (1975).
32. J. C. Giddings, H. C. Lin, K. D. Caldwell, and M. N. Myers, *Sep. Sci. Technol.*, **18**, 293 (1983).
33. K.-G. Wahlund, H. S. Winegarner,K. D. Caldwell, and J. C. Giddings, *Anal. Chem.*, **58**, 573 (1986).
34. J. C. Giddings, L. K. Smith, and M. N. Myers, *Sep. Sci. Technol.*, **13**, 367 (1978).
35. K.-G. Wahlund and J. C. Giddings, *Anal. Chem.*, **59**, 1332 (1987).
36. K. D. Caldwell, G. Karaiskakis, M. N. Myers, and J. C. Giddings, *J. Pharm. Sci.*, **70**, 1350 (1981).
37. J. J. Kirkland, C. H. Dilks, and W. W. Yau, *J. Chromatogr.*, **255**, 255 (1983).
38. K. D. Caldwell, S. L. Brimhall, Y. Gao, and J. C. Giddings, *J. Appl. Polym. Sci.*, **36**, 703 (1988).
39. P. S. Williams and J. C. Giddings, *Anal. Chem.*, **59**, 2038 (1987).
40. J. C. Giddings, *Anal. Chem.*, **58**, 735 (1986).
41. W. W. Yau and J. J. Kirkland, *Anal. Chem.*, **56**, 1461 (1984).
42. K. D. Caldwell and M. N. Myers, *Anal. Chem.*, **58**, 1583 (1986).
43. L. H. Tung, J. C. Moore and G. W. Knight, *J. Appl. Polym. Sci.*, **10**, 1261 (1966).
44. J. C. Giddings, M. N. Myers, F. J. F. Yang, and L. K. Smith, "Mass Analysis of Particle and Macromolecules by Field-Flow Fractionation," in *Colloid and Interface Science*, Vol. IV, M. Kerker, Ed., Academic Press, New York, 1976, p. 381.
45. J. C. Giddings, G. Karaiskakis, and K. D. Caldwell, *Sep. Sci. Technol.*, **16**, 725 (1981).
46. M. N. Myers, K. D. Caldwell, and J. C. Giddings, *Sep. Sci.*, **9**, 47 (1974).
47. R. Beckett, Z. Jue, and J. C. Giddings, *Environ. Sci. Technol.*, **21**, 289 (1987).
48. J. J. Kirkland and W. W. Yau, *J. Chromatogr.*, **353**, 95 (1986).
49. J. C. Giddings, K. D. Caldwell, and M. N. Myers, *Macromolecules*, **9**, 106 (1976).

50. S. L. Brimhall, M. N. Myers, K. D. Caldwell, and J. C. Giddings, *J. Polym. Sci., Polym. Lett. Ed.*, **22**, 339 (1984).
51. J. C. Giddings, F. J. F. Yang, M. N. Myers, *Anal. Chem.*, **46**, 1917 (1974).
52. J. Marra, H. A. van der Schee, G. J. Fleer, and J. Lyklema, in *Adsorption from Solution*, R. H. Ottewill, C. H. Rochester, and A. L. Smith, Eds., Academic Press, New York, London, 1983, p. 245.
53. K. D. Caldwell, "Polymer Analysis by Field-Flow Fractionation," in *Modern Methods of Polymer Analysis*, H. Barth, Ed., Wiley, New York, in press.
54. S. L. Brimhall, M. N. Myers, K. D. Caldwell, and J. C. Giddings, *J. Polym. Sci., Polym. Phys. Ed.*, **23**, 2445 (1985).
55. J. C. Giddings, M. N. Myers, and J. Janca, *J. Chromatogr.*, **186**, 37 (1979).
56. J. C. Giddings, M. Martin, and M. N. Myers, *J. Chromatogr.*, **158**, 419 (1978).
57. G. H. Thompson, M. N. Myers, and J. C. Giddings, *Anal. Chem.*, **41**, 1219 (1969).
58. F.-S. Yang, K. D. Caldwell, and J. C. Giddings, *J. Colloid Interface Sci.*, **92**, 81 (1983).
59. F.-S. Yang, K. D. Caldwell, M. N. Myers, and J. C. Giddings, *J. Colloid Interface Sci.*, **93**, 115 (1983).
60. J. C. Giddings, G. C. Lin, and M. N. Myers, *J. Liq. Chromatogr.*, **1**, 1 (1978).
61. S. L. Brimhall, M. N. Myers, K. D. Caldwell, and J. C. Giddings, *Sep. Sci. Technol.*, **16**, 671 (1981).
62. J. C. Giddings, L. K. Smith, and M. N. Myers, *Anal. Chem.*, **48**, 1587 (1976).

CHAPTER

13

SUPERCRITICAL FLUID CHROMATOGRAPHY

BRUCE E. RICHTER

Lee Scientific R&D Director, Salt Lake City, Utah

INTRODUCTION

What is supercritical fluid chromatography (SFC), and why do we need it when we have the other forms of chromatography?

Scientists have been asking these questions with more frequency in recent years, partly because of the increased number of publications and symposia that have dealt with the topic. There seems to be almost a battleline drawn between two factions. One side is denying that there is a need for SFC, while on the opposite extreme there appear those who say that SFC can solve nearly all the problems which cannot be solved by other chromatographic means. There are also two arguments as to which column configuration is better—packed or capillary—the same argument that has been going on for some time in gas chromatography. As is usually the case, the truth lies somewhere in between the two extremes.

In answer to the first part of the opening question, SFC is nothing more than chromatography using a supercritical fluid. In reference to the second part, the discussion of this chapter should show that SFC does have a place in analytical laboratories because problems have been solved with it which cannot be solved by other chromatographic methods. However, as powerful as it is, SFC has its limitations, as do all analytical techniques.

The unique solvating abilities of supercritical fluids were first reported in 1879 by Hannay and Hogarth (1). The solubilities of cobalt and ferric chlorides in supercritical ethanol were studied, and it was found that the concentrations of the salts were much higher than vapor pressures alone would predict. Hannay and Hogarth also demonstrated that the absorption spectra of the salts in the supercritical fluids were identical to the spectra of the same substances dissolved in the corresponding liquid. This indicated that no chemical reaction had taken place, but simply dissolution into the supercritical fluid.

The first report of the use of supercritical fluids as mobile phases in chromatography was in 1962 by Klesper et al. (2). Throughout the 1960s there were papers published in the field, but growth in this area of chromatography was slow until recent years. Fewer than six reports appeared annually in the literature until the early 1980s.

The 1980s have seen a renewed interest in SFC because of its advantages over other chromatographic techniques. Novotny and coworkers first reported the use of capillary columns in SFC in 1981 (3). The first commercial SFC instrument, a modified high performance liquid chromatography (HPLC) system for use with packed columns, was introduced in 1982. Commercial instrumentation for capillary SFC was introduced in 1986. Instrumentation for both capillary and packed column SFC is available from several vendors currently. Commercially available instrumentation has given more investigators easy access to SFC, and there were nearly 100 journal articles on SFC for 1987, which indicates how the level of interest in this technique is growing.

This chapter discusses the use of SFC for the analysis of oligomers. A brief overview of the development of SFC from a historical perspective is followed by a review of the physicochemical properties of supercritical fluids. Then instrumental aspects of the technique are treated, and finally some applications of the technique to various oligomer samples are discussed.

HISTORICAL PERSPECTIVE

Klesper and coworkers described the first use of supercritical fluids in chromatography, demonstrating the separation of various nickel porphyrins using supercritical chlorofluoromethanes. This work was later extended by Karayannis et al. with various porphyrins and metal chelates (4–7).

Sie and Rijnders (8–12) conducted various studies investigating the use of carbon dioxide, isopropanol, and pentane as mobile phases. They were probably the first to use the term "supercritical fluid chromatography." Their research was quite thorough and is very useful as reference material for SFC in general. In addition to demonstrating various applications of SFC, these investigators presented a good evaluation of many theoretical aspects of the technique. During this period, Giddings and coworkers were also active in dense-gas chromatography as it was called, with operating inlet pressures of up to 2000 atms (13–18).

Several useful reviews have appeared which are excellent treatments of packed column SFC. Van Wasen et al. (19) reviewed SFC in terms of physicochemical properties and instrumental considerations. Klesper (20) reviewed the principles of packed column SFC such as column performance, instrumentation, and applications. Reviews by Gouw and Jentoft (21–23) cover all aspects of SFC, applications and practical matters.

The first report of the use of capillary columns in SFC came in 1981 (31). Before this time, the growth of SFC had been slow, for three main reasons. First, instrumental difficulties were encountered in dealing with the pressure and temperature conditions required for maintaining the mobile phase in the supercritical state. Second, the attention of the chromatographic community was primarily focused on HPLC and gas chromatography (GC), both of which were experiencing rapid growth. Third, the early development of SFC was done using packed columns with 4.6 mm i.d., which were difficult to interface to detectors other than UV absorbance detectors. SFC did not show wide applicability because of the limited number of detectors possible and the column packing material used at that time. The unavailability of nonextractable stationary phases for open tubular columns made the development of capillary SFC extremely difficult.

The development of HPLC and GC provided the solutions for these obstacles to the growth of SFC. Supercritical fluid chromatography borrows much of its instrumentation and column technology from these two techniques. The use of capillary columns with their high permeability and low pressure drop allows the full potential of pressure or density programming of the fluid to be used. In addition, all types of chromatographic detectors, both GC and HPLC, are compatible with SFC, but generally they can be more easily interfaced to capillary SFC than to packed column SFC because of the low flow rates through the capillary columns. The use of highly inert fused silica and nonextractable stationary phases similar to those used in capillary GC allow the elution of compounds of larger molecular weight and greater polararity than is possible in packed column SFC. Polar organic modifiers are needed to elute polar compounds from packed columns because of the residual surface activity of the packing. The use of these organic mobile phases precludes the use of the flame ionization detector (FID). Since, however, there are many more phases available in packed columns than currently can be had in capillaries, more unique selectivities are possible in packed column SFC than with capillaries. The loadability of packed columns is higher because the columns have internal diameters in the range of 1.0–4.6 mm, while the capillaries used are in the range of 0.05–0.1 mm i.d. Speed of analysis favors packed columns as predicted by theory (24), but rapid analyses have been reported with capillaries, too (25). Both capillaries and packed columns have their distinct advantages and can be used in SFC depending on the particular application needed. The advantages of SFC over other chromatographic techniques have caused a renewed interest in the technique. Some of these advantages are discussed in more detail in the next section.

PHYSICOCHEMICAL PROPERTIES OF SUPERCRITICAL FLUIDS

If one compresses a gas, it will, at some pressure, liquefy; however, if that gas is held above its critical temperature and then compressed, it will not liquefy but will become a "supercritical fluid." In such a state, a fluid is generally two or more orders of magnitude more dense than it is in the gaseous state; yet the fluid is noncondensing. As can be expected, a supercritical fluid has some physical properties similar to a liquid and some similar to a gas. To be of use in SFC, it is not enough to have the mobile phase above its critical pressure and temperature; the fluid must have sufficiently strong intermolecular interactions to cause solute migration, or a solvating effect. Therefore, GC using helium as a carrier cannot be classified as SFC, even though the operating conditions may be above the critical conditions of helium ($T_c = -267.9°C$, $P_c = 2.26$ atm or 34 psi), because helium does not posses any solvating ability.

Supercritical fluids are of interest as chromatographic mobile phases because of the unique physical properties they exhibit. The three properties of most concern to chromatographers are viscosity, diffusion, and density. The values of these properties for supercritical fluids are intermediate between those of the gas and liquid states.

Supercritical fluid viscosities are less than those of liquids. Lower viscosities result in lower pressure drops across chromatographic columns in SFC than are observed with liquids in HPLC. Lower pressure drops mean that supercritical fluids can be pumped faster than liquids at the same operating pressure. This also signifies that within normal instrumental operating pressures (≤ 6000 psi), there is more latitude in the choice of mobile phase flow rates when using supercritical fluids.

The diffusion coefficients of solutes in supercritical fluids are smaller than in gases but still much greater than in liquids. The high diffusivity of supercritical fluids improves the analyte mass transfer between the mobile and stationary phases. The improved diffusion and viscosity properties of supercritical fluids versus liquids combine to produce higher optimum average linear velocities for SFC than for HPLC, which translates into faster analysis times for SFC than for HPLC (26).

Supercritical fluid densities are higher than those of gases and can approach those of liquids. The solvent strength of a fluid is related to its density, with density a function of pressure. If the fluid is pressure or density programmed (i.e., increased with time) the solvent strength will also increase. This increase in pressure, hence density, causes the elution of compounds of increasingly higher molecular weight from the chromatographic system, analogous to temperature programming in GC and gradient programming in HPLC. Supercritical fluid viscosity and diffusion properties allow fast, efficient analytical separations to be done (as in the case of GC), but with a mobile phase that has a strong solvating ability (as in HPLC). If the proper choice of mobile phase is made, then separations can be accomplished near room temperature, as in HPLC, with high efficiency and with sensitive, universal detectors like the FID, as in GC. In many ways, SFC has the advantages of both HPLC and GC, including high efficiency, mild operating temperatures, and detector compatibility. By controlling the temperature and pressure exerted on the carrier fluid, its solvent strength can be controlled or "fine-tuned" to achieve high resolution separations of similar compounds. Supercritical fluids exhibit high solvent strength—stronger than the corresponding liquid in many cases. An excellent review of the use of supercritical fluids for extraction has been written by Randall (27).

A variety of liquids can and have been used as mobile phases in SFC (see Table 1). Carbon dioxide is often the fluid of choice because it is compatible with an FID, making the analysis of nonabsorbing compounds more feasible. It can also be used with a UV absorbance detector because its its UV cutoff is about 190 nm, which is lower than many HPLC solvents. Carbon dioxide is colorless, nontoxic, odorless, nonflammable, and readily available in high purity. However, carbon dioxide is not the solvent of choice for all analyses. Currently, it is applicable to samples of moderately high polarity and with molecular weight below 10,000 daltons.

In addition to the use of pure mobile phases such as those listed in Table 1, there have many reports of the additions of modifiers to enhance solvent

Table 1. Physical Parameters of Selected Supercritical Fluids[a]

Fluid	T_c(°C)	P_c(atm)	ρ_c(g/mL)
Carbon dioxide	31.0	72.9	0.47
Nitrous oxide	36.5	71.7	0.45
Ammonia	132.5	112.5	0.24
n-Hexane	234.2	29.6	0.23
n-Pentane	196.6	33.3	0.23
n-Butane	152.0	37.5	0.23
n-Propane	96.8	42.0	0.22
Sulfur hexafluoride	45.5	37.1	0.74
Dichlorodifluoromethane	111.8	40.7	0.56
Trifluoromethane	25.9	46.9	0.52
Methanol	240.5	78.9	0.27
Ethanol	243.4	63.0	0.28
Isopropanol	253.3	47.0	0.27
Diethyl ether	193.6	36.3	0.27

[a]Data from Matheson Gas Data Book, East Rutherford, NJ, 1980 and *CRC Handbook of Chemistry and Physics*, CRC Press, Boca Raton, FL, 1984.

strength or change the selectivity of the mobile phase. For example, isopropanol or methanol can be added to carbon dioxide to improve the solubility of polar compounds and to change the retention characteristics of the columns (28, 29). Modifiers such as dioxane have been used with pentane as the primary fluid (30). As expected, when using organic mobile phases such as pentane or CO_2 containing organic modifiers, the use of the FID is precluded because of the unusable baseline offset produced by the organic materials. In these cases, different detection systems must be used, such as UV, fluorescence, or mass spectrometry.

INSTRUMENTAL ASPECTS

Reviews of the instrumental requirements of SFC have been published elsewhere (20, 31, 32). To understand the usefulness and applicability of SFC, it is good to understand some of the basic instrumental aspects of SFC. Until recently, the instrumentation used for SFC was home-built. Systems have been constructed by modifying and combining HPLC pumps and GC ovens. This approach, although acceptable in fundamental research laboratories, is arduous, uneconomical, and time-consuming for most industrial, contract research, and applications laboratories. The introduction of commercially available instrumentation has made SFC more reliable, and, as a result, this equipment is now being used routinely in many laboratories.

Supercritical fluid chromatographs, whether packed or capillary columns are used, consist of three basic components: a high pressure pumping system that delivers a fluid mobile phase under supercritical pressures, an oven

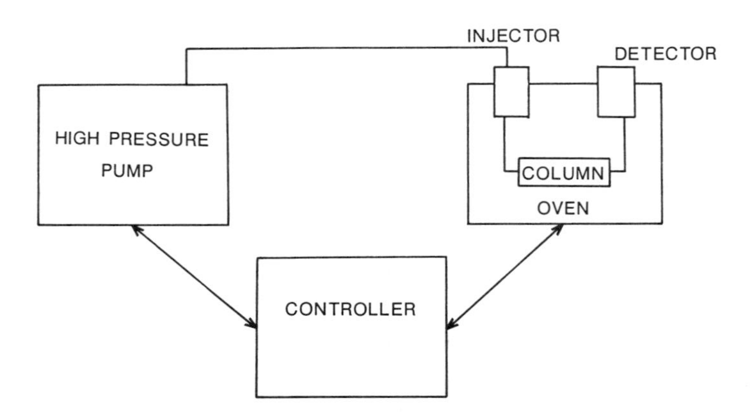

Figure 1. Schematic drawing of SFC instrumentation.

equipped with an injection system and detector(s) for temperature control of the column, and a controller to operate the entire system. Figure 1 illustrates schematically these instrumental components.

The early work on SFC was done using reciprocating pumps like those commonly used in HPLC. These pumps work well for packed column SFC with flow rates of 1–10 mL/min. However, with capillary columns, the flow rates are on the order of several microliters per minute, and reciprocating pumps cannot deliver these flow rates without pulses. The pump of choice for capillary SFC has been the syringe pump because it can deliver a nonfluctuating fluid supply under either pressure or flow control. Currently, all vendors of SFC instrumentation offer syringe pumps. To accommodate the larger flow rates of packed columns, syringe pumps with volumes close to 200 mL, or else pumps which have two syringe pumping chambers (one fills while the other is pumping) are used. These state-of-the-art pumping systems can be used with either capillary or packed columns.

Differential-pressure transducers are the control components of SFC pumps. Microprocessor control gives precise control of the pressure and density throughout the course of programmed runs. Accurate pressure delivery from 1 to 420 atm (15–6300 psi) with rapid programming up to 150 atm/min are possible with current pumping systems, which facilitate fast, high pressure analyses. These high pressures, which are necessary for the elution of large molecular weight compounds from the chromatograph, are not too far removed from those normally encountered in HPLC; thus standard compression fittings are usually adequate for SFC.

The most important aspect of the oven module is temperature uniformity. Precision temperature control should be 0.1°C, with thermal gradients between any two points in the oven, particularly in the column zone, being less than ±0.5°C. These features are crucial because slight changes in column temperature can affect the fluid density and set up density gradients along the length of the column, which would in turn affect the solvent strength of the mobile phase, hence the chromatography. These temperature effects are less drastic in GC or HPLC.

The injection valves used in SFC are of the same type as the valves used in HPLC because the injections are made under high pressure. Packed column SFC generally uses direct injection, in which the entire loop contents are loaded onto the column, as in HPLC. Three types of injection technique have been used in capillary SFC: split, timed-split, and direct injection (33). The small column diameters used in capillary SFC present special problems for sample injection. Well-designed injection systems must be used to prevent overloading and band broadening during injection. Split injection has been most commonly used to inject the small volumes (5–100 nL) used with these columns. Split injection gives sharp bands at the head of the column with the volumes needed, but this form of injection has its difficulties. The main problem seems to be discrimination between compounds of varying molecular weight, similar to what occurs in split injection in GC.

In an effort to overcome the problem of split injection discrimination, other injection procedures have been investigated. Harvey and Stearns (34) first investigated the timed-split or "moving-injection" technique for microbore HPLC. Fast pneumatics and electronics are used in this injection technique, which allows the valve to be rapidly moved from the load to the inject position and back again. This rapid actuation permits only a portion of the sample loop to be injected on the column. Quantities as small as picoliters have been injected reproducibly with this technique. Richter was the first to report the use of this injection method for capillary SFC (35). Recent work in the area has shown that this is the most reproducible injection procedure for capillary SFC (33, 36). Typical relative standard deviation for raw areas is in the range of 1–2%, and for internal standard methods it is between 0.5 and 1%.

Direct injection is a recent development for injection onto capillary columns in SFC. Richter et al. (33, 36) have shown that using a "retention gap" (a piece of bare fused silica on the front of the column to help concentrate the solutes), as has been reported in HPLC-GC multidimensional techniques (37–39) and on-column injection in capillary GC (40), can be used for the injection of sample sizes up to 1.0 μL directly onto capillary columns in SFC. As is expected, good reproducibility is achieved using these larger injection sizes, and larger injection volumes make trace analysis by capillary SFC easier.

One of the major advantages of SFC is flexibility in the choice of detectors available for use. Virtually all GC and HPLC detectors have been used in packed and capillary column SFC. In most cases, it is easier to interface capillary SFC to the detectors because of the low flow rates exiting these columns (1–10 mL/min gas flow). Novotny has reviewed detectors that have been used in conjunction with capillary SFC (41).

The most commonly used detector is the flame ionization detector. Other GC-type detectors, including thermionic, electron capture, and flame-photometric have been used in SFC. The UV-absorbance and fluorescence detectors have been reported as well. In addition, hyphenated techniques including SFC-MS (42) and SFC-FTIR (43) have proven to be useful.

The third component of an SFC system is the controller. The first pump controllers for SFC were simple analogue devices that allowed constant pressure control or crude programming. The oven was operated separately through the oven control itself. With the advent of microprocessors, much better control is achieved. All pump and oven operating functions, injection system operation, regulation of heated zones, and pressure and density ramping of the mobile phase are accomplished with a separate microcomputer or on-board microprocessor. Simultaneous temperature–density or temperature–pressure programming, which has been shown to be advantageous for some oligomer samples (44), can be done. Software for density–pressure calculations, multiple method storage, data acquisition, data storage, complete operation of the instrument, and manipulation of the detector signal output can be run using the instrument-control computer at the highest level of sophistication. On the other hand, the controller can simply operate the instrument and store a few methods, as is the case of the on-board microprocessor. In either case, this controller unit is all important because it determines, in part, the accuracy and precision of the method as well as the ease of use for routine analyses.

Mention should be made here of the analytical columns. Regardless of the sophistication of the analytical instrument, the real separation process is carried out in the chromatographic columns; if the columns are no good, meaningful data cannot be obtained from the instrument. In the case of packed columns, microbore HPLC columns (1.0 mm i.d.) are generally being used, although packed capillaries (0.1–0.75 mm i.d.) are gaining popularity. These columns use standard HPLC packings in the 3–10 μm range. The open tubular columns generally have small internal diameters (0.05–0.1 mm i.d.) and are coated with films of polysiloxanes similar to those used in capillary GC columns. The difference between the GC and SFC capillary columns is that the polymers used on the SFC columns must have a higher level of surface bonding and cross-linking to prevent stripping by the supercritical fluids themselves.

It is necessary to use some form of pressure restriction at the end of the columns to maintain the high pressures necessary for SFC uniformly throughout the columns and to control the flow rates for optimized chromatography. If a large pressure drop occurs along the length of the column, compounds may precipitate before entering the detector. For packed columns, back-pressure regulators are normally used to maintain constant pressures along the columns. For capillary columns, restrictors of various types have been used. These restrictors consist of some form of capillary tubing with regions of large pressure drops such that the return to atmospheric pressure occurs over a short region (0.5–2 cm). Crimped metal tubing, small bore capillaries, flame-tapered capillaries, and ceramic frits have all been used. Lee has reviewed the restrictor types which have been used (45), and some experimental and theoretical work on restrictor design has been reported recently (46, 47).

APPLICATIONS

In this section we explore some of the applications of SFC to the analysis of oligomer samples. The real usefulness of any technique is demonstrated by its ability to solve difficult, real-world problems. SFC lends itself to the analysis of some compounds that cannot easily be analyzed by GC, HPLC, or SEC because of thermal instability, nonvolatility, insufficient chromatographic efficiency, or detector incompatibility.

Polystyrenes

One of the first applications of SFC was the separation of polystyrene oligomers reported by Klesper and coworkers (48, 49) and also by Jentoft and Gouw (50–52) using packed columns. Polystyrenes have been also analyzed by capillary SFC (53). Gradient elution SFC with packed columns has been used for polystyrene separations by Klesper et al. (54, 55). In fact, polystyrenes are the compounds which have been most widely reported in the literature as applications of SFC. The most impressive separation of polystyrenes was done by Hirata and Nakata (56), who showed the separation of oligomers up to 70 monomer units (8000 approximate MW) using packed capillary columns. One advantage of SFC analysis is that the individual oligomers can be separated and quantified, whereas with SEC, only the molecular weight distribution is obtained.

Polyglycols

Polyglycols are widely used in the chemical industry—for example, as surfactants, heat transfer fluids, brake fluids, and raw materials in the production of other polymers such as polyurethanes. The physical and chemical properties of the final products depend, in part, on the molecular weight distribution of the polyglycols used as the starting materials. Many polygycols have no UV-absorbing groups, making HPLC analysis difficult, and most of these compounds have molecular weights beyond the molecular weight range of even high temperature GC($>400°C$). Some nonchromatographic techniques such as viscosity, cloud point, or freezing point measurements are often used to monitor the quality of these materials, but these techniques are not sensitive enough to detect the small differences in the average molecular weights which can cause differences in performance.

Figure 2 shows the chromatographic analysis of a poly(ethylene glycol) with an average molecular weight of 600 daltons. In this case, CO_2 at 100°C was used as the mobile phase with an FID as the detector. There is good separation of all the oligomers: the oligomers eluting at the end of the chromatogram have molecular weights near 1000 daltons. High resolution separations such as this are becoming more necessary as glycols are being more widely used in foodstuffs, cosmetics, drugs, and other formulations that come in contact with humans.

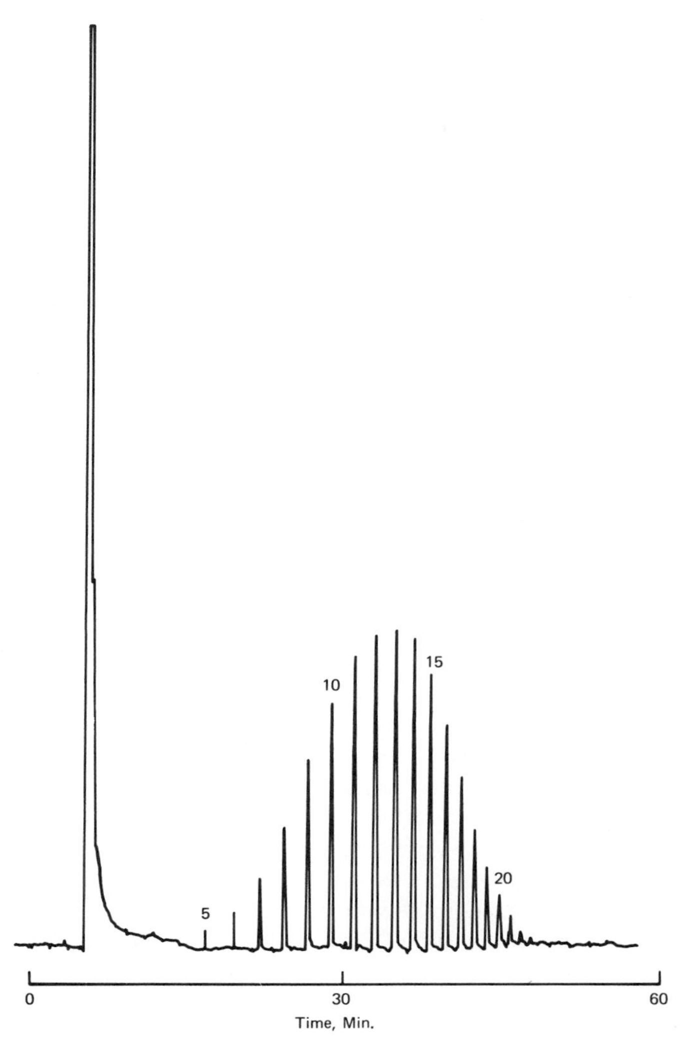

Figure 2. Supercritical fluid chromatogram of a poly(ethylene glycol) with an average molecular weight of 600. Conditions: 10 m × 50 μm i.d. fused silica column, CO_2 mobile phase at 100°C, FID at 350°C, multiramp density programmed. Numerals indicate number of ethylene oxide units.

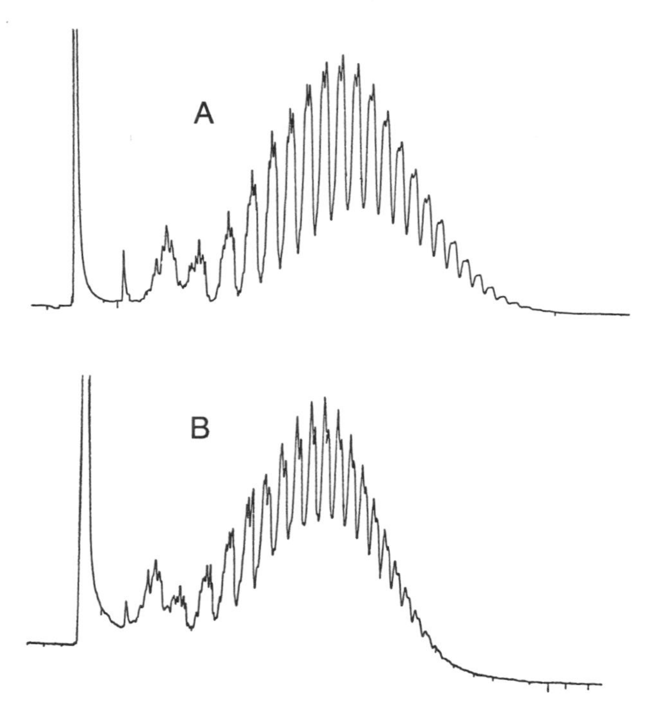

Figure 3. Supercritical fluid chromatograms of two lots of a polyglycol with an average molecular of 1200. Both chromatograms obtained under identical conditions: 15 m × 50 μm i.d. fused silica column, CO_2 mobile phase at 100°C, FID at 375°C, multiramp density programmed.

The usefulness of an analytical technique for routine analysis is demonstrated if this technique can be used to monitor lot-to-lot variability in chemical production. Figure 3 compares chromatograms of two lots of a polyglycol with an average molecular weight of 1200 daltons which was made by the addition of ethylene oxide to a fatty alcohol. The standard physical tests indicated that there were no differences between the two lots of material; however, the SFC chromatograms do show subtle differences. A pseudo-molecular-weight distribution plot can be made, by plotting area percent versus peak number or retention time for each of the peaks as shown in Figure 4. This information graphically illustrates that SFC can be used to detect the differences between these materials. Lot A is the material that did not perform according to the customer's specifications, and the reason for this is the shift in the molecular weight distribution evident from the SFC data.

Figure 5 shows a chromatogram obtained from the analysis of a glycol made by the addition of ethylene and propylene oxide to a fatty alcohol. When

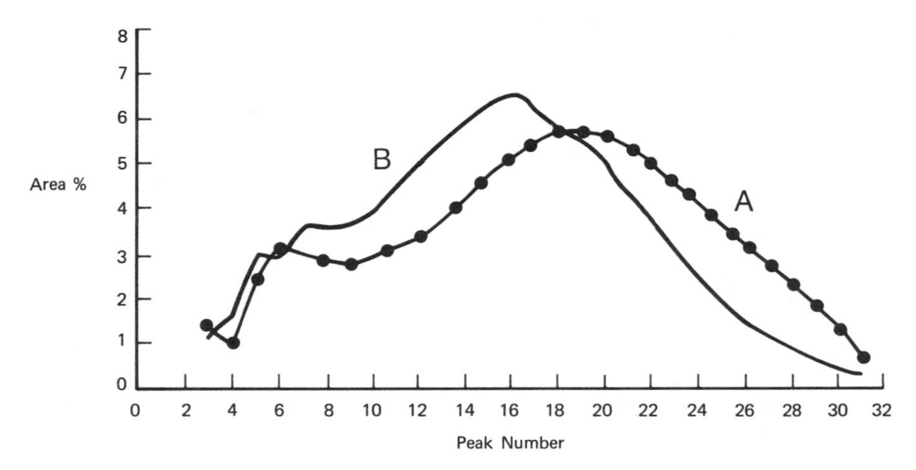

Figure 4. Pseudo-molecular-weight distribution plot of area percent versus peak number for the two lots of the polyglycol shown in Figure 3.

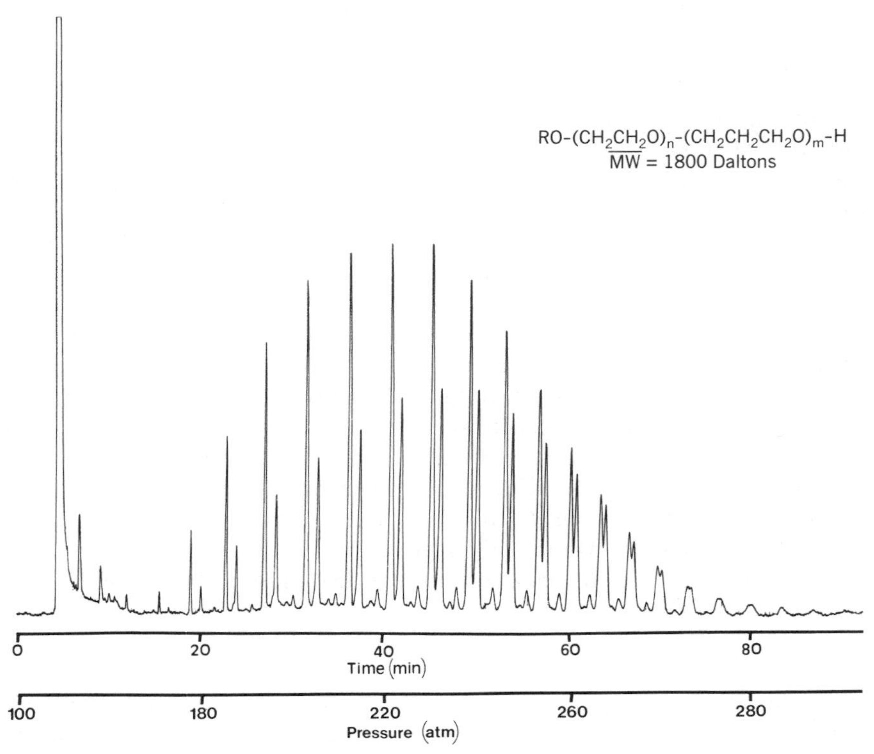

Figure 5. Supercritical fluid chromatogram of a polyglycol sample. Conditions: 25 m × 50 μm i.d. fused silica column, CO_2 mobile phase at 160°C, FID at 375°C, multiramp density programmed.

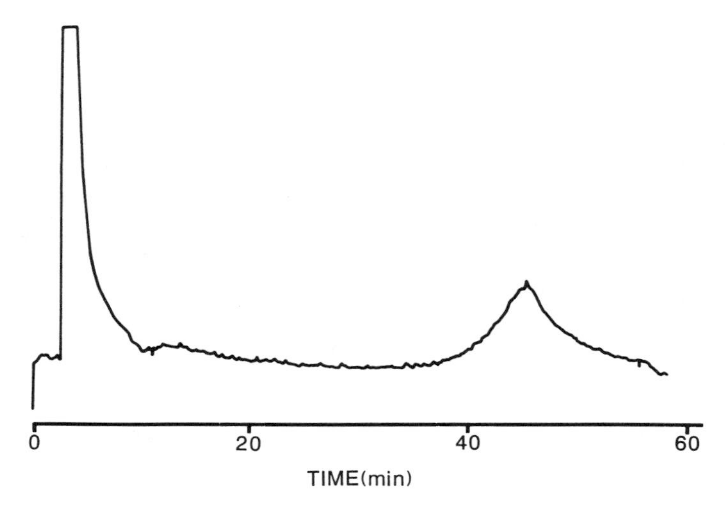

Figure 6. Supercritical fluid chromatogram of a glycerol-initiated polyglycol. Conditions: 5 m × 50 μm i.d. fused silica column, CO_2 mobile phase at 120°C, FID at 400°C, multiramp density programmed.

propylene oxide is used, there are many more isomers possible because addition can take place at either the first or second carbon of propylene oxide. Several separate distributions, due presumably to different isomers, are evident in this chromatogram. It should be noted that the average molecular weight of this material is 1800 and the compounds eluting at the last part of the chromatogram approach 2500; yet the different oligomers are still fairly well resolved.

SFC does have its limitations, however, and the type of compound analyzed will determine the molecular weight range that can be successfully chromatographed as well as the type of data obtained. Figure 6 shows a chromatogram of a polyglycol of average molecular weight 5000 synthesized by the addition of ethylene and propylene oxide to glycerol. This sample is much more complex, has a higher molecular weight, and is more polar than any of the other examples shown previously. All these factors combine to prevent the separation of the individual oligomers from being realized, but enough information is gained to characterize the distribution of the glycol.

Polysiloxanes

Polysiloxanes, also known as silicones, have many uses in industry as well as in consumer products. The unique properties of silicones give them the ability to perform a variety of functions. They can be used to lubricate, seal, bond, release, defoam, encapsulate, insulate, waterproof, and coat, to name a few uses. Basic to their performance is, of course, their molecular weight distribution.

Nieman and Rogers (57) were the first to report the analysis of polysiloxanes by packed column SFC. Fjeldsted and Lee (58) first demonstrated their analysis by capillary SFC. Other workers have shown improved results of the analysis of these compounds (44, 59).

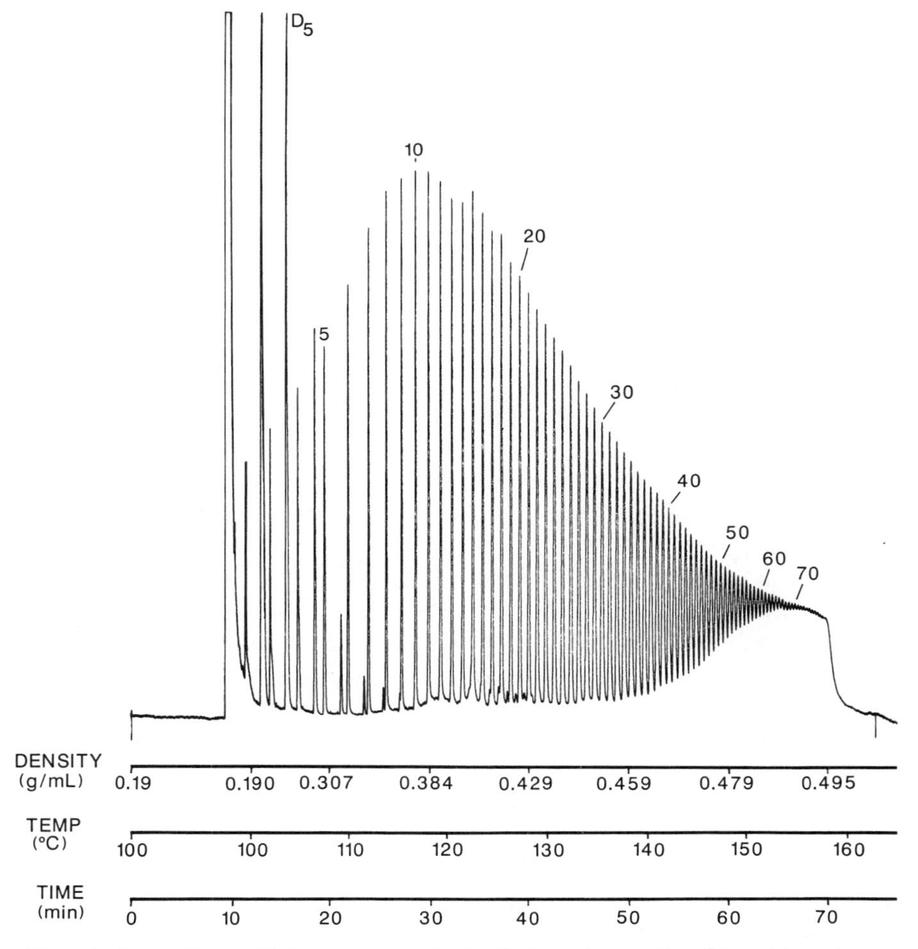

Figure 7. Supercritical fluid chromatogram of a dimethylpolysiloxane. Conditions: 10 m × 50 μm i.d. fused silica column, CO_2 mobile phase, FID at 400°C, simultaneous linear temperature–asymptotic density programmed.

Figure 7 shows the analysis of a dimethylpolysiloxane by capillary SFC using simultaneous density–temperature programming. Visible in this chromatogram are the distributions from the linear oligomers out to 70 monomer units in length and from the cyclic compounds as well. Hirata et al. (60) have shown the analysis of a phenylmethylpolysiloxane using packed capillary columns and mobile phase of hexane–ethanol with UV detection, and Richter et al. (61) have reported the analysis of a phenylhydropolysiloxane with CO_2 and UV detection. Clearly, SFC is well suited for the analysis of polysiloxanes because of their high solubility in supercritical fluids.

Other Applications

White and Houck demonstrated the use of capillary SFC for the analysis of chlorotrifluoroethylene polymers, polymeric alcohols, and fluorinated alkyl alkoxylates (59). Other oligomer samples that have been analyzed are polysulfides, polyterpenes, poly(α-olefins), polyethylenes, polymeric diisocyanates, polybutadienes, polyepichlorohydrins, epoxy resins, polymethacrylates, and polyether amines (62). There have been reports of the determination of residual monomers and polymer additives in oligomer and polymer samples by SFC (63, 64), although this is not directly related to oligomer analysis.

REFERENCES

1. J. B. Hannay and J. Hogarth, *Proc. R. Soc., London*, **29**, 324 (1879).
2. E. Klesper, A. H. Corwin, and D. A. Turner, *J. Org. Chem.*, **27**, 700 (1962).
3. M. Novotny, S. R. Springston, P. A. Peaden, J. C. Fjeldsted, and M. L. Lee, *Anal. Chem.*, **53**, 407A (1981).
4. N. M. Karayannis and A. H. Corwin, *Anal. Biochem.*, **26**, 34 (1968).
5. N. M. Karayannis and A. H. Corwin, *J. Chromatogr.*, **47**, 247 (1970).
6. N. M. Karayannis and A. H. Corwin, *J. Chromatogr. Sci.*, **8**, 251 (1970).
7. N. M. Karayannis, A. H. Corwin, E. W. Baker, E. Klesper, and J. A. Walker, *Anal. Chem.*, **40**, 1763 (1968).
8. S. T. Sie and G. W. A. Rijnders, *Sep. Sci.*, **2**, 699 (1967).
9. S. T. Sie and G. W. A. Rijnders, *Sep. Sci.*, **2**, 729 (1967).
10. S. T. Sie and G. W.A. Rijnders, *Sep. Sci.*, **2**, 755 (1967).
11. S. T. Sie, W. Van Beersum, and G. W. A. Rijnders, *Sep. Sci.*, **1**, 459 (1966).
12. S. T. Sie and G. W. A. Rijnders, *Anal. Chim. Acta*, **38**, 31 (1967).
13. M. N. Myers and J. C. Giddings, *Anal. Chem.*, **37**, 1453 (1965).
14. M. N. Myers and J. C. Giddings, *Sep. Sci.*, **1**, 761 (1966).
15. L. KcLaren, M. N. Myers, and J. C. Giddings, *Science*, **159**, 197 (1968).

16. J. C. Giddings, M. N. Myers, L. McLaren, and R. A. Keller, *Science*, **162**, 67 (1968).
17. J. C. Giddings, M. N. Myers, and J. W. King, *J. Chromatogr. Sci.*, **7**, 276 (1969).
18. M. N. Myers and J. C. Giddings, "High-Pressure Gas Chromatography," in *Progress in Separation and Purification*, Vol. 3, E. S. Perry and C. J. V. Oss, Eds., Wiley-Interscience, New York, 1970, p. 133.
19. U. Van Wasen, I. Swaid, and G. M. Schneider, *Angew. Chem., Int. Ed. Engl.*, **19**, 575 (1980).
20. E. Klesper, *Angew. Chem., Int. Ed. Engl.*, **17**, 738 (1978).
21. T. H. Gouw and R. E. Jentoft, *J. Chromatogr.*, **68**, 303 (1972).
22. T. H. Gouw and R. E. Jentoft, *Adv. Chromatogr.*, **13**, 1 (1975).
23. T. H. Gouw and R. E. Jentoft, *Chromatogr. Sci.* (*Chromatogr. Pet. Anal.*), **11**, 313 (1979).
24. P. A. Peaden and M. L. Lee, *J. Chromatogr.*, **259**, 1 (1983).
25. B. W. Wright and R. D. Smith, *J. High Resolut. Chromatogr. Chromatogr. Commun.*, **8**, 8 (1985).
26. D. R. Gere, R. Board, and D. McManigill, *Anal. Chem.*, **54**, 736 (1982).
27. L. G. Randall, *Sep. Sci. Technol.*, **17**, 1 (1982).
28. J. M. Levy and W. M. Ritchey, *J. Chromatogr. Sci.*, **24**, 242 (1986).
29. J. B. Crowther and J. D. Henion, *Anal. Chem.*, **57**, 2711 (1985).
30. F. P. Schmitz, H. Hilgers, and E. Klesper, *J. Chromatogr.*, **267**, 267 (1983).
31. P. A. Peaden, J. C. Fjeldsted, M. L. Lee, S. R. Springston, and M. Novotny, *Anal. Chem.*, **54**, 1090 (1982).
32. E. Klesper and W. Hartmann, *Eur. Polym. J.*, **14**, 77 (1978).
33. B. E. Richter, D. W. Knowles, M. R. Andersen, N. L. Porter, E. R. Campbell, and D. W. Later, in *Proceedings of the Eighth International Symposium on Capillary Chromatography*, Riva del Garada, Italy, May 19–21, 1987, P. Sandra, Ed., Hüthig, Heidelberg, 1987, p. 973.
34. M. C. Harvey and S. P. Stearns, *Anal. Chem.*, **56**, 837 (1984).
35. B. E. Richter, Paper No. 514, presented at the Pittsburgh Conference and Exposition on Analytical Chemistry and Applied Spectroscopy, March 10–14, 1986, Atlanitc City, NJ.
36. B. E. Richter, D. E. Knowles, M. R. Andersen, N. L. Porter, E. R. Campbell, and D. W. Later, *J. High Resolut. Chromatogr. Chromatogr. Commun.*, **11**, 29 (1988).
37. F. Munari, A. Trisciani, G. Mapelli, S. Trestianu, K. Grob, Jr., and J. M. Colin, *J. High Resolut. Chromatogr. Chromatogr. Commun.*, **8**, 602 (1985).
38. H. J. Cortes, C. D. Pfeiffer, and B. E. Richter, *J. High Resolut. Chromatogr. Chromatogr. Commun.*, **8**, 469 (1985).
39. H. J. Cortes, B. E. Richter, C. D. Pfeiffer, and D. E. Jensen, *J. Chromatogr.*, **344**, 55 (1985).
40. K. Grob, Jr., G. Karrer, and M. L. Riekkola, *J. Chromatogr. Rev.*, **334**, 129 (1985).
41. M. Novotny, *J. High Resolut. Chromatogr. Chromatogr. Commun.*, **9**, 137 (1986).

42. B. W. Wright, H. T. Kalinoski, H. R. Udseth, and R. D. Smith, *J. High Resolut. Chromatogr. Chromatogr. Commun.*, **9**, 145 (1986).
43. R. C. Wieboldt, Nicolet FTIR Application Note, AN-8705, Madison, WI.
44. D. W. Later, E. R. Campbell, and B. E. Richter, *J. High Resolut. Chromatogr. Chromatogr. Commun.*, **11**, 65 (1988).
45. M. L. Lee and K. E. Markides, *J. High Resolut. Chromatogr. Chromatogr. Commun.*, **9**, 652 (1986).
46. R. D. Smith, J. L. Fulton, R. C. Petersen, A. J. Kopriva, and B. W. Wright, *Anal. Chem.*, **58**, 2057 (1986).
47. R. W. Bally and C. A. Cramers, *J. High Resolut. Chromatogr. Chromatogr. Commun.*, **9**, 626 (1986).
48. E. Klesper and W. Hartman, *J. Polym. Sci., Polym. Lett. Ed.*, **15**, 9 (1977).
49. W. Hartman and E. Klesper, *J. Polym. Sci., Polym. Lett. Ed.*, **15**, 713 (1977).
50. R. E. Jentoft and T. H. Gouw, *J. Polym. Sci., Polym. Lett. Ed.*, **7**, 811 (1969).
51. R. E. Jentoft and T. H. Gouw, *J. Chromatogr. Sci.*, **8**, 138 (1970).
52. R. E. Jentoft and T. H. Gouw, *Anal. Chem.*, **44**, 681 (1972).
53. J. C. Fjeldsted, W. P. Jackson, P. A. Peaden, and M. L. Lee, *J. Chromatogr. Sci.*, **21**, 222 (1983).
54. F. P. Schmitz, H. Hilgers, and E. Klesper, *J. Chromatogr.*, **267**, 267 (1983).
55. F. P. Schmitz, H. Hilgers, B. Lorenschat, and E. Klesper, *J. Chromatogr.*, **346**, 315 (1985).
56. Y. Hirata and F. Nakata, *J. Chromatogr.*, **295**, 315 (1984).
57. J. A. Nieman and L. B. Rogers, *Sep. Sci.*, **10**, 517 (1975).
58. J. C. Fjeldsted and M. L. Lee, *Anal. Chem.*, **56**, 619A (1984).
59. C. M. White and R. K. Houck, *J. High Resolut. Chromatogr. Chromatogr. Commun.*, **9**, 4 (1986).
60. Y. Hirata, F. Nakata, and M. Kawasaki, *J. High Resolut. Chromatogr. Chromatogr. Commun.*, **9**, 633 (1986).
61. B. E. Richter, D. E. Knowles, M. R. Andersen, D. W. Later, and F. J. Yang, Paper No. 1021, presented at the pittsburgh Conference and Exposition on Analytical Chemistry and Applied Spectroscopy, March 9–13, 1986, Atlantic City, NJ.
62. D. E. Knowles and L. Nixon, personal commmunication, unpublished data, Lee Scientific Application Laboratory, Salt Lake City.
63. B. E. Richter, *Chromatogr. Forum*, **1**(4), 52 (1986).
64. M. W. Raynor, K. D. Bartle, I. L. Davies, A. Williams, A. A. Clifford, J. M. Chalmers, and B. W. Cook, *Anal. Chem.*, **60**, 427 (1988).

CHAPTER

14

MASS SPECTROMETRY

ROBERT P. LATTIMER and ROBERT E. HARRIS

The BFGoodrich Research and Development Center, Brecksville, Ohio

HANS-ROLF SCHULTEN

Fachhochschule Fresenius, Wiesbaden, Federal Republic of Germany

INTRODUCTION

Mass spectrometry involves the study of ions in the vapor phase. In the ideal situation, the polymer chemist would like to use mass spectrometry for direct analysis of polymer systems. Much can be learned about the composition of a polymer if the mass spectrometer can produce a spectrum of molecular or quasi-molecular ions for the various oligomers that are present. Mass spectrometry should, in principle, be a very effective method to determine molecular weight averages, since the technique can directly provide the information needed to perform the calculations (i.e., molecular weights and relative intensities or abundances of oligomers). The principal difficulty, of course, is the nonvolatility of polymers. To analyze any material by conventional mass spectrometry, the sample first must be vaporized and ionized in the instrument's vacuum system. Because of this rather severe limitation, most mass spectral methods of polymer analysis involve degradation of the material before the volatile fragments can be analyzed. The most common of these methods is pyrolysis, or thermal decomposition. Although degradative methods are very useful for many types of polymer characterization, they have no use in estimating molecular weight distributions, since the integrity of the macromolecule is destroyed before mass analysis can occur. Fortunately, during the past few years several newer methods have been developed for the vaporization–ionization of intact, higher molecular weight molecules into the mass spectrometer. The direct analysis of oligomeric mixtures using these newer desorption ionization methods is a fast-developing area of mass spectrometry that holds much promise for the future.

For the analysis of an oligomeric mixture, it is preferable that fragmentation and decomposition processes be absent, or at least minimal, in the mass spectrometer. The direct analysis of high polymers (MW $\sim 10^4$–10^7 amu) by mass spectrometry has not yet been effectively achieved, but some exciting developments in this quest have recently been reported. The intermediate molecular weight region ($\sim 10^3$–10^4 amu) is now accessible by a number of techniques (1–5). The applications to date have been mainly in the area of biomolecules, but several recent reports have described the characterization of synthetic polymer systems as well.

The necessity or advisability of pursuing techniques for mass spectral analysis of higher molecular weight polymers has at times been questioned, since a number of "classical" methods (e.g., gel permeation chromatography, vapor pressure osmometry, laser light scattering, magnetic resonance, infrared spectroscopy) are available. These are discussed in other chapters of this book. However, there are important reasons to pursue these developments other than scientific curiosity and desire for methodological improvements. Classical techniques, for example, are always averaging methods, that is, they measure the average properties of an oligomeric mixture and thus do not examine individual oligomers. Furthermore, classical techniques do not normally yield information on the different types of oligomer that may be present, nor do they distinguish and identify impurities in polymer samples. Copolymers, blends, and oligomers with different end groups often are not distinguished as to polymer type. Finally, classical methods in general do not provide absolute direct molecular weight distributions for polymers; instead they rely on calibrations made with accepted standards. Mass spectrometry clearly has great potential for direct analysis of polymers, and recent results from several laboratories are very encouraging. Nevertheless, the full realization of that potential must await the development of new and improved techniques that can overcome some of the formidable obstacles that currently exist. The following then is a progress report on molecular weight distributions using "macromass" spectrometry; one may expect further developments to occur rapidly during the next several years.

INSTRUMENTAL CONSIDERATIONS

For analysis of higher molecular weight materials, all components of the mass spectrometer system need to be considered: the inlet (vaporization) system, the ionization method, the mass analyzer, and the ion detector. All common types of analyzers have been adapted for higher mass transmission; these include high field or large radius magnets, low frequency quadrupoles, time-of-flight (TOF) analyzers, Fourier transform ion cyclotron resonance (FT-MS), and the plasma chromatograph (ion mobility analyzer). For higher mass operation, these analyzers are normally operated at rather low resolution to achieve maximum sensitivity. Quadrupole, time-of-flight, and ion mobility are inherently low resolution devices, but because of low ion abundances, the magnetic and FT-MS instruments are also often set to low resolution to gain maximum sensitivity for detection. Resolution is normally not of prime concern in oligomer analysis, however, since the objective is only to separate oligomers with relatively large mass differences ($\sim$ 30–100 amu).

Detection of higher mass ions can present special problems. Standard secondary electron multipliers become progressively less efficient as the mass of the ion increases (or conservely, as the ion velocity decreases). Beuhler and Friedman (6–8) have extensively studied the detection of higher mass ions and have established that there is a threshold velocity for secondary ion production when an ion strikes a surface. Practical limitations dictate a maximum of 10–20 keV for the energy to which ions can be accelerated in a mass spectrometer. Sundqvist et al. (9) have studied the secondary ion yield from insulin as a function of ion energy. They derived the following empirical formula, which estimates the minimum energy E needed to detect an ion of mass M:

$$E > 1.7 \times 10^{-3} M \,(\text{kev})$$

Thus at m/z 10,000, about 17 keV ion energy is required. This is readily achievable in many modern mass spectrometer systems with postacceleration detectors.

The vaporization–ionization of higher molecular weight molecules is currently the most restrictive area with respect to mass spectral analysis of polymers. Several desorption ionization approaches have been used with varying degrees of success, and the numerous techniques are discussed individually below. At present, no one technique has emerged as a dominant or preferred method. Often, however, one specific method is found to be optimal for a particular polymer system of interest.

CALCULATIONS

The calculations involved in determining molecular weight averages from mass spectral data are relatively straightforward. The analyst acquires mass spectra for the polymer using whatever vaporization–ionization method has been selected. It is always best to make several ($\sim$ 3–5) runs with the same material, so that the repeatability and precision of the measurments can be assessed. One then selects the molecular (or quasi-molecular) ions to be used for the intensity measurements. This includes an examination of the spectra to ascertain whether any fragment ions or ions due to decomposition products are present. If it is concluded that such interferences are absent or minimal, the intensities of the ions of interest are summed. This is most readily accomplished when the data are acquired by computer; in this case the area under the molecular ion envelop for each oligomer (which will include all the isotope peaks) is summed by the computer to represent the oligomer intensity. In some earlier studies (10–11), data were acquired by oscillograph, and peak height (rather than area) was used in the calculations. In this case corrections were made to account for the differing shapes of the peak envelopes which result from unresolved isotopic clusters. This type of correction is unnecessary if data are acquired by computer and total peak envelope areas are used.

The summed peak intensities N and mass spectrometric molecular weights M of the various oligomers are then used to calculate molecular weight averages as commonly defined:

$$\bar{M}_n = \sum N_i M_i / \sum N_i \qquad \text{and} \qquad \bar{M}_w = \sum N_i M_i^2 / \sum N_i M_i$$

Unfortunately, commercial mass spectral data systems do not yet provide subroutines to perform these calculations. It is to be hoped that this feature will be added in the future. Analysts to date have either written off-line (10) or on-line (12) computer programs to perform the calculations.

DIRECT PROBE INTRODUCTION (EI-MS)

A straightforward method of oligomer analysis is simply to heat the polymer in the direct probe of the mass spectrometer, with ionization by electron impact (EI). Under favorable conditions, oligomers of lower molecular weight polymers can be evaporated and detected before the polymer undergoes thermal decomposition. One must exercise caution, however, since the temperature required to evaporate oligomers is often close to, or overlaps with, the temperature at which the sample starts to decompose.

An interesting study of ethylene oxide (EO) and propylene oxide (PO) block copolymers by Lee and Sedgwick illustrates the utility of direct oligomer analysis by probe introduction with EI ionization (13). Polymer samples of low molecular weight (< 1700 amu) were heated in the direct probe of the mass spectrometer (200–320°C) with EI ionization at 70 eV. Since the samples were primary alcohols, no molecular ions were observed, but reasonably intense water-loss ions $(M - H_2O)^{+\cdot}$ could be monitored. The relative intensities of these ions were used to estimate number average molecular weights ($\bar{M}_n$) that were in good agreement with those determined by end-group titration. For example, Figure 1 represents the distribution of $(M - H_2O)^{+\cdot}$ ions found for a block copolymer of the type CD_3—O—$(EO)_n$—$(PO)_m$—H. The numbers on

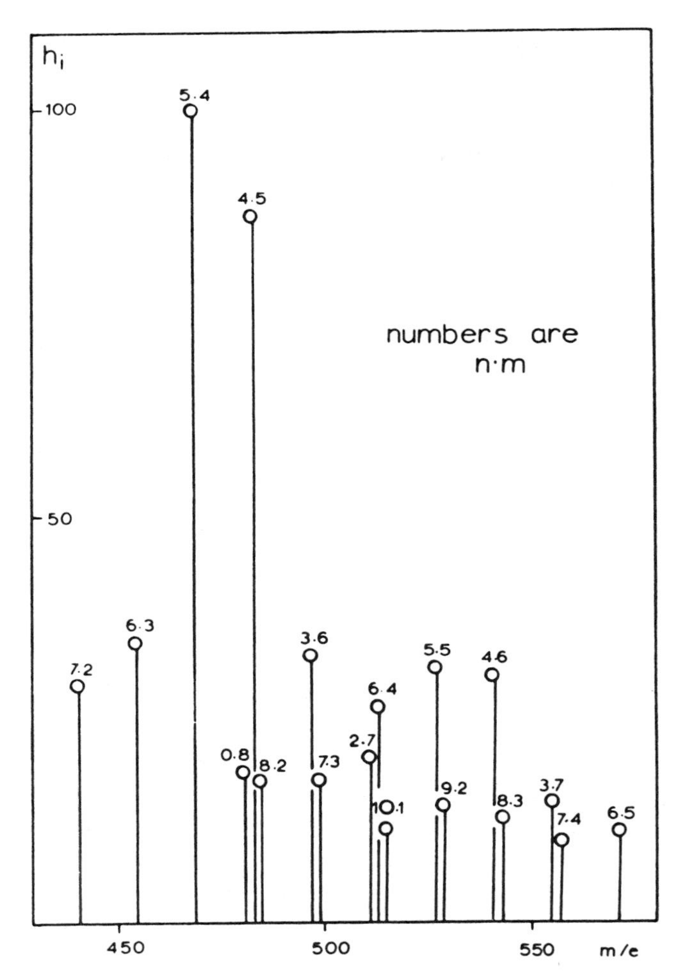

Figure 1. Distribution of ions $[CD_3O(EO)_n(PO)_mH—H_2O]^{+\cdot}$ from the block copolymer $CD_3O(EO)_n(PO)_mH$. (Reprinted with permission from Ref. 13.)

the diagram represent the block lengths (n—m) for each of the molecular species detected.

There are some potential problems associated with this type of analysis (14). The EI mass spectra are complex due to numerous fragmentation processes, and it seems likely that some thermal decompositon of the polymer takes place at the temperatures required for evaporation of oligomers into the ion source. EI fragmentation processes result in very complex mass spectra, and the fragmentation processes may not be the same for all oligomeric species. This makes the assignment of quasi-molecular species $(M - H_2O)^{+\cdot}$ difficult, and some components may be missed. Also, there will be some fractionation of components as a result of the different probe temperatures used for analysis (14). Nevertheless, Lee and Sedgwick were successful in obtaining useful quasi-molecular ion spectra for these rather fragile polymers. The good agreement of the mass spectral $\bar{M}_n$ values with those obtained by end-group titration indicates that their experimental procedure was essentially valid.

In another study, Wiley examined several oligomeric systems by direct probe introduction with EI ionization (15). In some cases, oligomer molecular ions were seen above 1000 amu. Unfortunately, it was often uncertain whether molecular ions originated from molecular species actually present in the sample, or whether the ions were due to thermal decomposition products of the sample. In general, EI molecular ions, when observed, were weak compared to fragment ions. In a series of experiments, Wiley isolated pure oligomers of 4,4′-isopropylidenediphenyl carbonate by gel permeation chromatography or prepared them by synthesis. The EI spectra showed molecular ions up to the tetramer (MW 1016), but fragmentation was extensive. It would not have been possible to determine molecular weight averages for these materials by direct probe EI analysis. Studies of this kind (13, 15) clearly show that direct probe EI-MS is not a viable method for determination of polymer molecular weight distributions.

RAPID HEATING (EI/CI-MS)

One method that may be used to facilitate electron impact or chemical ionization (CI) analysis of mixtures of oligomers of low volatility is to heat the sample so rapidly that the oligomers are evaporated before thermal decomposition can take place. "Rapid heating" [sometimes called flash desorption (16)] relies on a kinetic competition between evaporation and decomposition processes (17). At higher temperatures, vaporization processes may be favored over competitive decomposition processes (Fig. 2) (16). For the technique to be successful, samples must evaporated, ionized in close vicinity to the sample probe, and mass analyzed very quickly.

Udseth and Friedman have examined low molecular weight polystyrene ($\bar{M}_n$ 2100) using a rapid heating method (18). Samples were evaporated from a rhenium filament that was heated at a rate of ~1000°C/s. A special low frequency quadrupole system was used for mass analysis. The highest oligomer observed via EI ionization was the undecamer (MW 1202). Chemical ionization with argon or methane was much more successful, however, in terms of observing higher mass species. Both CI methods gave abundant ions

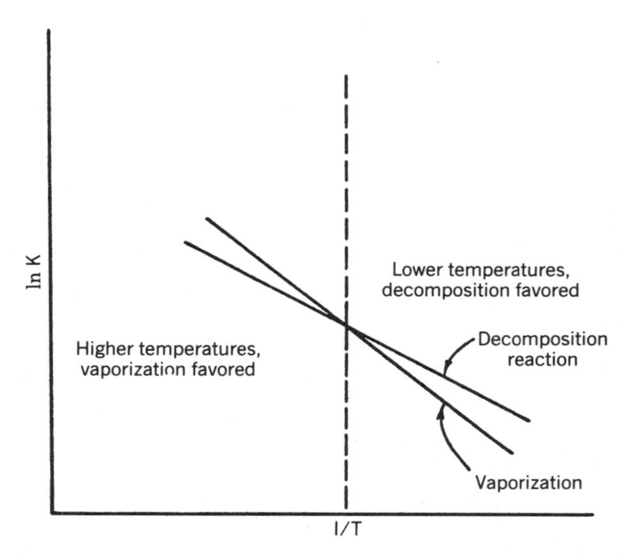

Figure 2. Relationship between the temperature dependencies of vaporization and molecular decomposition processes of an involatile, thermally labile compound. (Reprinted with permission from Ref. 16.)

in the 500–3000 amu range. Protonated molecular ions from the CI spectrum (Fig. 3) could be used to determine an average molecular weight ($\bar{M}_n = 1987$), in reasonably good agreement with the value supplied by the manufacturer ($\bar{M}_n = 2100$). The relatively large abundances of the lower mass oligomers ($n = 1$–3, Fig. 3) indicate that these species are primarily decomposition products. The distribution of oligomers (Fig. 3) also indicates that small quantities of higher mass oligomers (MW > 3000) may have been missed by the mass analyzer. This may be why the calculated $\bar{M}_n$ (1987 amu) is somewhat lower than the manufacturer's value (2100 amu).

The technique of "desorption" CI (DCI) or "in-beam" CI has gained considerable interest for the analysis of low volatility compounds (19). The general idea is to insert the sample directly inside the ionizing region via an extended probe tip made of some inert substance (e.g., glass, quartz, Teflon or Vespel). DCI has gained rapid popularity because it is relatively simple to adapt to almost any mass spectrometer. Although the technique is generally used with chemical ionization, it has also been used, albeit with less success, with electron impact ionization. The DCI method works at least partially because of the rapid heating that takes place when the probe tip is inserted into the ion beam. This is indicated by the fact that the best quasi-molecular ion spectra are obtained shortly after insertion of the sample probe into the ion source. The method may have some potential for use with synthetic polymers,

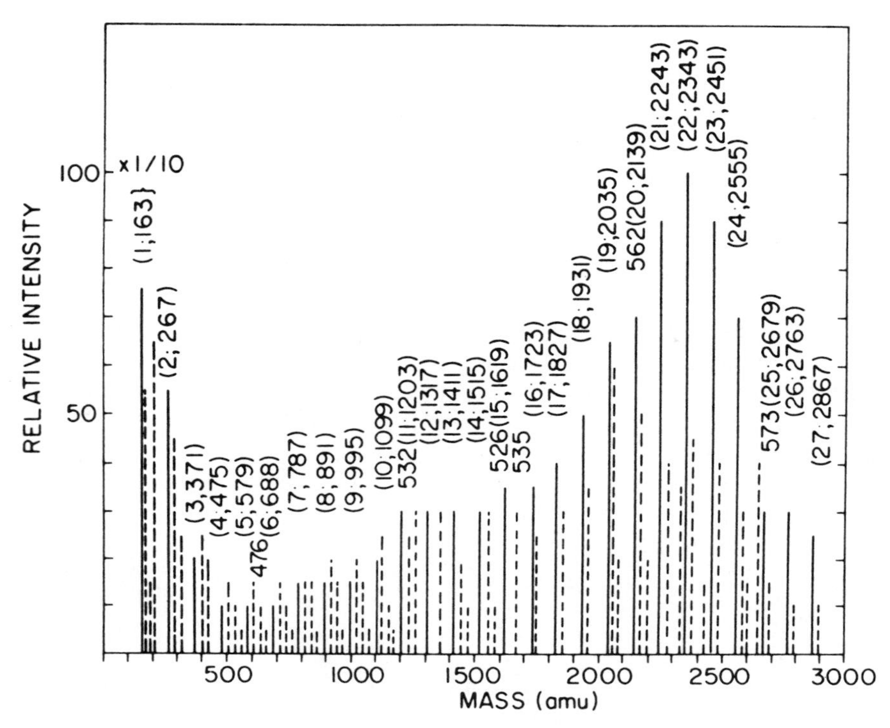

Figure 3. Partial methane chemical ionization mass spectrum of polystyrene: solid lines, oligomer sequence; dashed lines, solvation or fragment peaks. (Reprinted with permission from Ref. 18.)

but this and other rapid heating methods are in general not very satisfactory for polymer average molecular weight determinations.

CALIFORNIUM PLASMA DESORPTION

The californium-252 plasma desorption (PD-MS) technique uses high energy fission fragments from the decay of ^{252}Cf to vaporize and ionize nonvolatile samples (20, 21). It is essentially an ultrarapid heating technique. The sample in the form of a thin film is placed either in front of or behind the ^{252}Cf source. Fission fragments produce very rapid heating of the sample. Ionization of polar organic molecules occurs mainly via ion–molecule reactions or ion-pair formation. Positive ions produced are generally MH^+; cation attachment also occurs if alkali salts are present. Negative ions are also formed, generally $(M - H)^-$. Single-fission events are detected using a time-of-flight analyzer with channel plate detector. Accumulation of data is quite slow, often requiring several hours, and resolution is quite poor using the TOF system. Nevertheless, some very impressive results have been obtained by PD-MS on numerous biomolecules with molecular weights up to several thousand.

To date only a couple of reports have been issued describing the PD-MS analysis of synthetic polymers (22–23). Analysis of some poly(ethylene glycol) (PEG) samples yielded quasi-molecular ions (MH^+, MLi^+, MNa^+) for various oligomers (22). The distributions of quasi-molecular ions were reported to be indicative of the average molecular weights of the polymers, but no actual comparisons of molecular weight averages determined from PD-MS data with those determined by classical methods were given (22). It is evident that PD-MS deposits a great deal of energy into desorbed ions and neutrals. Thus it is likely to experience the same problems in polymer molecular weight analysis as the rapid heating techniques described in the preceding section. Its utility for determining polymer molecular weight distributions has not yet been assessed in any detail, however.

ELECTROSPRAY

Attempts to obtain mass spectra of high polymers ($MW > 10^4$) have thus far met with very limited success. The difficulties associated with getting stable, charged macromolecules of several thousand molecular weight into the vapor phase are indeed formidable. Without doubt the most ambitious project in this respect has been the work of Dole and coworkers over the past 20 years or so with electrospray mass spectrometry (ES-MS) (24–28). The concept is to spray very dilute solutions of macromolecules into a gas-filled chamber. The solution is forced through a syringe held at high potential (a few kiloelectron-volts), and the resulting aerosol droplets are charged. In theory, the droplets should break down, losing solvent and charge, so that eventually singly charged macromolecular ions will result. Earlier experiments using a time-of-flight mass analyzer had limited success because of difficulties in introducing the macroions into the mass spectrometer and in detecting the ions with the normal secondary electron multiplier (25). Later experiments used a plasma chromatography (or ion drift spectrometer) for mass analysis (26). This instrument has the advantages of atmospheric pressure operation and ion detection via a Faraday cage–vibrating reed electrometer system. The results

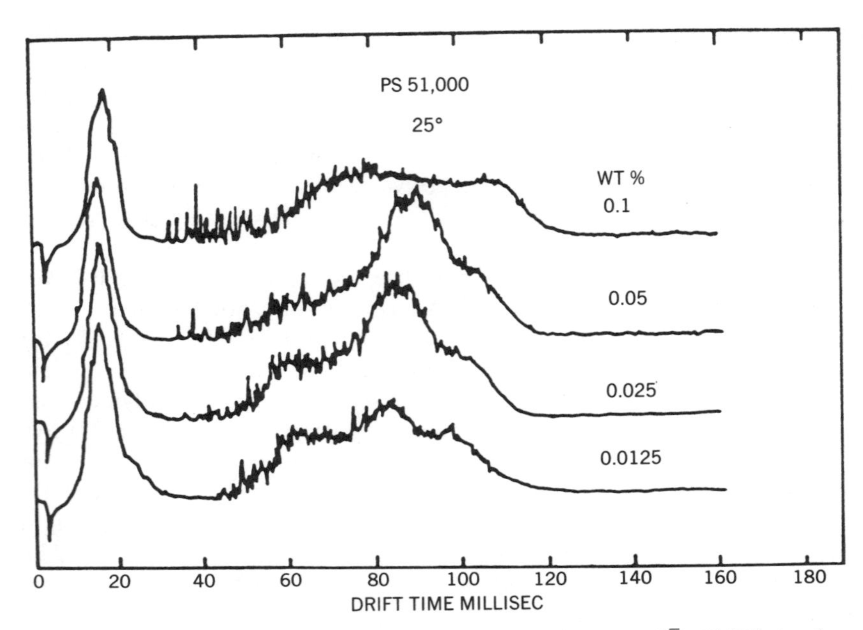

Figure 4. Electrospray ion signal versus drift time for polystyrene ($\bar{M}_n$ 51,000) in plasma chromatograph. (Reprinted with permission from Ref. 26.)

with this instrument were also disappointing, since it was not possible to clearly distinguish between polystyrene samples that differed widely in average molecular weight. This deficit was apparently due to cluster formation and multiply charged species.

In Figure 4, the signal intensity of polystyrene ($\bar{M}_n$ 51,000) is plotted against drift time in the plasma chromatograph (26). The ions observed at ~40–120 ms were believed to be rather large multiply charged aggregates of solvent and solute molecules. Dole has abandoned work on ES-MS and has concluded that accurate molecular weight averages of polymers cannot be determined using this technique (27). It is unfortunate that a method with so much potential has succumbed, at least temporarily, to the numerous experimental difficulties involved. It is encouraging that other workers are now experimenting with electrospray sources (29).

SECONDARY ION/FAST ATOM BOMBARDMENT

Secondary ion mass spectrometry (SIMS) has historically been used mainly for atomic (elemental) analysis of surfaces. Recently however, "organic" (or more generally "molecular") SIMS has found use as an analytical technique (30–35). The organic material is deposited as a thin film on a metal foil, sometimes with the addition of a salt. The sample is then bombarded with a primary ion beam, and secondary ions are sputtered from the surface and analyzed, usually with a quadrupole mass filter. Ions characteristic of molecular weight can form via several processes (32–35). If the molecule is already ionic, intact cations and/or anions can be produced by direct sputtering. Polar organics tend to form ions via cation attachment. Nonpolar organics can form odd-electron molecular ions ($M^{+\cdot}$ or $M^{-\cdot}$) via electron transfer, although ion attachment also occurs to some extent. So far SIMS spectra for molecular organic species have been obtained only on relatively low molecular weight materials (up to a few hundred amu). This is partially because the quadrupole mass filters normally used have a rather low mass range (< 1000 amu), but also because the SIMS process is relatively high in energy. Thus, thermal damage (pyrolysis) ensues, particularly with higher mass materials, and ions due to fragmentation–decomposition processes are prominent. SIMS has been used to study pyrolysis of high polymers (36, 37), and the new technique of time-of-flight SIMS (35) shows considerable facility for obtaining spectra of high mass polymer fragments. So far no reports have appeared describing the direct analysis of low molecular weight synthetic polymers, but SIMS may have some potential applications in this area.

The fast atom bombardment (FAB-MS) technique, introduced in 1981 (38), has become a very popular ionization technique for nonvolatile molecules. FAB is conceptually rather simple and is closely related (essentially identical) to organic SIMS. The sample, either a solid or a viscous liquid, is mixed with glycerol or other suitable matrix and placed on a flat metal probe. The probe is inserted into the ion source where it is bombarded with a stream of argon or xenon atoms (2–8 keV) from a gun. Both positive and negative ions are produced. Although very similar in concept to organic SIMS, FAB has certain advantages that make it more useful for analysis of large organic molecules. First, since the viscous matrix constantly resupplies fresh sample to the surface, spectra are intense and long-lasting. Second, since the primary beam consists of atoms rather than ions, ion source charging is reduced and essentially eliminated. Third, double focusing analyzers can easily be used, which extends the mass range and makes high resolution and metastable ion analysis possible. Fourth, the sample matrix (solvent) can be varied, with a view to extending the technique to include numerous classes of compounds. Finally, having the ion source at ambient temperature may provide an advantage over other techniques that rely, to a greater or lesser extent, on more direct thermal processes to effect vaporization.

Ions are formed in FAB primarily by proton attachment or abstraction; that is, for most organics, MH^+ is seen in positive ion mode and $(M - H)^-$ in

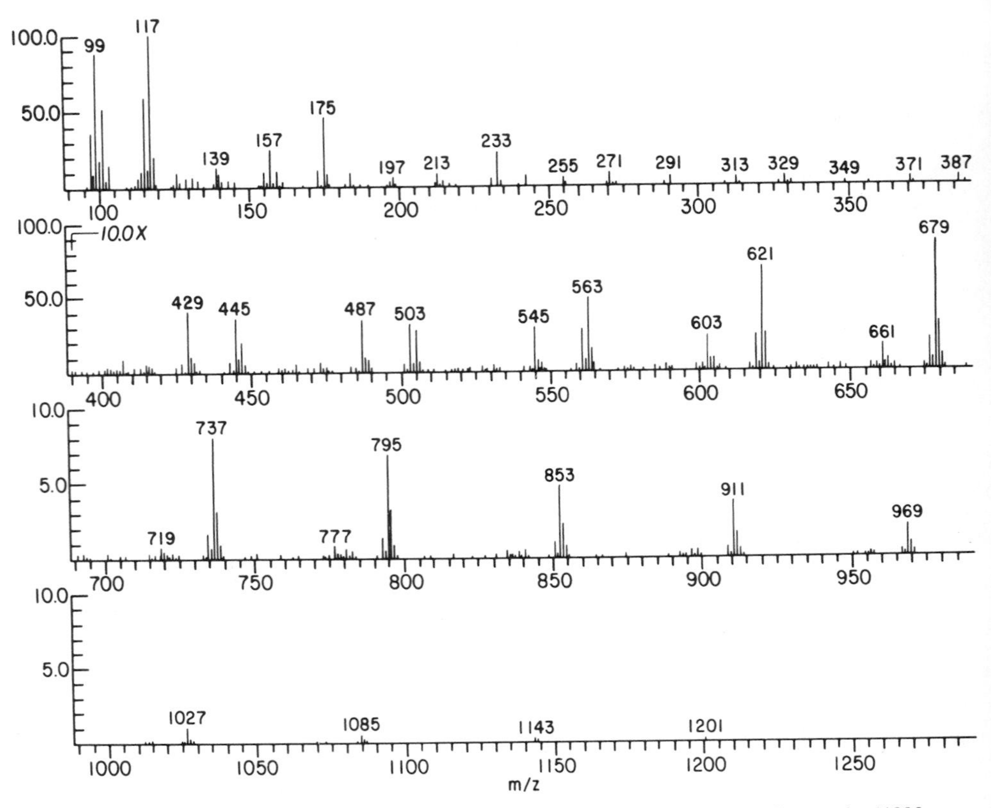

Figure 5. Fast atom bombardment mass spectrum of poly(propylene glycol) sample 41993. (Reprinted with permission from Ref. 39.)

negative mode. Cation attachment is also often observed. FAB seems to work best for highly polar organics, and as with other desorption ionization techniques, it has been used primarily for the analysis of large biomolecules. The FAB spectrum of a poly(propylene glycol) (PPG) with added NaBr salt is shown in Figure 5 (39). This sample gave high abundances of MNa^+ ions, among other species. Fragmentation was extensive, however, giving rise to intense low mass ions. Nevertheless, the spectra were easily interpretable in terms of protonated and cationized oligomers. FAB results for polyglycols (39) and poly(ethylene imine) (PEI) (40) show that oligomer quasi-molecular ion distributions can give approximate indications of molecular weight averages for the polymers. The molecular weight distributions are not quantitative in comparison with results determined by other methods, however. More specifically, lower mass oligomer intensities tend to be too high in FAB-MS as a result of extensive fragmentation reactions (39, 40). It thus appears that FAB (and probably organic SIMS as well) is not very useful for determination of molecular weight distributions of low mass polymers.

FIELD DESORPTION

Field desorption (FD-MS) (4, 5, 41, 42) has received considerable attention in terms of its ability for direct analysis of low molecular polymers. There are several reasons for this. First, FD-MS has been in existence longer than nearly all the other desorption ionization techniques, and commercial ion sources have been available for several years. Second, the fact that molecular or quasi-molecular ion intensities are normally very large compared to fragment ion intensities makes the technique extremely useful for analysis of oligomeric mixtures (43). Third, the fact that FD ion sources have been attached to numerous "high mass" analyzers has enabled several workers to directly examine polymers of fairly large molecular weight (up to several thousand amu) (10, 44, 45). Fourth, FD-MS has proved itself to be versatile enough to handle polar and nonpolar polymers of many types.

The principles of FD-MS are well documented elsewhere (41, 42). Basically, the compound to be analyzed is deposited on a specially prepared emitter consisting of a 10-μm tungsten wire on which microneedles of pyrolytic carbon have been grown. The commonest method of sample deposition is the dipping technique, in which the emitter wire is dipped into a solution of the sample to be analyzed. However, for better repeatability and in particular, quantitative work, the syringe technique should be used[42]. After the solvent has evaporated from the emitter wire, the sample is introduced into a special ion source, where a very high electric field ($\sim$10–100 MV/cm) causes the sample to be ionized and desorbed. Heating of the emitter is necessary to ensure effective ionization and desorption. For oligomeric mixtures in particular, all the molecular species will not desorb at the same emitter heating current (EHC). The more volatile oligomers will desorb first, and as the EHC is raised, the less volatile molecules will appear. Thus when using FD-MS data to determine molecular weight averages, spectra covering the entire desorption profile must be summed together to yield a composite mass spectrum.

Ionization mechanisms in the high electric field have been discussed in the literature (3, 41, 42), and in general two different cases can be distinguished. Nonpolar organics yield odd-electron molecular ions ($M^{+\cdot}$ or $M^{-\cdot}$) that are apparently formed by sublimation–field ionization (electron tunneling) at or near the fine microneedle tips on the emitter. Polar organics (e.g., aliphatic amines, alcohols, and acids) normally form ions by cation or anion attachment.

Cationization is efficient for alkali ions, so that alkali salts are sometimes added to samples to facilitate the formation of attachment ions (e.g., MLi^+). Preformed ions (from salts or organometallic complexes) are normally observed as the intact cation (anion) or as cluster ions.

The potential of FD-MS for examining higher molecular weight oligomers was demonstrated by Matsuo et al. in 1979 (46). These workers showed that oligomer molecular ions for polystyrene up to mass ~11,000 could be obtained. Subsequently, a fairly large number of reports has appeared describing the FD-MS characterization of low molecular weight polymer systems. It has been found in several cases that the relative intensities of FD molecular ions can be used directly to obtain accurate molecular weight averages for low mass polymers (< 10,000 amu) (10–12). For example, Figure 6 shows the distribution of oligomers for a particular batch of polystyrene (10). The $\bar{M}_n$ value obtained by FD-MS (1690 amu) compared very well with that determined by vapor pressure osmometry (VPO: 1710 amu). A number of other hydrocarbon polymers also have been examined by FD-MS—polybutadiene, polyisoprene, and polyethylene (12). For all these low polarity polymers, molecular weight averages derived from FD-MS data agreed well (within ~5%) with values obtained by conventional techniques.

A series of poly(ethylene glycol), poly(propylene glycol), and poly(tetrahydrofuran) samples also was examined by FD-MS (11). FD molecular weight averages for these polar polymers agreed quite well with

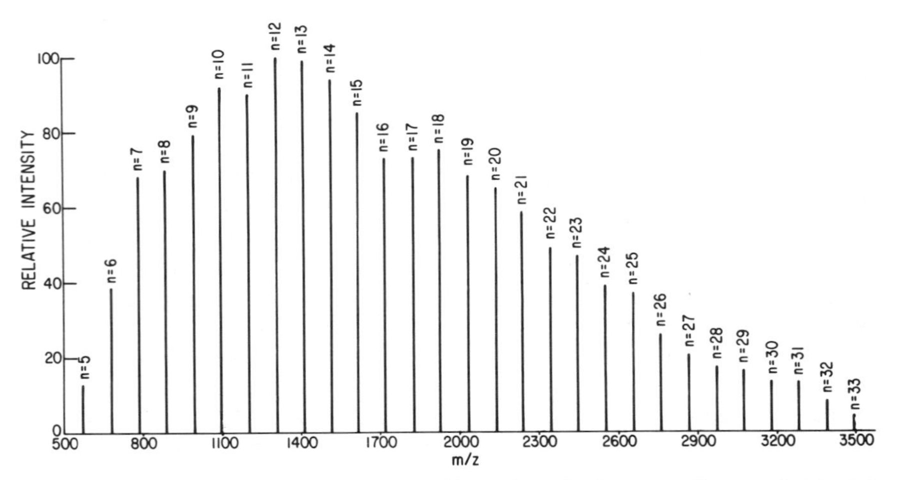

Figure 6. Relative field desorption mass spectral intensities of polystyrene oligomers (batch 12a). (Reprinted with permission from Ref. 10.)

values obtained by other methods. Polymers that are reasonably "well behaved" when studied by FD-MS should give similar good results. Samples that are "poor desorbers" may not work so well, however. For example, molecular weight averages determined by FD-MS for alkylated phenol–formaldehyde resins gave consistently lower values than those determined by gel permeation chromatography or VPO (47). This was due to the poor desorption characteristics of the higher mass oligomers. Poly(ethylene imine) (PEI) has also been studied by FD-MS (40). In this case the molecular weight averages determined from FD-MS data were also somewhat lower than the values reported by the manufacturer. For these PEIs, unfortunately, it was not known which values for $\bar{M}_n$ and $\bar{M}_w$ were correct. At a minimum, FD-MS was able to establish lower limits for the average molecular weight values (40).

ELECTROHYDRODYNAMIC IONIZATION

The technique of electrohydrodynamic ionization (EH-MS) appears to have considerable potential for polymer analysis. In EH-MS, organic molecules are dissolved in a solvent of low volatility (e.g., glycerol) with an added electrolyte (e.g., an alkali halide) (48). The solution is supported in a syringe held at high potential (a few kilovolts), and charged droplets (sometimes solvated) are extracted into the vacuum system of a double focusing mass spectrometer. Ions are generated directly from the liquid surface. EH-MS is limited to polar molecules that are inherently ionic or can form quasi-molecular ions readily via ion attachment or proton abstraction. EH-MS has been applied successfully to the analysis of several poly(ethylene glycol) (49–51) and poly(ethylene imine) (52) samples. In the first study, sodium attachment ions were detected for PEG up to ~1900 amu, and reliable molecular weight distributions were determined by averaging the spectra from several scans (49). The data indicated that oligomer ions were formed and extracted with roughly equal efficiencies, regardless of mass. No significant fragmentation was observed, although multiply charged ions were formed to some extent. In a later report, multiply charged ions were used to detect oligomers with molecular weights in excess of 3500 amu (50). The molecular weight averages determined for PEI samples (52), on the other hand, were much lower than values reported by the manufacturer as well as those estimated from FD-MS data (40). The lower EH-MS values may be attributed to a sampling bias induced by strong hydrogen bonding interactions in the EH experiment (40, 52).

Ion formation via EH-MS appears to have some analogies to FD-MS, since a high electric field is used in both techniques to facilitate the ionization–vaporization process (48). EH-MS appears to have several advantages which may apply to the analysis of polar macromolecules: (1) the ion beam can be intense and long lasting, (2) ionization–vaporization occurs at the ambient temperature of the ion source, (3) both positive and negative ions can be formed, often with equal ease, (4) the solvent and electrolyte can be varied to make the technique more generally applicable to polymers of different molecular structures, and (5) fragmentation–decomposition of desorbed ions is minimal. This latter point is especially advantageous in the determination of polymer molecular weight averages. It is clear that applications for EH-MS are not yet well developed. The technique is limited in the types of molecule that can be examined (nonpolar macromolecules, e.g., apparently cannot be examined). The speed of analysis via EH-MS is likely to be slow, since one needs to find an appropriate solvent and electrolyte. Also, the ion source may become contaminated with the relatively large amounts of solvent that are involved. One would hope that in the future EH ion sources will be attached to high field magnet spectrometers so that the mass range can be extended to higher mass polymers than have been studied to date. Results thus far certainly indicate that EH-MS can be a very useful technique for determining molecular weight distributions of oligomeric mixtures.

LASER DESORPTION

A very interesting and fast-developing technique for polymer analysis is laser desorption (LD) mass spectrometry (53). The sample is placed on a metal surface, sometimes with an added salt (e.g., an alkali halide), and is then exposed to a submicrosecond laser pulse. Ionization occurs mainly by cation attachment to the desorbed molecules, which apparently limits the technique to polar molecules which can provide a site for the attachment. Ion currents from a single pulse last from less than a second to at most a few seconds, which makes it mandatory to be able to acquire and record spectra quickly.

Two types of mass analyzer are used in LD-MS, time-of-flight and FT-MS. Mattern and Hercules (54) have described the use of a commercial LD-TOF system (LAMMA-1000) for low molecular weight polyglycol analysis. Several PEG and PPG batches were found to yield intense cation attachment mass spectra that could be used to calculate accurate molecular weight averages. In general, fragmentation–decomposition ions were of low abundance compared to quasi-molecular ions. Mattern and Hercules found that postacceleration (from 6.0 to 9.5 kV) was necessary for optimal detection of the higher mass oligomers (54). It was found that the LAMMA system was limited to an effective mass range of only $\sim$1000 amu.

Cotter et al. have designed a specialized LD-TOF system (55, 56) that has been used for low molecular weight polymer analysis (57). This system uses a time-delayed drawout pulse which allows some control over the experimental conditions and the resulting LD-TOF spectra. Several batches of PEG, PPG, and PEI have been examined, and quasi-molecular ions (generally MK^+ with added KCl) were observed in some cases up to nearly *m/z* 5000. Although accurate molecular weight averages could be readily determined for the PEG and PPG samples, the averages obtained for three PEI samples were somewhat lower than expected (57). These were the same PEI samples that were also examined by FD-MS (40) and EH-MS (52). The $\bar{M}_n$ and $\bar{M}_w$ values were similar to, but a little higher than, the values calculated from FD-MS data (40). As mentioned in an earlier section, unambiguous $\bar{M}_n$ and $\bar{M}_w$ values were not available for these polymers.

LD-FT-MS (58) appears to have considerable utility for polymer analysis. Two attractive advantages for FT-MS are its wide mass range with little or no mass discrimination and its ability to operate at much higher resolution compared to the LD-TOF systems. Cody et al. have examined several poly(*p*-phenylene) (PPP) samples using a Nicolet FTMS-1000 LD-MS system (59). PPP is insoluble and thus difficult to handle by most desorption ionization methods. Molecular ions were the predominant species observed for PPP by LD-FT-MS, and it was found that molecular weight averages could be determined from these data (59).

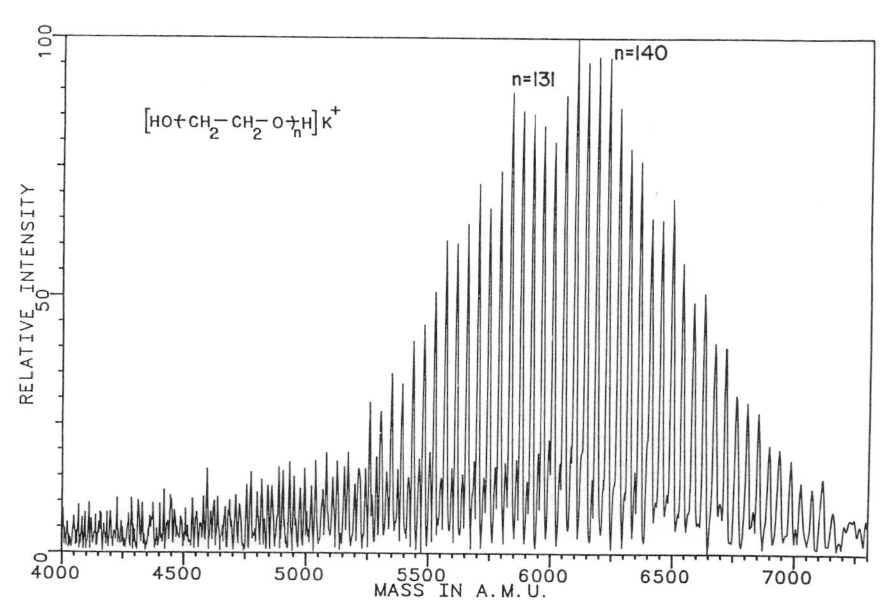

Figure 7. Laser desorption FT mass spectrum of poly(ethylene glycol) 6000. (Reprinted with permission from Ref. 61.)

Wilkins et al. (160, 61) have studied several low molecular weight polymers by LD-FT-MS (Nicolet FTMS-1000 system): PEG, PPG, PEI, polystyrene, and polycaprolactone. In some samples quasi-molecular ions were observed in excess of m/z 7000 (Fig. 7). In all cases accurate molecular weight averages could be determined from the data (61). The data for PEI were particularly interesting, since the $\bar{M}_n$ values determined for this polymer were considerably higher than those reported by any other mass spectral method; $\bar{M}_n$ was 685 by LD-FT-MS for PEI 600 and 1137 for PEI 1200. Since these values are close to the values reported by the manufacturer, the LD-FT-MS technique at first glance appears to give results superior to those obtained by other desorption ionization methods (40, 52, 57). It is not clear, however, why the LD-FT-MS system (61) should be inherently any better than the LD-TOF-MS approach (57). The method used for laser desorption is very similar in both instruments. It is to be hoped that continued experimentation will be able to sort out these inconsistencies.

THERMOSPRAY

Perhaps the newest desorption ionization technique is Vestal's thermospray (TSP-MS) system (3, 62). Developed as an LC-MS interface, this technique calls for a continuous flow of sample in solution. The liquid passes through a capillary tube, where it becomes superheated. The effluent from the capillary is in the form of a hot mist that is carried in a supersonic vapor jet. Some of the

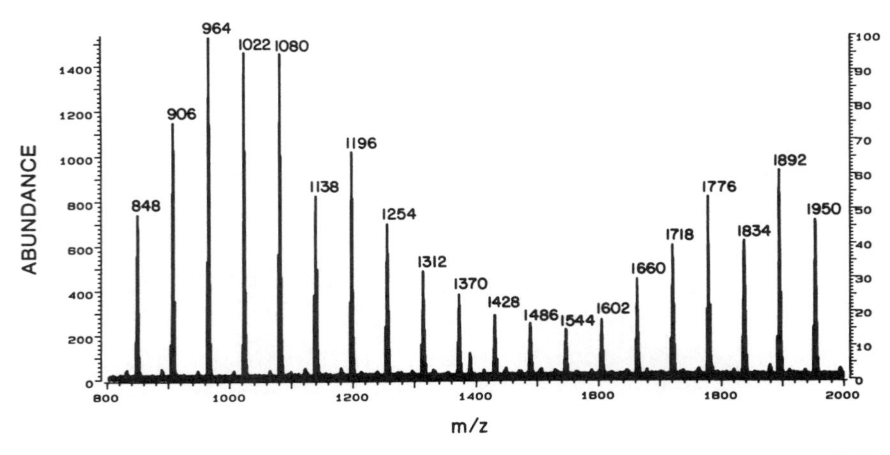

Figure 8. Thermospray mass spectrum of a mixed poly(propylene glycol) (MW 1000 + 2000). (Reprinted with permission from Ref. 63.)

vapor droplets are charged, and as solvent evaporates, charged sample ions result. The ions are sampled into a mass spectrometer via a small orifice. It is obvious that to be analyzed by TSP-MS, molecules must be reasonably polar.

Thus far only preliminary studies on the use of TSP-MS for low molecular weight polymer analysis have been conducted (63). Goodley has analyzed PEG and PPG using a thermospray source attached to a quadrupole mass spectrometer. Figure 8 is the TSP spectrum of a mixed PPG sample (MW 1000 + 2000). The solvent was methanol–water (25:75) with added ammonium acetate. The solution was injected directly into the TSP ion source, and the ions observed are $(M + NH_4)^+$. Molecular weight averages were not calculated in this preliminary study (63), but the intense quasi-molecular ions with minimal fragmentation suggest that TSP-MS has excellent potential for analysis of polar polymers.

Table 1. Mass Spectral Molecular Weight Averages $\bar{M}_n$ for Poly(ethylene glycol)

		$\bar{M}_n$	
MS Method	Ref.	MS	Other[a]
FD-MS	11	444	406
		601	605
		1010	1041
		1360	1396
EH-MS	49	406	406
		572	605
		962	1041
		1365	1396
LD-TOF-MS (LAMMA)	54	420	406
		618	605
		1007	1041
		1266	1396
LD-TOF-MS	57	1350	1349
		3130	3297
LD-FT-MS	61	661	(600)
		1088	(1000)
		1384	1349
		3160	3297
		6033	(6000)

[a]Determined by end-group titration, except for numbers in parentheses, which are values supplied by the manufacturer.

SUMMARY

The use of desorption ionization mass spectral methods for determining molecular weight averages in low molecular weight polymers is still a relatively new field. As such, there is much to be learned in terms of determining the optimal conditions for analysis. Results to date suggest that very accurate $\bar{M}_n$ and $\bar{M}_w$ values can be determined for many types of polymers with average molecular weights of around 5000 amu or less. The desorption ionization methods that have thus far yielded the best results are field desorption, electrohydrodynamic ionization, and laser desorption. As an example of typical results, Table 1 lists $\bar{M}_n$ values for poly(ethylene glycol) which were determined by several different mass spectral methods.

It is clear that as the methodology improves, higher molecular weight polymers and a wider variety of polymer types will be handled in the future. Mass spectrometry has the rather unique advantage (compared with most other methods) that absolute molecular weight averages can be determined, since individual oligomers can be examined directly by the instrument. This points to a very promising future for mass spectrometry as a tool for use in polymer science.

ACKNOWLEDGMENTS

Appreciation is expressed to The BFGoodrich Company for support of this work. We are indebted to K. D. Cook, R. J. Cotter, P. C. Goodley, and C. L. Wilkins for supplying preprints of work not yet published for inclusion in this chapter.

REFERENCES

1. C. C. Fenselau, *Anal. Chem.*, **54**, 105A (1982).
2. K. L. Busch and R. G. Cooks, *Science*, **218**, 247 (1982).
3. M. L. Vestal, *Mass Spectrom. Rev.*, **2**, 447 (1983).
4. H.-R. Schulten and R. P. Lattimer, *Mass Spectrom. Rev.*, **3**, 231 (1984).
5. R. P. Lattimer, R. E. Harris, and H.-R. Schulten, *Rubber Chem. Technol.*, **58**, 577 (1985).
6. R. J. Beuhler and L. Friedman, *J. Appl. Phys.*, **48**, 3928 (1977).
7. R. J. Beuhler and L. Friedman, *Int. J. Mass Spectrom. Ion Phys.*, **23**, 81 (1977).
8. R. J. Beuhler and L. Friedman, *Nucl. Instrum. Methods*, **170**, 309 (1980).
9. B. Sundqvist, A. Hedin, P. Hakansson, I. Kamensky, M. Salehpour, and G. Sawe, *Int. J. Mass Spectrom. Ion Proc.*, **65**, 69 (1985).

10. R. P. Lattimer, D. J. Harmon, and G. E. Hansen, *Anal. Chem.*, **52**, 1808 (1980).
11. R. P. Lattimer and G. E. Hansen, *Macromolecules*, **14**, 776 (1981).
12. R. P. Lattimer and H.-R. Schulten, *Int. J. Mass Spectrom. Ion Phys.*, **52**, 105 (1983).
13. A. K. Lee and R. D. Sedgwick, *J. Polym. Sci., Polym. Chem. Ed.*, **16**, 685 (1978).
14. B. Weibull, *J. Polym. Sci., Polym. Chem. Ed.*, **18**, 1633 (1980).
15. R. H. Wiley, *J. Polym. Sci., Macromol. Rev.*, **14**, 379 (1979).
16. G. D. Daves, Jr., *Acc. Chem. Res.*, **12**, 359 (1979).
17. R. J. Beuhler, E. Flanigan, L. J. Greene, and L. Friedman, *J. Am. Chem. Soc.*, **96**, 3990 (1974).
18. H. R. Udseth and L. Friedman, *Anal. Chem.*, **53**, 29 (1977).
19. R. J. Cotter, *Anal. Chem.*, **52**, 1589A (1980).
20. R. D. Macfarlane and D. F. Torgerson, *Science*, **191**, 920 (1976).
21. R. D. Macfarlane, *Anal. Chem.*, **55**, 1247A (1983).
22. B. T. Chait, J. Shpungin, and F. H. Field, *Int. J. Mass Spectrom. Ion Proc.*, **58**, 121 (1984).
23. R. C. Robbins, M. Alai, P. Demirev, and R. J. Cotter, *Proceedings of the 33rd Annual Conference on Mass Spectrometry and Allied Topics*, San Diego, May 26–31, 1985, p. 403.
24. M. Dole, L. L. Mack, R. L. Hines, R. C. Mobley, L. D. Ferguson, and M. B. Alice, *J. Chem. Phys.*, **49**, 2240 (1968).
25. M. Dole, H. L. Cox, Jr., and J. Gieniec, *Adv. Chem. Ser.*, **125**, 73 (1973).
26. M. Dole, C. V. Gupta, L. L. Mack, and K. Nakamae, *Polym. Prepr. Am. Chem. Soc., Div. Polym. Chem.*, **18**(2), 188 (1977).
27. K. Nakamae, V. Kumar, and M. Dole, *Proceedings of the 29th Annual Conference of Mass Spectrometry and Allied Topics*, Minneapolis, May 24–29, 1981, p. 517.
28. J. Gieniec, L. L. Mack, K. Nakamae, C. V. Gupta, V. Kumar, and M. Dole, *Biomed. Mass Spectrom.*, **11**, 259 (1984).
29. C. M. Whitehouse, R. N. Dreyer, M. Yamashita, and J. B. Fenn, *Anal. Chem.*, **57**, 675 (1985).
30. H. Grade and R. G. Cooks, *J. Am. Chem. Soc.*, **100**, 5615 (1978).
31. R. J. Day, S. E. Unger, and R. G. Cooks, *Anal. Chem.*, **52**, 557A (1980).
32. A. Benninghoven, D. Jaspers, and W. Sichtermann, *Appl. Phys.*, **11**, 35 (1976).
33. A. Benninghoven and W. Sichtermann, *Org. Mass Spectrom.*, **12**, 595 (1977).
34. A. Eicke, W. Sichtermann, and A. Benninghoven, *Biomed. Mass Spectrom.*, **15**, 289 (1980).
35. A. Benninghoven, Ed., *Ion Formation from Organic Solids*, Springer-Verlag, New York, 1983.
36. J. A. Gardella, Jr., and D. M. Hercules, *Anal. Chem.*, **52**, 226 (1980).
37. I. V. Bletsos, D. M. Hercules, D. Greifendorf, and A. Benninghoven, *Anal. Chem.*, **57**, 2384 (1985).

38. M. Barber, R. S. Bordoli, R. D. Sedgwick, and A. N. Tyler, *J. Chem. Soc., Chem. Commun.*, **1981**, 325.
39. R. P. Lattimer, *Int. J. Mass Spectrom. Ion Proc.*, **55**, 221 (1983/1984).
40. R. P. Lattimer and H.-R. Schulten, *Int. J. Mass Spectrom. Ion Proc.*, **67**, 277 (1985).
41. H. D. Beckey, *Principles of Field Ionization and Field Desorption Mass Spectrometry*, Pergamon Press, Elmsford, NY, 1977.
42. H.-R. Schulten, *Int. J. Mass Spectrom. Ion Phys.*, **32**, 97 (1979).
43. R. P. Lattimer and K. R. Welch, *Rubber Chem. Technol.*, **53**, 151 (1980).
44. H. Matsuda, *At. Masses Fundam. Constants, Proc. 5th Int. Conf.*, **5**, 185 (1976).
45. P. G. Cullis, G. M. Neumann, D. E. Rogers, and P. J. Derrick, *Adv. Mass Spectrom.*, **8B**, 1729 (1980).
46. T. Matsuo, H. Matsuda, and I. Katakuse, *Anal. Chem.*, **51**, 1329 (1979).
47. R. P. Lattimer, E. R. Hooser, H. E. Diem, and C. K. Rhee, *Rubber Chem. Technol.*, **55**, 442 (1982).
48. B. P. Stimpson, D. S. Simons, and C. A. Evans, Jr., *J. Phys. Chem.*, **82**, 660 (1978).
49. S.-T. F. Lai, K. W. S. Chan, and K. D. Cook, *Macromolecules*, **13**, 953 (1980).
50. K. W. S. Chan and K. D. Cook, *Org. Mass Spectrom.*, **18**, 423 (1983).
51. K. W. S. Chan and K. D. Cook, *Macromolecules*, **16**, 1736 (1983).
52. J. H. Callahan and K. D. Cook, *Proceedings of the 32nd Annual Conference on Mass Spectrometry and Allied Topics*, San Antonio, May 27–June 1, 1984, p. 552.
53. M. A. Posthumus, P. G. Kistemaker, H. L. C. Meuzelaar, and M. C. Ten Noever de Brauw, *Anal. Chem.*, **50**, 985 (1978).
54. D. E. Mattern and D. M. Hercules, *Anal. Chem.*, **57**, 2041 (1985).
55. R. B. van Breemen, M. Snow, and R. J. Cotter, *Int. J. Mass Spectrom. Ion Phys.*, **49**, 35 (1983).
56. J.-C. Tabet and R. J. Cotter, *Anal. Chem.*, **56**, 1662 (1984).
57. R. J. Cotter, J. P. Honovich, J. K. Olthoff, and R. P. Lattimer, *Macromolecules*, **19**, 2996 (1986).
58. D. A. McCrery, E. B. Ledford, Jr., and M. L. Gross, *Anal. Chem.*, **54**, 1435 (1982).
59. C. E. Brown, P. Kovacic, C. A. Wilkie, R. B. Cody, Jr., and J. A. Kinsinger, *J. Polym. Sci., Polym. Lett. Ed.*, **23**, 453 (1985).
60. C. L. Wilkins, D. A. Weil, C. L. C. Yang, and C. F. Ijames, *Anal. Chem.*, **57**, 520 (1985).
61. R. S. Brown, D. A. Weil, and C. L. Wilkins, *Macromolecules*, **19**, 1255 (1986).
62. C. R. Blakley and M. L. Vestal, *Anal. Chem.*, **55**, 750 (1983).
63. P. C. Goodley, private communication.

CHAPTER

15

PHYSICAL AND CHEMICAL METHODS FOR CHARACTERIZING INSOLUBLE POLYMERS

A. MARIE ZAPER and JACK L. KOENIG

Department of Macromolecular Science, Case Western Reserve University, Cleveland, Ohio

INTRODUCTION

The polymerization of many different monomers has led to formulations for a great variety of polymeric materials, available for technological applications. Further innovative uses for polymeric materials will require improvements in such physical properties as temperature and size stability, adhesion, and mechanical strength of these systems. Since it is known that the molecular structure of a polymer influences its macroscopic physical behavior, the synthesis of a polymer can be designed so that the chemical structure of the system will allow the attainment of specific improved material properties.

Certain polymers have the ability to form network structures which exhibit the excellent engineering behavior that is desired for special materials. Network formation can occur with the curing of a multifunctional resin or by the cross-linking of chains of a linear polymer. Therefore network polymers are the result of nonlinear polymerization and contain chemical cross-links having more than two functionalities. A three-dimensional network structure, which is also referred to as a gel, has an infinite molecular weight and is insoluble and intractable. A complete characterization of such a system entails determining the reactions that occur during cross-linking and resolving the final structure of the material. In addition, it is desirable to determine the basic network parameters, including molecular weight before gelation, sol fraction, gel point, cross-link density, and the molecular weight between cross-links.

In the 1940s Flory (1, 2) and Stockmayer (3) developed relationships between the extent of reaction and the final polymer structure for nonlinear polymerizations, to derive expressions for the molecular weight distribution of network materials. Since then several methods and models have been described for deriving properties such as molecular weight, gel point, and cross-link density for network systems. This chapter discusses the important cross-link theories and models and gives a comprehensive review of the experimental methods available for characterizing insoluble polymers.

THEORIES FOR NETWORKS

Flory used a probabilistic approach to determine the distribution of molecular weight for a linear polymerization. For a polymer sample the number distribution of molecules of length n is given by the equation

$$N_n = N_0(1-p)^2 p^{n-1}$$

where p is defined as the extent of reaction and N_0 is the total number of monomer units present initially. The weight fraction W_n of polymer molecules that are n units long is given by

$$W_n = np^{n-1}(1-p)^2$$

The number and weight distribution functions are used to obtain the respective degrees of polymerization, which reflect the number or weight average of the repeat units in the chain. In final form, the number average degree of polymerization is found with the equation

$$\bar{X}_n = \frac{1}{1-p}$$

[The same relationship between the degree of polymerization and the extent of reaction was established previously by Carothers (4).] The weight average degree of polymerization is determined to be

$$\bar{X}_w = \frac{1+p}{1-p}$$

The molecular weight of the polymer is then obtained by multiplying the degree of polymerization by the molecular weight of the structural units. The measure of the breadth of a polymer distribution is determined by the ratio of the weight average distribution to the number average distribution, given by

$$\bar{X}_w/\bar{X}_n = 1+p$$

When more than two functional groups are present on the monomer units, the potential for more complicated polymer structures arises. Polymerization

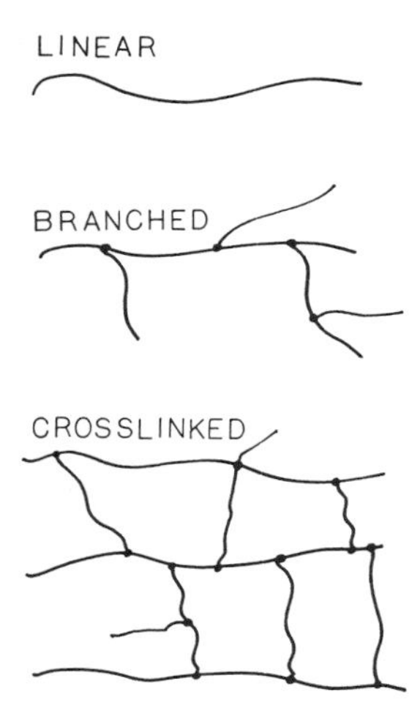

Figure 1. Schematic representation of the structures of linear, branched, and cross-linked polymers.

of monomers of these types results in branched polymers, which in many cases lead to the formation of three-dimensional network systems. Chemical networks are formed as a result of cross-linking reactions between branch units of polymer chains. For a monodisperse polymer sample, gelation or infinite network formation occurs when there is one cross-link for every two polymer molecules, resulting in a macroscopic molecule. Some of the possible structures of polymers are represented in Figure 1.

The theoretical concepts for the formation of three-dimensional polymer networks were also established in the classical work of Flory (1,2) and Stockmayer (3) using statistical branching theory. A relationship was developed between gelation and the extent of reaction or composition for nonlinear polymerization. Their treatment of polymer network systems was based on the assumption of an ideal network formation, where (1) all functional groups of the same type are equally reactive, (2) all groups react independently of one another, and (3) no intramolecular reactions occur in finite species. Their statistical approach is based on determining $\bar{X}_n$ and $\bar{X}_w$ at the limit of infinite size of the polymer network.

In this method, the branching coefficient α is defined as the probability that a given functional group of a branch unit in a polymer chain leads to another branch unit. For gel formation, the critical branching coefficient $[\alpha_c]$ is found to be

$$[\alpha_c] = \frac{1}{f-1}$$

where f is the functionality (> 2) of the branch units.

To determine the extent of reaction, the probability α must first be evaluated. Flory derived the following equation for a polymerization of A—A with B—B, where both A—A and B—B are reactants with two identical functional groups, and A_f is a monomer having a functionality greater than 2

$$\alpha = \frac{p_A p_B \rho}{1 - p_A p_B(1 - \rho)}$$

Assuming equal reactivity of functional groups, the probability p_A that an A group has undergone reaction equals the fraction of A groups which have reacted, p_B is the extent of reaction of the B functional groups, and ρ is the ratio of all A groups, both reacted and unreacted, belonging to branch units, to the total number of A functional groups in the entire system.

If r is defined as the ratio of all A groups to all B groups, then $p_B = rp_A$, and either p_A or p_B can be eliminated from the expression above. The resulting equation is combined with the critical branching coefficient equation to yield

$$p_c = \frac{1}{\{r[1 + \rho(f - 2)]\}^{1/2}}$$

which describes the extent of reaction at the gel point for the A functional groups. At this point the reaction mixture is divided between the gel and the sol portions. The sol is the soluble portion of the reaction mixture, which can be extracted from the gel. When the extent of reaction is greater than p_c, an infinite network exists. Flory and Stockmayer give several examples for special cases of reactions, including a polymerization in which both A and B types of branch units occur, and systems using monomers which have a functionality greater than 2. The uses and examples of the theoretical relationships which have been described can be found in the literature with experimental data in the papers mentioned.

For the types of polymerization systems discussed above, further derivations lead to expressions for the number and weight average degrees of polymerization (5):

$$\bar{X}_n = \frac{1}{1 - \alpha f/2}$$

$$\bar{X}_w = \frac{1 + \alpha}{1 - (f - 1)\alpha}$$

$$\bar{X}_w/\bar{X}_n = \frac{(1 + \alpha)(1 - \alpha f/2)}{1 - (f - 1)\alpha}$$

In general, much narrower molecular weight distributions are expected for nonlinear polymerizations than for linear polymerizations. The approach of Flory and Stockmayer, which used the primary assumption of equal and independent reactivity of all functional groups and no loop formation, is also known as the combinatorial method.

Marsh et al. (6) developed a mathematical model to characterize the postgelation state of cross-linked condensation polymers containing trifunctional branch units. The model deals with the distribution of branch units and is based on Flory's branching coefficient. It is centered around a hierarchy of branching coefficients and derived functions. The hierarchy is based on the observation that chain-terminating structural elements negate the cross-linking function of branch units and transform their function, one for one, into chain extension units. Parameters including composition, ingredient functionality distributions, and the extent of reaction are needed for the calculations. With this particular model it is possible to determine the sol and gel fractions of the reaction system, concentrations of the effective chains and of the cross-linking sites, average effective chain lengths, and the distribution of a hierarchy of pendant chains. Good correlation has been found between the calculated and measured values of the gel fractions and of the effective chain concentrations in simple polyesters.

The method of Flory and Stockmayer requires calculating the distributions of all species present and then using the distributions to calculate average properties. Gordon (7, 8) has also applied statistical branching theory to the investigations of structure and physical properties of polymer network systems. Gordon and coworkers have used the methods of stochastic branching processes, which require the use of probability generating functions to directly calculate the molecular weight average, gel point, and the sol fraction of a polymer network (9–11). With this technique, monomers are expressed in terms of their reactive functionalities using vectorial probability-generating functions which yield equations in matrix form. The generating functions include a parameter to measure the conversion of the particular functionality in the system under study. This methodology uses very little probability theory and does not require distribution functions for the different species but does involve a great deal of mathematics.

Dusek and coworkers have recently studied the curing of epoxy resins using the theory of stochastic branching processes (12, 13). In their investigations, epoxy resins were reacted with diamines. It was assumed that both the polymerization of the epoxy groups themselves and the cyclization reactions are negligible. First it was necessary to derive the expressions for molecular weight averages in the sol and in the gel fractions; then the conditions for the gel point of the reaction had to be determined. The required parameters are a function of the reaction stoichiometry for the epoxy system and of the reaction rate constants of the hydrogen atoms in the primary and secondary amino groups, $k_2/k_1 = r$.

According to the theoretical calculations for a stoichiometrically equivalent ratio of functionalities, the critical conversion at the gel point is found to be $\frac{1}{2}$ for r approaching infinity and increases to $(\frac{1}{2})(5^{1/2} - 1)$ for r approaching zero. Experimental studies, particularly titrations up to the gel point and gel fraction analysis by extraction in postgelation, were carried out, and the results were compared with the theoretical calculations. Acceptable agreement was found between the theoretical calculations and the experimental data, and it was concluded that the assumptions made initially were reasonable. Dusek and Ilavsky extended this particular approach for network characterization to study the occurrence of cyclization reactions in cross-linking polymerization (14, 15). These investigators have calculated the probability of cyclization by taking into account the ability of unsaturated monomers to form chains that contain repeat units having unreacted double bonds, are part of a branch point, or have cyclized. It was determined that the probability of cyclization depends on the conformation of the polymer chain between the reactive functional groups.

Macosko and Miller have used the recursive nature of the branching process and elementary probability laws to characterize nonlinear stepwise polymerizations (16–18). For ideal nonlinear polymerization, where it is assumed that all functional groups are equally and independently reactive and that there is no loop formation, expressions have been developed for the direct calculation of average molecular weight up to the gel point of a network system. Expressions for postgel properties such as the soluble fraction and the cross-link density have also been derived. In addition the consequences of nonideality have been studied with this method. Miller and Macosko have also treated the effects of both unequal reactivity of functional groups (19), as well as changes in reactivity of functional groups upon reaction (20). The method of Macosko and Miller will be described for the condensation homopolymerization of A_f, which is schematically represented below.

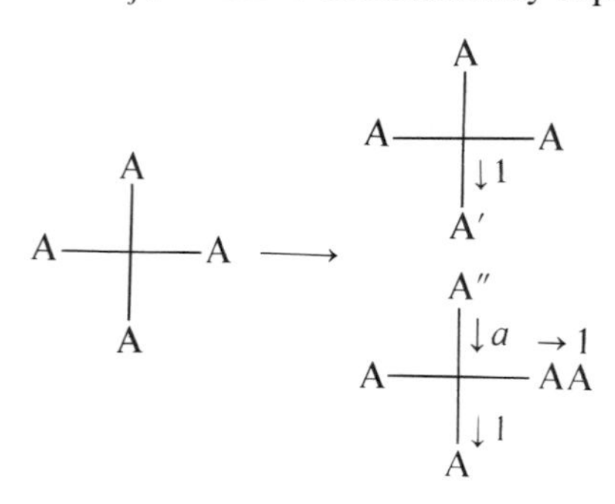

It was recognized that the use of the law of conditional expectation is a simple approach to the problem of calculating the average molecular weight of nonlinear polymers (16). During the onset of stepwise polymerization, finite branches containing residual unreacted functionalities are formed. As the reaction progresses, the fraction of unreacted functionalities decreases and the molecular weight increases as a direct result of branch formation; therefore the molecular weight determination of the system at hand becomes more complex. The elementary law of conditional expectation is given here in a general form and is used in conjunction with the recursive method. In this law, Y is a random variable, A is an event, and B is its complement; $E(Y)$ is the expectation or average value and $E(Y/\mathrm{A})$ is the conditional expectation given that the event A has occurred. Then the law of total probability for expectation is

$$E(Y) = E(Y/\mathrm{A})P(\mathrm{A}) + E(Y/\mathrm{B})P(\mathrm{B})$$

The schematic above represents the stepwise condensation reaction of A_f moles of monomer having f similar groups. The polymerization is allowed to proceed until a fraction p of the A groups have reacted. Extent of reaction p is

$$p = \frac{\mathrm{A} - \mathrm{A}_t}{\mathrm{A}}$$

where the number of initial moles of the A-type functional groups is denoted by A, and A_t is the number of moles after a reaction time t. With this method an A group is picked at random (e.g., A′ is chosen), and then it is necessary to determine the weight attached to A′ looking in the 1 direction—that is, out from its parent molecule. The symbol for this particular weight is $W_{\mathrm{A}'}^{\mathrm{out}}$ and it is zero if A′ has not reacted. If A′ has reacted, in this case with A″, then $W_{\mathrm{A}'}^{\mathrm{out}}$ equals $W_{\mathrm{A}''}^{\mathrm{in}}$, which is the weight attached to A″ looking into the parent molecule of A″.

To compute the average weight, the law of total probability for expectation is used:

$$\begin{aligned} E(W_{\mathrm{A}'}^{\mathrm{out}}) &= E(W_{\mathrm{A}''}^{\mathrm{in}}/\mathrm{A}\text{ reacts })P(A\text{ reacts}) \\ &\quad + E(W_{\mathrm{A}'}^{\mathrm{out}}/\mathrm{A}\text{ does not react})P(\mathrm{A}\text{ does not react}) \\ &= E(W_{\mathrm{A}''}^{\mathrm{in}})p + 0(1 - p) \\ &= pE(W_{\mathrm{A}''}^{\mathrm{in}}) \end{aligned}$$

The weight $E(W_{\mathrm{A}''}^{\mathrm{in}})$ is the molecular weight of A_f plus the sum of the expected weights on each of the remaining $f - 1$ arms; $E(W_{\mathrm{A}''}^{\mathrm{in}})$ is the average weight on any A looking into its parent molecule. The expression for expected weight is given by

$$E(W_{\mathrm{A}}^{\mathrm{in}}) = M_{\mathrm{A}_f} + (f - 1)E(W_{\mathrm{A}}^{\mathrm{out}})$$

Because of the repetitive nature of the branched molecule, the system ends up the same way as in the starting case. The approach assumes that the reaction of a functional group with a polymer chain end depends only on the functional group found at the end of that chain and does not depend on any functional groups further back on the chain.

The total molecular weight of the molecule W_{A_f} is the weight attached to one of the arms of A_f taking both directions into account:

$$W_{A_f} = W_A^{in} + W_A^{out}$$

Therefore the expected or average molecular weight attached to a random A_f is given by

$$\bar{M}_w = E(W_{A_f}) = E(W_A^{in}) + E(W_A^{out})$$

Upon appropriate substitution, the molecular weight can be calculated directly from the expression

$$\bar{M}_w = M_{A_f} \frac{1+p}{1-p(f-1)}$$

$$\bar{X}_w = \frac{1+p}{1-p(f-1)}$$

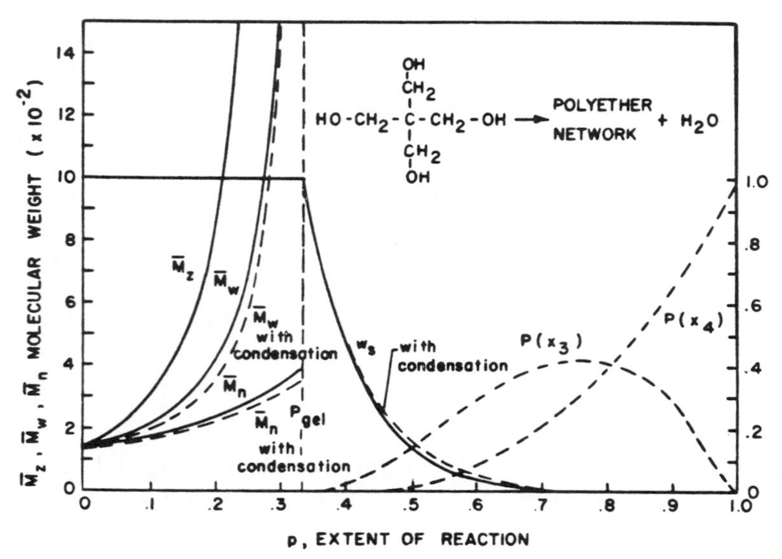

Figure 2. Calculated properties for the polyether network formation by the stepwise polymerization of pentaeythritol. (From Ref. 19).

Picking an A_f group at random is equivalent to picking a unit mass and then determining the expected weight of the molecule of which it is a part. Therefore the molecular weight that is determined in this way is a weight average value as opposed to a number average value. Figure 2 plots the average molecular weight for a tetrafunctional network as a function of the extent of reaction. The critical degree of conversion is $p_c = 1/(f-1)$ because infinite network formation occurs when the weight average molecular weight diverges—that is, $p(f-1) \geqslant 1$.

Next, attention turns to the characterization of the postgel region. Before the gel point, only finite molecules or branched structures exist and they tend to be soluble; therefore the weight fraction of the soluble material W_s is one. At the onset of gelation a fraction of the branched molecules become part of the network structure and the weight fraction of the soluble material decreases. The state of the ntework at this point can be characterized by W_s. The weight fraction of the soluble material can be calculated in a manner similar to the approach used to find molecular weight because the randomly chosen A_f is part of the sol fraction if all f of its arms lead to finite chains. For this particular calculation it is necessary to determine the probability that looking out from an A group, in direction 1, leads to a finite chain. This particular probability is denoted as $P(F_A^{out})$. The reaction of an A unit corresponds to an event occurring with a probability p. In this case $P(F_A^{out})$ is equivalent to $P(F_A^{in})$, the event of seeing a finite chain when looking toward an A unit in direction 2. If A does not react, signifying chain termination, which is an event occurring with a probability $1-p$, then $F_A^{out} = 1$. Using the law of total probability, this situation can be written in the following form:

$$\begin{aligned} P(F_A^{out}) &= P(F_A^{out}/\text{A reacts})P(\text{A reacts}) \\ &\quad + P(F_A^{out}/\text{A does not react})P(\text{A does not react}) \\ &= P(F_A^{in})p + 1(1-p) \\ &= pP(F_A^{in}) + (1-p) \end{aligned}$$

In accordance with the repetitive nature of the branched molecule, all the $f-1$ arms must be simultaneously finite, which requires multiplying the probability that each arm (looking out) is finite:

$$P(F_A^{in}) = P(F_A^{out})^{f-1}$$

It is now possible to solve for $P(F_A^{out})$:

$$pP(F_A^{out})^{f-1} - P(F_A^{out}) - p + 1 = 0$$

For postgel relations, $P(F_A^{out})$ has a root between 0 and 1. If $P(F_A^{out})$ is 1, the reaction system has not gelled. For homopolymerization there are two cases of interest. The solutions for these two cases are:

$$\text{for } f=3, \qquad P(F_A^{out}) = \frac{1-p}{p}$$

$$\text{for } f=4, \qquad P(F_A^{out}) = (1/p - 3/4)^{\frac{1}{2}} - \tfrac{1}{2}$$

The weight fraction of solubles W_s can be calculated if the probability $P(F_A^{out})$ is known. A randomly chosen A_f molecule is part of the sol if it has f finite arms:

$$W_s = P(F_A^{out})f$$

As a result, when $f = 3$,

$$W_s = \left(\frac{1-p}{p}\right)^3$$

In Figure 2, W_s is plotted as a function of extent of conversion for the case of $f = 4$. At high conversion only very small species such as monomers and oligomers are present in the sol fraction.

Macosko and Miller have also calculated the cross-link density or concentration of effective junction points in the infinite network (17). An effective junction point is defined for A_f chosen randomly if three or more arms lead to the infinite network. In general the probability that an A_{fi} monomer will be an effective cross-link of degree M is

$$P(X_{M_i f_i}) = \binom{f_i}{M} P(F_A^{out})^{f_i - M}[1 - P(F_A^{out})]^M$$

The calculated value of A_f as a cross-link of degree 3 (trifunctional cross-link) $[P(X_3)]$ and as a tetrafunctional cross-link $[P(X_4)]$ is shown in Figure 2. Initially trifunctional cross-links are formed, which then give way to tetrafunctional cross-links. At complete reaction, assuming that no side reactions or small loops occur, the tetrafunctional groups become unity.

The cross-link density is the internal concentration of $[A_{f_i}]_0$ multiplied by the probability $P(X_{M_i f_i})$, which is summed over $f_i = M$ to the highest functionality in the reaction system:

$$[X_M] = [A_{f_i}]_0 P(X_{M_i f_i})$$

The total cross-link density [X] for the system is the sum of all the $[X_M]$

from $M = 3$ to f_k. For the limit of complete reaction where p goes to 1, $[X_M] = [A_{fm}]_0$.

As was mentioned previously, Macosko and Miller have calculated the property relations for nonlinear polymerizations with unequal reactivities (19) and also for the effect of change in functionality as a result of the state of the other groups (20). In addition, they have considered systems of networks which are formed by cross-linking long polymer chains after polymerization (17). The derivations of these calculations and examples of their treatment can be found in the literature in the references noted.

Gelation has also been studied using lattice percolation concepts. For a random system, percolation theory deals with the effects of variations in the pairing of points on a lattice (21). Then the properties of the resulting system such as the size, shape, density, and appearance of the clusters of the paired points that are formed are studied. In 1976 de Gennes (22) and Stauffer (23) independently analyzed the connection between percolation theory and the sol-to-gel transition. In the percolation model, polyfunctional monomer units occupy the sites of a periodic lattice. A bond is formed between the two nearest neighbors of the lattice sites with a probability p. The linking of the lattice sites leads to the formation of clusters of bonds, and then the molecular weight determination of the system corresponds to analyzing the cluster size distributions. The importance of ring structures and steric hindrance effects near the gel point are handily studied with this method. This is in contrast to the Flory–Stockmayer model, which assumes that intramolecular reactions do not occur, in which case the cyclic structures are avoided. Stauffer et al. (24) have compared percolation theory to classical gelation theory and have reviewed some experimental results to check the validity of percolation theory. Thus far experimental difficulties have prevented the clear resolution of the question of validity.

Analyzing the percolation process involves simulating polymerization on a lattice which can be modeled by a Monte Carlo method. The Monte Carlo procedure uses random numbers to generate sample configurations of a partially filled lattice (25, 26). Many examples of computer simulations of the gelation process can be found in the literature. For the kinetic gelation of a random mixture of bifunctional and tetrafunctional monomers placed on a cubic lattice, Herrmann et al. (27) have calculated the gel fraction and the degree of polymerization. The gel fraction was identified with the largest of the clusters formed by the percolation process, while the average cluster size corresponded to the weight average degree of polymerization. The researchers found that their particular computer simulation of a model for irreversible gelation belongs to a class of its own. Because of differences in results, it was established that the model differed from both random percolation and classical gelation theories. Jan et al. (28) extended the same method and polymerization system to a two-dimensional lattice, which is defined for a system that does not require a solvent and lacks mobility. In that particular study the gel point, gel fraction, and weight average degree of polymerization were determined as a function of time.

The percolation process on a lattice has also been used to model the cross-linking of paraffins induced by gamma radiation (29). For this study the cluster size distribution for the percolation process on a lattice was determined by using the perimeter polynomial method to calculate the probability of each cluster size. (A cluster is a set of sites joined by occupied bonds.) This method allows for an exact calculation of the cluster probabilities by using the number and probability of occupied and empty bonds on the cluster perimeter. The perimeter polynomial is obtained by summing the expressions of cluster probabilities for clusters with a given number of bonds. This polynomial provides the number of clusters of a particular size per lattice site and is then multiplied by the size of the cluster itself. The result corresponds to the fraction of sites found in the clusters of the particular size. Since this value corresponds to the weight fraction of paraffin molecules with a particular degree of polymerization, the molecular weight distribution is also described by this expression. The calculated results obtained were compared to the molecular weight distribution of the cross-linked paraffins as determined by gel permeation chromotography. The comparison showed that more cross-linking occurs in the paraffin system than is predicted by the percolation model for this study.

Recently Leung and Eichenger (30, 31) have also used Monte Carlo methods to simulate network formation. Their model was developed to permit consideration of the spatial arrangement of reactive groups; that is, intramolecular reactions can also be accounted for using this method. In this study the end-linking of elastomers was simulated by computer. An algorithm was developed to place a random array of primary chains in an image container on the computer. The ends of the chains are joined together with the possibility of both intramolecular and intermolecular reactions occurring. Large random graphs are formed from which the connected components are sorted; thus the sol and gel components are identified. Because molecules of different shapes are sorted and counted, this procedure also allows for the determination of cyclic species in the sol and the proportion of defects in the gel.

Several other theoretical approaches and models have been used to characterize polymer network systems. Stafford has used kinetic reaction schemes to derive the molecular size distributions and molecular weight averages for multifunctional polycondensation systems (32). The principle of equal reactivities of functional groups and the absence of the occurrence of intramolecular reactions were assumed in the theoretical developments. Some of the results were found to be equivalent to the classical distributions of Flory and Stockmayer. When gelation was considered, it was found that none of the previously established expressions, having the same inherent assumptions, were very accurate when compared to experimental data. The deviations were attributed to the neglect of intramolecular reactions for reaction systems of these types. Further uses of statistical theory have allowed Klonowski to predict the gel point and the gel fraction using only one variable, which represents the fraction of reacted functionalities (33–35). Durand and Bruneau retained the assumptions of an ideal network and used propagation expectation expressions to calculate the pregel and postgel properties (36, 37).

Now that the primary theories and several models of three-dimensional networks have been presented, we turn to the experimental characterization of these polymeric network systems.

CHARACTERIZATION TECHNIQUES

Chemical Methods

Since the early days of polymer chemistry, the chemical analysis of polymer samples has provided valuable information on the structure and purity of materials. Typically, a representative sample is prepared for actual analysis by extraction, solution, and separation techniques (38). Extraction methods allow for the removal of impurities from polymer samples. Separation procedures also remove impurities and, in addition, separate from the polymer itself the low molecular weight fragments produced by chemical reactions. The techniques used require dissolution of the polymer in suitable solvents. The chemical analysis of the material is then carried out using the standard methods of chemistry. Conventional chemical analyses include elemental analysis and functional group analysis using a variety of experimental methods. The details of the various identification techniques for polymers can be found in several books (39–47). The application of chemical analysis to polymer systems is also covered in the literature (48–51).

Network polymers are insoluble and as a result are not as easily characterized by the conventional chemical methods, which require dissolution. Identification tests, based primarily on decomposition techniques, have been extended to several insoluble polymers (42). For example, phenols can be detected and identified as they are released from phenolic resins and cross-linked epoxy resins upon pyrolysis. Melamine–formaldehyde resins are detected by the presence of melamine chloride. Such chemical tests are used primarily to identify unknown polymeric materials.

In the *Handbook of Analysis of Synthetic Polymers and Plastics*, Urbanski et al. include chemical analysis as a characterization technique for some of the cured resins (42). Cured epoxy resins have been analyzed for the degree of cure which is obtained by determining the quantity of epoxy groups that have reacted. A sample is placed into a solution containing hydrogen chloride in which the cured resin swells but does not dissolve, while the unreacted epoxy groups react with the hydrogen chloride. A quantification of the excess hydrogen chloride present in the solution allows for the determination of the degree of conversion. For unsaturated polyester resins, degree of cure or cross-linking has been determined by hydrolyzing the resins followed by an extraction of the hydrolyzate. Urbanski et al. also include procedures for determining the resin content, filler content, and amount of low molecular weight components present in specific polymeric systems. Their book contains many identification tests, elemental analysis experiments, and estimation procedures for chemical characteristics, which are applied mainly to linear polymers.

Cross-linked polymers have been characterized by gel content analysis, which is dependent on the inherent insolubility of the gel portion of the sample. The gel content method requires isolation of the gel from the sol, followed by a determination of the weight fraction of the gel in the sample. In one particular case the cross-linking efficiency of a system was estimated from the gel content for known molecular weight distributions of the base resin (52). Linear polyethylene was cross-linked with dicumyl peroxide (DCP), after which the gel content of the samples was determined by an experimental analysis. The gel content was also predicted with the aid of a computer program, which required the input of several parameters, including the concentration of cross-linking agent and the molecular weight distribution of the resin. The program assumed 100% cross-linking efficiency by means of intermolecular cross-links only. Both the experimental gel content and the calculated gel content were plotted versus the proportion of DCP cross-linking agent used. From the difference of the two slopes of the curves, a measure of the cross-linking efficiency is obtained. For the case of linear polyethylene and DCP, it was found that the system had a cross-linking efficiency of approximately 43%.

Bell et al. (53, 54) have developed a chemical technique for measuring the cross-link density of amine-cured epoxy resins. The method is based on the stoichiometry of the curing reaction and on the amount of primary amino groups and epoxy groups remaining in the polymer at a particular time. A complete understanding of the reaction mechanism is necessary for success in the method. The unreacted epoxy is potentiometrically titrated to determine its concentration, while the primary and secondary amine concentrations are measured spectrophotometrically by derivatizing the amines with a chromophore sensitive to the amine structure. The values for molecular weight between cross-links calculated by this method were consistent with the values obtained by equilibrium swelling measurements for their epoxy cure system.

Chemical probes were one of the first methods used to characterize sulfur-vulcanized natural rubber networks, which are comprised of monosulfidic, disulfidic, and polysulfidic cross-links. The different types of cross-link may be preferentially cleaved with the use of chemical reagents, whereupon the resulting structures are identified (55). The polysulfidic cross-links in a vulcanizate sample can be cleaved with the use of propane-2-thiol and piperidine in a solution of *n*-heptane, while both the polysulfidic and disulfidic cross-links are cleaved when *n*-hexanethiol is used in a piperidine solution. These chemical probe methods must then be used in conjunction with a physical or mechanical technique to quantify the number and nature of cross-links. Chemical analysis provides basic structural information, which is greatly extended with the use of physical methods of characterization.

Thermal Analysis

Thermal analysis has provided important contributions in the characterization of polymeric materials, and a great deal has been written on this subject (56–62). For insoluble network polymers, thermal techniques have been used to establish the degree and rate of cure, to study the chemical kinetics of curing reactions and the curing behavior itself, and to study degradation reactions. We cover the following thermal analysis techniques and their applications: differential scanning calorimetry (DSC), differential thermal analysis (DTA), and thermogravimetry (TG).

DSC and DTA are techniques which monitor either the heat evolution or absorption for any reactions that are occurring in a sample. The DTA experiment involves the measurement of the temperature difference between a sample and a reference material which are being either heated or cooled at a linear rate. Then the thermal effects accompanying physical or chemical changes in the sample are detected. In DSC the heat flow into or out of a sample and into or out of a reference material is measured as a function of time or temperature; therefore it is a useful technique with which to measure the heats of transition. The basic differences between DSC and DTA are found in the heating system and in the method of operation of the instrument. In DSC, the sample and the reference are heated separately by individually controlled elements. To maintain the same temperature for the sample and reference materials, the power to the heaters is adjusted continually in response to any thermal effects in the sample. The differential power to achieve this condition is recorded as the ordinate on an x–y recorder, while the abscissa is the programmed temperature of the system. In DTA, the sample forms a significant part of the thermal conduction path. The thermal conductivity of a sample changes in a way that is generally unknown during a transition and as a result the proportionality between temperature differences and energy changes is also unknown. Because the conversion of peak area to energies is somewhat uncertain, DTA is generally a qualitative technique. In contrast, the peak area in a DSC curve is a true measure of electrical energy input that is necessary to maintain the same sample and reference temperatures independent of the instrument's thermal constants or any changes in the thermal behavior of the sample. The calibration constant that relates the DSC peak area to calories is known and allows for a quantitative analysis of the data.

DSC is the more frequently used technique in the measurement of heats of reactions for the types of polymerization under discussion because it does give a quantitative measure of the heat and the rate of the curing reaction. As will be demonstrated, DSC provides a considerable amount of insight into the chemical mechanism of the curing process, but thus far a relationship between the T_g (glass transition temperature) determined from the DSC method and the cross-link density has not been found. A brief description of the type of information that can be obtained is given here; for a more comprehensive overview of the thermal analysis of thermosets, before and after curve, one can refer to Chapter 5 of Reference 58.

The degree of cure in a network system is determined by temperature variations during the actual cure, which depend on the heat of the reaction. Fava (63) described kinetic measurements involving the heat of reaction for measuring the extent of cure in an epoxy resin. The three methods for obtaining isothermal cure curves using the DSC technique are isothermal operation, analysis of thermograms with different scan rates (a dynamic method), and scans on partly cured resins. From DSC curves, the state of cure can be monitored and the kinetic parameters of cure can be determined.

DSC scans with isothermal operation cannot be used for reaction systems that have very fast cure rates. This method requires a time period in which the sample and reference material must heat up to the desired temperature. As a result, the beginning section of the isotherm is lost during the scan. The experimental difficulty can be alleviated by obtaining an analysis of thermograms with different scan rates. Figure 3 shows schematic set of displaced thermograms on a temperature axis. From this method one can construct an isothermal cure curve at temperature T_0 as shown in Figure 3*b*. An ordinate at T_0 is drawn in each of the thermograms. The state of the resin is described by three parameters at the point at which it dissects the curve: temperature T_0, heat generation rate dH/dt, and heat of reaction H. We calculate H from the shaded area and find the total heat of reaction H_0 by the area enclosed by the complete curve on a time axis. The eight isothermal states shown in Figure 3*a* are plotted in reduced form as $[d(H/H_0)/dt]^{-1}$ versus H/H_0 as shown in Figure 3*b*. The integral of the curve from zero to H/H_0 is equal to the time necessary to reach the degree of cure H/H_0, at temperature T_0. As a result, the isothermal cure curve of H/H_0 versus time can be plotted. It is possible to

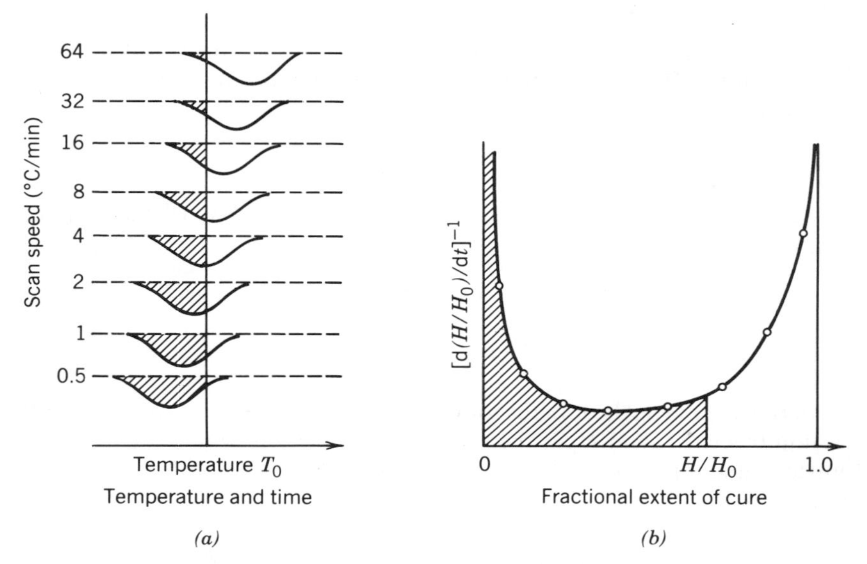

Figure 3. (*a*) Set of displaced thermograms. (*b*) Curve deduced from (*a*) and used to obtain isothermal cure curve at temperature T_0. (From Ref. 63.)

repeat this process for any temperature greater than T_0. For temperatures that are lower than T_0, the final segment of the cure curve will be missing. Therefore, one can expect reliable results from the isothermal method at low temperatures, while the scan method does not provide enough data in this region. The different scan rate method must be used for systems having a fast cure or very high curing temperatures. There is also a third method by which partially cured resins are scanned. For these samples, the area of the thermogram gives the residual heat of reaction ($H_0 - H$) from which H can be obtained. This method is useful when the rate of heat evolution is too small for isothermal detection.

Two methods are commonly used to analyze the DSC curves obtained for a curing reaction. In the first method a DSC thermogram is studied by means of a differential–integral analysis, which yields values for activation energy, order of reaction, and total heat of reaction (64). The general kinetic equation is of the Arrhenius type and is given by

$$dx/dt = k(1 - x)^n$$

where dx/dt is the rate of reaction, k is the reaction rate constant, and x is the fractional extent of reaction or the extent of cure.

With the Arrhenius relationship, $k = A\exp(-E/RT)$, where E is an activation energy, R is the gas constant, T is the absolute temperature, and A is the preexponential or the frequency factor, the equation above can be rewritten as

$$dx/dt = A\exp(-E/RT)(1-x)^n$$

It is assumed that the evolution of heat is proportional to the extent of reaction:

$$x = H/H_0$$

and

$$dx/dt = dH/dH_0$$

where H is the partial heat of reaction, which varies in direct proportion to the fraction reacted x, and H_0 is the total heat of reaction, which is equal to the total area under the curve. When the equations above are combined, the result gives

$$dH/dt = A(H_0)\exp(-E/RT)(1-H/H_0)^n$$

In a dynamic DSC run, the heating rate or the scan rate is denoted as $G = dT/dt$; thus the equation above can be written in a form which shows the change of degree of cure with temperature:

$$dH/dT = A(H_0/G)\exp(-E/RT)(1-H/H_0)^n$$

Crane et al. (65) extended these calculations to a study of the curing reaction of an epoxy resin system. To determine the kinetic parameters for their study, they derived the equation

$$\left(\frac{d^2H/dT^2}{dH/dt}\right)T^2 = E/R - [nT^2/(1-\alpha)H_0](dH/dT)$$

where dH/dt is the ordinate scale of a DSC curve and is converted to the curve height $h = dH/Gdt = dH/dT$ by the scan rate given above; d^2H/dT^2 is the slope of the curve S at temperature T. Using the definition $x = H/H_0$, the expression $(1-x)H_0$ is equal to $(H_0 - H)$, which is the remaining area A, found under the DSC curve. If these substitutions are made in the equations above, we have

$$(S/h)T^2 = E/R - nT^2(h/A)$$

After the values of the slope S, the curve height h, and the remaining area A, have been taken at various temperatures, a plot of $(S/h)T^2$ versus $T^2(h/A)$ will be a straight line whose slope defines the order of reaction n, while the activation energy E is obtained from the intercept at $T^2(h/A) = 0$.

Kissinger (66) developed a second method to analyze the DSC curves of curing reactions which uses multiple DSC scans obtained at various rates of heating. In this method it is assumed that the reaction rate is a maximum at the peak temperature T of a DSC curve. At the peak temperature, $h = dH/dT$ has reached its maximum value and the slope $S = d^2H/dT^2$ is equal to zero. From the assumptions of this method the following equation is obtained:

$$(d \ln G)/[d(1/T)] = -E/R - 2T$$

where G is the scan rate and T is the peak temperature of the DSC curve. A plot of $\ln G$ versus $1/T$ from a series of DSC curves should be linear, and the activation energy E is obtained from the slope when $E/R \gg 2T$. This method allows for an evaluation of the effects of scan rate on the cure kinetics. Generally one would want to use both methods for complete analysis of the kinetics occurring during a curing reaction.

Extensive research on the cure kinetics of various network systems has been completed. DSC and the three methods discussed have been widely used in the study of the curing kinetics and reaction mechanisms of epoxy resin systems (67–76). Prime (69) addressed the question of equivalency of cure kinetics derived from dynamic experiments with published isothermal data for epoxy cure systems. It was determined that the dynamic kinetic parameters measured on epoxies are significantly larger than for parameters determined by isothermal methods. Horie et al. (67) investigated the reaction rates of an epoxy resin with aliphatic diamines over the entire range of conversion. The various kinetic parameters were measured for the system in addition to the determination of extent of conversion of the epoxide group and the cross-linking density of the postcured resin. Detailed experimental techniques and procedures can be found in the literature, in addition to the analyses of the data obtained. The importance of correct baselines is stressed for obtaining accurate measurements (71). DSC methods have been used to study phenolic resins (77–84), melamine–formaldehyde resins (85), and vulcanizates (86), among other cross-linked systems (87–90). In general, the DSC technique measures the extent and the rate of chemical conversion in the curing process. The technique does not actually detect gelation because it is insensitive to the physical changes occurring at the gel point. A DSC cure study also allows for a determination of various glass transition temperatures such as that of thermosets at zero, one, or the highest attainable degree of conversion. For a discussion of these measurements, see Prime's chapter on thermal characterization of thermosets (58).

Differential thermal analysis is a less often used calorimetric technique for such measurements, but useful data have been obtained. Recently it was shown that there is reasonable correlation between the gelation time and the temperature corresponding to the peak of the isotherm on DTA curves for an epoxy resin system (91). The DTA method has yielded accurate kinetic results for a phenol–formaldehyde resin system, in addition to providing an insight to the actual chemistry of the curing process (92). In another study dealing with phenolic resins, sealed-cell or pressure DTA was used to analyze the reaction system (93). This type of cell was necessary to retain the volatile products generated in the experiment because removal of the volatiles produced large endotherm peaks, which prevented the observance of the curing exotherm peak. The experiment allowed for the detection of the heating effects associated with the curing, and it was possible to correlate the analysis of the thermograms with the actual curing behavior.

Thermogravimetry is a technique in which the mass of a substance is monitored as a function of temperature or time as a sample is subjected to a programmed rate of heating in a regulated atmosphere. TG has been used to measure the extent of reaction for thermosets that cure with the formation of condensation products such as water. This measurement is possible because the weight loss of a sample is analogous to the enthalpy loss for such a system under ideal reaction conditions. Using the methods of TG analysis, Greenberg and Kamel (94) followed the curing kinetics for a polyacrylic acid system which forms anhydride during the course of the reaction. Three separate peaks were identified in the analysis and attributed to three different phenomena. The first peak was correlated to the removal of bound water. The second stage of curing, which had the highest weight loss, was determined to be the actual cross-linking reaction and produced a peak which was attributed to anhydride formation. The degree of conversion of the acid groups was determined with the data obtained in this particular region of the analysis. The lower weight loss at the highest temperature, corresponding to the third peak of the analysis, was identified with polymer degradation. The authors also studied the kinetics of the anhydride reaction and determined the order of the reaction mechanism and the activation energy using the TG data.

A common application of thermogravimetric analysis is in polymer degradation kinetics (95, 96). From TG studies of an epoxide resin system, it was found that higher curing temperatures apparently lead to higher activation energies for degradation (97). In the same study it was observed that at a constant rate of heating, the samples cured at higher temperatures have more thermal stability. TG experiments have also been used to identify oxidation reactions in epoxy resins (98). The primary degradation route for the phenolic resin system has been determined to be oxidation (96).

Spectroscopic Techniques

The use of spectroscopic methods for the characterization of polymeric systems has provided important molecular level descriptions of these systems. Two spectroscopic techniques which have prove to be especially powerful for the analysis of cross-linked network polymers are solid state nuclear magnetic resonance (NMR) and Fourier transform infrared (FTIR) spectroscopy. Both methods allow for solid sample analysis (i.e., dissolution is not necessary) and provide important structural information. The basis of these two methods is described, followed by a review of the application of these techniques to the studies of polymeric network systems.

HIGH RESOLUTION SOLID STATE NMR SPECTROSCOPY

High resolution NMR has proved to be a significant analytical technique for the characterization of polymeric systems. The use of NMR spectroscopy for solid polymers has been reviewed by McBrierty (99–101), who has covered molecular motion studies in addition to the structural characterization of these systems in great detail. Jelinski addressed the subject of chemical information and problem solving for both solution and solid state polymer studies (102). Relevant information can be obtained from modern NMR methods including cross-polarization (CP), magic angle spinning (MAS), and two-dimensional spectroscopy. Havens and Koenig reviewed the application of exclusively solid state carbon-13 (^{13}C) CP-MAS NMR to polymer characterization (103). Recently Harrison et al. examined the use of NMR spectroscopy, among other physical and mechanical techniques, for the investigation of cross-linked polymers (104). They found a considerable amount of data in the literature relating the cross-link density to the molecular level relaxation properties of network systems using the pulsed proton NMR techniques. Carbon-13 nuclear relaxation measurements have not been used as frequently, but some studies have shown virtually no changes in the spin–lattice (T_1) relaxation times between cross-linked and unaltered rubber systems, while small changes were noted for the spin–spin (T_2) relaxation times of these systems (105, 106). Carbon-13 high resolution NMR spectroscopy has been used to directly detect cross-links.

This section focuses primarily on network characterization using the techniques developed for solid state magic angle NMR. Solution NMR, with carbon-13 and hydrogen-1 the most commonly observed nuclei, has been a very effective characterization method for macromolecular systems for many years (107, 108). If a ^{13}C NMR spectrum is obtained for a solid material under solution conditions with a pulsed Fourier transform NMR experiment, one obtains broad resonances which may extend into the kilohertz range. This makes the assignment of resonance peaks to individual carbon types in the sample impossible. In contrast, the line widths of resonances in solution spectra are several hertz, which is a direct result of the rapid molecular motion, which averages out strong dipolar interactions between spins. Techniques have been established to eliminate the line-broadening problem and to obtain the high resolution NMR spectra of dilute spins such as ^{13}C in solid samples.

In the subsections that follow, high power decoupling, magic angle spinning, and cross-polarization are explained with ^{13}C as the observed nucleus. Then we review the application of these techniques and the types of structural and dynamic information obtained for cross-linked network systems. The solid state NMR studies of these systems have demonstrated that high resolution spectra are attainable and that this method of characterization with its capabilities and relative ease of interpretation is a valid and useful one.

High Power Decoupling

Two line-narrowing techniques are used to solve resolution difficulties that formerly arose from $^{1}H/^{13}C$ dipole–dipole interactions and chemical shift anisotropy. The first method is high power decoupling, which removes the dipolar broadening and is the critical factor in obtaining high resolution spectra. If the proton spins undergo transitions at a rate faster than the frequency of the $^{1}H/^{13}C$ interactions, then the dipolar fields are averaged to zero. This procedure can be accomplished by means of a strong radiofrequency field applied at the Larmor frequency of the protons (109). This high power decoupling requires a radiofrequency field of about 10 gauss, which is the magnitude of the proton line widths (110).

Magic Angle Spinning

The anisotropy of the chemical shift, which is controlled by the electronic structure surrounding a carbon, also induces line broadening and in addition creates unsymmetrical line shapes. Solids have a dispersion of chemical shifts resulting from the anisotropy of the electronic environment in the static field, and the chain motion is insufficiently rapid to average these shifts as occurs in a liquid. The magnitude of the chemical shift may be shown to be proportional to $3\cos^2\theta - 1$, where θ is the angle between the sample and the magnetic field direction (111). To reduce this factor to zero, the "magic angle" of 54.7° is employed as the second line-narrowing technique (112). Sample spinning at this angle with speeds that are slightly greater than the dispersion of chemical shifts, known as magic angle spinning (MAS), reduces the resonances to their isotropic averages and thereby achieves substantial improvement in spectral resolution.

Cross-Polarization

Even with the improvement in resolution achieved by magic angle spinning and high power decoupling, a pulsed NMR experiment on a solid system is inconvenienced by a time consideration. Repetition of the pulse sequence is necessary for coaddition of the free induction decays for the concomitant signal-to-noise enhancement. However, a delay of several ^{13}C spin–lattice relaxation times must be tolerated before the nuclei return to their equilibrium distribution. Such delays, which must be on the order of tens of seconds for solid polymers (113), are avoided by transferring polarization from nearby protons to the carbons by means of static dipolar interactions. This procedure, called cross-polarization (CP) (114), requires a match of the resonant frequencies of the carbon and protons ($\omega_C = \omega_H$) known as the Hartmann–Hahn condition (115). The major advantage of the technique lies in the relative speed of the repolarization of the protons, which is considerably more efficient than that of the carbons; thus the pulse sequence may be repeated with less delay. In addition, it may be shown theoretically that the transfer of polarization from the protons to the carbons improves the sensitivity and resolution of the spectra (114). The pulse sequence used for the cross-polarization experiment is shown in Figure 4*A* and a typical pulse sequence, which depends on carbon magnetization, is shown in Figure 4*B*. The CP pulse

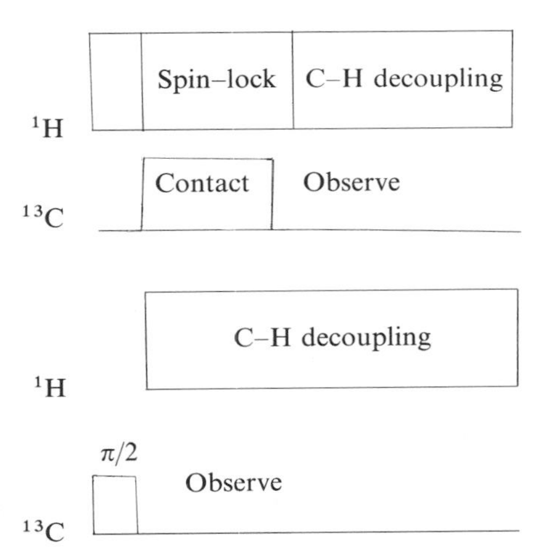

Figure 4. Carbon-13 NMR pulse sequences. (*A*) Cross-polarization (CP) experiment. (*B*) Gated high power decoupling (GHPD) experiment.

sequence consists of applying a 90-degree radiofrequency (rf) pulse to the proton spins, followed by spin locking the proton rf field to direct it collinear with the proton magnetization. Next, the carbon spins are brought into contact with the proton spins by means of matching the carbon and the proton rf power levels. Transfer of magnetization from the ^{1}H spins to the ^{13}C spins occurs at this point, after which the carbon pulse is turned off and the resultant free induction decay is observed.

By means of these three techniques, high resolution NMR spectra are obtainable for solid, rigid samples. The CP-MAS NMR method provides structural and dynamic information on the molecular level for solid polymeric materials. For additional information on the techniques and applications, the reader is referred to the literature (116–121). The discussion that follows details the range of information attainable for cross-linked network systems using the solid state NMR techniques. We cover four types of network system: cross-linked elastomers, curing of epoxies and other thermosetting resins, curing of acetylene-terminated resins, and radiation-induced cross-linking.

Cross-linked Elastomers

Network formation can occur by cross-linking the chains of a linear polymer or by the curing of multifunctional thermosetting resins. Elastomers such as polybutadiene and polyisoprene (natural rubber) are polymer chains which can be cross-linked by free-radical cross-linking or by sulfur vulcanization. The formation of such networks improves the macroscopic physical properties of these systems, including mechanical performance, temperature, and size stability. A highly cross-linked material is generally insoluble and intractable, which does not allow for a simple spectroscopic analysis of the network formation and final structure. Solid state ^{13}C NMR has been used to address this particular problem of structure elucidation.

The use of dicumyl peroxide as a curing agent for natural rubber and *cis*-polybutadiene produces a network that contains carbon-to-carbon cross-links. Although ^{13}C NMR spectroscopy in the solution state has been used to study the reaction mechanisms for these systems (122, 123), the information obtainable from this work is minimal because of the resonance broadening that arises upon cross-linking. The techniques of solid state ^{13}C NMR have been applied to this problem (124, 125). Figure 5 shows ^{13}C MAS spectra of *cis*-polybutadiene cured for 2 hours at 149°C with various levels of dicumyl peroxide. The spectra in Figure 5A were obtained under normal FT (NFT) conditions with magic angle spinning, while the spectra in Figure 5*B* were obtained with cross-polarization in addition to magic angle spinning. Solid samples without modification were used in obtaining both sets of spectra. The free radicals that promote the cross-linking process are generated at various

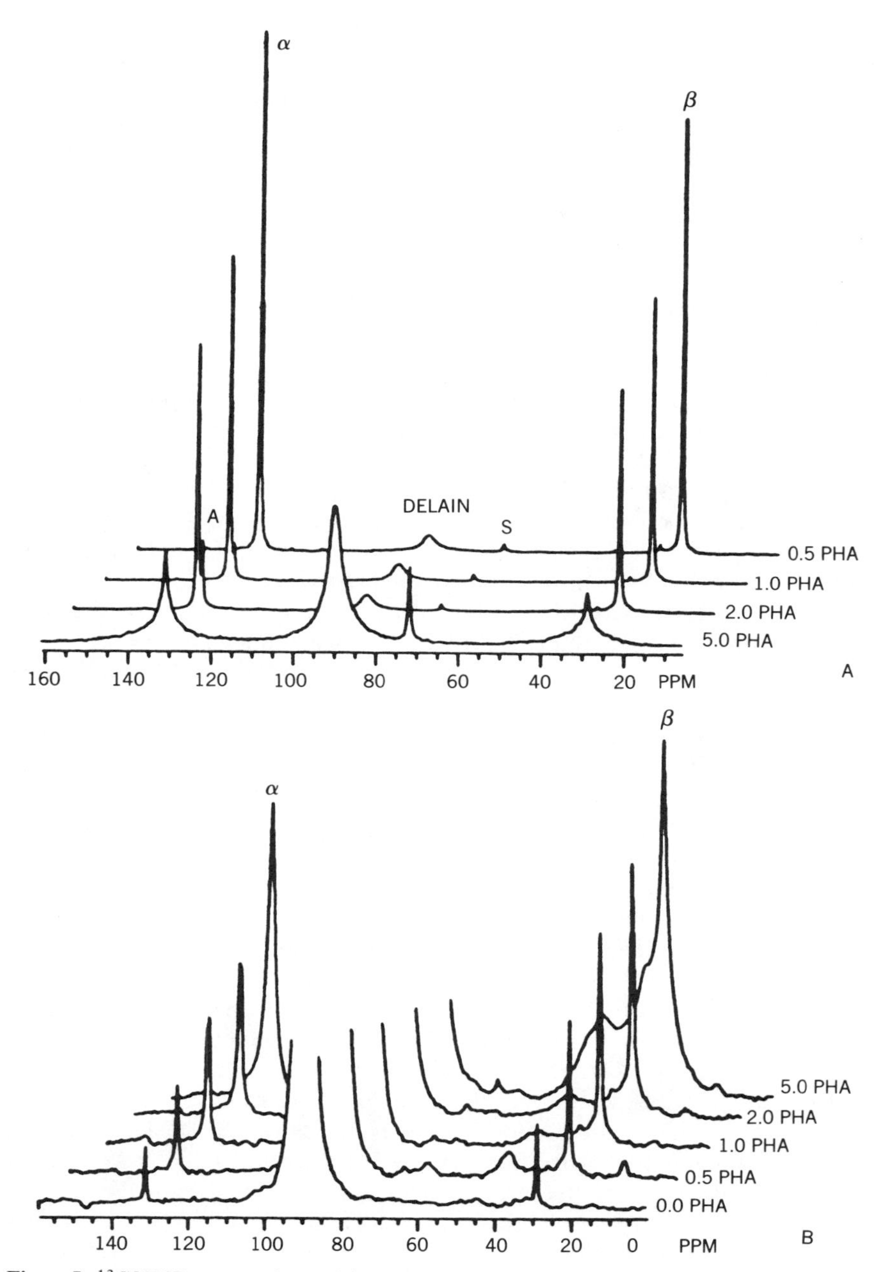

Figure 5. ^{13}C NMR spectra of *cis*-polybutadiene cured for 2 hours at 149°C with various levels of dicumyl peroxide [given in parts per hundred (phr)]. The spectra were recorded at 38 MHz. (*A*) Spectra obtained under normal FT conditions with magic angle spinning. (*B*) Spectra obtained with cross-polarization and magic angle spinning.

concentrations by the dissociation of dicumyl peroxide. Both sets of spectra show appreciable resonance broadening with increase in degree of cross-linking. The effects on the carbon resonances in the NFT spectra arise from the decrease in segmental motional freedom of the carbon chain backbone, in addition to line broadening due to the dipolar interactions. In the CP spectra, the broad resonance peaks indicate a wider distribution of isotropic chemical shifts arising from the differences in bonding and molecular packing associated with network formation.

With the increase in cross-linking it is also observed that the total integrated area of the CP spectrum increases. An inherent feature of cross-polarization is exclusion of highly mobile regions of a sample in an NMR spectrum because of the intensity dependence on the static dipolar coupling between carbon and proton nuclei. Cross-linking promotes rigidity in the polybutadiene samples, which is reflected by the increase in the resonances in the CP spectra. Another significant observation in the CP spectra is the presence of a broad quaternary carbon resonance at 45 ppm, which arises from the carbons directly attacked by the radicals leading to the formation of cross-links. Patterson and Koenig (125) also found spectral evidence of cis-to-trans isomerization during cross-linking for both natural rubber and polybutadiene systems using the solid state ^{13}C MAS NMR techniques.

Sulfur-cured natural rubber systems are also being studied by solid state ^{13}C NMR methods (126). The network structure of sulfur-vulcanized natural rubber is highly influenced by the curing additives and by the conditions used in the processing procedures. The curing process may induce monosulfidic, disulfidic, and polysulfidic cross-links in addition to other main chain structural modifications such as cyclic sulfides, pendant thiol groups, and cis-to-trans isomerization of the natural rubber chain. The ^{13}C NMR spectra of two different formulations of sulfur-vulcanized natural rubber are shown in Figure 6. The resonances due to the natural rubber repeat unit are labeled. An efficient rubber formulation is one in which the ratio of curing accelerator to elemental sulfur is high, whereas a conventional rubber system contains a high ratio of sulfur to accelerator. Both spectra were obtained by magic angle spinning of solid samples using a gated high power decoupling pulse sequence instead of the cross-polarization pulse sequence.

Considerable differences are observed for the two cases. The efficient vulcanized system yields a simpler, nonoverlapping ^{13}C NMR spectrum (Figure 6*A*), compared with the ^{13}C NMR spectrum of the conventional system (Figure 6*B*). The five prominent resonances are due to the natural rubber carbons themselves. Evidence of cis-to-trans isomerization is observed by means of the resonance appearing at 40.3 ppm, which is assigned to the methylene carbons of *trans*-natural rubber and another resonance at 16.1 ppm assigned to the methyl carbons of the trans isomer. Resonances due to carbon–

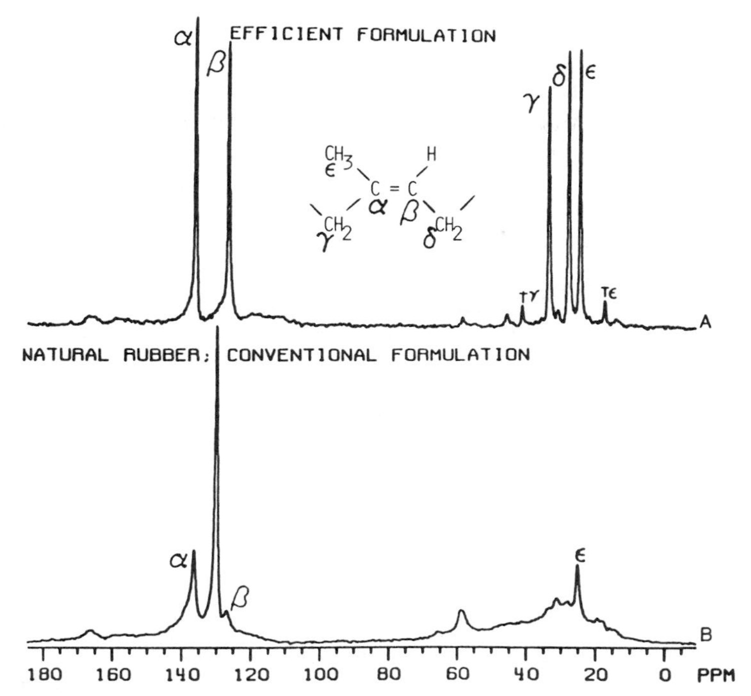

Figure 6. ^{13}C NMR spectra of sulfur-cured natural rubber systems obtained with the gated high power decoupling pulse sequence and magic angle spinning. (*A*) An efficient formulation (high ratio of accelerator to sulfur). (*B*) A conventional formulation (high ratio of sulfur to accelerator).

sulfur cross-link points at different sites on the natural rubber chain appear at 45.0 and 57.6 ppm. The resonance peak at 45.0 ppm is probably also due to pendant accelerator groups on the natural rubber main chain. In this efficient formulation the accelerator controls the use of elemental sulfur during the vulcanization process.

Conventional rubber vulcanizate systems containing high ratios of sulfur to cross-linking accelerator are known to produce complex network systems (127) with considerable main chain structural modification, as is evident from Figure 6*B*. The curing induces a loss of spectral resolution, particularly in the aliphatic carbon region between 10 and 60 ppm. The increase in line width is due to a distribution of chemical shifts that arise from the structural changes that occur during the cure process, including resonances due to cross-linking and other modifications of the natural rubber chain. In addition, network formation establishes a more rigid structure; as a result, the decrease in molecular motion is also contributing to spectral broadening. The resonance

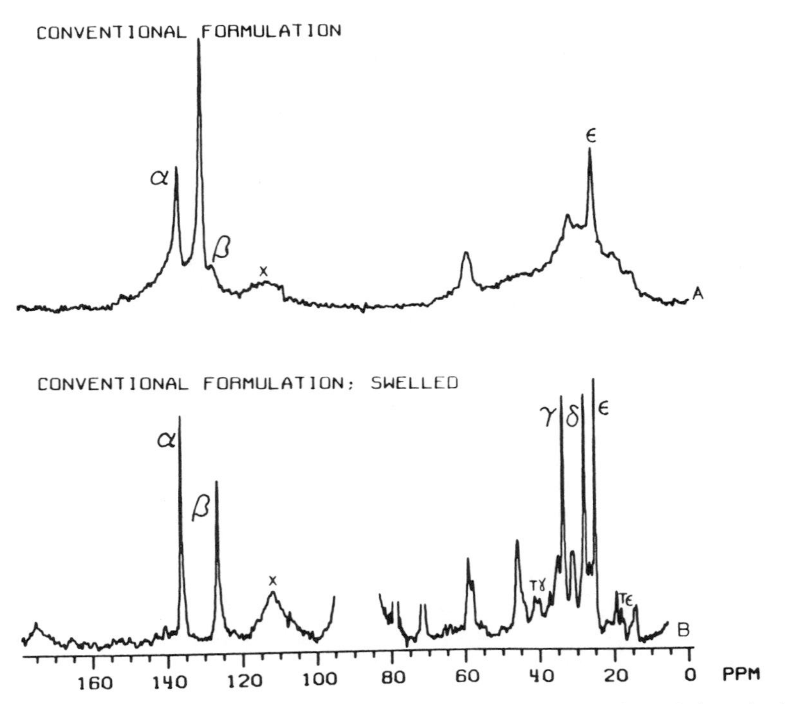

Figure 7. ^{13}C NMR spectra of natural rubber cured with a conventional formulation, obtained with the gated high power decoupling pulse sequence and magic angle spinning. (*A*) Unswollen sample. (*B*) Sample swollen in benzene.

appearing at 58 ppm is assigned to cross-link points. The prominent resonance found at 128 ppm is due to trapped benzene in the vulcanizate. Benzene was used for network swelling measurements, which preceded the NMR measurements.

To obtain further spectral information about the actual network structure using ^{13}C NMR experiments, the conventional sample was swollen in chloroform to induce more mobility in the cross-linked sample, hence to obtain additional spectral resolution (126). The resulting ^{13}C NMR spectrum is shown in Figure 7. There is considerably more resolution in this particular spectrum of the conventional formulation; therefore it is assumed that the solvent molecules in the swollen, cross-linked sample allow for the necessary mobility for resolution. The five prominent resonances that appear are due to carbon groups of the natural rubber repeat unit itself. The enhanced resolution obtained by swelling the sample allows for a closer examination of the network structure. Various cross-linked resonances are observed at 57.6, 56.9, and

44.5 ppm, in addition to evidence of cis-to-trans isomerization at 40.3 and 16.1 ppm. The resonances appearing between 30 to 45 ppm may be due to cyclic sulfides and pendant thiol groups. In summary, the solid state ^{13}C NMR technique has proved to be a significant method for the detection of cross-links and other structural modifications that occur in sulfur-cured natural rubber systems.

Curing of Epoxies and Other Thermosetting Resins

The network analysis of thermosetting resins by the conventional chemical and physical characterization methods has been limited because of the

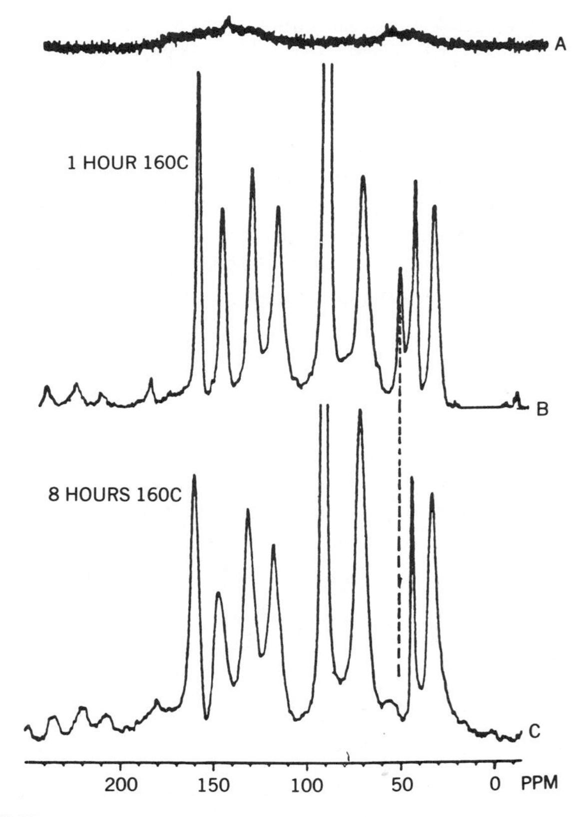

Figure 8. ^{13}C NMR spectra recorded at 38 MHz of epoxy resin systems based on DGEBA and cured with BDMA. (*A*) Spectrum of the resin cured for 1 hour at 160°C, obtained under typical solution NMR conditions. (*B*) Spectrum of the same sample, obtained with cross-polarization and magic angle spinning. (*C*) CP-MAS NMR spectrum of the resin cured for 8 hours at 160°C.

insolubility of cross-linked polymers. The CP-MAS NMR techniques are an obvious application for the studies of these polymeric systems in their solid state, and several investigators have studied the curing of epoxies with the solid state ^{13}C NMR techniques (128–136). Epoxy resins have been cured by various reactions involving the use of amines, anhydrides, or other accelerators (137–143), which lead to the formation of specific reaction products. In all cases network formation is the result of the curing reaction, which has proved to be a difficult product to characterize because of its insolubility. The work that has been reported using solid state NMR techniques shows the usefulness of the method for structural network characterization.

Cholli, Ritchey, and Koenig have investigated the curing of the diglycidyl ether of bisphenol A (DGEBA) with 2% benzyldimethyl amine (BDMA) (128). The ^{13}C CP-MAS NMR spectra of this system cured at 160°C for 1 and for 8 hours are shown in Figure 8 (*B* and *C*, respectively). Conventional solution NMR techniques were used to obtain the spectrum shown in Figure 8*A* for the sample cured for 1 hour at 160°C. By comparing the spectra of Figure 8*A* and *B*, the advantage over conventional NMR techniques of using the magic angle spinning and cross-polarization techniques for this network system is clearly demonstrated. This epoxy cure study followed the disappearance of the intensity of the resonances assigned to the epoxy ring carbons and the increase in the intensity of the resonances due to the methylene carbons attached to the oxygen atoms. Upon extending the cure time from 1 to 8 hours, the methine

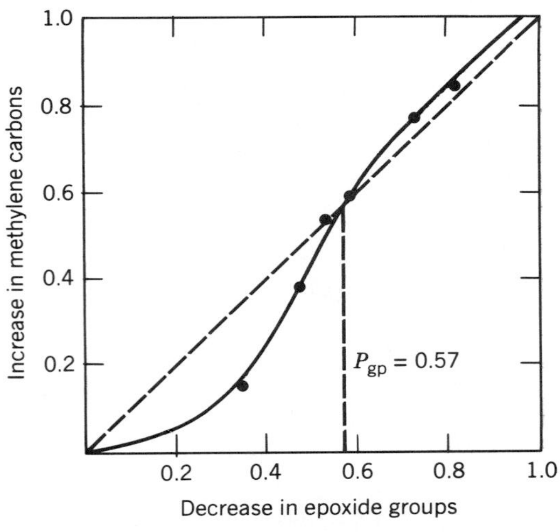

Figure 9. Data corresponding to the ^{13}C NMR data of Figure 8 (gp = gel point). Plot of increase in methylene resonance intensity against the decrease of the epoxide group peak intensity.

resonance of the epoxide group at 50 ppm shifts downfield to approximately 70 ppm as the ring opens. A similar effect is observed for the methylene carbon of the epoxide group at 44 ppm.

The increase in oxymethylene carbon resonance is plotted against the decrease in the epoxide carbon intensities in Figure 9. The conversion of one epoxide group should produce one additional oxymethylene unit, shown by means of a straight line having a slope of 1 as a function of conversion. The experimental data of Figure 9 deviate from ideality in that the curve decreases as the extent of reaction proceeds and follows ideal behavior after the extent of reaction reaches 0.57. This reaction behavior was analyzed on the basis of other chemical reactions which may occur during cure, including side reactions with water or alcohol which consume epoxy groups without the formation of oxymethylene units. The deviation in the experimental curve was also interpreted as an effect of the change in the physical state of the system induced during cure, which would in turn affect the dynamics of the CP NMR experiment. The transition at the extent of reaction of 0.57 may indicate the gelation point.

Resing and Moniz (129) used the techniques of cross-polarization and high power decoupling to obtain the ^{13}C NMR spectra of DGEBA cured with three different curing agents: piperidine, hexahydrophthalic anhydride (HHPA), and nadic methyl anhydride (NMA). Because magic angle spinning was not used, line broadening due to chemical shift anisotropy obscured spectral resolution. Line diagrams were constructed to represent the approximate chemical shift anisotropy (CSA) patterns expected for the various carbon types, including quaternary aromatic carbons, unsubstituted aromatic carbons, methylene carbons, and methyl carbons of the cured epoxy systems. By comparing the diagrams to the NMR spectra, it was possible to assign the resonances in the CP spectra of the cured systems. Spectral differences for the three systems were associated with the rigidity of the polymers as well as the differences in the chemistry of the curing agents.

Garroway, Moniz, and Resing characterized four epoxy polymers based on DGEBA, using the techniques of solid state ^{13}C NMR (130–132). Two of the systems were cured with amines, and the other two were cured with anhydrides. In this particular study high power decoupling, cross-polarization, and magic angle spinning techniques were used, and the solid state spectra obtained were compared to the solution spectra of the unreacted epoxy systems dissolved in a solvent. These spectra (shown in Figure 10 and identified by main curing agent) clearly reveal the potential for chemical identification of cured epoxy polymers. Resonance peaks due to the epoxide ring carbons were not present in the polymerized DGEBA of Figure 10*a*, thus giving a rough indication of the degree of polymerization of the reaction system. Further evidence of reaction in the spectrum of Figure 10*a* can be

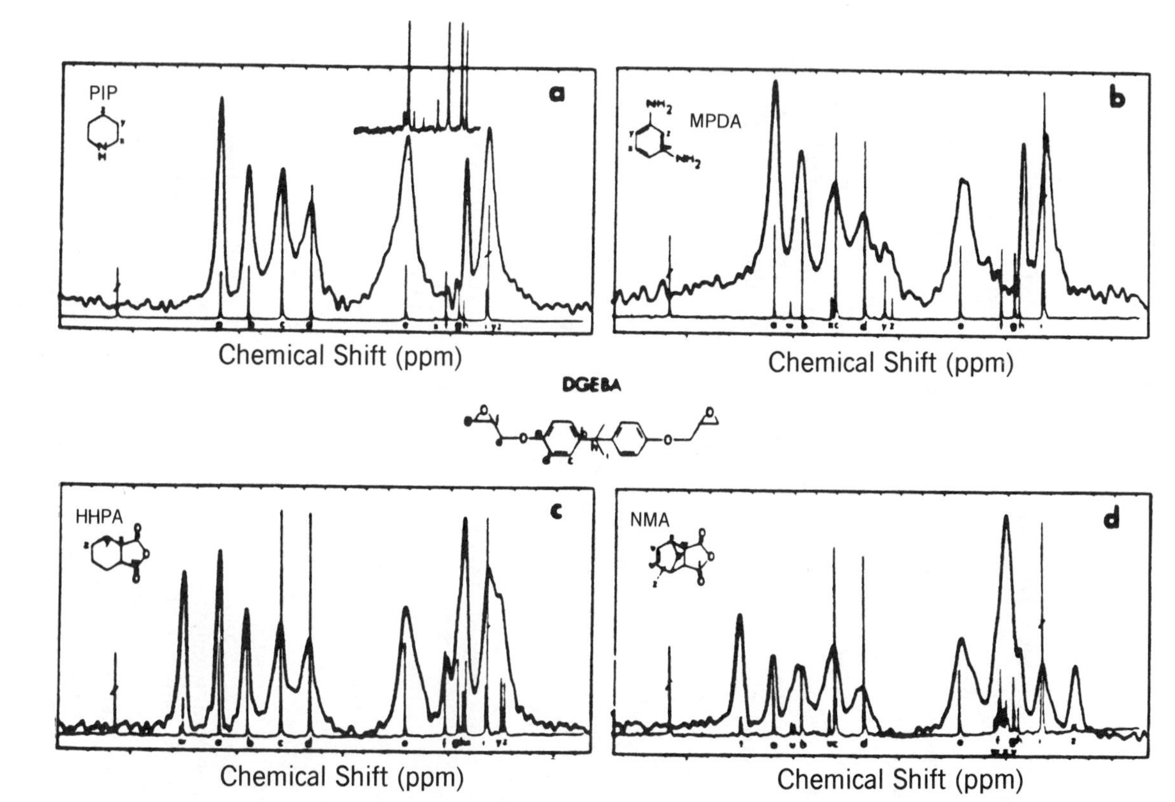

Figure 10. Comparison of solid state ^{13}C NMR spectra of four different epoxies (based on the resin diglycidyl ether of bisphenol A) with the liquid state spectra of their respective unreacted components. The epoxies are identified by their main curing agent. (*a*) PIP, piperidine, (*b*) MPDA, metaphenylene diamine. (*c*) HHPA, hexahydrophthalic anhydride. (*d*) NMA, nadic methyl anhydride. (From Ref. 132.)

observed by the resonances near peak e, which are assigned to the carboxyl–methine ether carbon and to the methylene carbon found near the chemical reaction site.

The researchers were particularly interested in determining whether all the carbons were counted and what the limits of resolution were in these solid state NMR studies. It was resolved that for moderate magic angle spinning rates and for spin–lock cross-polarization in radiofrequency fields far removed from the $T_{1\rho}$ (rotating frame spin–lattice relaxation) minimum, the CP spectrum represents all the carbon-13 nuclei which come into contact with the proton spin reservoir. In the epoxy systems, spectral line broadening is caused by the distribution of isotropic chemical shifts and inadequate proton decoupling. For the highest possible resolution, it is required to use thermal activation to eliminate any conformational anisotropy, and to work near the point of the glass transition temperature of the sample. Solid state NMR spectral resolution studies have been carried out for a variety of polymeric systems in addition to epoxies (133, 134).

Garroway et al. studied the ^{13}C rotating frame relaxation rates on an epoxy system of DGEBA cured with piperidine (131). The objective was to measure the ^{13}C $T_{1\rho}$ values for the network system and to estimate the contribution of the proton spin–spin relaxation process to the ^{13}C $T_{1\rho}$ nuclear relaxation rates. When strong proton–proton coupling is absent in such a polymeric system, the ^{13}C $T_{1\rho}$ relaxation rates reflect molecular motional fluctuations in the 25–100 kHz frequency range in addition to reflecting local motions. The dependence on the magic angle spinning rate of the cross-polarization contact times was found for the resolved resonances of this epoxy system. The radiofrequency field for the experiment was 38 kHz, and the proton $T_{1\rho}$ of the system was found to be 2.6 ms. The nonprotonated carbons showed an averaging relaxation behavior as a result of magic angle spinning. On the other hand, protonated carbons relaxed in less time than was necessary for one revolution of the sample specimen. In other words, as a result of the large interaction between the carbon and the proton nuclei, the cross-polarization process was completed within this time period.

Figure 11 presents the rf dependence of the ^{13}C $T_{1\rho}$ values, normalized by the cross-polarization rates obtained at a 1 kHz spinning speed, from the same study (131). A weak rf field dependence was observed for the different carbon types, indicating that the $T_{1\rho}$ of this system was not dominated by spin–spin effects. It was concluded that at room temperature and at rf fields above 40 kHz, the ^{13}C $T_{1\rho}$ values are affected mainly by spin–lattice effects. The temperature dependence of the ^{13}C $T_{1\rho}$ values at a 55 kHz rf field strength was also investigated for this system. The relaxation values at these conditions should indicate molecular motions. The data for the temperature range between 242 and 324 K are shown in Figure 12. The relaxation times of the

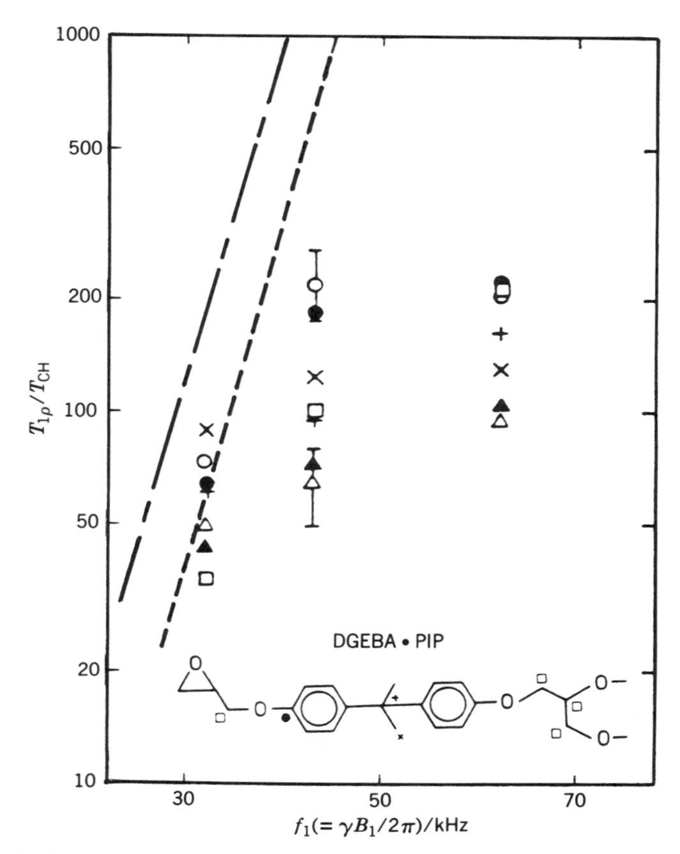

Figure 11. Carbon-13 rotating frame spin–lattice (^{13}C $T_{1\rho}$) relaxation measurements for a cured epoxy polymer system at three different radiofrequency field strengths. The epoxy system is DGEBA cured with piperidine. The relaxation times at 43 and 66 kHz are shorter than those predicted for purely spin–spin effects; thus the high field results indicate molecular motion. (From Ref. 131.)

methyl group showed the only prominent temperature dependence in this range. An unresolved peak containing methylene and methine resonances showed a weak temperature variation. Garroway and coworkers concluded that the determination of the ^{13}C $T_{1\rho}$ in solid samples is complicated by the presence of a strongly coupled proton system. For this particular epoxy system at rf fields greater than 40 kHz, the observed ^{13}C $T_{1\rho}$ values are not dominated by spin–spin interaction effects and therefore reflect molecular motion.

Further molecular motion studies have been completed on the DGEBA

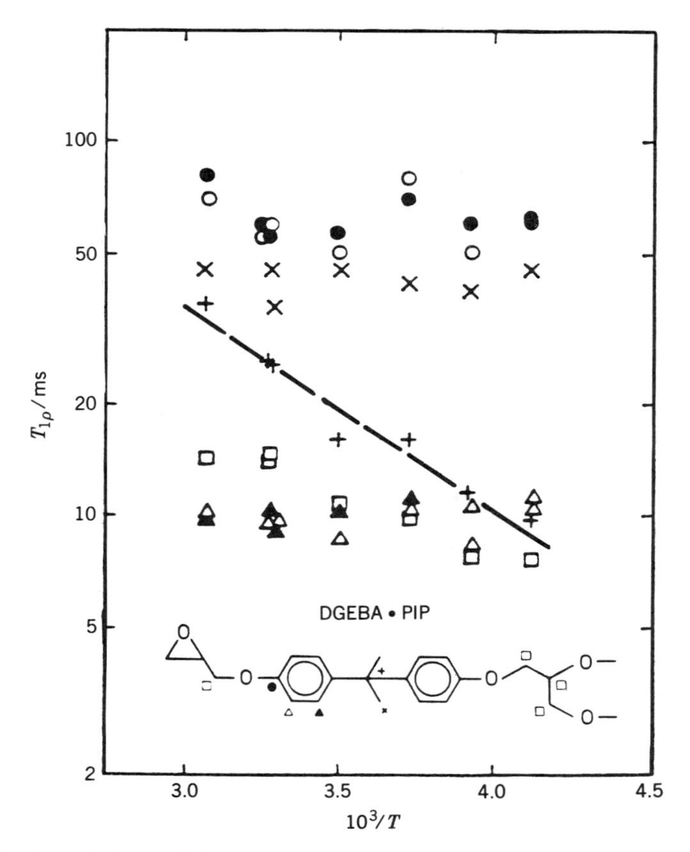

Figure 12. The variation of ^{13}C $T_{1\rho}$ values with temperature of the epoxy system in Figure 11. Only the methyl resonance shows a well-defined temperature dependence. (From Ref. 131.)

epoxy system using variable-temperature solid state ^{13}C NMR. Garroway, Ritchey, and Moniz (136) obtained the variable-temperature spectra of DGEBA cured with amines and anhydrides over the temperature range of 150–350 K. These spectra were compared to the NMR spectra of four different phases of the DGEBA resin (liquid, crystalline, amorphous, and polymerized). The ^{13}C NMR study of the crystalline DGEBA resin produced evidence, which was corroborated by X-ray studies, of more than one conformation available to the epoxide groups. The piperidine-cured epoxy system gave the best resolved NMR spectra and was therefore studied in greatest detail. In the low temperature spectra of this system, the resonances from the carbons found in positions which are ortho and meta to the oxygen atom are observed as partly resolved doublets. With an increase in the temperature, these resonance lines individually coalesce. The researchers noted that the observed coalescence of the resonance lines in this epoxy polymer system indicated a 180° rotation of the phenylene ring, which may proceed by small diffusive steps. Garroway et al. proceeded to extract the dynamics of the phenylene reorientation for two models of molecular motion. It was found that a broad distribution of correlation times is necessary to describe the molecular motion over the full temperature range. The NMR data then were compared to dynamic mechanical results for the piperidine-cured epoxy.

Recently Jelinski et al. used solid state deuterium NMR to investigate the nature of the water–epoxy interaction (144, 145). The epoxy samples were based on the diglycidyl ether of bisphenol A, cured with nadic methyl anhydride, followed by an exchange with D_2O (water). Some of the samples that had been previously exchanged with D_2O were dried to provide reference samples. The deuterium NMR spectroscopic results provided information on the plasticization of epoxy resins by water. In particular it was shown that (1) although the absorbed water is very mobile, it is not identical with bulk water in this respect, (2) the water is distributed homogeneously throughout the sample, (3) approximately one water molecule is entrained for every six OH/OD sites on the polymer backbone, and (4) the water exchanges with the OH/OD groups on the polymer backbone (144). The deuterium NMR study was extended to provide further insight into the molecular details of the interaction of epoxy resins with water (145). By means of T_1 relaxation data, it was established that (1) the water is impeded in its movement, with the molecules hopping from site to site with an approximate residence time per site of 7×10^{-10} s, (2) there is no free water, (3) there is no evidence for tightly bound water, and (4) it is unlikely that the water disrupts the hydrogen-bound network with the epoxy resin. These results are in agreement with the idea that water is a plasticizer for these polymeric systems that contain hydroxyl groups, just as it is for polymers which contain no exchangeable protons.

The curing reaction of epoxy resins has been studied by the solution ^{13}C NMR techniques. In one particular study, variable-temperature NMR spectroscopy was used to follow the disappearance of the monomer in a piperidine–DGEBPA system (146). In another ^{13}C NMR study, the spectra of epoxy resins were run under quantitative conditions in which the loss of epoxides at the end groups were detected by reactions such as hydrolysis (147). The products of acid hydrolysis of the DGEBPA resin were observed and assigned.

Various other thermosetting resins have been studied with CP-MAS NMR. Fyfe et al. have studied the curing of phenolic resins (148). The reaction of phenol and formaldehyde under alkaline conditions yields a solid resol-type material which contains methylene bridges, dibenzyl ether linkages, and free methylol groups. With the ^{13}C NMR data, the researchers were able to elucidate the substitution patterns on the phenol rings by observing chemical shifts. The relative peak intensities of substituted and unsubstituted aromatic carbons indicate a high overall degree of substitution in the system which is controlled by the cure conditions. It was observed that higher degrees of cure cause conversion of methylol residues to methylene bridges. Fyfe et al. also used ^{13}C CP-MAS NMR to investigate the mechanism of the thermal decomposition of cured phenolic resins (149). Degradation induces structural changes, with the first step involving the loss of methylol groupings. Oxidized functionalities including aldehydes, ketones, carboxylic acids, and anhydrides appear upon further degradation, with a simultaneous loss in methylene groupings. At higher temperatures, the NMR spectra indicate the loss of almost all linking groups and suggest the formation of a highly aromatic structure. The CP-MAS NMR spectra of the same samples degraded under vacuum show no evidence of oxidation. The data show a loss of methylol functionalities and the formation of methyl groups, but the basic chemical structure of the sample does not appear to change significantly upon heating.

Other researchers have also studied both the Novolak type and the resol type of phenol–formaldehyde resins (150–152). In one particular study of the Novolak phenolic resins, the ^{13}C CP-MAS NMR spectra obtained were assigned on the basis of comparisons with solution state ^{13}C NMR data (150). The spectral regions due to the unsubstituted aromatic carbons lost intensity upon curing, while the region due to methylene-substituted aromatics increased in intensity. These results confirm that cross-linking does accompany the curing process. In their investigation of resol-type phenol–formaldehyde resins, Maciel et al. found spectroscopic evidence for the direct involvement of the hydroxyl group of the phenol ring and the methylene bridges in the curing process (151). So and Rudin used ^{13}C CP-MAS measurements on cured resols to follow the degree of cure quantitatively (152). They obtained spectra under the optimal CP conditions for a quantitative data analysis for the particular cured system at hand. The degree of cross-linking of cured phenolics was estimated by measuring the ratios of methylene carbons to the aromatic carbons to which the hydroxyl group is attached. This ratio was measured as 0.65 with the CP-MAS NMR spectrum, while a value of 0.63 was obtained with the fully relaxed spectrum obtained with MAS and high power decoupling for the same system. These results indicate that CP-MAS NMR spectra can provide quantitative data, in a time which is much shorter than that required to obtain a fully relaxed ^{13}C NMR spectrum.

Maciel et al. have also used the ^{13}C CP-MAS technique for the structural characterization of cured furfuryl alcohol resins (153, 154) and of urea–formaldehyde resins (155). The urea–formaldehyde resins gave broad ^{13}C CP-MAS spectra due to the dipolar coupling of ^{13}C to ^{14}N. It was necessary to use higher static magnetic field strengths (i.e., increase from a 1.4 T magnet to a 4.7 T magnet) to obtain the resolution of individual carbon types. The researchers deconvoluted the complex regions of the spectra and made tentative peak assignments based on comparisons with solution NMR data. The CP-MAS spectra allowed for the characterization of the chemical structures present in the solid urea–formaldehyde resin. Methylol end groups and ether groups decreased with cure, while branching tertiary nitrogens increased with cure. Spectra for the urea–formaldehyde resins have also been obtained by ^{15}N NMR methods (156). CP and MAS techniques were used on samples prepared with ^{15}N enriched to 99% and on model compounds with ^{15}N in natural abundance to supplement the ^{13}C CP-MAS NMR data. Even though there was a problem with overlap of peak positions for some of the chemical structural types of interest for urea–formaldehyde resins, it was possible to distinguish between secondary and tertiary amide nitrogens according to their ^{15}N chemical shifts. The ^{15}N NMR data complemented the ^{13}C NMR data that had been obtained previously.

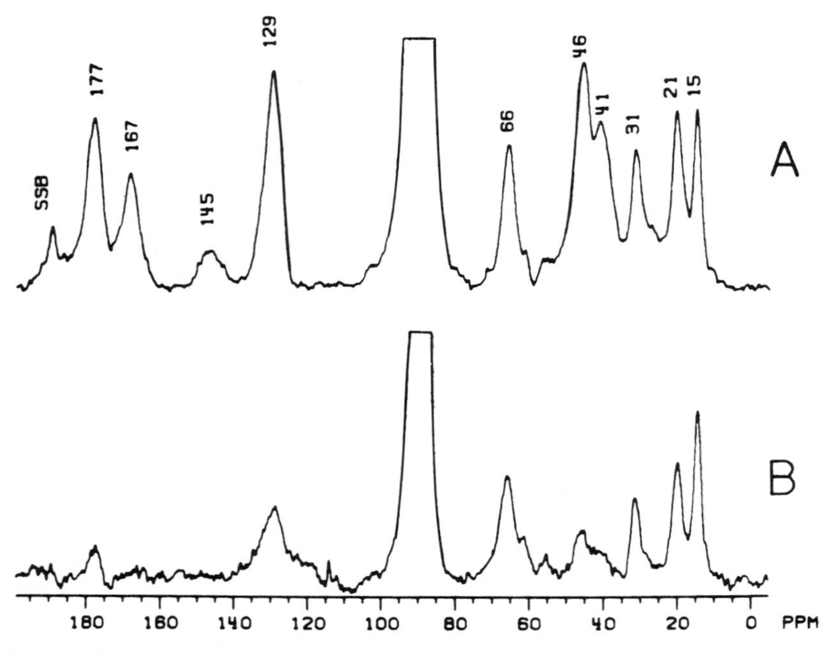

Figure 13. ^{13}C NMR spectra of acrylic copolymer melamine–formaldehyde coating obtained at 38 MHz with magic angle spinning. The broad resonance at 89 ppm is due to the polyxymethylene rotor used. (*A*) Spectrum obtained with the cross-polarization pulse sequence. (*B*) Spectrum obtained with the gated high power decoupling pulse sequence.

Bauer, Dickie, and Koenig have obtained the MAS ^{13}C NMR results for cured and degraded acrylic melamine formaldehyde coatings (157, 158). The CP and the gated high power decoupling NMR spectra of one particular coating formulation are shown in Figure 13. The structures found in the coating have been assigned, and it was determined that the intensities of the resonances correlate well with the known stoichiometry of the coating. The gated high power decoupling experiments were used to determine the relative mobilities of the different carbon resonances as a function of the acrylic copolymer composition and as a function of degradation. The researchers found that the mobilities of the main chain carbons of the acrylic copolymer decrease with increasing glass transition temperature of the acrylic copolymer, while the side chain carbons are insensitive to both the chemical composition and the extent of degradation. The CP spectra provided information on changes in coating composition after ultraviolet weathering. The hydrolysis of acrylic–melamine cross-links was observed, followed by the formation of melamine–melamine cross-links. Photooxidation in the carbonyl region of the ^{13}C NMR spectra was also observed.

Curing of Acetylene-Terminated Resins

Acetylene-terminated (AT) resins polymerize thermally without the evolution of volatile by-products and form a rigid and highly cross-linked network structure (159). The curing reaction is thought to proceed by the formation of substituted aromatic rings from the acetylenic end groups. Because of the complexity of the possible reactions that can occur with thermal cure and the intractable nature of the cured material, the mechanism of curing has not been verified. Solid state ^{13}C NMR was an appropriate method for the characterization of this system. Figure 14*B* presents the ^{13}C CP-MAS NMR spectrum of 4-bis[(3-ethynylphenoxy)phenyl]sulfone (160, 161). The spectrum of this AT monomer shows well-resolved resonances for each individual carbon type present in the structure. It is possible to obtain a spectrum of the same system in which the protonated carbons are suppressed, by the use of an interrupted decoupling pulse sequence (162). The ^{13}C NMR spectrum of the nonprotonated carbons of the AT monomer is shown in Figure 14*A*. This pulse sequence and its resulting spectrum is very useful for aiding in peak assignment, especially for the closely spaced aromatic resonances between 140 and 115 ppm.

Figure 15*A* shows the ^{13}C CP-MAS NMR spectrum of the acetylene-terminated sulfone resin cured for 30 minutes at 200°C; the spectrum of the monomer (Fig. 15*B*) is shown for comparison. The curing conditions cause spectral changes that are attributed to the structure of the cross-linked material. The resonance at 82 ppm decreases significantly in intensity, which

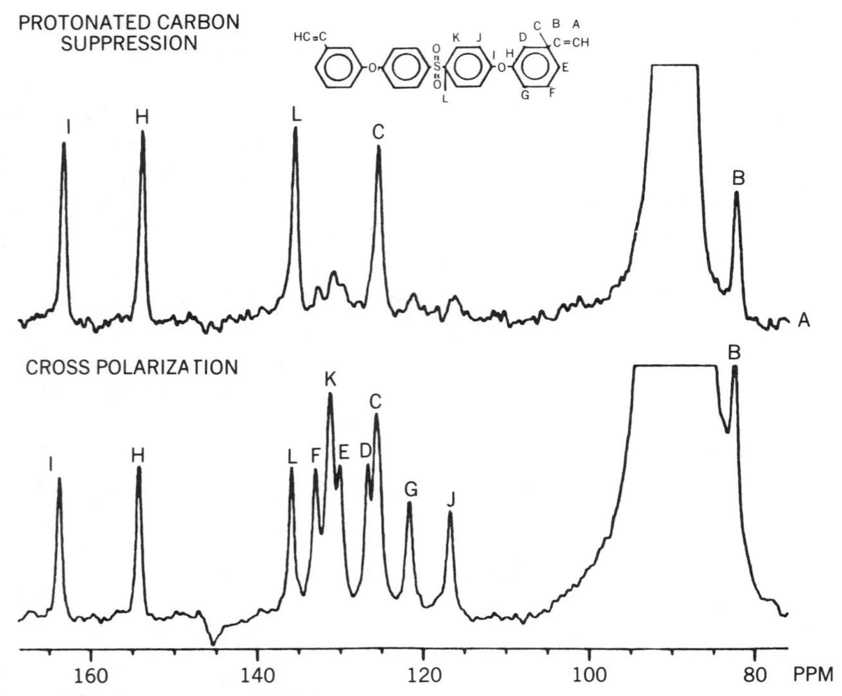

Figure 14. ^{13}C NMR spectra of 4-bis[(3-ethynylphenoxy)phenyl]sulfone obtained at 38 MHz with magic angle spinning. (*A*) Spectrum obtained with cross-polarization and interrupted decoupling for 30 μs prior to acquisition. (*B*) Normal cross-polarization spectrum. The truncated resonance at 89 ppm arises from the polyoxymethylene rotor. The assignments are indicated.

corresponds to the consumption of the acetylene carbons during the cure process. Considerable broadening of the resonances is observed because of the wider distribution of isotropic chemical shifts expected for cross-linked materials. Also, contributions to the broadening are expected from variations in molecular packing and conformation, which are in effect frozen into this solid polymeric material. It is the rapid molecular tumbling in solutions that results in isotropic chemical shifts for different carbon types. A variety of products can be expected with thermal cure such as substituted vinylacetylene, benzenes, and linear polyene chains (163). The chemical shifts of these expected reaction products give rise to resonances near the existing resonances of the aromatic carbons in the resin structure. Because the cured material gives broad resonances in the aromatic carbon region, it is not simple to detect small structural changes. But the changes in the 150–110 ppm region of the spectrum

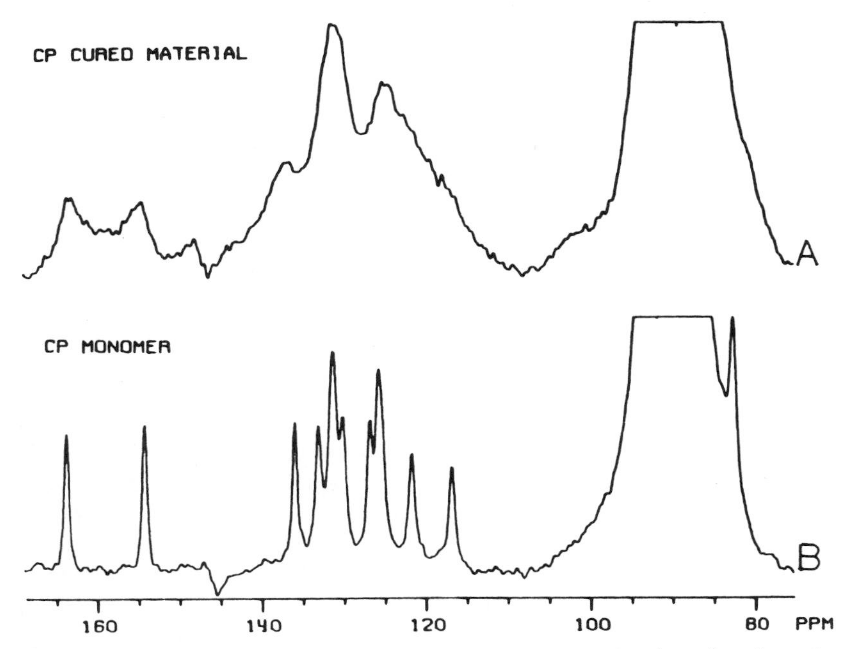

Figure 15. ^{13}C CP-MAS NMR spectra obtained at 38 MHz, illustrating the curing of acetylene-terminated sulfones. (*A*) The prepolymer cured for 30 minutes at 200°C. (*B*) The prepolymer 4-bis[(3-ethynylphenoxy)phenyl]sulfone.

correspond well to those calculated for trisubstituted benzene structures or linear polyene chains.

Sefcik et al. used CP-MAS ^{13}C NMR to study the curing and structure of acetylene-terminated polyimide resins (164). Upon curing of the prepolymer, the acetylene carbon resonance peak which appears at 84 ppm in the ^{13}C NMR spectrum decreases in intensity; with postcuring, the resonance completely disappears. The results indicate that cyclotrimerization of the end groups to form trisubstituted benzenes may be part of the cross-linking process. But it was established that not more than 30% of the acetylenes undergo cyclotrimerization or other condensation reactions. Instead, a variety of addition reactions probably are involved in the consumption of the remainder of the acetylenic end groups.

Radiation-Induced Cross-Linking

Various polymer chains can be cross-linked by means of radiation treatment. With the future objective of studying radiation-induced cross-linking in polyethylene, some preliminary work for the direct detection of cross-links with ^{13}C NMR has been carried out. Bennett et al. have synthesized 1, 1, 2, 2-tetra(tridecyl)ethane, which is an H-shaped model compound representing cross-linked polyethylene (165). During a ^{13}C NMR investigation, the researchers detected, by means of a resonance peak appearing at 39.5 ppm, a tertiary carbon atom associated with the cross-link point (166). Other hydrocarbons, *n*-hexadecane and *n*-eicosane, were also examined with ^{13}C NMR before and after irradiation. A new resonance was found at 39.5 ppm for these two systems also and was associated with cross-linking.

In their CP-MAS ^{13}C NMR studies of solid polyethylenes, Earl and VanderHart studied a linear polyethylene which had been radiation cross-linked with an ionizing radiation dosage of 624 Mrad (167). The spectral resolution allowed for a clear identification of methyl groups and methylene carbons, on the high field end of the NMR spectrum, characteristic of chain ends in the sample. In spite of the insufficient signal-to-noise ratio in the ^{13}C NMR spectrum, a small indication of intensity on the low field side of the central methylene resonance was observed. The researchers expect to find the tertiary and quaternary carbons of the branch points due to cross-linking in that particular region of the NMR spectrum. Cholli and Koenig and their coworkers studied a similar irradiated system on an NMR instrument having a higher magnetic field (161). The intent was to characterize the branching in greater detail with the enhanced sensitivity of the higher field instrument.

Figure 16 shows the CP-MAS ^{13}C NMR spectrum of peroxide-cured,

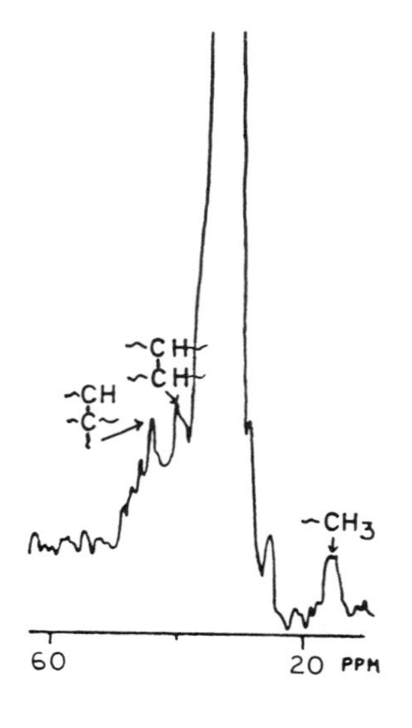

Figure 16. ^{13}C NMR spectra of gamma-irradiated high density polyethylene (600 Mrad) obtained at 38 MHz with cross-polarization and magic angle spinning. The resonance assignments are indicated. The truncated resonance is due to the methylene units of the main chain of polyethylene.

gamma-irradiated (600 Mrad), high density polyethylene. The broad resonance in the center of the spectrum is due to the methylene carbons of linear polyethylene itself. The resonance appearing at 39.3 ppm is assigned to the carbons involved in direct cross-linking between two chains, while the resonance peak observed at 43.0 ppm is assigned to the quaternary carbon of a cross-link which also has a long chain attached to it. The resonance assignments were aided by chemical shift calculations using the additivity relationships of Grant and Paul (168).

Sohma et al. used CP-MAS ^{13}C NMR data to study the chemistry of cross-links induced by gamma irradiation of ethylene–propylene rubber (169). Ethylene–propylene was irradiated to 640 kGy, which provided for a new resonance at 24 ppm and a shoulder on an existing peak appearing at 37 ppm, in addition to changes in the intensities of the methyl and polyethylene–methylene peaks. The relative concentration of each species was semiquantitatively determined. The peak at 24 ppm was attributed to methyl groups at the end of short branches and contributed 5% to the total area of the NMR spectrum. The shoulder at 37 ppm was assigned to tertiary and quaternary carbon atoms at cross-link sites occupying 3% of the area. The researchers also observed a 3% area loss in the methyl group resonance, which arose because of cross-linking at this site. Such NMR data indicated a 1% effective cross-linking density, compared with 0.5% predicted from a Charlesby–Pinner analysis. O'Donnell and Whitaker also used ^{13}C NMR to study the radiation degradation in ethylene–propylene copolymer samples subjected to a 10-MGy radiation dose (170). They found a new peak at 14.3 ppm, which they attributed to methyl groups arising from chain scission. An increase in intensity for the spectral region between 39 and 47 ppm, observed when compared with a simulated NMR spectrum, was assigned to the presence of H cross-links. A quantitative analysis of the NMR results allowed for the determination of G(scission) and G(cross-link) values which were in close agreement with the same types of values obtained by extraction procedures. The data tend to indicate that the cross-links are not clustered in the sample under study. If the cross-links were clustered, the cross-link density as determined by extraction would be reduced.

Diacetylenes can polymerize in the solid state when exposed to high energy radiation. The resultant polymers have interesting electrical properties, especially when doped with suitable inorganic compounds. Havens et al. have studied the cross-polymerization of the macromonomer poly(1,11-dodecadiyne) (171). This cross-polymerization reaction is schematized in Figure 17. The intent was to use CP-MAS ^{13}C NMR to study the details of the reaction and in particular to discern whether the polymerization leads to acetylenic or butatriene structures. The ^{13}C NMR spectra of solid poly(1,11-dodecadiyne) were examined before and after the gamma-irradiation

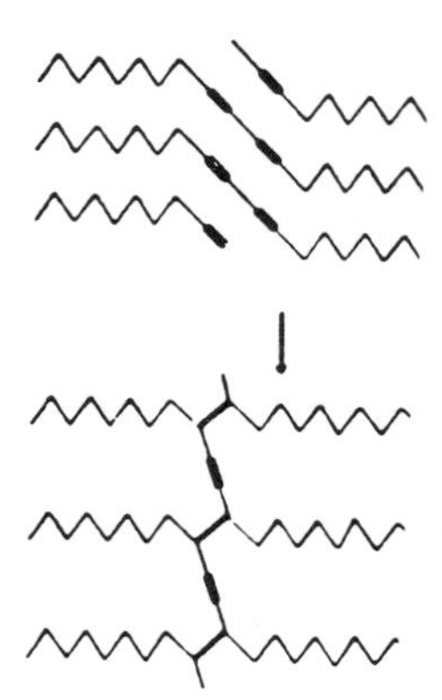

Figure 17. A schematic representation of the cross-polymerization of poly(1, 11-dodecadiyne) initiated by gamma-irradiation.

treatment. The spectroscopic evidence showed that the cross-polymerization reaction results in a conjugated backbone characterized by the acetylenic structure instead of by the butatriene possibility. Using solid state NMR techniques, the researchers were also able to resolve amorphous and crystalline contributions in the NMR spectra.

FOURIER TRANSFORM INFRARED SPECTROSCOPY

Infrared spectroscopy is a familiar and very powerful technique for the characterization of polymeric systems. The basis of the method depends on the absorption of radiation in the infrared frequency range, which occurs as a result of the molecular vibrations of the various functional groups found in the polymer chain. Fourier transform infrared (FTIR) spectroscopy, with its speed and improved signal-to-noise ratio, has allowed for further applications of IR spectroscopy to polymer research, including the characterization of cross-linked systems. Before the advent of FTIR, a dispersive instrument which depends on gratings and prisms to geometrically disperse the infrared radiation was necessary for IR spectroscopy. The dispersed infrared radiation was passed over a slit system, by means of a scanning device, and thus the frequency range falling on the detector was isolated. The data indicated the amount of energy transmitted through a sample as a function of frequency, and as a result an infrared spectrum could be obtained. However the sensitivity of the technique is relatively low, because a large percentage of the available energy from the source of radiation does not fall on the open slits and is lost to the technique. FTIR uses a more efficient optical system, which allows for increased sensitivity; as a result, more spectral applications have been

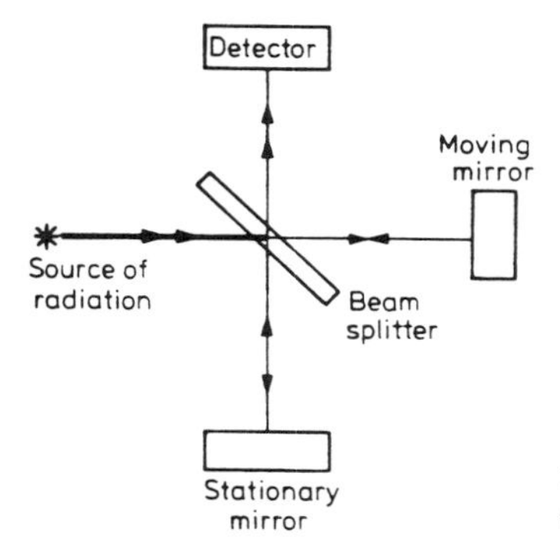

Figure 18. Optical diagram of the Michelson interferometer.

developed. The FTIR method is described next, followed by a review of this spectroscopic technique for network studies.

Description of the FTIR Method

The FTIR techniques make use of the Michelson interferometer (Fig. 18) to examine the transmitted energy through a sample at all times. Instrumentation theory and practice have been discussed to a great extent in the literature (172–175); thus only a brief description is given here. The Michelson interferometer consists of two perpendicular plane mirrors. One mirror is stationary, while the other moves at constant velocity. Between the two mirrors is a beam splitter, where the incident beam is divided and later, after a path difference has been introduced between the two beams, recombined. The two recombined light beams will interfere constructively or destructively depending on their path differences and the wavelengths of the light. When the optical path of the two arms is identical, all the frequencies of the radiation constructively interfere and add coherently, which allows the detector to view a maximum signal. As the movable mirror is shifted from this particular point, a difference in the path lengths of the two arms of the interferometer will result. Initially all the wavelengths of the light will exhibit some destructive interference, causing a drop in intensity. As the path length difference increases, some of the shorter wavelengths will start to constructively interfere and the longer wavelengths will destructively interfere. The resulting flux, which is the sum of energies from all the wavelengths comprising the incoming light, is sampled at the detector, which detects an interferogram. For a monochromatic source of frequency v, the interferogram is given by

$$I(x) = 2RTI(v)(1 + \cos 2\pi v x)$$

where R is the reflectance of the beam splitter, T is the transmittance of the beam splitter, $I(v)$ is the input energy at frequency v, and x is the path difference. The interferogram consists of two parts, a constant (dc) component equal to $2RTI$, and a modulated (ac) component. The ac component is called the interferogram and is given by

$$I(x) = 2RTI(v)\cos(2\pi v x)$$

An infrared detector and the ac amplifier convert this flux into an electrical signal:

$$V(x) = \mathrm{re}I(x) \qquad \text{volts}$$

where re is the response of the detector and the amplifier.

The interferogram is obtained in the time domain and is expressed as a function of the rate of the mirror drive; it must be transformed into a spectrum in the frequency domain. The interferogram for a polychromatic source $A(v)$ is

$$I(x) = \int_0^\infty A(v)(1 + \cos 2\pi v x)\, dv$$

The methods of evaluating these integrals involve a determination of the values at zero path length and very long or infinite path length.

At zero difference

$$I(\overset{0}{x}) = 2\int_0^\infty A(v)\, dv$$

For large path differences

$$I(\overset{\infty}{x}) = \int_0^\infty A(v)\, dv = \frac{I(0)}{2}$$

The actual interferogram $F(x)$ is

$$F(x) = I(x) - I(\infty) = \int_0^\infty A(v)\cos(2\pi v x)\, dv$$

The interferogram is transformed into the spectrum by performing a Fourier transformation of the data. From Fourier transform theory (176–178)

$$A(v) = 2\int_0^\infty F(x)\cos(2\pi v x)\, dx$$

The computational difficulty of this transformation prevented the application of this interferometric technique to spectroscopy. An important advance was made with the discovery by Cooley and Tukey of the fast Fourier transform algorithm (179). The field of spectroscopy, using interferometers, was revived because the calculation of the Fourier transform could be carried out rapidly. With the improvements in computers that permit real-time spectra to be obtained with the transformation time of the data on the order of fractions of seconds, FTIR has become a routine technique.

The advantages for the use of the interferometer over the grating devices arise from several sources. First, all the frequencies are observed all the time, which means that all the information from all frequencies is gathered simultaneously. This is known as the multiplex advantage [Fellgett's advantage (180)], which shows that measurements taken at equal resolution with the same optical throughput and with the same detector can result in a complete interferogram in the time that it takes to measure one resolution unit by a dispersive spectrometer. For a frequency range of 4000–400 cm^{-1} with 1-cm^{-1} resolution, the FTIR measurement time is 3600 times faster than with a dispersive instrument, which means a signal-to-noise improvement of 60.

The Jacquinot or throughput advantage (181) arises because the interferometer has no gratings or slits to limit the optical throughput. The advantage is a result of the more effective use of the power of the radiation source in the interferometer instead of losing the energy in the slits. The Conne or frequency advantage (182) provides for increased accuracy relative to a dispersive spectrometer. The frequencies of an FTIR instrument are internally calibrated by a laser, which monitors the position of the moving mirror in the interferometer. Wavenumber precision and accuracy are necessary for the coaddition of spectra for signal averaging.

Data Processing Techniques Using Digitized Infrared Spectra

Spectral processing methods can significantly contribute to the structural information obtained from the spectra of polymeric systems. The digital subtraction of absorbance spectra is widely used to isolate or enhance small differences or changes in the analysis between two samples or a sample and a reference material (183). Koenig et al. have applied the absorbance subtraction method to a variety of polymeric systems and have extracted a great deal of structural information (184–188). Other data processing techniques include the ratio method, factor analysis, and least-squares curve fitting for quantitative analysis. These techniques are covered by Koenig in a review which details the theory and application of Fourier transform infrared spectroscopy to the characterization of polymers (189). Factor analysis is based on expressing a property as linear sum of terms called factors. For the case of FTIR, factor analysis allows one to extract information from mixture spectra. In particular, it is possible to determine the number of components and the relative concentration of each component for a series of mixture spectra and then to reproduce the absorbances. Least-squares curve fitting analyses may be used in the determination of multicomponent mixtures with overlapping spectral features. These data processing techniques have aided significantly in the upgrading type and quality of spectral information obtained in the infrared studies of cross-linked systems.

Sampling Techniques for FTIR

A range of sampling techniques have been developed for infrared spectroscopy and may be used both for FTIR instrumentation and for dispersive spectrometers. Diagrams for the various experimental configurations appear in Figure 19. The most commonly used spectroscopic sampling techniques for the studies of cross-linked systems are transmission, internal reflectance (IRS), diffuse reflectance, and photoacoustic. The use of one spectroscopic method rather than another depends on the sample geometry involved. For example, powdered samples are recommended for diffuse reflectance spectroscopy, while in IRS the spectrum is obtained with the sample in optical contact with another material.

Transmission spectroscopy remains the most commonly used infrared measurement for samples that can be prepared in a transparent form. Polymer samples can be prepared as KBr pellets, mulls, or self-supporting thin films, or pressed into transparent wafers (190). For suitable samples, the transmission technique produces spectra with high signal-to-noise ratios and given the nature of the method, the spectra are quantitative in the infrared region. Internal reflection spectroscopy also known as attenuated total reflection (ATR), which is schematically represented in Figure 19*e*, requires contact of the material to be characterized with the internal reflection element. Radiation passes through the element and forms a total internal reflection at the surface of the sample. The optics of this phenomenon are quite involved and have been discussed in detail elsewhere (191–194). Qualitatively an IRS spectrum resembles a transmission spectrum. For powdered samples, diffuse reflectance spectroscopy appears to be particularly advantageous, since no sample preparation (which could change the morphology of the sample) is required. With this technique, a study of the diffuse reflected light can be used to measure the amount of light which is absorbed in the sample. This is possible because some of the light directed on the sample is absorbed and the remainder is reflected. Diffuse reflectance spectroscopy requires specially designed cells, which allow the measurement of diffuse reflectance spectra using FTIR instrumentation. The experimental design of this technique (195), in addition

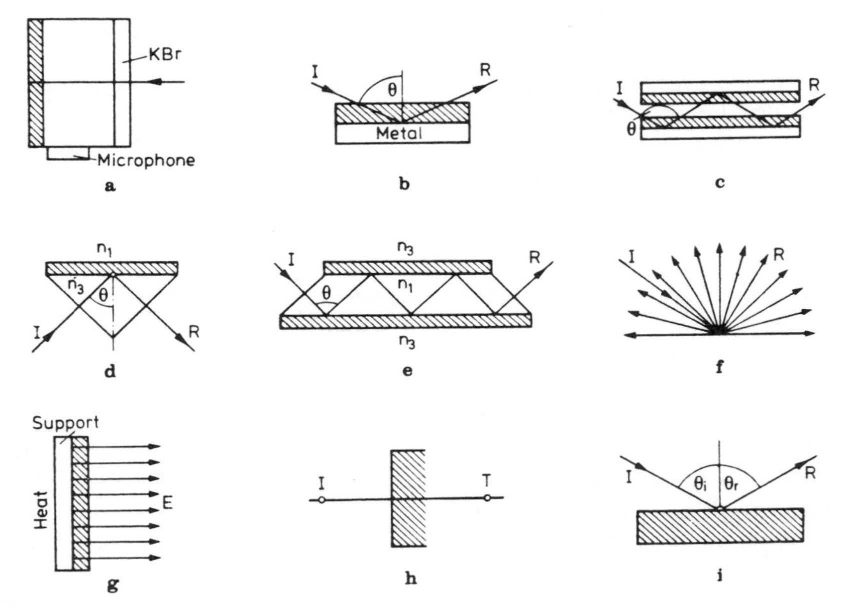

Figure 19. (*a*) Photoacoustic cell: the incident light produces pressure fluctuations, which are detected by a sensitive microphone. (*b*) Single-reflection [reflection-absorption (RA)] setup: light penetrates the sample first and is reflected by the metal mirrors (θ should be 70–89.5°). (*c*) Multiple-reflection RA setup: light penetrates the sample first and is reflected by the metal mirrors (θ should be 70–89.5°). (*d*) Single-reflection IRS setup: light passes through the internal reflection element first and is totally reflected at $\theta > \theta_c$; $n_1 \sin \theta_c = n_2 \sin 90$; $\sin \theta_c = n_2/n_1$. (*e*) Multiple-reflection IRS setup. (*f*) Diffuse reflectance: the scattered light is collected by mirrors and directed to the detector. (*g*) Emission technique: the sample is heated and the emitted radiation is analyzed. (*h*) Transmission spectroscopy. (*i*) Spectral reflection (mirrorlike): angle of incidence equals angle of reflection.

to the theory and principles (196–199), can be found in the literature.

Photoacoustic spectroscopy (PAS) is a new FTIR technique for solid samples (200, 201). A sample is placed in a cell under a KBr window as in Figure 19*a* and the cell is filled with a coupling gas, such as helium, argon, or air. The sample is exposed to modulated light and is heated to the extent that it absorbs the incident light. The energy is lost to heat through nonradiative processes. An acoustic signal is generated by the sample and detected by a sensitive microphone. The absorption that is detected is due to the infrared beam that penetrates the sample as opposed to the component of the beam that is reflected. The photoacoustic technique requires no sample preparation before a spectrum can be obtained. However low sensitivity may be a problem, making long scan times necessary to obtain good spectra. FTIR PAS has yielded high quality IR spectra of various polymeric systems (200, 201) and appears to be a useful method to study cross-linked systems.

Next we cover five areas of network characterization using FTIR: curing kinetics and mechanisms, composition studies, degradation processes, chemical reactions of epoxies with coupling agents, and vulcanized rubber systems.

Curing Kinetics and Mechanisms

For every new and useful polymerization system that is discovered, there is an attempt to obtain a complete characterization of the system, including a knowledge of the polymerization kinetics and reaction mechanism. Elucidating the reaction mechanism and the order of the reaction for cross-linking systems tends to be a difficult process. Infrared spectroscopy has been a common approach used for the characterization of cured epoxy systems (202, 203). Studies with IR spectroscopy are based on the absorption intensity of the various functional groups found in the particular epoxide–anhydride–amine system at hand. Typically it is possible to monitor the epoxy, amine, anhydride, or hydroxyl functional group in the infrared spectrum of the reaction system. Using a thermally controlled cell installed in an IR instrument, kinetic studies using FTIR can be carried out (204). Antoon and Koenig analyzed solutions of epoxy resin, nadic methyl anhydride isomers, and dimethylbenzylamine as thin films between NaCl disks. Reaction kinetics of the system were studied by obtaining spectra at the reaction temperature (80°C) of the stoichiometric mixture. The kinetics were followed by means of the 1858-cm^{-1} absorption band, which is assigned to the carbonyl functional group of the anhydride molecule. The results are shown in Figure 20, where the data are plotted as a zeroth-order process. Up to the gelation point found at 55%, the data are linear, slowing, however, above 55% and eventually stopping at 71% conversion. A great reduction in polymerization rate is expected when the kinetic processes are slowed by increased viscosity near the point of gelation. Since all the frequencies of a system are being recorded for every data point when FTIR techniques are used, it is possible to generate plots of this type for several of the functional groups in a curing system.

The rapid scanning capabilities of FTIR allow for the study of reactions with short half-lives. Buckley and Roylance (205) demonstrated this technique by studying the kinetics of a sterically hindered amine-cured epoxy resin system. FTIR spectra were recorded at one-minute intervals during cure at several temperatures for fresh mixtures of the reactants. Variation in an epoxide absorbance (915 cm^{-1}) caused by differences in specimen thickness can be eliminated by normalizing the epoxide peak height to an internal reference peak (appearing at 1510 cm^{-1} due to the phenyl groups of the

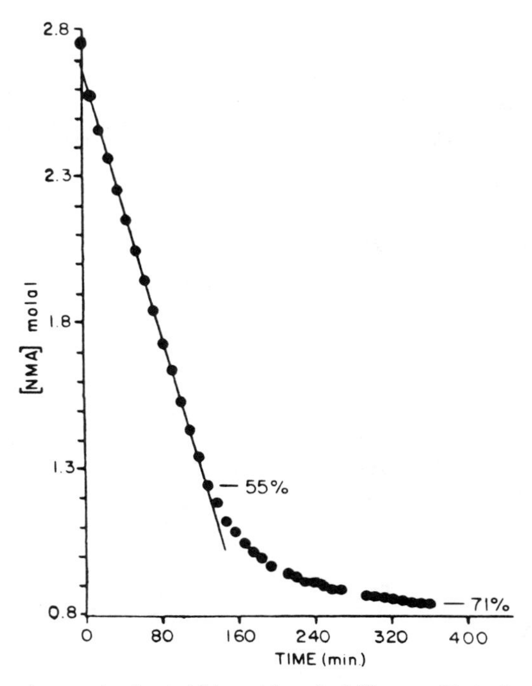

Figure 20. Zeroth-order graph of a stoichiometric anhydride–epoxide tertiary amine-catalyzed copolymerization at 80°C.

system):

$$A_{915}(t) = \frac{A_{915,s}}{A_{1510,s}} \frac{A_{1510,ref}}{A_{915,ref}}$$

where $A_{915}(t)$ = fraction unreacted epoxide at time t

$A_{915,s}$ = specimen absorbance at $915\,cm^{-1}$ at time t

$A_{1510,s}$ = specimen absorbance at $1510\,cm^{-1}$ at time t

$A_{915,ref}$ = initial absorbance at $915\,cm^{-1}$

$A_{1510,ref}$ = initial absorbance at $1510\,cm^{-1}$

The data enabled the researchers to calculate the activation energies for the curing process. The values were found to be in close agreement with values obtained by others and by other experimental methods, such as torsional braid analysis.

Other studies have been reported on the epoxy curing process studied by FTIR. In one particular epoxy resin system, the interest was to determine the cure rate of tetraglycidyl-4,4′-diaminodiphenylmethane (TGDDM) and diaminodiphenylsulfone (DDS) and/or boron trifluoride monoethylamine (BF_3-MEA) (206). In an earlier study it had been shown that BF_3-MEA is quickly hydrolyzed to fluoroboric acid (HBF_4) at the elevated temperatures used to cure epoxy resins (207). It was determined that in this particular curing reaction system, HBF_4 was acting as the catalyst, not the BF_3-MEA. FTIR was used to follow the rates of cure of TGDDM and TGDDM plus DDS, which was found to be slower that the rate of hydrolysis of BF_3-MEA. The addition of HBF_4 to the reaction mixture noticeably accelerated the epoxy resin cure.

It is useful to compare the results obtained in the study of an epoxy resin cure by FTIR with other techniques such as differential scanning calorimetry (DSC) and torsional braid analysis (TBA). Donnellan and Roylance used all three techniques in their curing study of a bisphenol A-type epoxy resin with diamino-*p*-methane (208). The FTIR and DSC data gave very close activation energy values. In general, the resin was found to react to a partially cured (52%) state at room temperature and then to vitrify. The combination of characterization techniques for comparing and contrasting data has been used by other researchers (71, 75). In addition to epoxy resin systems, polyimide materials, which are obtained by the reaction of 4,4′-bis(maleimidodiphenylmethane) with diamine, have been characterized by FTIR (209). In this study it was possible to characterize the cross-linking of the materials using the maleimide and amine absorption bands. The researchers observed that the amine group reaction with double bonds occurs readily and without regard to the temperature of curing, while the decrease of the maleimide double bonds depends on the reaction temperature. FTIR has been very useful for interpreting the cross-linking reactions of curing resins.

Composition Studies

The analysis of cross-linked resins may require determining the composition and the degree of curing. This type of information is generally necessary for materials qualification. Infrared methods have been developed for the problem of verifying the purity of the initial resin components and mixtures of those components. The purity of a reactant is commonly tested by comparing its spectrum to the spectrum of a known standard. Because of the nature of reactions and interactions, the composition analysis of blended and cured resins is considerably more complicated. In several cases, characteristic absorption bands have been isolated for the different components of a cured system, whereupon the composition was estimated by comparison to standard spectra of known composition (210, 211).

FTIR difference spectra have been used for investigating the composition of neat epoxy resin, hardener, and catalyst, as well as the composition and degree of cross-linking of the cured resins (212). Precision was achieved by using a least-squares curve fitting of the digitized spectra of the neat resin (213). The technique involves fitting, by a least-squares criterion, the spectra of one or more pure or standard components to the spectrum of a mixture of these components, followed by calculation of the fractional amount of each component. The method provides for the determination of spectral scaling coefficients that are necessary prior to digital subtraction. A least-squares analysis uses all the spectral data in the region of interest and accounts for spectral overlapping features. It is also possible to use weighting factors to maximize the use of the spectral data available and to minimize regions containing high noise (214); baselines also can be calculated (215).

Antoon, Starkey, and Koenig used FTIR difference spectra to detect and define slow chemical reactions in the initial reactants of an epoxy resin system as a method of quality control (212). The intent was to look for spectroscopic changes induced by storage as well as for the presence and effect of water and other impurities in the system. The system was comprised of EPON-828 epoxy resin and cured by nadic methyl anhydride (NMA) using *N,N*-dimethylbenzylamine (NNDB) curing accelerator. Digital subtraction showed an increase in the water content of EPON-828 during the storage period, as well as indications of polyetherification. An example of the least-squares curve-fitting method is shown in Figure 21. The partial infrared spectrum of a stoichiometric mixture of EPON-828 and NMA in the region from 3100 to 2800 cm^{-1} (solid line) is shown, and superimposed on it is the spectrum created by a least-squares curve fit of the pure EPON and NMA spectra to the actual mixture spectrum. The closeness of the fit is extremely good.

The degree of cure and the differences in chemical processes occurring during various stages of the cross-linking reaction have been examined (204, 212). For these studies the degree of cure was obtained by means of the

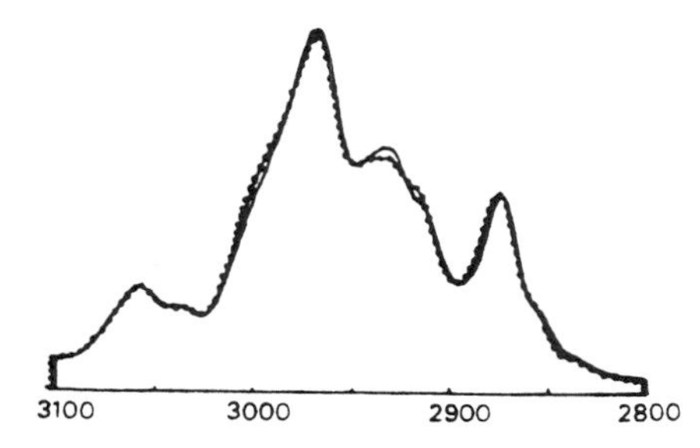

Figure 21. Least-squares, curve-fitting method for composition determination of uncured epoxy resin mixture:——, experimental FTIR spectrum of a 1:1 stoichiometric mixture of EPON-828 and NMA; ···, best, least-squares fit of the spectra of pure EPON-828 and pure NMA to the experimental mixture spectrum.

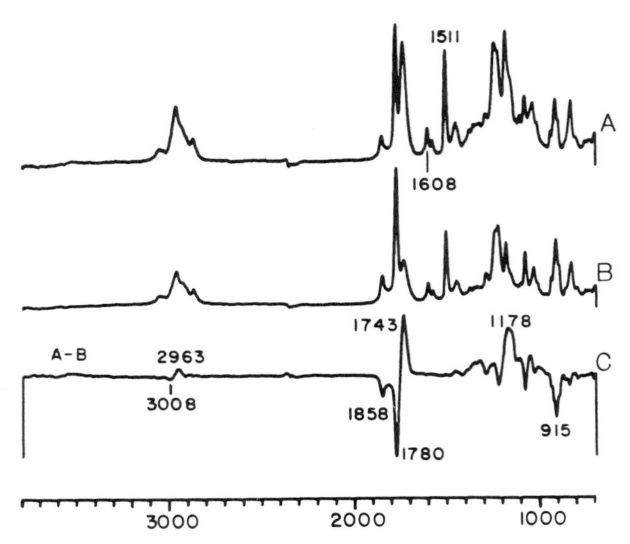

Figure 22. Tertiary-amine-catalyzed epoxide isomerization. (*A*) Absorbance spectrum of stoichiometric anhydride/epoxide mixture (2.0 wt % BDMA catalyst) polymerized 121 minutes at 80°C. (*B*) Absorbance spectrum of the system above after polymerization for 26 minutes at 80°C. (*C*) Difference spectrum (*A* − *B*) showing spectral changes occurring as a function of cure.

number of anhydride groups remaining, as measured by the intensities of two bands (I_{1780}/I_{1608}). The 1780 cm^{-1} band is assigned to the carbonyl of the anhydride ring and the 1608 cm^{-1} band is assigned to the invariant benzene ring modes. The use of the 918 cm^{-1} band to measure the epoxy content (216) and the hydroxy content at 3450 cm^{-1} (217) for epoxy systems is well documented. Difference spectroscopy allows for the study of the reactions taking place at various stages of the curing reaction. A difference spectrum as shown in Figure 22*C* is characteristic of the cross-linking mechanism. In this case the spectrum was generated by subtracting the absorbance spectrum of a 1 : 1 anhydride/epoxide reaction mixture (2.00 wt% BDMA) at 13% conversion (Fig. 22*B*) from the absorbance spectrum at 52% conversion (Fig. 22*A*). The criterion for subtraction is elimination of the 1508-cm^{-1} band presumed to be associated with the benzene ring and therefore independent of the degree of cure but proportional to the mass of the epoxy resin in the beam. The remaining bands in the difference spectrum are characteristic of the curing reaction. Absorption decreases due to the reacting species are observed at 1858, 1780, and 1082 cm^{-1}, indicating reaction of the anhydride groups and at 3008, 1034, and 915 cm^{-1}, indicative of reduction in epoxy groups. Intensity increases from formation of the ester group and nearby structures are found at 2963, 1778, and 1743 cm^{-1}.

Problems arise when the spectra of pure components are used to determine the composition of the cured resins with a least-squares curve-fitting procedure. The curing process introduces extensive spectral changes so that the spectra of the initial reactants do not represent their spectra in the cured state (212). A factor analysis procedure has been presented for determining the number of spectroscopically distinguishable components in a cross-linked epoxy matrix. Then the spectra of those components are fitted by a least-squares criterion to spectra of the multicomponent matrix (218). The least-squares coefficients yield the matrix composition in terms of the initial reactant composition and the extent of cross-linking.

Factor analysis, as described in detail elsewhere (219, 220), is based on the concept that the analysis of any multicomponent resin is simplified if the spectrum of that material is expressed by a linear combination of a finite set of "pure component" spectra. The entire process may be separated into three steps: calculation of the number of species present, identification of each of those species, and curve fitting of the spectra of these species to the spectra of the multicomponent resin (218). The epoxy–anhydride system was investigated by this procedure.

Factor analysis of the absorbance spectra of nine mixtures of EPON-828 and NMA (0.5 wt% BDMA in each mixture also) was performed in the 2000–1400 cm^{-1} spectral region. In each case the mixture was cross-linked arbitrarily between 0 and 65% by heating at 80°C. The plot of log eigenvalue which was generated by the analysis is shown in Figure 23. Theory shows that the eigenvalues can be grouped together with an error contribution and a

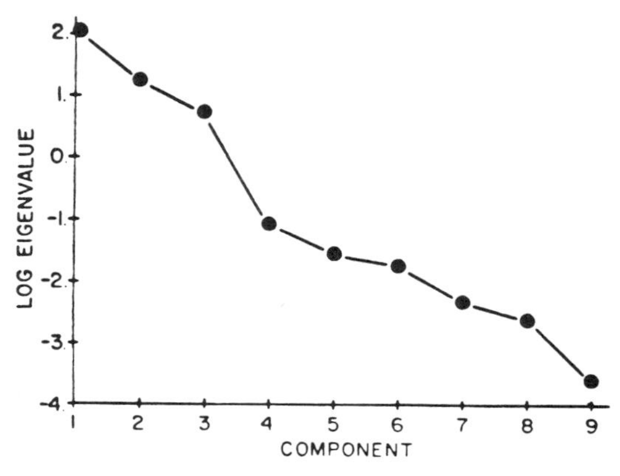

Figure 23. Factor analysis results of a cross-linked epoxy/anhydride resin FTIR study. Plot of log eigenvalue versus the number of components, showing the number of components being 3.

secondary set composed entirely of error. If the logs of the eigenvalues are plotted versus the number of the eigenvalues in descending order, a break will occur between the real and noise eigenvalues. Figure 23 reveals a break in the eigenvalue magnitude between the third and fourth eigenvalue. Therefore the spectrum of the cross-linked epoxy matrix may be approximated by a linear combination of only three linearly independent component spectra.

The three spectra chosen to represent the components were those of pure EPON-828 and pure NMA, and a difference spectrum characteristic of the cross-linking reaction. The difference spectrum was calculated by subtracting the spectrum of a stoichiometric mixture of EPON-828 and NMA (with 2.0 wt% BDMA catalyst) cross-linked for 37 minutes at 80°C from the spectrum of the same reactant mixture cross-linked for 83 minutes at 80°C. The procedure is illustrated for the 2000–600 cm^{-1} spectral region in Figure 24. The results for the stoichiometric mixtures of epoxy and anhydride cross-linked to various extents were calculated, and they compared very favorably with the expected results. For each case the weight percent compositions were derived from the least-squares coefficients after scaling the EPON-828 and NMA spectra to give the exact result for a standard sample of known composition as described elsewhere (212). The actual weight percent

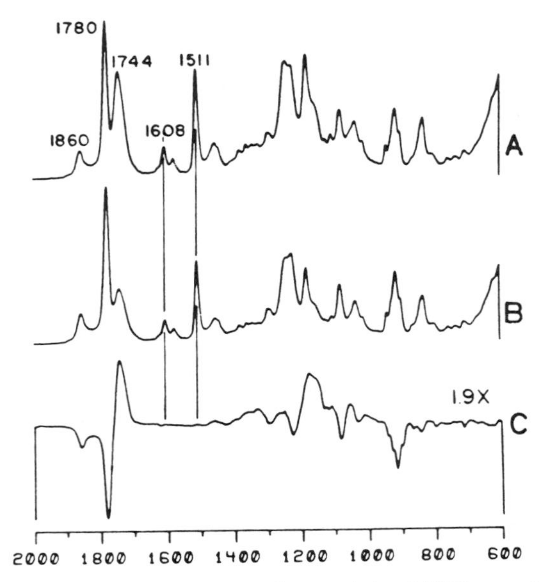

Figure 24. Generation of difference spectrum characteristic of 80°C cross-linking of stoichiometric mixture of NMA and EPON-828. (*A*) Cross-linked for 83 minutes. (*B*) Cross-linked for 37 minutes. (*C*) Difference spectrum (*A*–*B*) showing spectral changes occurring as a function of cure.

compositions were obtained by weighing the samples. The accuracy as indicated by calculated standard error was within 2%.

The thermal polymerization of the acetylene-terminated resin bis[4-(3-ethynylphenoxy)phenyl]sulfone was characterized by FTIR spectroscopy (212). Spectral processing techniques, including the method of factor analysis, were used in the analysis of the infrared data. In this study, two series of spectra were generated, the first following the cure at 150°C and the second at 180°C at various intervals over a 12 hour period. Each series of spectra was considered to be a series of "mixture" spectra which, with cure time, change in respect to the number and quantity of components present. In this case, the infrared spectrum at zero cure time is composed of only one component, AT monomer. As the polymerization proceeds, the monomer content decreases while one or more reaction products appear as additional components in the mixture. It was assumed that FTIR absorbance spectra would reflect these changes in the sample by corresponding spectral changes and that Beer's law was obeyed; therefore it was proposed that the factor analysis could be applied successfully to this system. It was expected that the results would distinguish a decreasing component (corresponding to the AT monomer) and the number of products that are formed upon the thermal cure. Factor analysis indicated the presence of two components in both series of mixture spectra. The calculated pure component spectra indicated that the monomer was one of the two components and that there was only one spectroscopically distinguishable product component in this particular reaction system.

Other data processing techniques were used in addition to factor analysis for this acetylene-terminated sulfone study (221). The degree of cure was quantitatively monitored by observing bands assigned to the terminal acetylene groups. Using least-squares curve fitting, it was possible to fit the 3350–3150 cm^{-1} region and to monitor the degree of cure by calculating the percent decrease in the amount of acetylene functional groups present in the sample. An intense asymmetrical band at approximately 3300 cm^{-1} due to the C—H stretching mode characterized the terminal acetylene groups. Figure 25 plots degree of cure versus cure time for both 150 and 180°C. The FTIR spectra of this system after postcure at 300°C showed a new band at 948 cm^{-1}. The new absorbance was broad and was overlapped both by a band at 942 cm^{-1} (attributed to the C—H deformation mode of a terminal acetylene moiety) and by a band at 965 cm^{-1} (associated with a trans-unsaturated moiety). This region of the FTIR spectrum has been shown to be characteristic of C—H deformation modes of trans-unsaturated ethylene structures, and the presence of bands in this region has been cited for differentiating trans isomers from their cis counterparts for several arylacetylenes. These observations indicated that of the possible structures that may result upon curing, a linear *trans*-polyene structure best represents the FTIR data. There was no spectroscopic

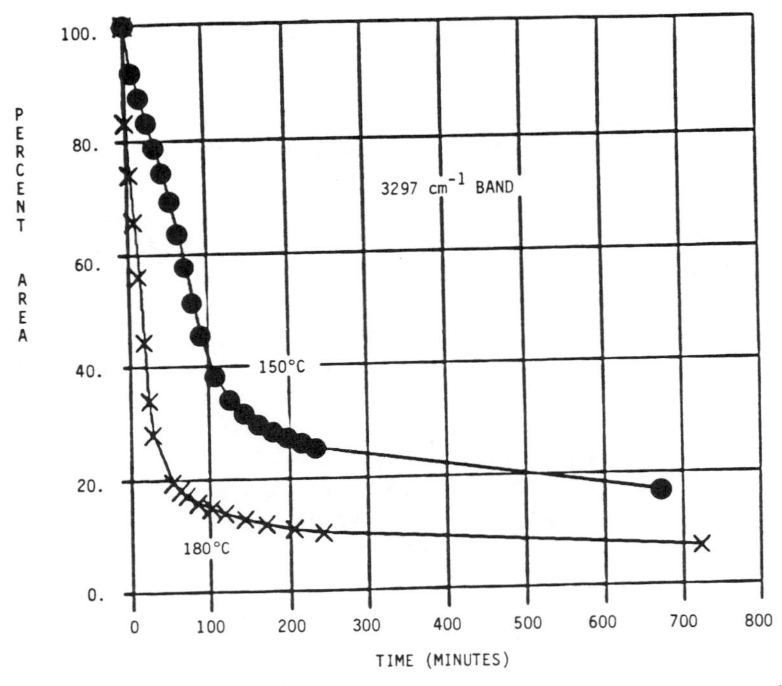

Figure 25. Decrease in acetylene content as measured from area of band at 3300 cm^{-1} for acetylene-terminated sulfone cure at 150 and 180°C.

evidence for inter- or intramolecular cyclization reactions, which previously had been proposed to be the polymerization mechanisms.

Degradation Processes

FTIR spectroscopy may be used to determine the effects of moisture, hydrolysis, and degradation on cross-linked polymers. To design reliable structural materials, it is critical to understand the reversible and irreversible effects of moisture and degradation processes on a molecular level. FTIR is especially advantageous for this purpose, since spectra can be obtained on the same or similar samples as a function of exposure time and very small spectral differences can be detected by absorbance subtraction.

Antoon and Koeing, who applied FTIR spectroscopy to characterize the interaction of a cross-linked epoxy matrix with sorbed water (222), studied the vibrations of sorbed water as well as the altered polymer vibrations. The vibrational bending mode of water in epoxy was observed at 1638 cm^{-1}

compared to the band in liquid water at 1648 cm^{-1}. Since the bending mode of sorbed water at 1628 cm^{-1} is intermediate in frequency between the bands of liquid water and free water, it was proposed that the sorbed water is held within the resin by hydrogen bonding.

The FTIR spectra also demonstrated that several vibrational modes of the epoxy polymer were influenced by the presence of sorbed moisture. The spectral effects indicated an interaction such as hydrogen bonding between the ester groups and the water molecules. Other spectral changes occurred which further implied that the water molecules interact with other polar species such as the phenyl ether linkage and with the phenyl rings themselves. For the most part, it was observed that the polymer–water interactions were reversible for extended periods.

The irreversible effects of moisture exposure on these anhydride-cross-linked epoxy resin films were also investigated (223). The exposure of this type of epoxy film to a 80°C liquid water environment causes hydrolysis of the unreacted anhydride groups to the diacid, and then the hydrolyzed material is leached from the resin. The long-term irreversible observations show that the alkaline hydrolysis of the matrix is accelerated by the application of high tensile stress. Matrix hydrolysis is also enhanced in the presence of organic fibers.

The surface of a glass-fiber-reinforced epoxy composite is degraded rapidly upon outer exposure unless it is protected by a UV absorber or paint. This photooxidation phenomenon is difficult to study because of the intractability of the cured composite. George, Sacher, and Sprouse investigated the photoprotection of the surface resin in a reinforced composite system by means of FTIR spectroscopy using a single-pass internal reflectance attachment for the surface study (224). The resin used in the composite was a mixture of Novolak epoxy and bisphenol A epoxy. The infrared spectrum showed a strong ester carbonyl band appearing at 1735 cm^{-1}, which was used as an index of the extent of photooxidation after UV exposure. It was observed that the oxidation rate for the Novolak is eight times that of the bisphenol A epoxy. The high photooxidation rate of the cured epoxy Novolak is related to the cure process itself. Changes in the IR spectrum during cure under vacuum or in an inert atmosphere show evidence for epoxide ring opening. Oxidation bands were observed in the spectra of both the commercial composite panels and laboratory-cured resin films, only when the cure was not carried out under vacuum. It was shown that the photooxidation rate can be largely inhibited by curing and postcuring under vacuum, while the incorporation of an antioxidant in the resin before cure should inhibit aromatic and aliphatic carbonyl formation.

The problem of thermal and oxidative thermal degradation of cross-linked materials is an important issue because of the significant loss of polymer properties when exposed to certain environments. Lin, Bulkin, and Pearce used FTIR to investigate three cured epoxy resins under various degradation conditions in an effort to further their understanding of the degradation mechanisms (225). New approaches to the stabilization of cross-linked systems may arise if the deterioration process is better understood. The epoxy resins that were studied included diglycidyl ethers of bisphenol A (DGEBA), phenolphthalein (DGEPP), and 9, 9-bis(4-hydroxyphenyl)fluorene (DGEBF). The researchers used the subtraction of the absorbance spectra to compare the functional group stability of the epoxy resin. In the thermal degradation study, the stability order of the functional groups in DGEBA was:

$$\text{total methyl group} = \text{total benzene group} > \text{methylene} > p\text{-phenylene} > \text{ether linkage} > \text{isopropylene}$$

The FTIR results showed that the oxidative thermal and photodegradation processes were related to the classical autocatalytic oxidation of aliphatic hydrocarbon segments. The Wieland rearrangement, a Noorish-type reaction, the Claisen rearrangement, and other possible degradation mechanisms were suggested by the data.

Chemical Reactions of Epoxies with Coupling Agents

Coupling agents are an important component of reinforced plastics and composites. Surface treatment of glass fibers with silane coupling agents to improve the mechanical performance of a composite is a common practice. When the treated glass fibers are mixed with resin and cured to form a composite, a region known as the silane interphase is formed (226, 227). It is believed that the silane interphase can be divided into two regions: the chemisorbed and the physisorbed. Physisorbed silane can be washed from the glass fiber surface by organic solvents; the chemisorbed cannot. FTIR has been used to study the nature of the interfacial reactions between the cured matrix and silane coupling agents in fiber-reinforced composites (228–231).

FTIR spectroscopy offers the capability of subtracting out the absorbances of the bulk resin and the glass fiber from the composite system being studied, allowing the spectra of interfaces to be obtained. Consequently, the difference spectrum allows the analysis of reactions occurring at the interface between the treated glass and the resin matrix material. It has been spectroscopically determined that nadic methyl anhydride can react with γ-methylaminopropyltriethoxysilane and N-methylaminopropyltrimethoxysilane (228). In comparing the relative reactivities of the two coupling agents to the epoxy resin, the secondary aminosilane has a higher reactivity than the primary aminosilane.

Chiang and Koeing have investigated the molecular structure at the resin–coupling agent interface of an anhydride-cured epoxy resin on fiberglass surfaces treated with an *N*-methylaminopropyltrimethoxysilane coupling agent using FTIR spectroscopy (229). The results showed that the copolymerization of the resin and the silane generates amide and tertiary amine groups and consumes the secondary amine groups of the silane in the composite system. Therefore the structure of the interface of the silane and the resin-fiber-reinforced composites is composed of copolymers of the epoxy resin with the organofunctionality of the deposited silanes. The researchers also determined that the number of interfacial bonds formed depends on the amount of silane coupling agent deposited on the fiberglass and on the specific reaction conditions of the system. The spectroscopic study indicated that the silane induces additional esterification in the resin and increases the curing density of the epoxy matrix near the fiber surface by roughly 5–10% relative to the bulk resin.

The chemical reaction between γ-aminopropyltriethoxysilane on E-glass fibers in a composite system with a difunctional epoxy resin has been characterized (231). The spectroscopic data verified the occurrence of the reaction between the amine and the epoxide groups, with the resulting formation of the alcohol and the secondary amine, followed by a further reaction of the alcohol groups to form ether linkages. It was observed that the reactivity of the silane coupling agent with the epoxy resin varies with the drying conditions of the silane. The reactions between the coupling agent and the epoxy resin were minimized when the silane was most condensed. A bicarbonate salt, which forms with the primary amines of the coupling agent, slightly decreases the reactivity of the coupling agent with the epoxy resin. The researchers concluded that the penetration of the resin into the silane interphase and the chemical reaction of the two components must be optimized if optimum mechanical properties of the composite materials are to be achieved.

Vulcanized Rubber Studies

There has been a long-standing interest in obtaining a complete characterization of sulfur-vulcanized natural rubber systems. The ongoing research motives are designed to obtain a further understanding of the relationship between the chemical structure and the physicomechanical properties of the resulting materials. It is known that the vulcanization process produces several different cross-link structures including mono-, di, and polysulfidic linkages and that the network structure varies with the time and temperature of vulcanization (232–234). The reaction mechanisms of unaccelerated and sulfur-accelerated vulcanizations are still not entirely understood. Additional insight into the reaction mechanisms involved in the process may lead to the ability to predict the resulting network structure under a specific set of vulcanization conditions. Some of the approaches that have been available for the structural elucidation of vulcanizate networks have included structural resolution by low molecular weight analogues of natural rubber, determination of cross-linking efficiency, treatment of the vulcanizate network with chemical agents, independent physical evaluation of cross-links using equilibrium swelling and stress–strain measurements, and instrument analysis of networks, including spectroscopic treatments. Recently sulfur-vulcanized natural rubber has been studied by Fourier transform infrared spectroscopy (235–237).

Chen et al. have studied a series of natural rubber materials compounded with a sulfur-accelerator curing system. To determine the structural basis for the reversion process to which some rubber systems are subject, the investigators used FTIR spectroscopy (235). Reversion occurs when continued curing causes the polymers to exhibit a loss of such physical and mechanical properties as tensile strength, stiffness, resilience, and wear resistance. It is speculated that reversion occurs when the desulfurization reaction is faster than the curing reaction (238), but the structural explanation for this

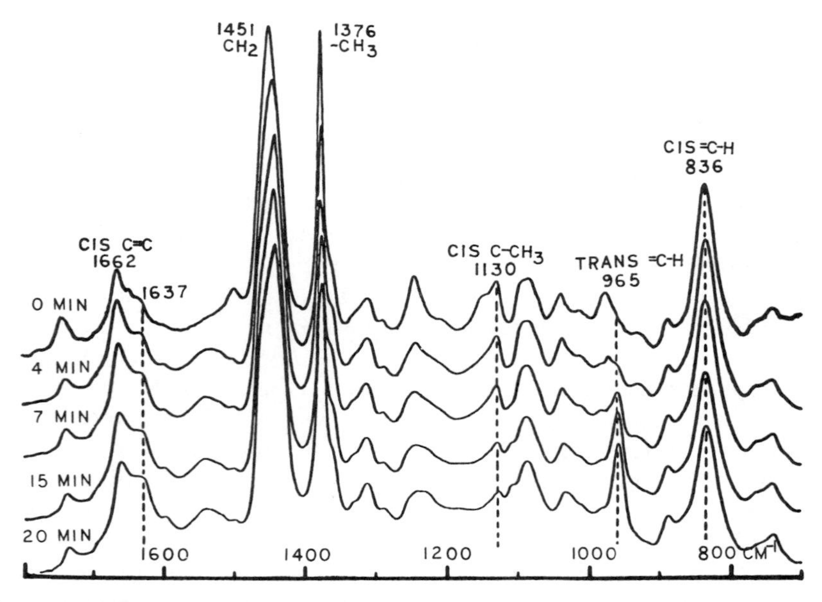

Figure 26. FTIR spectra of sulfur-accelerated natural rubber cured at 150°C. The bands are labeled.

phenomenon is not clear (232, 239). Isothermal curing studies of the natural rubber were carried out within the infrared spectrometer itself, using a heating cell. Solvent swelling and dynamic mechanical analysis were used in addition to the FTIR spectroscopy.

Figure 26 presents a series of infrared spectra of natural rubber cured with a sulfur accelerator at 150°C with increasing cure times. The principal band assignments are indicated on the spectra. Significant spectral changes are observed with curing time in the region between 800 and 1000 cm^{-1}. The reversion process was found to be associated with the formation of a *trans*-methine structure detected by the band appearing at 965 cm^{-1}. By using the three different techniques, a correlation was found between the microscopic chemical structure of the network and the macroscopic physicomechanical properties of the vulcanizate. The researchers consistently noted that when there was no reversion observable in the actual vulcanization curves, there was no detectable *trans*-methine structure with the infrared analysis. This *trans*-methine structure is generated by the main chain modification of the natural rubber chains through the desulfurization process.

In summary, it was determined that the final network structure is the result of the overlapping of two competing reactions—namely, cross-linking and desulfurization. The same researchers also investigated the influence of compounding variables on the reversion process (236) and the effects of carbon black on the reversion process (237) in the vulcanized natural rubber systems.

RAMAN SPECTROSCOPY

Since the advent of lasers, Raman spectroscopy has become an important technique in polymer research. Raman scattering occurs for those vibrational motions which produce a polarization or distortion of the electron charge of the chemical bonds (38). Thus the stretching motion of a homonuclear diatomic molecule is active in the Raman effect. The Raman effect provides more information about the nonpolar portions of the molecules, while the IR effect yields information about the polar motions of the molecule. Because of the complementary nature of the two types of spectroscopy, the techniques should be used on the same system whenever possible. Raman spectroscopy enhances the effectiveness of IR for solving chemical structure problems, and vice versa. Since the Raman effect is a light-scattering rather than a light-transmitting process, the transparency, size, and shape of the samples are relatively unimportant—one can run large samples or extremely small samples with comparable ease. Filled polymer composites present some difficulty for IR investigations, since fillers such as glass, clay, and silica are strong IR absorbers that block the IR spectrum of the polymer. Since these fillers are poor Raman scatterers, the Raman spectrum of the polymer is obtainable without removal of the filler.

Raman spectroscopy has been used to a small extent for the studies of cross-linked systems, having served to complement infrared spectroscopy in the characterization of rubber networks (240–242). In particular, the mechanism of accelerated sulfur vulcanization was probed. The Raman spectroscopic studies indicated a mixed free-radical and ionic mechanism for this type of rubber vulcanization. Lu and Koenig used Raman spectroscopy to study the curing of epoxy resins (243). Strong Raman lines due to the bisphenol A skeleton, which are characteristic of the epoxy resin and independent of the state of cure for the system, are found at 640, 823, 1114, 1188, 1232, and 1608 cm^{-1}. The epoxy group has spectral lines which appear at 768, 809, and 1156 cm^{-1} and are sensitive to the degree of cross-linking.

Myers et al. used laser Raman spectroscopy to analyze the structures of urea–formaldehyde resins (244). Initial band assignments were based on the analysis of model compounds allowing for the study of the spectra of the resins after various stages of cure. The trends that were noted in the spectra agreed with previous kinetic and mechanistic studies of the cure process. The principal limitation of Raman spectroscopy is the fluoresence which results from absorption of the laser beam and subsequent emission in the same spectral region as the Raman scattering.

SUMMARY

It has been demonstrated that there are significant characterization methods currently available for the analysis of polymeric network materials. The techniques that have been presented have been used successfully to study these network systems in spite of the inherent complexity of cross-linked polymers. These experimental methods are particularly valuable because of their ability to overcome the problems of insolubility and intractability, allowing the characterization of the materials in their solid state. It is preferable to study such materials in their solid and unaltered end-use state. Each technique provides important information for a more comprehensive understanding of any one cross-linked system.

REFERENCES

1. P. J. Flory, *Principles of Polymer Chemistry*, Cornell University Press, Ithaca, NY, 1953, Chap. 9.
2. P. J. Flory, *J. Am. Chem. Soc.* **63**, 3083, 3091, 3096 (1941).

3. W. H. Stockmayer, *J. Chem. Phys.* **11**, 45 (1943); **12**, 125 (1944).
4. W. H. Carothers, *Trans. Faraday Soc.* **32**, 39 (1936).
5. G. Odian, *Principles of Polymerization*, Wiley New York, 1981, Chap. 2.
6. H. E. Marsh, S. Y. Chung, G. C. Hsu, and C. J. Wallace, "Prediction of the Structure of Crosslinked Polymers," in *Chemistry and Properties of Crosslinked Polymers*, S. S. Labana, Ed., Academic Press, New York, 1977.
7. M. Gordon, *Proc. R. Soc. London, Ser. A*, **268**, 240 (1962).
8. M. Gordon and M. Judd, *Nature*, **234**, 96 (1971).
9. M. Gordon, G. N. Malcolm, and D. S. Butler, *Proc. R. Soc. London, Ser. A*, **295**, 29 (1966).
10. M. Gordon and G. R. Scantlebury, *J. Chem. Soc. B*, 1 (1967).
11. M. Gordon, T. C. Ward, and R. S. Whitney, in *Polymer Networks*, A. J. Chompf and S. Newman, Eds., Plenum, New York, 1971.
12. K. Dusek, M. Ilavsky, and S. Lunak, *J. Polym. Sci., Polym. Symp.*, **53**, 29 (1975).
13. S. Lunak and K. Dusek, *J. Polym. Sci., Polym. Symp.*, **53**, 45 (1975).
14. K. Dusek and M. Ilavsky, *J. Polym. Sci., Polym. Symp.*, **53**, 57 (1975).
15. K. Dusek and M. Ilavsky, *J. Polym. Sci., Polym. Symp.*, **53**, 75 (1975).
16. C. W. Macosko and D. R. Miller, *Macromolecules*, **9**, 199 (1976).
17. C. W. Macosko and D. R. Miller, *Macromolecules*, **9**, 206 (1976).
18. D. R. Miller, E. M. Valles, and C. W. Macosko, *Polym. Eng. Sci.*, **19**, 272 (1979).
19. D. R. Miller and C. W. Macosko, *Macromolecules*, **11**, 656 (1978).
20. D. R. Miller and C. W. Macosko, *Macromolecules*, **13**, 1063 (1980).
21. R. Zallen *The Physics of Amorphous Solids*, Wiley New York, 1983, Chap. 4.
22. P. G. de Gennes, *J. Phys. Lett. (Paris)*, **L37**, 1 (1976).
23. D. Stauffer, *J. Chem. Soc., Faraday Trans. II*, **72**, 1354 (1976).
24. D. Stauffer, A. Conglio, and M. Adam, *Adv. Polym. Sci.* **44**, 103 (1982).
25. W. Bruns, I. Motoc, and K. F. O'Driscoll, "Monte Carlo Applications in Polymer Science," in *Lecture Notes in Chemistry*, Vol. 27, Springer-Verlag, Berlin, 1981.
26. "Applications of the Monte Carlo Method," in *Topics in Current Physics*, K. Binder, Ed., Springer-Verlag, Berlin, 1984.
27. H. J. Herrmann, D. Stauffer, and D. P. Landau, *J. Phys. A: Math. Gen.*, **16**, 1221 (1983).
28. N. Jan, T. Lookman, and D. Stauffer, *J. Phys. A, Math. Gen.*, **16**, L117 (1983).
29. J. Stejny, *Macromolecules*, **17**, 2055 (1984).
30. Y. Leung and B. E. Eichenger, *J. Chem. Phys.*, **80**, 3877 (1984).
31. Y. Leung and B. E. Eichenger, *J. Chem. Phys.*, **80**, 3885 (1984).
32. J. W. Stafford, *J. Polym. Sci., Polym. Chem. Ed.*, **19**, 3219 (1981).
33. W. Klonowski, *Rheol. Acta*, **18**, 442 (1979).
34. W. Klonowski, *Rheol. Acta*, **18**, 667 (1979).
35. W. Klonowski, *Rheol. Acta*, **18**, 673 (1979).

36. D. Durand and C. M. Bruneau, *Polymer*, **23**, 69 (1982).
37. D. Durand and C. M. Bruneau, *Makromol. Chem.*, **183**, 1007 (1982).
38. J. L. Koenig, *Chemical Microstructure of Polymer Chains*, Wiley, New York, 1980, Chap. 6.
39. N. M. Bikales, Ed., *Characterization of Polymers—Reprints of Encyclopedia of Polymer Science and Technology*, Wiley, New York, 1971.
40. J. Haslam and H. A. Willis, *Identification and Analysis of Plastics*, Iliffe, London, 1965.
41. G. Kline, Ed., *Analytical Chemistry of Polymers*, Part III: *Identification Procedures and Chemical Analysis*, Wiley-Interscience, New York, 1962.
42. J. Urbanski, W. Czerwinski, K. Janicka, F. Majewska, and H. Zowall, *Handbook of Analysis of Synthetic Polymers and Plastics*, Ellis Horwood, Chichester, 1977.
43. A. D. Jenkins, *Polymer Science—A Materials Science Handbook*, North-Holland, Amsterdam, 1972.
44. A. Krause and A. Lange, *Introduction to the Chemical Analysis of Plastics*, Iliffe, London, 1969.
45. C. P. A. Kappelmeier, Ed., *Chemical Analysis of Resin-Based Coating Materials*, Wiley-Interscience, New York, 1959.
46. A. L. Smith, Ed., *Analysis of Silicones*, Wiley, New York, 1974.
47. W. C. Wake, *The Analysis of Rubber and Rubberlike Polymers*, Wiley-Interscience, New York, 1958.
48. T. P. Gladstone Shaw, *Ind. Eng. Chem.*, **16**, 541 (1944).
49. H. Mark, *Anal. Chem.*, **20**, 104 (1948).
50. F. E. Critchfield and D. P. Johnson, *Anal. Chem.*, **33**, 1834 (1961).
51. H. Nechamkin, *J. Chem. Educ.*, **28**, 97 (1951).
52. A. J. Peacock, *Polym. Commun.*, **25**, 169 (1984).
53. J. P. Bell, J. *Polym. Sci., A2*, **8**, 417 (1970).
54. C. J. Lin and J. P. Bell, *J. Appl. Polym. Sci.*, **16**, 1721 (1972).
55. B. Saville and A. A. Watson, *Rubber Chem. Technol.*, **40**, 100 (1967).
56. H. H. G. Jellinek, Ed., *Aspects of Degradation and Stabilization of Polymers*, Elsevier Scientific Publishing, New York, 1978.
57. E. A. Turi, *Thermal Analysis in Polymer Characterization*, Heyden, Philadelphia, 1981.
58. E. A. Turi, *Thermal Characterization of Polymeric Materials*, Academic Press, New York, 1981.
59. P. E. Slade, Jr., and L. T. Jenkins, *Thermal Characterization Techniques*, Dekker, New York, 1970.
60. J. Runt and I. R. Harrison, "Thermal Analysis of Polymers," in *Methods of Experimental Physics: Polymers*, Vol. 16B, R. A. Fava, Ed., Academic Press, New York, 1980.

61. B. Ke, *Newer Methods of Polymer Characterization*, Wiley-Interscience, New York, 1964.
62. V. V. Korshak, *The Chemical Structure and Thermal Characteristics of Polymers*, Israel Program for Scientific Translation, Jerusalem, 1971.
63. R. A. Fava, *Polymer*, **9**, 137 (1968).
64. S. M. Ellerstein, in *Analytical Calorimetry*, R. S. Porter and J. F. Johnson, Eds., Plenum, New York, 1968.
65. L. W. Crane, P. J. Dynes, and D. H. Kaelble, *J. Polym. Sci., Polym. Lett. Ed.*, **11**, 533 (1973).
66. H. E. Kissinger, *Anal. Chem.*, **29**, 1702 (1957).
67. K. Horie, H. Hiura, M. Savada, I. Mita, and H. Kambe, *J. Polym. Sci.: A1*, **8**, 1357 (1970).
68. M. A. Acitelli, R. B. Prime, and E. Sacher, *Polymer*, **12**, 335 (1971).
69. R. B. Prime, *Polym. Eng. Sci.*, **13**, 365 (1973).
70. S. Sourour and M. R. Kamal, *Thermochim. Acta*, **14**, 41 (1976).
71. N. S. Schneider, J. F. Sprouse, G. L. Hagnauer, and J. K. Gillham, *Polym. Eng. Sci.*, **19**, 304 (1979).
72. C. C. Ricardi, H. E. Adabbo, and R. J. J. Williams, *J. Appl. Polym. Sci.*, **29**, 2481 (1984).
73. T. Olcese, O. Spelta, and S. Vargiu, *J. Polym. Sci., Polym. Symp.*, **53**, 113 (1975).
74. J. M. Barton, *Br. Polym. J.*, **11**, 115 (1979).
75. J. A. Mikroyannidis and D. A. Kourtides, *J. Appl. Polym. Sci.*, **29**, 197 (1984).
76. P. P. Shah, P. H. Parsania, and S. R. Patel, *Br. Polym. J.*, **17**, 64 (1985).
77. Z. Katovic, *J. Appl. Polym. Sci.*, **11**, 85 (1967).
78. V. I. Kurachenkov and L. A. Igonin, *J. Polym. Sci., A1*, **9**, 2283 (1971).
79. A. Sebenik, I. Vizovisek, and S. Lapanje, *Eur. Polym. J.*, **10**, 273 (1974).
80. P. W. King, R. H. Mitchell, and A. R. Westwood, *J. Appl. Polym. Sci.*, **18**, 1117 (1974).
81. A. Siegmann and M. Narkis, *J. Appl. Polym. Sci.*, **21**, 2311 (1977).
82. S. Chow and P. R. Steiner, *J. Appl. Polym. Sci.*, **23**, 1973 (1979).
83. J. A. Koutsky and R. Ebewele, "Differential Scanning Calorimetry of Phenolic Resins," in *Chemistry and Properties of Crosslinked Polymers*, S. S. Labana, Ed., Academic Press, New York, 1977.
84. A. W. Christiansen and L. Gollob, *J. Appl. Polym. Sci.*, **30**, 2279 (1985).
85. R. Kay and A. R. Westwood, *Eur. Polym. J.*, **11**, 25 (1975).
86. D. W. Brazier, "Calorimetric Studies of Rubber Vulcanisation and Vulcanisates," in *Developments in Polymer Degradation*, Vol. 3, N. Grassie, Ed., Applied Science Publishers, Barking, 1981, Chap. 2.
87. P. E. Willard, *Polym. Eng. Sci.*, **12**, 120 (1972).

88. D. W. Sundstrom and M. F. English, *Polym. Eng. Sci.*, **18**, 728 (1978).
89. K. Horie, I. Mita, and H. Kambe, *J. Polym. Sci., A1*, **8**, 2839 (1970).
90. G. R. Tryson and A. R. Shultz, *J. Polym. Sci., Polym. Phys. Ed.*, **17**, 2059 (1979).
91. A. G. Ulukhanov, V. A. Lapitskii, M. S. Akutin, and L. D. Skokova, *Plast. Massy*, **9**, 59 (1981).
92. Z. Katovic, *J. Appl. Polym. Sci.*, **11**, 95 (1967).
93. M. Ezrin and G. C. Claver, *Appl. Polym. Symp.*, **8**, 159 (1969).
94. A. R. Greenberg and I. Kamel, *J. Polym. Sci., Polym. Chem. Ed.*, **15**, 2137 (1977).
95. B. Dickens and J. H. Flynn, "Thermogravimetry Applied to Polymer Degradation Kinetics," in *Polymer Characterization* (ACS Advances in Chemistry Series No. 203), C. D. Craver, Ed., American Chemical Society, Washington, DC 1983.
96. R. T. Conley, "Thermosetting Resins," in *Thermal Stability of Polymers*, R. T. Conley, Ed., Dekker, New York, 1970, Chap. 11.
97. H. T. Lee and D. W. Levi, *J. Appl. Polym. Sci.*, **13**, 1703 (1969).
98. T. R. Manley, *J. Macromol. Sci.-Chem.*, **A8**, 53 (1974).
99. V. J. McBrierty, *Polymer*, **15**, 503 (1974).
100. V. J. McBrierty and D. C. Douglass, *Phys. Rep.*, **63**, 61 (1980).
101. V. J. McBrierty, *Magn. Reson. Rev.*, **8**, 165 (1983).
102. L. W. Jelinski, *Chem. Eng. News*, **62**(45), 26 (1984).
103. J. R. Havens and J. L. Koenig, *Appl. Spectrosc.*, **37**, 226 (1983).
104. D. J. P. Harrison, W. R. Yates, and J. F. Johnson, *J. Macromol. Sci., Rev. Macromol. Chem. Phys.*, **C25**, 481 (1985).
105. J. Schaefer, *Macromolecules*, **6**, 882 (1973).
106. R. A. Komoroski and L. Mandelkern, *J. Polym. Sci., Polym. Lett. Ed.*, **14**, 253 (1976).
107. J. C. Randall, *Polymer Sequence Determination Carbon-13 NMR Method*, Academic Press, New York, 1977.
108. F. A. Bovey, *High Resolution NMR of Macromolecules*, Academic Press, New York, 1972.
109. F. Bloch, *Phys. Rev.*, **111**, 841 (1958).
110. J. Schaefer, E. O. Stejskal, and R. Buchdahl, *Macromolecules*, **8**, 291 (1975).
111. E. R. Andrew, *Prog. Nucl. Magn. Reson. Spectrosc.*, **8**, 1 (1971).
112. E. R. Andrew, *Phil. Trans. R. Soc. London*, **A299**, 505 (1981).
113. W. R. Earl and D. L. Vanderhart, *Macromolecules*, **12**, 762 (1979).
114. A. Pines, M. G. Gibby, and J. S. Waugh, *J. Chem. Phys.*, **59**, 569 (1973).
115. S. R. Hartmann and E. L. Hahn, *Phys. Rev.*, **12B**, 2042 (1962).
116. J. Schaefer and E. O. Stejskal, *J. Am. Chem. Soc.*, **98**, 1031 (1976).
117. J. Schaefer, E. O. Stejskal, and R. Buchdahl, *Macromolecules*, **10**, 384 (1977).
118. J. Schaefer and E. O. Stejskal, "High-Resolution C-13 NMR of Solid Polymers,"

in *Topics in Carbon-13 NMR Spectroscopy*, Vol. 3, G. C. Levy, Ed., Wiley, New York, 1979.

119. J. R. Lyerla, "High-Resolution Carbon-13 NMR Studies of Bulk Polymers," in *Contemporary Topics in Polymer Science*, Vol. 3, M. Shen, Ed., Plenum, New York, 1979.
120. M. Mehring, *High Resolution NMR in Solids*, Springer-Verlag, New York, Berlin, 1983.
121. C. A. Fyfe, *Solid State NMR for Chemists*, C.F.C. Press, Ontario, 1983.
122. M. W. Duch and D. M. Grant, *Macromolecules*, **3**, 165 (1970).
123. V. D. Mochel, *J. Macromol. Sci., Rev. Macromol. Chem.*, **8**, 289 (1972).
124. D. J. Patterson, J. L. Koenig, and J. R. Shelton, *Rubber Chem. Technol.*, **56**, 971 (1983).
125. D. J. Patterson and J. L. Koenig, "Peroxide Cross-linked Natural Rubber and *cis*-Polybutadiene," in *Characterization of Highly Cross-Linked Polymers*, S. S. Labana and R. A. Dickie, Eds. (ACS Symposium Series No. 203), American Chemical Society, Washington, DC, 1984.
126. A. M. Zaper and J. L. Koenig, to be published.
127. W. Hofmann, *Vulcanization and Vulcanizing Agents*, Maclaren, London, 1967.
128. A. Cholli, W. M. Ritchey, and J. L. Koenig, "Magic Angle NMR of an Epoxy Resin," in *Characterization of Highly Cross-Linked Polymers*, S. S. Labana and R. A. Dickie, Eds. (ACS Symposium Series No. 203), American Chemical Society, Washington, DC, 1984.
129. H. A. Resing and W. B. Moniz, *Macromolecules*, **8**, 560 (1975).
130. A. N. Garroway, W. B. Moniz, and H. A. Resing, *Am. Chem. Soc., Coatings and Plast. Prepr. 172nd Meeting*, **36**, 133 (1976).
131. A. N. Garroway, W. B. Moniz, and H. A. Resing, *Faraday Symp. Chem. Soc.*, **13**, 63 (1978).
132. A. N. Garroway, W. B. Moniz, and H. A. Resing, *ACS Symp. Ser.*, **103**, 67 (1979).
133. A. N. Garroway, D. L. Vanderhart, and W. L. Earl, *Phil. Trans. R. Soc. Lond.*, **A299**, 609 (1981).
134. D. L. Vanderhart, W. L. Earl, and A. N. Garroway, *J. Magn. Reson.*, **44**, 361 (1981).
135. G. E. Balimann, C. J. Groombridge, R. K. Harris, K. J. Packer, B. J. Say, and S. F. Tanner, *Phil. Trans. R. Soc. London*, **A299**, 643 (1981).
136. A. N. Garroway, W. M. Ritchey, and W. B. Moniz, *Macromolecules*, **15**, 1051 (1982).
137. R. F. Fischer, *J. Polym. Sci.*, **44**, 155 (1960).
138. Y. Tanaka and H. Kakiuchi, *J. Appl. Polym. Sci.*, **7**, 1063 (1963).
139. R. B. Prime and E. Sacher, *Polymer*, **13**, 455 (1972).
140. S. A. Sojka and W. B. Moniz, *J. Appl. Polym. Sci.*, **20**, 1977 (1976).
141. R. Peyser and W. D. Bascom, *J. Appl. Polym. Sci.*, **721**, 2359 (1977).

142. J. Luston, Z. Manasek, and M. Kulickova, *J. Macromol. Sci. Chem.*, **A12**, 995 (1978).
143. M. K. Antoon, Ph.D. Dissertation, Case Western Reserve University, Cleveland, 1980.
144. L. W. Jelinski, J. J. Dumais, R. E. Stark, T. S. Ellis, and F. E. Karasz, *Macromolecules*, **16**, 1019 (1983).
145. L. W. Jelinski, J. J. Dumais, A. L. Cholli, T. S. Ellis, and F. E. Karasz, *Macromolecules*, **18**, 1091 (1985).
146. S. A. Sojka and W. B. Moniz, *J. Appl. Polym. Sci.*, **20**, 1977 (1976).
147. W. W. Fleming, *J. Appl. Polym. Sci.*, **30**, 2853 (1985).
148. C. A. Fyfe, A. Rudin, and W. Tchir, *Macromolecules*, **13**, 1320 (1980).
149. C. A. Fyfe, M. S. McKinnon, A. Rudin, and W. J. Tchir, *Macromolecules*, **16**, 1216 (1983).
150. R. L. Bryson, G. R. Hatfield, T. A. Early, A. R. Palmer, and G. E. Maciel, *Macromolecules*, **16**, 1669 (1983).
151. G. E. Maciel, I. Chuang, and L. Gollob, *Macromolecules*, **17**, 1081 (1984).
152. S. So and A. Rudin, *J. Polym. Sci., Polym. Lett. Ed.*, **23**, 403 (1985).
153. G. E. Maciel, I. Chuang, and G. E. Myers, *Macromolecules*, **15**, 1218 (1982).
154. I. Chuang, G. E. Maciel, and G. E. Myers, *Macromolecules*, **17**, 1087 (1984).
155. G. E. Maciel, N. M. Szevernyi, T. A. Early, and G. E. Myers, *Macromolecules*, **16**, 598 (1983).
156. I. Chuang, B. L. Hawkins, G. E. Maciel, and G. E. Myers, *Macromolecules*, **18**, 1482 (1985).
157. D. R. Bauer, R. A. Dickie, and J. L. Koenig, *J. Polym. Sci., Polym. Phys. Ed.*, **22**, 2209 (1984).
158. D. R. Bauer, R. A. Dickie, and J. L. Koenig, *Ind. Eng. Chem. Prod. Res. Dev.*, **24**, 121 (1985).
159. P. M. Hergenrother, *J. Macromol. Sci., Rev. Macromol. Chem.*, **C19**, 1 (1980).
160. C. Shields and J. L. Koenig, unpublished results.
161. D. J. Patterson, C. M. Shields, A. Cholli, and J. L. Koenig, *Polym. Prepr., ACS Div. Polym. Chem.*, **25**, 358 (1984).
162. S. J. Opella and M. N. Frey, *J. Am. Chem. Soc.*, **101**, 5854 (1979).
163. H. G. Viehe, Ed., *Chemistry of Acetylenes*, Dekker, New York, 1969.
164. M. D. Sefcik, E. O. Stejskal, R. A. McKay, and J. Schaefer, *Macromolecules*, **12**, 423 (1979).
165. R. L. Bennett, A. Keller, and J. Stejny, *J. Polym. Sci., Polym. Chem. Ed.*, **14**, 3021 (1976).
166. R. L. Bennett, A. Keller, J. Stejny, and M. Murray, *J. Polym. Sci., Polym. Chem. Ed.*, **14**, 3027 (1976).
167. W. L. Earl and D. L. Vanderhart, *Macromolecules*, **12**, 762 (1979).
168. D. M. Grant and E. G. Paul, *J. Am. Chem. Soc.*, **86**, 2984 (1964).

169. J. Sohma, M. Shiotani, and S. Murakami, *Radiat. Phys. Chem.*, **21**, 413 (1983).
170. J. H. O'Donnell and A. K. Whitaker, *Br. Polym. J.*, **17**, 51 (1985).
171. J. R. Havens, M. Thakur, J. B. Lando, and J. L. Koenig, *Macromolecules*, **17**, 1071 (1984).
172. P. R. Griffiths, *Chemical Infrared Fourier Transform Spectroscopy*, Wiley, New York, 1975, Chap. 2.
173. P. Griffiths, "FTIR: Theory and Instrumentation," in *Transform Techniques in Chemistry*, P. Griffiths, Ed., Plenum, New York, 1978.
174. R. J. Bell, *Introductory Fourier Transform Spectroscopy*, Academic Press, New York, 1982.
175. J. Chamberlain, *Principles of Interferometric Spectroscopy*, Wiley-Interscience, New York, 1979.
176. L. Mertz, *Transformation in Optics*, Wiley, New York, 1965.
177. R. Bracewell, *The Fourier Transform and Its Applications*, McGraw-Hill, New York, 1965.
178. D. C. Champeney, *Fourier Transforms and Their Physical Applications*, Academic Press, New York, 1973.
179. J. W. Cooley and J. W. Tukey, *Math. Comput.*, **19**, 297 (1965).
180. P. B. Fellgett, *J. Phys. Radium*, **19**, 187 (1958).
181. P. Jacquinot, *Rep. Prog. Phys.*, **13**, 267 (1960).
182. J. Conne and P. Conne, *J. Opt. Soc. Am.*, **56**, 896 (1966).
183. J. L. Koenig, *Appl. Spectrosc.*, **29**, 293 (1975).
184. L. D'Esposito and J. L. Koenig, "Applications of Fourier Transform Infrared to Synthetic Polymers and Biological Macromolecules," in *Fourier Transform Infrared Spectroscopy*, Vol. 1, J. R. Ferraro and L. J. Basilie, Eds., Academic Press, New York, 1978, Chap. 2.
185. D. L. Tabb, J. J. Sevcik, and J. L. Koenig, *J. Polym. Sci., Polym. Phys. Ed.*, **13**, 815 (1975).
186. L. D'Esposito and J. L. Koenig, *J. Polym. Sci., Polym. Phys. Ed.*, **14**, 1731 (1976).
187. P. C. Painter and J. L. Koenig, *J. Polym. Sci., Polym. Phys. Ed.*, **15**, 1885 (1977).
188. D. L. Tabb and J. L. Koenig, *Macromolecules*, **8**, 929 (1975).
189. J. L. Koenig, "Fourier Transform Infrared Spectroscopy of Polymers," in *Advances in Polymer Science*, Vol. 54, Springer-Verlag, Berlin, 1983.
190. J. Haslam, H. A. Willis, and D. C. M. Squirrell, *Identification and Analysis of Plastics*, 2nd ed., Heyden, Philadelphia, 1972.
191. J. Fahrenfort, *Spectrochim. Acta*, **17**, 698 (1961).
192. J. Fahrenfort and W. M. Visser, *Spectrochim. Acta*, **18**, 110 (1962).
193. N. J. Harrick, *Phys. Rev.*, **125**, 1165 (1962).
194. N. J. Harrick, *Internal Reflection Spectroscopy*, Wiley, New York, 1967.
195. M. P. Fuller and P. R. Griffiths, *Anal. Chem.*, **50**, 1906 (1978).

196. G. Kortum, *Reflectance Spectroscopy Principles, Methods, and Applications*, Springer-Verlag, Berlin, 1964.
197. P. Kubelka and F. Munk, *Z. Tech. Phys.*, **12**, 593 (1931).
198. P. Kubelka, *J. Opt. Soc. Am.*, **38**, 448 (1948).
199. H. G. Hecht, *J. Res. Natl. Bur. Stand.*, **80A**, 567 (1976).
200. D. W. Vidrine, *Appl. Spectrosc.*, **34**, 314 (1980).
201. K. Krishnan, *Appl. Spectrosc.*, **35**, 549 (1981).
202. J. Luston, A. Manasek, and M. Kulickova, *J. Macromol. Sci. Chem.*, **12**, 995 (1978).
203. G. C. Stevens, *J. Appl. Polym. Sci.*, **26**, 4279 (1981).
204. M. K. Antoon and J. L. Koenig, *J. Polym. Sci., Polym. Chem. Ed.*, **19**, 549 (1981).
205. L. Buckley and D. Roylance, *Polym. Sci. Eng.*, **22**, 166 (1982).
206. R. E. Smith, F. N. Larsen, and C. L. Long, *J. Appl. Polym. Sci.*, **29**, 3713 (1984).
207. R. E. Smith, F. N. Larsen, and C. L. Long, *J. Appl. Polym. Sci.*, **29**, 3697 (1984).
208. T. Donnellan and D. Roylance, *Polym. Eng. Sci.*, **22**, 821 (1982).
209. C. D. Giulio, M. Gautier, and B. Jasse, *J. Appl. Polym. Sci.*, **29**, 1771 (1984).
210. R. B. Barnes, in *Applications of Instruments in Chemistry*, R. E. Burk and O. Gummitt, Eds., Wiley-Interscience, New York, 1945.
211. W. J. Murphy, *Ind. Eng. Chem.*, **15**, 659 (1943).
212. M. K. Antoon, K. M. Starkey, and J. L. Koenig, "Applications of FTIR to Quality Control of the Epoxy Matrix," Composite Materials: Fifth Conference of the American Society for Testing and Materials on Testing and Design, ASTM, Philadelphia, 1979, STP 674, 541.
213. M. K. Antoon, J. H. Koenig, and J. L. Koenig, *Appl. Spectrosc.*, **31**, 518 (1977).
214. D. M. Haaland and R. G. Easterling, *Appl. Spectrosc.*, **34**, 539 (1980).
215. D. M. Haaland and R. G. Easterling, *Appl. Spectrosc.*, **36**, 665 (1982).
216. H. Dannenberg and W. R. Harp, Jr., *Anal. Chem.*, **28**, 86 (1956).
217. W. A. Patterson, *Anal. Chem.*, **26**, 823 (1954).
218. M. K. Antoon, B. E. Zehner, and J. L. Koenig, *Polym. Comp.*, **2**, 81 (1981).
219. E. R. Malinowski and D. G. Lowery, *Factor Analysis in Chemistry*, Wiley, New York, 1980.
220. M. K. Antoon, L. D'Esposito, and J. L. Koenig, *Appl. Spectrosc.*, **33**, 351 (1979).
221. J. L. Koenig and C. M. Shields, *J. Polym. Sci., Polym. Phys. Ed.*, **23**, 845 (1985).
222. M. K. Antoon, J. L. Koenig, and T. Serafini, *J. Polym. Sci., Polym. Phys. Ed.*, **19**, 1567 (1981).
223. M. K. Antoon and J. L. Koenig, *J. Polym. Sci., Polym. Phys. Ed.*, **19**, 197 (1977).
224. G. A. George, R. E. Sacher, and J. R. Sprouse, *J. Appl. Polym. Sci.*, **21**, 2241 (1977).
225. S. C. Lin, B. J. Bulkin, and E. M. Pearce, *J. Polym. Sci., Polym. Chem. Ed.*, **17**, 3121 (1979).
226. H. Ishida and J. L. Koenig, *J. Polym. Sci., Polym. Phys. Ed.*, **18**, 1931 (1980).
227. H. Ishida and J. L. Koenig, *J. Colloid Interface Sci.*, **64**, 565 (1978).

228. C. Chiang and J. L. Koenig, *Polym. Comp.*, **2**, 192 (1981).
229. C. Chiang and J. L. Koenig, *J. Polym. Sci., Polym. Phys. Ed.*, **20**, 2135 (1982).
230. R. T. Graf, J. L. Koenig, and H. Ishida, *Anal. Chem.*, **56**, 773 (1984).
231. S. R. Culler, H. Ishida, and J. L. Koenig, *J. Colloid Interface Sci.*, **109**, 1 (1986).
232. L. Bateman, *The Chemistry and Physics of Rubber-Like Substances*, Maclaren, London, 1963.
233. C. G. Moore and M. Porter, *J. Appl. Polym. Sci.*, **11**, 2227 (1967).
234. B. Saville and A. A. Watson, *Rubber Chem. Technol.*, **40**, 100 (1967).
235. C. H. Chen, J. L. Koenig, J. R. Shelton, and E. A. Collins, *Rubber Chem. Technol.*, **54**, 734 (1981).
236. C. H. Chen, E. A. Collins, J. R. Shelton, and J. L. Koenig, *Rubber Chem. Technol.*, **55**, 1221 (1982).
237. C. H. Chen, J. L. Koenig, J. R. Shelton, and E. A. Collins, *Rubber Chem. Technol.* **55**, 103 (1982).
238. W. Hoffmann, *Vulcanization and Vulcanizing Agents*, Maclaren, London, 1967.
239. J. I. Cunneen, G. M. C. Higgins, and R. A. Wilkes, *J. Polym. Sci.*, **A3**, 3503 (1965).
240. J. L. Koenig, M. M. Coleman, J. R. Shelton, and P. H. Starmer, *Rubber Chem. Technol.*, **44**, 71 (1971).
241. J. R. Shelton, J. L. Koenig, and M. M. Coleman, *Rubber Chem. Technol.*, **44**, 904 (1971).
242. M. M. Coleman, J. R. Shelton and J. L. Koenig, *Rubber Chem. Technol.*, **45**, 173 (1972).
243. C. S. Lu and J. L. Koenig, *Am. Chem. Soc. Prepr., Div. Org. Coatings Plast. Chem.*, **32**(1), 112 (1972).
244. C. G. Hill, Jr., A. M. Hedren, G. E. Myers, and J. A. Koutsky, *J. Appl. Polym. Sci.*, **29**, 2749 (1984).

CHAPTER

16

MECHANICAL AND SOLVENT-SWELLING METHODS FOR CHARACTERIZING INSOLUBLE POLYMER

JEAN-PIÉRRE QUESLEL

Manufacture Michelin, Clermont-Ferrand, France

J. E. MARK

Department of Chemistry and the Polymer Research Center, The University of Cincinnati, Cincinnati, Ohio

INTRODUCTION

Linear chain molecules of any molecular weight are completely soluble in good solvents (1, 2), as are branched molecules such as stars or combs. When such primary molecules are joined intermolecularly by means of a chemical reaction, the first stage can be viewed as formation of giant tree devoid of cyclic (intramolecular) connections. Such a molecule would also be soluble in some solvents. However, additional cross-links would engage units intramolecularly into a three-dimensional network (3–10) which is no longer soluble in any solvent, although good solvents would swell the network structure (1, 11–15).

The functionality ϕ of the cross-links or junctions—that is, the number of chains emanating from each of them—depends on the chemical process used to link the chain molecules. The network junctions ordinarily are provided by covalent bonds. In some instances, however, physical association of chain segments (e.g., by crystallites) may serve the same function as permanent chemical interlinkages (16). Difunctional junctions are of no interest, since they lead only to chain extension. The two most important types of networks are (1) the tetrafunctional ($\phi = 4$, almost invariably obtained upon joining two segments from different chains, as in the case of sulfur vulcanization or peroxide or irradiation cross-linking), and (2) the trifunctional [$\phi = 3$, obtained, e.g., when forming a polyurethane network by end-linking hydroxyl-terminated chains with a triisocyanate (17)].

The pore or mesh size is a fundamental quantity which characterizes the structure of the insoluble polymer network, and can be taken to be the chain molecular weight M_c between two consecutive cross-links. It controls all the properties of the network, in particular its mechanical properties. For perfect networks obtained by chain end-linking with a multifunctional agent, values of M_c are essentially equal to the values of the number average molecular weight $\bar{M}_n$ of the chains before their end-linking, and values of ϕ are equal to the functionality of the end-linking agent (18, 19). In this approach, chemical analysis, and viscometric and gel permeation chromatographic (GPC) measurements, are generally carried out on the chains before they are cross-linked, to determine their average value of $\bar{M}_n$ and the distribution about this average. However, randomly cross-linked networks contain no information related to the precursor chain molecular weight, except for the number of dangling chains per unit volume, which is $2\rho/\bar{M}_n$, where ρ is the polymer density (20). Another important way of characterizing a network is unquestionably in terms of its degree of cross-linking, that is, the number density of junctions μ/V joining the chains into a permanent structure, where V is the reference volume of the isotropic unswollen network.

Polymer networks are conveniently characterized in the elastomeric state, which is exhibited at temperatures above the glass-to-rubber transition temperature T_g. In this state, the large ensemble of configurations accessible to flexible chain molecules by Brownian motion is very amenable to statistical mechanical analysis (3, 21, 22). Polymers with relatively high values of T_g, like polystyrene (23) or elastin (24, 25), are generally studied in the swollen state to lower their values of T_g to below the temperature of investigation. It is also advantageous to study network behavior in the swollen state because this facilitates the approach to elastic equilibrium, which is required for application of rubberlike elasticity theories based on statistical thermodynamics (26–28).

Above T_g, networks composed of long, flexible chains, with weak interchain interactions, exhibit rubberlike elasticity, that is, high extensibility under stress coupled with capacity of full recovery when subsequently unloaded (29). Network topology, cross-link density, and junction functionality affect all the elastomeric properties, including (1) equilibrium properties such as the modulus, ultimate strength, maximum extensibility, and degree of swelling, and (2) dynamic mechanical properties such as viscoelastic losses. Consequently, it may be possible to describe network topology from mechanical and solvent swelling measurements. This requires, however, a model of elasticity relating macroscopic stress and macroscopic strain or swelling ratio to molecular network structure and molecular deformation. The most useful such model is discussed below.

NETWORK TOPOLOGY

Flory has shown (3, 4, 30) that the cycle rank ξ of the network, or the number of independent circuits it contains as defined in graph theory (31), is the appropriate measure of the network connectivity, regardless of network imperfections. The cycle rank may be defined alternatively as the minimum number of scissions required to reduce the network to a spanning tree—that is, a unified structure comprising all the chains and containing no closed circuits or loops. Subsidiary quantities called the number of effective chains and junctions designated, respectively, ν_e and μ_e, can be defined by the relationships

$$\xi = \nu_e - \mu_e + 1 \simeq \nu_e - \mu_e \tag{1}$$

In a perfect network, μ_e is equal to the total number of junctions, whereas in a randomly cross-linked network, μ_e is the number of junctions required to transform the acyclic structure or tree into the final network (32). The value of ν_e depends on μ_e and on the (average) junction functionality ϕ, the specific relationship being

$$\nu_e = \phi\mu_e/2 \tag{2}$$

Scanlan (33) and Case (34) have defined an active junction as one joined by at least three paths to the gel network and an active chain as one terminated by an active junction at both ends. It can be shown (4, 35) that

$$\nu_e - \mu_e = \nu_a - \mu_a \tag{3}$$

where ν_a and μ_a are, respectively, the numbers of active chains and junctions.

Perfect Networks

As mentioned in the introduction, perfect model networks are networks obtained in a controlled manner—for example, by end-linking functionally terminated chains (of number average molecular weight $\bar{M}_n$) with multifunctional reagents. The molecular weight between cross-link points is then $M_c = \bar{M}_n$, and the functionality of the network is the functionality of the cross-linking agent [or the average functionality (36) if several types of cross-linking agents are used]. The following relationships now hold for a perfect end-linked network (no chain ends or loops):

$$\xi/V = \frac{(1 - 2/\phi)\rho}{M_c} \tag{4}$$

$$\mu_e V = \mu_a/V = 2\rho/\phi M_c \tag{5}$$

Randomly Cross-Linked Networks

Although model networks are very useful to test elasticity theories, common applications involve statistically cross-linked polymers (e.g., those irradiated, or those cured by dicumyl peroxide). Randomly cross-linked networks have indeed a complex topology, and analytical methods currently used to determine parameters of the network structure do not permit one to determine rigorously the concentration of cross-link units, the junction functionality, or the numbers of loops and free chains in the network. Several approaches based on probabilistic formations and soluble fraction measurements have been proposed (5–8, 10, 37), but it is difficult to draw reliable conclusions without assumptions on the statistics of network formation.

Recent developments in rubberlike elasticity models based on statistical mechanical analysis have shown that mechanical properties (such as the modulus) and swelling extent of elastomeric networks depend only on one topological entity, namely the cycle rank (3, 12, 32, 38). Only a relationship between the cycle rank and the molecular weight M_c is then necessary for network characterization through a set of swelling and mechanical property experiments. In the case of random cross-linking of linear chains, Queslel and Mark (9, 11) have established the following simple relationship between M_c, $\bar{M}_n$, and ξ, the cycle rank for regular networks having no defects other than chain ends. Specifically:

$$\xi/V = (\rho/2M_c)(1 - 3M_c/\bar{M}_n) \tag{6}$$

A similar relationship holds for the number of active junctions μ_a assumed to be equal to the total number of junctions μ:

$$\mu_a/V = (\rho/2M_c)(1 - M_c/\bar{M}_n) \tag{7}$$

The topology of networks formed by random tetrafunctional cross-linking of star polymers having A arms has also been investigated (39). It was shown that the following relationships exist between the molecular weight M_c between junctions, the number average molecular weight $\bar{M}_n$ of the primary star arms, the cycle rank ξ, the total number density μ/V of junctions, and the average junction functionality ϕ_{av}:

$$\xi/V = (\rho/2M_c)[1 - (A + 2)M_c/A\bar{M}_n] \tag{8}$$

$$\mu/V = (\rho/2M_c)[1 - (A - 2)M_c/A\bar{M}_n] \tag{9}$$

$$\phi_{av} = 4 + 2(A - 4)(2 - A + A\bar{M}_n/M_c)^{-1} \tag{10}$$

MOLECULAR ELASTICITY THEORY OF REAL NETWORKS

Relationships between structure of elastomeric networks and their mechanical properties have been investigated, from the very first, by statistical mechanics (1, 3, 40–50). The statistical mechanical theories of rubber elasticity are based on two fundamental postulates: (1) molecular chain configurations are random in undeformed amorphous polymers, and (2) the elastic response of the network originates within the chains and not to a significant extent from interactions between them. The high deformability of an elastomer and the elastic force generated by deformation stem from the configurations accessible to long molecular chains. Thus, the stress–strain behavior of elastomeric networks can be explained only if the relationship between molecular chain deformation and macroscopic strain is understood. Theoretical investigations on typical polymeric chains have shown that the Gaussian representation of the chain end-to-end distance distribution function should be quite satisfactory for chains consisting of 100 bonds or more (3, 51, 52). Basically, two models of Gaussian networks were constructed. In the affine network model, displacements of junctions and of chain vectors are assumed to be affine (linear) in the macroscopic strain (49). In the phantom network, chains are devoid of material properties (53), and macroscopic constraints operate only on the periphery of the network sample. Other junctions fluctuate without restraints from neighboring chains. In both cases, the reduced stress measured in uniaxial extension is predicted to be independent of deformation. This quantity is defined as (54)

$$[f^*] \equiv \frac{f v_2^{1/3}}{A_v(\alpha - \alpha^{-2})} \tag{11}$$

where α is the extension ratio (defined relative to the undeformed swollen state), f the measured equilibrium force, v_2 the volume fraction of polymer in the swollen network, A_v the cross-sectional area of the isotropic unswollen sample (of volume V). Departures from this behavior have long been reported (27, 55). Specifically, the experimental reduced force decreases markedly with elongation and with dilation by swelling. These observations have suggested that the limiting value of the reduced force at high elongation or high dilation is a fundamental characteristic of a given network. It became apparent later that this quantity was the phantom modulus $[f^*]_{ph}$ proportional to the cycle rank density (30)

$$[f^*]_{ph} = (\xi kT/V) v_{2S}^{2/3} \tag{12}$$

where k is the Boltzmann constant and T the absolute temperature and v_{2s} the volume fraction of polymer present during the cross-linking process.

Real networks present strong overlapping of chain configurations and intermolecular steric hindrances of chain motion commonly termed entanglements. These complications have been recognized as the origin of departures from both phantom and affine predictions. Different formalisms were proposed to include these intermolecular effects in rubberlike elasticity analysis at thermodynamic equilibrium (4). In the one due to Flory and Erman (38), entanglements are embodied as domains of constraints acting as restrictions on junction fluctuations. The Flory–Erman theory was successful in accounting for the stress–strain relationships for all varieties of strains of typical elastomers throughout ranges accessible to experiments (56–60). Applied to uniaxial extension, this model effectively predicts a decrease in the reduced force with increasing deformation, interpreting it as a gradual transition from nearly affine behavior (at small deformations) to the very nonaffine behavior chatacteristic of the phantom state.

The contraints due to entanglements are introduced in the Flory–Erman model by means of springlike forces acting on the fluctuating junctions (42, 61–63). A center of constraints is defined for each junction. The deviation of the real network from the corresponding phantom analogue is represented by the parameter κ defined as

$$\kappa = \frac{\langle (\Delta R)^2 \rangle_{\mathrm{ph}}}{\langle (\Delta s)^2 \rangle_0} \tag{13}$$

where $\langle (\Delta R)^2 \rangle_{\mathrm{ph}}$ and $\langle (\Delta s)^2 \rangle_0$ are respectively the mean-square fluctuation of junctions from their mean positions in the phantom network and from their centers of constraints in the real network.

On the plausible grounds that the constraints on junctions are determined by the degree of interpenetration of the chains about the junctions in the network, the parameter κ and the number of junctions μ should be related according to (32, 41, 56, 64)

$$\kappa = I \langle r^2 \rangle_0^{3/2} \mu / V \tag{14}$$

The quantity I is an interpenetration parameter, which is frequently ~ 0.5 (9, 13, 56, 65), and the quantity $\langle r^2 \rangle_0$ is the mean-square, end-to-end length of the (unperturbed) network chain. For sufficiently long chains,

$$\langle r^2 \rangle_0 = C_\infty M_c N l^2 / m_0 \tag{15}$$

where C_∞ is the characteristic ratio, N the average number of bonds per repeat unit, l^2 the mean-square skeletal bond length, and m_0 the monomer molecular weight.

The elastic free energy change is written as the sum of the elastic free energy chainge ΔA_{ph} for a phantom network and a term ΔA_c, which accounts for entanglement constraints:

$$\Delta A = \Delta A_{ph} + \Delta A_c \tag{16}$$

with

$$\Delta A_{ph} = \tfrac{1}{2}\xi kT\left(\sum_{t=x,y,z} \lambda_t^2 - 3\right) \tag{17}$$

where $\lambda_x, \lambda_y, \lambda_z$ are the principal extension ratios relative to the isotropic state of volume V_0 in which polymer was cross-linked. The other free energy change is

$$\Delta A_c = (\mu_e kT/2) \sum_{t=x,y,z} \{(1 + g_t)B_t - \ln[(B_t + 1)(g_t B_t + 1)]\} \tag{18}$$

where

$$B_t = (\lambda_t - 1)(1 + \lambda_t - \zeta\lambda_t^2)(1 + g_t)^{-2} \tag{19}$$

$$g_t = \lambda_t^2[\kappa^{-1} + \zeta(\lambda_t - 1)] \tag{20}$$

The quantity ζ is a parameter of marginal importance which accounts for the possible nonaffine transformation of the constraint domains with strain. It increases with increase in the number of network inhomogeneities (65). The response of stress to strain of an elastomeric network and the swelling extent in a solvent are related to the derivative of the free energy change with respect to deformation (29, 66–68). Therefore, analysis of mechanical or swelling measurements through equations (14)–(20), in conjunction with equations (4) and (5) for model networks or equations (6) and (7) for random networks permits the characterization of the network structure (9, 10, 13, 56, 65, 69).

APPLICATIONS TO NETWORK CHARACTERIZATION

Mechanical Methods

The mechanical experimental techniques most commonly used for network characterization are uniaxial elongation and compression (70–73). The apparatus typically used to measure the force at thermodynamic equilibrium in uniaxial elongation is shown in Figure 1 (74). Specifically, an elastomeric strip is mounted between two clamps, the lower one fixed and the upper one attached to a movable force gage. A recorder is used to monitor the output of

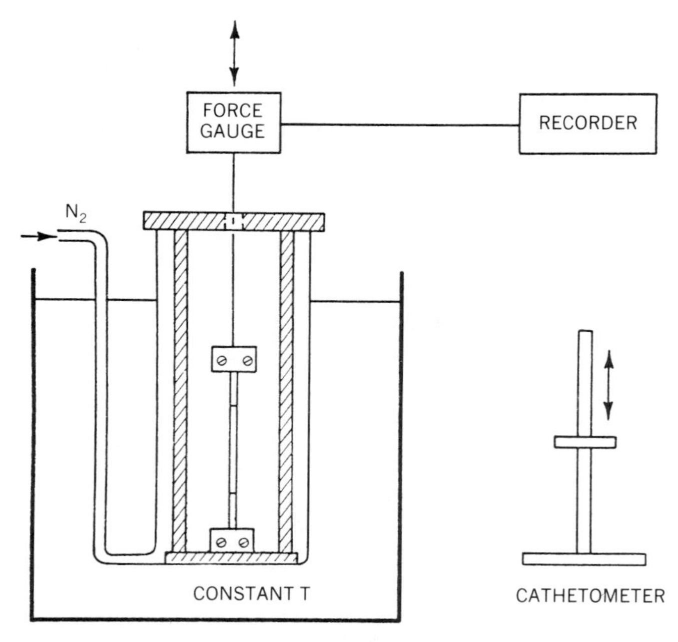

Figure 1. Schematic diagram of a typical apparatus used to measure the uniaxial stress as a function of strain for an elastomer. (From Ref. 74.)

the gage as a function of time, to obtain equilibrium values of the force f suitable for theoretical analysis. To prevent network degradation, the sample is generally protected with an inert atmosphere such as nitrogen, particularly in the case of measurements carried out at elevated temperatures. Both the sample cell and surrounding constant-temperature bath are glass, thus permitting use of a cathetometer or traveling microscope to obtain values of the strain α (see eq. 11) by measuring the distance between two lines marked on the central portion of the test sample. The cross-sectional area A_v is generally obtained by means of a micrometer. Stress–strain measurements are usually made using a sequence of increasing values of the elongation, with some inclusions of values out of sequence to test for reversibility (75, 76). For networks studied in the swollen state, a final variable is the volume fraction v_2 of polymer in the swollen network (28, 77).

Compression measurements are much more difficult than elongation measurements, since friction between the sample and the compression plates generally causes serious sample distortion ("barreling") (78–80). For this reason, equivalent experiments using equibiaxial extension, including inflation of sheets (81), are frequently carried out. Simultaneous prediction of elongation and compression mechanical behaviors is a stringent test which has been passed by the Flory–Erman model (60, 81, 82), which effectively accounts for the maximum in the reduced force found as the compression increases. Furthermore, this model has been extended to birefringence (83) and chain orientation measurements (84, 85). The elastic free energy expression, as in equation (16), yields the following relationship between the tensile reduced force $[f^*]$ (defined in equ. 11) and the uniaxial deformation α along the x axis:

$$[f^*] = (\xi kT/V) v_{2S}^{2/3} \{1 + (\mu_e/\xi)[\alpha K(\lambda_x^2) - \alpha^{-2} K(\lambda_y^2)](\alpha - \alpha^{-2})^{-1}\} \quad (21)$$

According to equations (1) and (2),

$$\mu_e/\xi = 2/(\phi - 2) \quad (22)$$

The function K is defined as

$$K(\lambda_t^2) \equiv B_t[\dot{B}_t(B_t + 1)^{-1} + g_t(\dot{B}_t g_t + \dot{g}_t B_t)(g_t B_t + 1)^{-1}] \quad (23)$$

with

$$\dot{B}_t \equiv \partial B_t/\partial \lambda_t^2 = B_t\{[2\lambda_t(\lambda_t - 1)]^{-1} + (1 - 2\zeta\lambda_t) \times [2\lambda_t(1 + \lambda_t - \zeta\lambda_t^2)]^{-1} - 2\dot{g}_t(1 + g_t)^{-1}\} \quad (24)$$

and

$$\dot{g}_t \equiv \partial g_t/\partial \lambda_t^2 = \kappa^{-1} - \zeta(1 - 3\lambda_t/2) \quad (25)$$

The deformation ratios λ_t ($t = x, y, z$) characterizing the deformation of the swollen network relative to the reference state of volume V_0 are related to α by

$$\lambda_x = \alpha v_2^{-1/3} v_{2S}^{1/3} \quad (26)$$

$$\lambda_y = \lambda_z = \alpha^{-1/2} v_2^{-1/3} v_{2S}^{1/3} \quad (27)$$

In the limit of vanishing constraints, $\kappa = 0$, $\kappa\zeta = 0$, and $K(\lambda_t^2) = 0$. The reduced force given by equation (21) then becomes equal to the phantom modulus $[f^*]_{ph}$. The opposite extreme corresponds to the affine transformation of the distribution of chain vectors, which is reached in the limit $\kappa \to \infty$, $\zeta = 0$. Then $K(\lambda_t^2) = 1 - \lambda_t^{-2}$. The nature of real network deformation is between affine and phantom behaviors and is characterized by a set of parameters κ and ζ, κ being itself linked to the network topology by the interpenetration concept (eqs. 14 and 15). For example, in the case of a perfect network, combination of equations (5), (14), and (15) leads to

$$\kappa \propto M_c^{1/2} \quad (28)$$

Table 1. Characterization of *cis*-1, 4-Polyisoprene Vulcanizates

Sample	$10^{-5}\bar{M}_n$ (g/mol)	Dicup[a]	$[f^*]_{ph}$ (N/mm^2)[b]	κ	ζ	$10^{-3}M_c$ (g/mol)	$10^5\mu/V$ (mol/cm^3)
1IR360	3.60	1.0	0.2350	4.6	0.000	4.69	9.58
2IR360	3.60	0.8	0.2113	4.9	0.011	5.19	8.63
1IR245	2.45	0.8	0.1967	4.9	0.017	5.44	8.17
2IR245	2.45	0.6	0.1592	5.4	0.011	6.62	6.68
3IR245	2.45	0.35	0.0930	6.8	0.000	10.72	4.06

[a]Weight percent in bulk polymer.
[b]At $T = 303$ K.
From Ref. 65.

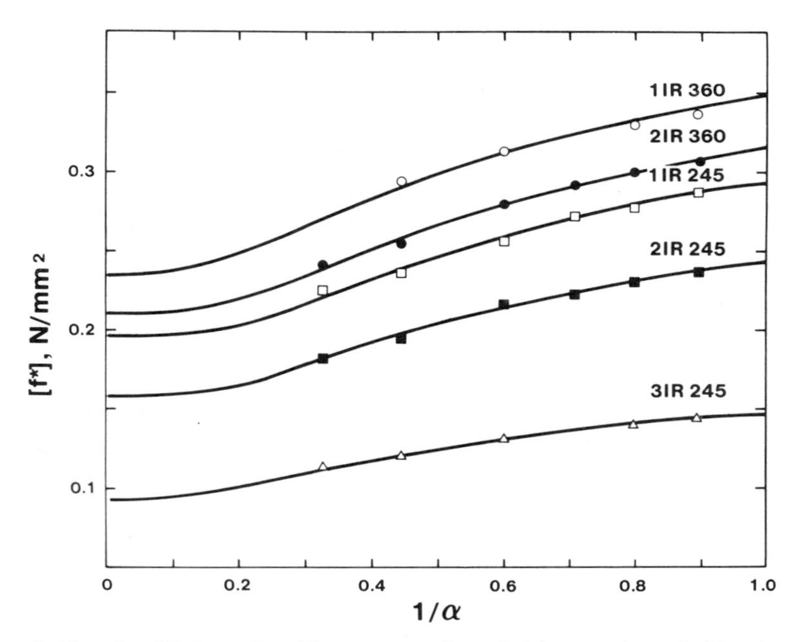

Figure 2. Plot of equilibrium reduced force versus reciprocal of the extension ratio. Experimental points were obtained by extrapolation of relaxation measurements to infinite time. Continuous curves are theoretical predictions calculated from equation (21) with parameters listed in Table 1. (From Ref. 65.)

The proportionality constant depends on polymer microstructure through the quantities C_∞, N, l^2, and m_0 of equation (15). The other parameter ζ is the result of the relationship between κ and network inhomogeneities (65), and its magnitude is estimated by experiment.

An illustrative analysis of typical elongation is taken from a study by Queslel et al. (65). Precursor polymer was an anionic commercial polyisoprene (Shell IR307) with a high *cis*-1,4 stereochemical structure. Two batches of varying number average molecular weight, namely $\bar{M}_n = 3.60 \times 10^5$ and 2.45×10^5 g/mol, were prepared by degradative working on a two-roll mill. Each precursor was mixed in bulk with several amounts of pure dicumyl peroxide, molded, and cured (30 minutes at 170°C). See Table 1 for designation of vulcanizates, number average molecular weight of precursor polymer, and quantity of peroxide. The equilibrium reduced forces, which were obtained by a method involving relaxation and extrapolation to infinite time, are plotted versus the inverse of the deformation ratio for all five vulcanizates in Figure 2. They were fitted by least-squares analysis with theoretical curves calculated through equations (6), (7), (14), (15), and (21). For *cis*-1,4-polyisoprene, the relationship between $\langle r^2 \rangle_0$ and the number average molecular weight between cross-links M_c is

$$\langle r^2 \rangle_0 = 3.8 \times 10^7 M_c/N_A \tag{29}$$

where N_A is Avogadro's number. The quantities M_c and $\langle r^2 \rangle_0$ are expressed in grams per mole and square centimeters, respectively. Equation (29) was obtained by choosing a characteristic ratio C_∞ equal to 5.0, a mean-square bond length of 2.18 Å^2, and an average number of bonds per repeat unit of 3.94; these are typical values for IR307 polymer (92% cis, %5 trans) (86, 87).

The mesh size M_c and the cross-link density μ/V are directly obtained from the least-squares analysis and are reported in Table 1. The results seem to show that 1 mole of dicumyl peroxide creates approximately 3 moles of C—C cross-links in this polymer; κ increases with M_c and, at constant κ, ζ increases with the number density of dangling chains $2\rho/\bar{M}_n$.

The Swelling Method

Samples to be used in swelling equilibrium experiments are generally extracted first to remove any soluble materials and then dried and weighed. They are next placed in a thermodynamically good solvent (1) and periodically reweighed swollen until constant weight is observed (75), at which point equilibrium has been achieved. This extent of swelling is conveniently characterized by v_{2m}, the volume fraction of polymer in the network at maximum (equilibrium) swelling. It is generally calculated assuming simple additivity of volumes.

Swelling measurements are extremely simple. They require for their interpretation, however, information on the interactions between the polymer and the swelling solvent at the temperature of equilibration. Generally, all solvents and temperatures chosen for such investigations are those for which such thermodynamic parameters are available (88–91). The polymer–solvent interaction parameter χ can be obtained by several techniques including osmometry, vapor sorption, gas–liquid chromatography, freezing-point depression of solvent, intrinsic viscosity, swelling equilibrium, and critical solution temperatures (91). Figure 3 is a schematic diagram of the apparatus designed to measure, by differential solvent vapor sorption, χ and the activities of solvent in un-cross-linked and cross-linked polymers (90). A Cahn RG electrobalance is housed in a stainless steel case B. Polymer samples on quartz sample pans are suspended on both sides of the balance with 32-gage nichrome wires. A strip-chart recorder is used to display portions of the electrobalance readout. The un-cross-linked and cross-linked samples of approximately equal weights are placed on the two sides of the electrobalance. A glass weighing chamber K communicates with the balance case, and a quartz spring is attached to a removable chamber cap D. An un-cross-linked polymer sample is suspended on the quartz spring, and all dry sample weights are accurately determined. Reservoir C, which serves to cushion against sudden pressure

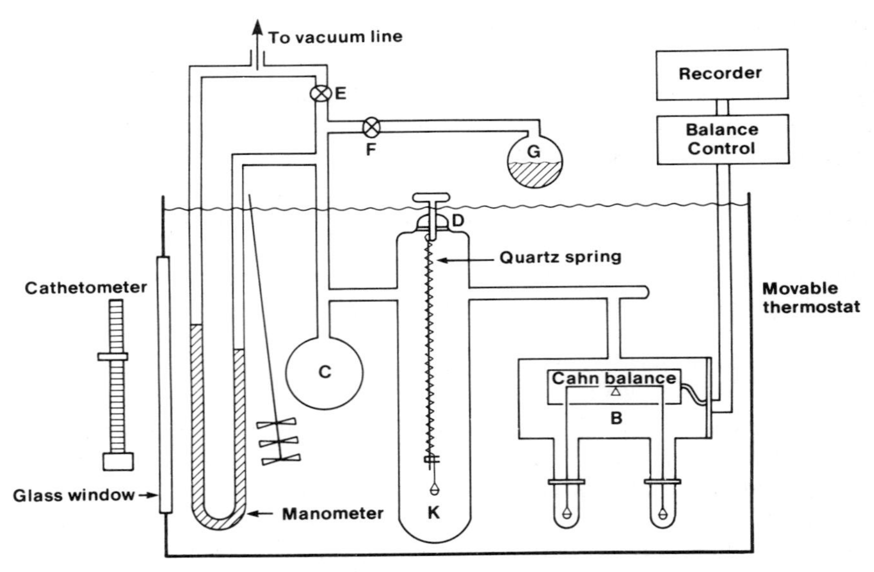

Figure 3. Schematic diagram of differential vapor sorption measurement apparatus. (From Ref. 90.)

Table 2. Results of Swelling Measurements for v_{2m} in Toluene and Calculations of the Phantom Modulus

Sample	v_{2m}	κ	$\xi KT/V$ (N/mm^2)[a]
1	0.067	10	0.0163
2	0.056	12	0.0110

[a]At $T = 298$ K.
From Ref. 13.

changes and acts as a mercury trap, is connected to the chamber K and to one arm of the mercury manometer, which is conneted to a vacuum line. A solvent reservoir G can be opened to the system via a mercury float valve F. A similar valve E connects all parts of the apparatus to the vacuum line, and a water bath is used as a thermostat for the apparatus. Measurements of the displacement of the quartz spring and manometer are made with a cathetometer, through a glass window. Valve F is opened to allow solvent vapor into the system. The amount of solvent introduced is controlled by controlling the temperature of the solvent reservoir. Valve F is closed after each addition of vapor, and solvent vapor and polymer samples are then allowed to equilibrate.

The integral and differential sorptions are observed on the respective balances and the pressure of solvent vapor is recorded. It is then possible to calculate the volume fraction of polymer in the cross-linked and un-cross-linked swollen samples, v_2^c and v_2^u, respectively, at a solvent pressure $p_1^c = p_1^u$ and thus to know p_1^c and p_1^u at equal volume fractions $v_2 = v_2^c = v_2^u$ and to calculate the ratio of activities $a_1^c/a_1^u = p_1^c/p_1^u$. If the ratio of molar volumes of polymer and solvent is high, χ at volume fraction v_2 is obtained by solving

$$RT \ln\left(\frac{p_1^u}{p_1^0}\right) = RT[\ln(1 - v_2) + \chi v_2^2 + v_2] \tag{30}$$

where p_1^0 is the pure solvent vapor pressure.

The ratio of solvent activities over the network and over the un-cross-linked polymer at constant volume fraction v_2 is obtained by differentiation of the elastic free energy change ΔA with respect to the number of moles of solvent n_1:

$$RT \ln\left(\frac{a_1^c}{a_1^u}\right) = \left(\frac{\partial \Delta A}{\partial n_1}\right)_{T,p} \tag{31}$$

Swelling is an isotropic dilation with deformation ratios given by

$$\lambda = \lambda_x = \lambda_y = \lambda_z = v_2^{-1/3} v_{2S}^{1/3} \tag{32}$$

where V_1 is the solvent molar volume.

A plot of $\lambda \ln(a_1^c/a_1^u)$ versus λ^2 exhibits a maximum, which is qualitatively predicted if the Flory–Erman elastic free energy change (eq. 16) is used in equation (31) (12, 15, 92–94). At equilibrium (maximum) swelling,

$$RT[\ln(1 - v_{2m}) + \chi v_{2m}^2 + v_{2m}] + \left(\frac{\partial \Delta A}{\partial n_1}\right)_{T,p} = 0 \tag{33}$$

Use of equation (16) for ΔA leads to the following relationship, which can be applied to network characterization from an equilibrium swelling experiment:

$$\ln(1 - v_{2m}) + \chi v_{2m}^2 + v_{2m} = -(\xi/V) V_1 v_{2m}^{1/3} v_{2S}^{2/3} \{1 + (\mu_e/\xi) K(v_{2m}^{-2/3} v_{2S}^{2/3})\} \tag{34}$$

An example of such an application is given by Erman and Baysal (13). Two cross-linked polystyrene networks were prepared by polymerization of mixtures of styrene, benzoyl peroxide, and *p*-divinylbenzene. Samples with dimensions of $\sim 8 \times 3 \times 1\,\text{mm}^3$ were machined from the dry, glassy cross-linked networks. Two pieces of thin wires were placed approximately 6 mm apart on each sample, forming fiducial marks from which linear deformation upon swelling was measured. Samples were stored horizontally in a constant-temperature bath (± 0.02°C), and length changes were measured in the immersed state with a traveling microscope. The degree of cross-linking was obtained by measuring the degree of equilibrium swelling of the networks in toluene. The χ parameter of polystyrene–toluene at 22–30°C is obtained from the data compiled by Orwoll (91). Specifically,

$$\chi = 0.455 - 0.155 v_2 \tag{35}$$

For polystyrene (at 25°C),

$$\kappa = 1.334(\xi kT/V)^{-1/2} \tag{36}$$

where the numerical coefficient is in units of newtons and millimeters ($\text{N}^{1/2}$/mm). Equation (36) was obtained from equations (14) and (15), assuming a perfect tetrafunctional network: that is, $\xi/V = \mu/V = \rho/2M_c$, with $C_\infty = 8.5$, $m_0 = 100$ g/mol, $l = 1.53 \times 10^{-8}$ cm, and $N = 2$.

The results of swelling measurements of v_{2m} in toluene and calculations of the phantom modulus $\xi kT/V$ by combination of equations (23) and (34)–(36) are reported in Table 2 with values of κ calculated through equation (36) from cycle rank values. Subsequent measurements of swelling ratios were carried out in a mixture of toluene and methanol (75/25 wt%) mixture. By making use of the cycle rank values determined above, the interaction parameter for the polystyrene–toluene–methanol system was calculated at different temperatures through equation (34). Least-squares analysis of the data points gave the following relationship for the dependence of χ on v_2 and T:

$$\chi = 0.4103 + 16.03/T + (0.5054 + 5.60/T)v_2 \tag{37}$$

Another example of swelling equilibrium analysis was given by Queslel et al. (95), who showed that swelling equilibrium and stress–strain characterization of networks were consistent if the results were interpreted through the constrained junction model proposed by Flory and Erman.

REFERENCES

1. P. J. Flory, *Principles of Polymer Chemistry*, Cornell University Press, Ithaca, NY, 1953, Chaps. 12 and 13.
2. P. J. Flory, *J. Chem. Phys.*, **10**, 51 (1942).
3. P. J. Flory, *Proc. R. Soc. London.*, **A351**, 351 (1976).
4. P. J. Flory, *Macromolecules*, **15**, 99 (1982).
5. A. Charlesby, *Proc. R. Soc. London*, **A222**, 542 (1954).
6. A. Charlesby and S. H. Pinner, *Proc. R. Soc. London*, **A249**, 367 (1959).
7. C. W. Macosko and D. R. Miller, *Macromolecules*, **9**, 199 (1976).
8. D. R. Miller and C. W. Macosko, *Macromolecules*, **9**, 206 (1976).
9. J. P. Queslel and J. E. Mark, *J. Chem. Phys.*, **82**, 3449 (1985).
10. N. R. Langley, *Macromolecules*, **1**, 348 (1968).
11. J. P. Queslel and J. E. Mark, *Adv. Polym. Sci.*, **71**, 229 (1985).
12. P. J. Flory, *Macromolecules*, **12**, 119 (1979).
13. B. Erman and B. M. Baysal, *Macromolecules*, **18**, 1696 (1985).
14. M. Gottlieb and R. J. Gaylord, *Macromolecules*, **17**, 2024 (1984).
15. R. W. Brotzman and B. E. Eichinger, *Macromolecules*, **16**, 1131 (1983).
16. A. Y. Coran, in *Science and Technology of Rubber*, F. R. Eirich, Ed., Academic Press, New York, 1978.
17. J. E. Mark and P. H. Sung, *Eur. Polym. J.*, **16**, 1223 (1980).
18. J. E. Mark, *Rubber Chem. Technol.*, **54**, 809 (1981).
19. J. E. Mark, *Adv. Polym. Sci.*, **44**, 1 (1982).
20. P. J. Flory, *Chem. Rev.*, **35**, 51 (1944).

21. P. J. Flory, *Statistical Mechanics of Chain Molecules*, Wiley-Interscience, New York, 1969.
22. M. V. Volkenstein, *Configurational Statistics of Polymeric Chains*, translated from the Russian edition by S. N. Timasheff and M. J. Timasheff, Wiley-Interscience, New York, 1963.
23. J. Bastide, C. Picot, and S. Candau, *J. Polym. Sci., Polym. Phys. Ed.*, **17**, 1441 (1979).
24. C. A. J. Hoeve and P. J. Flory, *Biopolymers*, **13**, 677 (1974).
25. J. E. Mark, *Biopolymers*, **15**, 1853 (1976).
26. G. Gee, *Trans. Faraday Soc.*, **42**, 585 (1946).
27. G. Allen, M. J. Kirkham, J. Padget, and C. Price, *Trans. Faraday Soc.*, **67**, 1278 (1971).
28. B. Erman, *Br. Polym. J.*, **17**, 140 (1985).
29. L. R. G. Treloar, *The Physics of Rubber Elasticity*, 3rd ed., Oxford University Press Clarendon, Oxford, London/New York, 1975.
31. F. Harari, *Graph Theory*, Addision-Wesley, Reading MA, 1971.
32. P. J. Flory, *Br. Polym. J.*, **17**, 96 (1985).
33. J. Scanlan, *J. Polym. Sci.*, **43**, 501 (1960).
34. L. C. Case, *J. Polym. Sci.*, **45**, 397 (1960).
35. D. S. Pearson and W. W Graessley, *Macromolecules*, **11**, 528 (1978).
36. J. P. Queslel and J. E. Mark, *Polym. Bull.*, **12**, 311 (1984).
37. M. Hoffmann, *Makromol. Chem.*, **183**, 2191 (1982).
38. P. J. Flory and B. Erman, *Macromolecules*, **15**, 800 (1982).
39. J. P. Queslel and J. E. Mark, *Polym. J.*, **18**, 263 (1986).
40. B. E. Eichinger, *Annu. Rev. Phys. Chem.*, **34**, 359 (1983).
41. P. J. Flory, *Polym. J.*, **17**, 1 (1985).
42. G. Ronca and G. Allegra, *J. Chem. Phys.*, **63**, 4990 (1975).
43. S. F. Edwards, *B. Polym. J.*, **17**, 122 (1985).
44. W. W. Graessley and D. S. Pearson, *J. Chem. Phys.*, **66**, 3363 (1977).
45. K. Dusek and W. Prins, *Adv. Polym. Sci.*, **6**, 1 (1969).
46. G. Gee, *Polymer*, **7**, 373 (1966).
47. M. C. Wang and E. Guth, *J. Chem. Phys.*, **20**, 1144 (1952).
48. W. R. Krigbaum and R.-J. Roe, *Rubber Chem. Technol.*, **38**, 1039 (1965).
49. F. T. Wall, *J. Chem. Phys.*, **11**, 527 (1943).
50. W. Kuhn and F. Grun, *J. Polym. Sci.*, **1**, 183 (1946).
51. D. Y. Yoon and P. J. Flory, *J. Chem. Phys.*, **61**, 5366 (1974).
52. P. J. Flory and V. W. C. Chang, *Macromolecules*, **9**, 33 (1976).
53. H. M. James and E. Guth, *J. Chem. Phys.*, **15**, 669 (1947).
54. M. Mooney, *J. Appl. Phys.*, **11**, 582 (1940).
55. G. Gee, *Trans. Faraday Soc.*, **50**, 881 (1954).
56. B. Erman and P. J. Flory, *Macromolecules*, **15**, 806 (1982).

57. B. Erman, *J. Polym. Sci., Polym. Phys. Ed.*, **19**, 829 (1981).
58. L. R. G. Treloar, *Br. Polym. J.*, **14**, 121 (1982).
59. J. P. Queslel and J. E. Mark, *Adv. Polym. Sci.*, **65**, 135 (1984).
60. M. Gottlieb and R. J. Gaylord, *Polymer*, **24**, 1644 (1983).
61. P. J. Flory, in *Contemporary Topics in Polymer Science*, Vol. 2, E. M. Pearce and J. R. Schaefgen, Eds., Plenum, New York, 1977, p. 1.
62. P. J. Flory, *J. Chem. Phys.*, **66**, 5720 (1977).
63. B. Erman and P. J. Flory, *J. Chem. Phys.*, **68**, 5363 (1978).
64. B. Erman and P. J. Flory, *Macromolecules*, **16**, 1607 (1983).
65. J. P. Queslel, P. Thirion, and L. Monnerie, *Polymer*, **27**, 1869 (1986).
66. P. J. Flory and J. R. Rehner, *J. Chem. Phys.*, **11**, 512 (1943).
67. P. J. Flory and J. R. Rehner, *J. Chem. Phys.*, **11**, 521 (1943).
68. P. J. Flory and Y.-I. Tatara, *J. Polym. Sci., Polym. Phys. Ed.*, **13**, 683 (1975).
69. J. P. Queslel and J. E. Mark, "Elasticity," in *Encyclopedia of Polymer Science and Engineering*, Vol. 5, 2nd ed., Wiley, New York, 1986, p. 365.
70. J. E. Mark, *Makromol. Chem.*, Suppl. 2, 87 (1979).
71. J. E. Mark, *J. Chem. Educ.*, **58**, 898 (1981).
72. J. E. Mark, *Rubber Chem. Technol.*, **55**, 762 (1982).
73. R. P. Brown, *Physical Testing of Rubbers*, Applied Science Publishers, Barking, 1979.
74. J. E. Mark, *J. Polym. Sci., Macromol. Rev.*, **11**, 135 (1976).
75. J. E. Mark and J. L. Sullivan, *J. Chem. Phys.*, **66**, 1006 (1977).
76. M. A. Llorente and J. E. Mark, *J. Chem. Phys.*, **71**, 682 (1979).
77. B. Erman, W. Wagner, and P. J. Flory, *Macromolecules*, **13**, 1554 (1980).
78. M. J. Forster, *J. Appl. Phys.*, **26**, 1104 (1955).
79. F. P. Wolf, *Polymer*, **13**, 347 (1972).
80. R. Y. S. Chen, C. U. Yu, and J. E. Mark, *Macromolecules*, **6**, 746 (1973).
81. H. Pak and P. J. Flory, *J. Polym. Sci., Polym. Phys. Ed.*, **17**, 1845 (1979).
82. B. Erman and P. J. Flory, *J. Polym. Sci., Polym. Phys. Ed.*, **16**, 1115 (1978).
83. B. Erman and P. J. Flory, *Macromolecules*, **16**, 1601 (1983).
84. B. Erman and L. Monnerie, *Macromolecules*, **18**, 1985 (1985).
85. J. P. Queslel, B. Erman, and L. Monnerie, *Macromolecules*, **18**, 1991 (1985).
86. J. W. Mays, Ph.D. Thesis, University of Akron, Akron, OH, 1984.
87. J. W. Mays, N. Hadjichristidis, and L. J. Fetters, *Macromolecules*, **17**, 2723 (1984).
88. J. Brandrup and E. H. Immergut, Eds., *Polymer Handbook*, 2nd ed., Wiley, New York, 1975.
89. B. E. Eichinger and P. J. Flory, *Trans. Faraday Soc.*, **64**, 2035, 2053, 2061, 2066 (1968).
90. L. Y. Yen and B. E. Eichinger, *J. Polym. Sci., Polym. Phys. Ed.*, **16**, 117, 121 (1978).

91. R. A. Orwoll, *Rubber Chem. Technol.*, **50**, 451 (1977).
92. R. W. Brotzman and B. E. Eichinger, *Macromolecules*, **15**, 531 (1982).
93. R. W. Brotzman and B. E. Eichinger, *Macromolecules*, **14**, 1445 (1981).
94. G. Gee, J. B. M. Herbert, and R. C. Roberts, *Polymer*, **6**, 541 (1965).
95. J. P. Queslel, F. Fontaine, and L. Monnerie, *Polymer*, **29**, 1086 (1988).

CHAPTER

17

STANDARD POLYMERS

PETER H. VERDIER and LESLIE E. SMITH

Polymers Division National Bureau of Standards Gaithersburg, Maryland

STANDARD POLYMER DATA

A variety of polymer samples with stated values of chemical or physical properties and available to the world at large are widely employed in polymer science and technology. Such "standard polymers" are used in three ways. First, they are used to calibrate instruments and techniques which determine physicochemical properties of polymers in a relative sense. Chromatographic methods of molecular weight determination are well-known examples of relative measurements that require calibration. Second, they are used to check the accuracy of instruments and techniques which in principle are absolute but are in practice complex enough to need verification. Finally, such polymers are often used as well-characterized starting materials for research.

This chapter surveys commercially available synthetic polymers for which molecular weight (and sometimes other properties as well) is given. Standards which merely happen to be made of polymeric materials are not included. (Some examples of such incidental polymer standards, all of which are issued by the National Bureau of Standards, are polystyrene latex spheres used as size standards; polyester films used as gas transmission standards, fluoropolymers used as dielectric standards; and elastomer copolymers used as compounding standards.) In addition, the techniques and problems encountered with biopolymers are sufficiently different from those for synthetic polymers to warrant their exclusion from this survey. An exception has been made for some polysaccharide standards whose nature and characterization appear to be comparable with those of synthetic polymers.

The number of standard polymers available has increased greatly since the last edition of the Encyclopedia of Polymer Science and Engineering (Second Edition, 1988) was published, and is still increasing. The tables which follow list standard samples for more than four dozen kinds of polymer. Because of the rapid increase in the availability of these samples, the listings of standard polymers in this chapter should be regarded as exemplary

Table 1. Standard Samples of Polybutadiene

Range in Molecular Weight	Number of Samples	Range of $\bar{M}_w/\bar{M}_n$	Supplier Codes[a]
2×10^2–5×10^2	1	1.07–1.17	G
5×10^2–1×10^3	8	1.07–1.2	C, D, G, P, Q, R
1×10^3–2×10^3	6	1.06–1.2	C, D, P, Q, R, W
2×10^3–5×10^3	8	1.08–1.2	C, D, G, P, Q, R, W
5×10^3–1×10^4	3	1.07–1.3	C, G, R
1×10^4–2×10^4	0		
2×10^4–5×10^4	5	1.04–1.11	C, G, P, R
5×10^4–1×10^5	0		
1×10^5–2×10^5	3	1.26–1.87	C, P, R
2×10^5–5×10^5	8	1.32–4.5	C, P, R

[a]Suppliers are listed in Table 15.

Table 2. Standard Samples of Poly(dimethyl siloxane)

Range in Molecular weight	Number of Samples	Range of $\bar{M}_w/\bar{M}_n$	Supplies Codes[a]
1×10^3–2×10^3	1	1.6	C
2×10^3–5×10^3	2	1.06–1.1	C
5×10^3–1×10^4	2	1.2–1.4	C
1×10^4–2×10^4	2	1.4–1.7	C
2×10^4–5×10^4	2	1.7–1.8	C
5×10^4–1×10^5	4	2.2–2.5	A, C, P, R
1×10^5–2×10^5	5	2.4–3.5	A, C, P, R
2×10^5–5×10^5	2	3.5	C
5×10^5–1×10^6	3	3.2–4.6	A, C, P
1×10^6–2×10^6	1	5.6	R

[a]Suppliers are listed in Table 15.

Table 3. Standard Samples of Linear Polyethylene

Range in Molecular Weight	Number of Samples	Range of $\bar{M}_w/\bar{M}_n$	Supplier Codes[a]
5×10^2–1×10^3	4	1.11–1.15	O, Q
1×10^3–2×10^3	2	1.20	O, Q
2×10^3–5×10^3	2	1.14	O, Q
5×10^3–1×10^4	0		
1×10^4–2×10^4	5	1.3–2.0	N, O, Q, R
2×10^4–5×10^4	13	1.1–6.0	N, O, Q, R
5×10^4–1×10^5	3	2.9	N, O, P
1×10^5–2×10^5	2	1.2	N, O

[a]Samples of hydrogenated polybutadiene are not included in this table; see polybutadiene hydrogenated Table 13.
[b]See Table 13 for branched polyethylene.
[c]Suppliers are listed in Table 15.

Table 4. Standard Samples of Poly(ethylene glycol)[a]

Range in Molecular Weight	Number of Samples	Range of $\bar{M}_w/\bar{M}_n$	Supplier Codes[b]
5×10^1–1×10^2	1	1.0	C
1×10^2–2×10^2	9	1.0–1.11	B, C, O, P, Q, R
2×10^2–5×10^2	13	1.07–1.2	B, C, O, P, Q, R
5×10^2–1×10^3	9	1.07–1.10	B, C, O, P, Q, R
1×10^3–2×10^3	11	1.05–1.10	B, C, O, P, Q, R
2×10^3–5×10^3	11	1.03–1.10	B, C, O, Q, R
5×10^3–1×10^4	6	1.03–1.10	B, C, O, P
1×10^4–2×10^4	12	1.04–1.34	B, C, O, P, Q, R
2×10^4–5×10^4	3	1.06–1.13	B, O, P

[a]Although poly(ethylene oxide) and poly(ethylene glycol) have the same basic repeating unit, [$—CH_2CH_2O—$], the name poly(ethylene glycol) implies that there are hydroxyl groups on both ends of the chain (which is important for some applications of these materials), whereas poly(ethylene oxide) implies nothing about the nature of the end groups. We have therefore tabluated these materials with the names used by their suppliers.
[b]Suppliers are listed in Table 15.

rather than exhaustive. Nearly all the available standards are linear or essentially linear homopolymers, though a few branched polymers and a few copolymers are available. Almost without exception, the primary certified or otherwise stated properties of these materials are molecular weights of various kinds. (A few polymers of specified tacticity are also available; these are

Table 5. Standard Samples of Poly(ethylene oxide)[a]

Range in Molecular Weight	Number of Samples	Range of $\bar{M}_w/\bar{M}_n$	Supplier Codes[b]
1×10^4–2×10^4	3	1.10	B, C, O
2×10^4–5×10^4	9	1.03–1.14	B, C, O, T, V, W
5×10^4–1×10^5	6	1.02–1.1	B. C, O, T, V, W
1×10^5–2×10^5	6	1.03–1.1	B, C, O, T, V, W
2×10^5–5×10^5	6	1.04–1.1	B, C, O, T, V, W
5×10^5–1×10^6	9	1.04–1.10	B, C, O, T, V, W
1×10^6–2×10^6	3	1.1–1.12	T, V, W

[a]Although poly(ethylene oxide) and poly(ethylene glycol) have the same basic repeating unit, [$—CH_2CH_2O—$], the name poly(ethylene glycol) implies that there are hydroxyl groups on both ends of the chain (which is important for some applications of these materials), whereas poly(ethylene oxide) implies nothing about the nature of the end groups. We have therefore tabulated these materials with the names used by their suppliers.
[b]Suppliers are listed in Table 15.

Table 6. Standard Samples of Polyisoprene

Range in Molecular Weight	Number of Samples	Range of $\bar{M}_w/\bar{M}_n$	Supplier Codes
5×10^2–1×10^3	4	1–1.2	D, G, W
1×10^3–2×10^3	4	1.05–1.2	B, O, P, R
2×10^3–5×10^3	4	1.05–1.10	B, G, O, P
5×10^3–1×10^4	4	1.04–1.2	D, G, O, R
1×10^4–2×10^4	11	1.04–1.2	B, D, G, O, P, R
2×10^4–5×10^4	6	1.04–1.34	B, D, O, P, R
5×10^4–1×10^5	5	1.04–1.16	B, G, O, P, R
1×10^5–2×10^5	3	1.04–1.1	B, O, P
2×10^5–5×10^5	5	1.05–1.1	B, O, P
5×10^5–1×10^6	2	1.05–1.10	O, P
1×10^6–2×10^6	2	1.08	B, O
2×10^6–5×10^6	2	1.15	B, O

[a]Suppliers are listed in Table 15.

Table 7. Standard Samples of Poly(methyl methacrylate)[a]

Range in Molecular Weight	Number of Samples	Range of $\bar{M}_w/\bar{M}_n$	Supplier Codes[b]
2×10^3–5×10^3	2	1.15	B, O
5×10^3–1×10^4	2	1.15	N, Q
1×10^4–2×10^4	4	1.10–1.8	B, C, O
2×10^4–5×10^4	12	1.07–2.4	B, C, N, O, P, Q, R
5×10^4–1×10^5	14	1.08–2.0	A, B, C, O, P, Q, R
1×10^5–2×10^5	12	1.04–2.44	B, N, O, P, Q, R
2×10^5–5×10^5	16	1.09–4.1	B, C, O, P, Q, R
5×10^5–1×10^6	12	1.10–3.3	B, C, O, P, Q, R
1×10^6–2×10^6	4	1.3–4.0	B, O, P
	Deuterated poly(methyl methacrylate)		
1×10^4–2×10^4	1	1.1	O
2×10^4–5×10^4	1	1.1	O
5×10^4–1×10^5	2	1.1	O
1×10^5–2×10^5	2	1.1	O
2×10^5–5×10^5	1	1.1	O

[a]See Table 14 for stereoregular poly(methyl methacrylate).
[b]Suppliers are listed in Table 15.

Table 8. Standard Samples of Poly(α-methylstyrene)

Range in Molecular Weight	Number of Samples	Range of $\bar{M}_w/\bar{M}_n$	Supplier Codes[a]
1×10^3–2×10^3	1	1.15	O
2×10^3–5×10^3	1	1.04	O
5×10^3–1×10^4	5	1.04–1.20	O, P, Q, R
1×10^4–2×10^4	2	1.04–1.10	O, P
2×10^4–5×10^4	5	1.04–1.15	O, P, Q, R
5×10^4–1×10^5	5	1.05–1.10	O, P, Q, R
1×10^5–2×10^5	5	1.05–1.10	O, P, Q, R
2×10^5–5×10^5	4	1.05–1.10	O, P, Q, R
5×10^5–1×10^6	4	1.05–1.10	O, P, R
1×10^6–2×10^6	1	1.10	O

[a]Suppliers are listed in Table 15.

tabulated separately.) The rapid increase in samples of specified molecular weights is surely a result of the widespread use of gel permeation chromotography (GPC) (also known as size exclusion chromatography, or SEC), with its attendant need for calibrants of known molecular weights.

A variety of techniques are used to obtain molecular weights for standard polymers. Absolute techniques employed include end-group analysis and membrane osmometry for number average molecular weight $\bar{M}_n$; and light scattering and sedimentation equilibrium ultracentrifugation for weight average molecular weight $\bar{M}_w$. Nonabsolute methods, which require calibration with samples of known molecular weight, include vapor phase osmometry for $\bar{M}_n$; limiting viscosity number (LVN) for viscosity average molecular weight $\bar{M}_v$ (usually close to $\bar{M}_w$); and GPC for $\bar{M}_n$, $\bar{M}_w$, peak molecular weight, and at least qualitatively, distribution in molecular weight. General references (1–3) may be consulted for the definitions of these quantities and descriptions of the techniques.

The present (mid-1987) availability of standard polymers is summarized in the 15 tables that follow. Tables 1–12 list standard samples available for the polymers for which there are four or more suppliers, with the ranges of molecular weight available and the ranges of polydispersity $\bar{M}_w/\bar{M}_n$, a measure of the breadth of the distribution in molecular weight. The ranges in molecular weight are chosen to give three ranges per decade.

Table 13 lists similar information for polymers with fewer than four suppliers, except that only the total range in molecular weight is shown. It should be noted that since many identical samples are sold by more than one supplier, the number of independent samples is frequently less than the total

Table 9. Standard Samples of Polystyrene[a,b]

Range in Molecular Weight	Number of Samples	Range of $\bar{M}_w/\bar{M}_n$	Supplier Codes[c]
1×10^2–2×10^2	1	1.0	O
2×10^2–5×10^2	3	1.1–1.15	P, T, V
5×10^2–1×10^3	17	1.03–1.3	B, D, G, O, P, Q, R, S, T, V
1×10^3–2×10^3	8	1.10–1.3	B, G, O, P, R
2×10^3–5×10^3	20	1.04–1.3	B, D, M, O, P, Q, R, S, T, V, W
5×10^3–1×10^4	22	1.04–1.19	B, C, D, G, M, O, P, Q, R, S, T, V
1×10^4–2×10^4	19	1.02–1.15	B, C, D, G, M, O, Q, R, S, T, V, W
2×10^4–5×10^4	25	1.01–1.4	B, C, D, G, M, N, O, P, Q, R, S, T, V
5×10^4–1×10^5	21	1.03–1.16	B, C, G, O, P, Q, R, S, W
1×10^5–2×10^5	21	1.02–1.18	C, D, G, M, N, O, P, Q, R, S, T, V, W
2×10^5–5×10^5	29	1.04–3.8	A, B, C, D, M, N, O, P, Q, R, S, T,V, W
5×10^5–1×10^5	20	1.01–1.12	B, C, D, G, M, O, P, Q, R, S, T, V
1×10^6–2×10^6	20	1.05–1.85	B, C, D, G, M, N, O, P, Q, R, S, T, V, W
2×10^6–5×10^6	17	1.04–1.2	B, O, P, R, T, V, W
5×10^6–1×10^7	9	1.14–1.2	B, O, P, T, V
1×10^7–2×10^7	5	1.2–1.3	B, O, P
2×10^7–5×10^7	4	1.2–1.3	P, T, V
		Deuterated Polystyrenes	
5×10^3–1×10^4	2	1.05	O
1×10^4–2×10^4	2	1.05–1.1	O
2×10^4–5×10^4	1	1.05	O
5×10^4–1×10^5	2	1.05–1.1	O
1×10^5–2×10^5	2	1.05–1.1	O
2×10^5–5×10^5	2	1.10	O
5×10^5–1×10^6	1	1.10	O
1×10^6–2×10^6	1	1.30	O
		Polystyrene Oligomer Mixture	
13 peaks in distribution in molecular weight, from 208 to 1458			Q, R

[a]See Table 13 polystyrene star molecules.
[b]See Table 14 for stereoregular polystyrene.
[c]Suppliers are listed in Table 15.

Table 10. Standard Samples of Sodium Polystyrene Sulfonate

Range in Molecular Weight	Number of Samples	Range of $\bar{M}_w/\bar{M}_n$	Supplier Codes[a]
1×10^3–2×10^3	4	1.25	O, P, Q, R
2×10^3–5×10^3	5	1.10	B, O, P, Q, R
5×10^3–1×10^4	5	1.10	B, O, P, Q, R
1×10^4–2×10^4	5	1.10	B, O, P, Q, R
2×10^4–5×10^4	5	1.10	B, O, P, Q, R
5×10^4–1×10^5	10	1.10	B, O, P, Q, R
1×10^5–2×10^5	5	1.10	B, O, P, Q, R
2×10^5–5×10^5	6	1.10	B, O, P, Q, R
5×10^5–1×10^6	5	1.10	B, O, P, Q, R
1×10^6–2×10^6	6	1.10	B, O, P, Q, R

[a]Suppliers are listed in Table 15.

Table 11. Standard Samples of Poly(vinyl acetate)

Range in Molecular Weight	Number of Samples	Range of $\bar{M}_w/\bar{M}_n$	Supplier Codes[a]
1×10^5–2×10^5	9	2.4–9.1	A, C, P, R
2×10^5–5×10^5	3	2.6–3.2	A, C, R

[a]Suppliers are listed in Table 15.

Table 12. Standard Samples of Poly(vinyl chloride)

Range in Molecular Weight	Number of Samples	Range of $\bar{M}_w/\bar{M}_n$	Supplier Codes[a]
2×10^4–5×10^4	1	1.8	C
5×10^4–1×10^5	6	1.8–2.5	A, C, P, R
1×10^5–2×10^5	6	2.1–2.4	C, R

[a]Suppliers are listed in Table 15.

Table 13. Standard Samples of Miscellaneous Polymers

Polymer	Range in Molecular Weight	Number of Samples	Range of $\bar{M}_w/\bar{M}_b$ (if given)	Supplier Codes[a]
Poly(hexamethylenediamine-*co*-adipic acid) (nylon-6, 6)	2×10^4–1×10^5	8	2.2–2.8	C
Phenoxy resin	5×10^4–1×10^5	1	3.8	R
Polyacrylamide	2×10^4–1×10^7	19	15–2.6	C, P
Poly(acrylic acid)	2×10^3–5×10^6	16	22	A, P, R
Poly(acrylic acid), Na salt	1×10^3–2×10^6	23	1.3–2.4	C
Poly(acrylonitrile)	5×10^4–5×10^5	3	3.8–5.3	A, P, R
Polybutadiene, hydrogenated	5×10^2–5×10^5	10	1.16–2.66	O, R
Poly(butyl acrylate)	5×10^4–1×10^5	1	3	A
Poly(butyl methacrylate)	2×10^5–5×10^5	1	4.4	A
Poly caprolactone	2×10^4–5×10^4	2	3.1–3.2	P, R
Poly carbonate	2×10^4–1×10^5	9	2.5–4.5	A, P, R
Poly(2, 6-dimethyl-*p*-phenylene oxide)	5×10^4–5×10^5	2	2.5–7.6	P, R
Poly epichlorohydrin	1×10^3–5×10^3	2	1.4–1.58	P
Poly(ethyl acrylate)	5×10^4–1×10^5	3	2.5	A, P, R
Polyethylene, branched	5×10^4–5×10^5	5	4–16	C
Poly(ethylene terephthalate)	2×10^4–5×10^4	1	1.9	C
Poly(2-ethylhexyl acrylate)	1×10^5–2×10^5	3	2.6	A, P, R
Poly(ethyl methacrylate)	2×10^5–5×10^5	3	2.7	A, P, R
Poly(isobutyl acrylate)	1×10^5–2×10^5	1	3.7	R
Poly isobutylene	5×10^2–2×10^6	7	1.03–3.5	C
Poly(isobutyl methacrylate)	2×10^5–5×10^5	2	2.1	A, R
Poly(lauryl methacrylate)	1×10^5–5×10^5	3	1.4–3.2	A, P, R
Poly(methyl acrylate)	2×10^4–5×10^4	3	2.9	A, P, R
Poly(*n*-butyl acrylate)	5×10^4–2×10^5	2	3.0–3.6	P, R

Poly(*n*-butyl methacrylate)	2×10^5–5×10^5	1	4.4	P, R
Poly(octadecyl acrylate)	2×10^4–5×10^4	1	1.8	R
Poly(octadecyl methacrylate)	2×10^5–5×10^5	2	2.2–3.3	A, R
Polypropylene	2×10^4–5×10^5	14	2.1–8.0	C
Poly(propylene glycol)	5×10^1–1×10^4	10	1.0–1.2	C, W
Polysaccharide	5×10^3–1×10^6	16	1.06–1.14	B, O
Polystyrene, star-branched	5×10^3–2×10^5	4	—	P
Polysulfone resin	5×10^4–1×10^5	2	3.3	A, R
Poly(tetrahydrofuran) (Poly(butylene oxide))	1×10^3–1×10^6	27	1.03–1.20	B, O, P
Poly(vinyl alcohol)	5×10^3–5×10^5	7	1.7–2.2	C
Poly(vinyl butyral)	1×10^5–2×10^5	2	6.1	P, R
Poly(vinyl carbazole)	5×10^5–1×10^6	3	3.6	A, P, R
Poly(vinyl fluoride)	1×10^5–2×10^5	2	3.2	P, R
Poly(vinyl formal)	2×10^4–5×10^4	2	2.8	P, R
Poly(vinyl methyl ether)	5×10^4–1×10^5	2	2.1	P, R
Poly(2-vinylpyridine)	2×10^3–2×10^6	20	1.05–2.05	P, Q, R
Poly(vinyl stearate)	5×10^5–1×10^6	1	9.4	A

[a] Suppliers are listed in Table 15.

Table 14. Standard Samples of Stereoregular Polymers

Polymer	Supplier Codes[a]
Poly(butene-1), isotactic	P
Poly(methyl methacrylate), isotactic	O, P, Q
Poly(methyl methacrylate), syndiotactic	O, P
Polypropylene, isotactic	P, Q
Polystyrene, isotactic	O, Q
Poly(2-vinylpyridine), isotactic	Q

[a]Suppliers are listed in Table

Table 15. Suppliers of Standard Polymers

Code	Supplier
A	Aldrich Chemical Company, Inc. 940 West St. Paul Ave. Milwaukee, WI 53233
B	Alltech Associates, Inc. Applied Science Laboratories 2051 Waukegan Road Deerfield, IL 60015
C	American Polymer Standards Corporation P.O. Box 901 Mentor, OH 44061-0901
D	Arro Laboratories, Inc. P.O. Box 686 Caton Farm Road Joliet, IL 60434
G	Goodyer Tire & Rubber Company Corporate Research Building 142 Goodyear Boulevard Akron, OH 44316
M	Morton Thiokol, Inc. Alta Products 152 Andover Street Danvers, MA 01923
N	Office of Standard Reference Materials U.S. Department of Commerce National Bureau of Standards Gaithersburg, MD 20899

Table 15. *(Continued)*

Code	Supplier
O	Polymer Laboratories Ltd. Church Stretton Shropshire SY6 6AX United Kingdom U.S. branch: Polymer Laboratories, Inc. Amherst Fields Research Park 160 Old Farm Road Amherst, MA 01002
P	Polysciences, Inc. 400 Valley Road Warrington, PA 18976
Q	Pressure Chemical Company 3419 Smallman Street Pittsburgh, PA 15201
R	Scientific Polymer Products, Inc. 6265 Dean Parkway Ontario, NY 14519
S	Supelco, Inc. Bellefonte, PA 16823-0048
T	Toyo Soda Manufacturing Company, Ltd. Toso Building 1-7-7 Akasaka, Minato-ku Tokyo 107 Japan
V	Varian Associates Instrument Group Service Center 220 Humboldt Court Sunnyvale, CA 94089
W	Waters Chromatography Division Millipore Corporation 34 Maple Street Milford, MA 01757

number of samples shown in these tables. The ranges in molecular weight shown are based on $\bar{M}_w$ values where these are given by the supplier, otherwise on peak molecular weight or "nominal" molecular weight. Range A–B means $A \leqslant M < B$. Deuterated polymers, where available, are listed with the corresponding undeuterated materials. Branched polyethylene and star-branched polystyrene are listed separately in Table 13, rather than with their linear counterparts in Tables 3 and 9, respectively.

Table 14 shows the available samples of stereoregular polymers. Finally, Table 15 lists the suppliers of standard samples and the letter codes with which Tables 1–14 refer to them.

REFERENCES

1. P. J. Flory, *Principles of Polymer Chemistry*, Cornell University Press, Ithaca, NY, 1953.
2. H. Morawetz, *Macromolecules in Solution*, Wiley-Interscience, New York, 1965.
3. H. Yamakawa, *Modern Theory of Polymer Solutions*, Harper & Row, New York, 1971.

INDEX

(*continued from front*)

Vol. 65. **The Interpretation of Analytical Chemical Data by the Use of Cluster Analysis.** By D. Luc Massart and Leonard Kaufman

Vol. 66. **Solid Phase Biochemistry: Analytical and Synthetic Aspects.** Edited by William H. Scouten

Vol. 67. **An Introduction to Photoelectron Spectroscopy.** By Pradip K. Ghosh

Vol. 68. **Room Temperature Phosphorimetry for Chemical Analysis.** By Tuan Vo-Dinh

Vol. 69. **Potentiometry and Potentiometric Titrations.** By E. P. Serjeant

Vol. 70. **Design and Application of Process Analyzer Systems.** By Paul E. Mix

Vol. 71. **Analysis of Organic and Biological Surfaces.** Edited by Patrick Echlin

Vol. 72. **Small Bore Liquid Chromatography Columns: Their Properties and Uses.** Edited by Raymond P. W. Scott

Vol. 73. **Modern Methods of Particle Size Analysis.** Edited by Howard G. Barth

Vol. 74. **Auger Electron Spectroscopy.** By Michael Thompson, M. D. Baker, Alec Christie, and J. F. Tyson

Vol. 75. **Spot Test Analysis: Clinical, Environmental, Forensic and Geochemical Applications.** By Ervin Jungreis

Vol. 76. **Receptor Modeling in Environmental Chemistry.** By Philip K. Hopke

Vol. 77. **Molecular Luminescence Spectroscopy: Methods and Applications** (*in two parts*). Edited by Stephen G. Schulman

Vol. 78. **Inorganic Chromatographic Analysis.** Edited by John C. MacDonald

Vol. 79. **Analytical Solution Calorimetry.** Edited by J. K. Grime

Vol. 80. **Selected Methods of Trace Metal Analysis: Biological and Environmental Samples.** By Jon C. VanLoon

Vol. 81. **The Analysis of Extraterrestrial Materials.** By Isidore Adler

Vol. 82. **Chemometrics.** By Muhammad A. Sharaf, Deborah L. Illman, and Bruce R. Kowalski

Vol. 83. **Fourier Transform Infrared Spectrometry.** By Peter R. Griffiths and James A. de Haseth

Vol. 84. **Trace Analysis: Spectroscopic Methods for Molecules.** Edited by Gary Christian and James B. Callis

Vol. 85. **Ultratrace Analysis of Pharmaceuticals and Other Compounds of Interest.** Edited by S. Ahuja

Vol. 86. **Secondary Ion Mass Spectrometry: Basic Concepts, Instrumental Aspects, Applications and Trends.** By A. Benninghoven, F. G. Rüdenauer, and H. W. Werner

Vol. 87. **Analytical Applications of Lasers.** Edited by Edward H. Piepmeier

Vol. 88. **Applied Geochemical Analysis.** by C. O. Ingamells and F. F. Pitard

Vol. 89. **Detectors for Liquid Chromatography.** Edited by Edward S. Yeung

Vol. 90. **Inductively Coupled Plasma Emission Spectroscopy: Part I: Methodology, Instrumentation, and Performance; Part II: Applications and Fundamentals.** Edited by J. M. Boumans

Vol. 91. **Applications of New Mass Spectrometry Techniques in Pesticide Chemistry.** Edited by Joseph Rosen

Vol. 92. **X-Ray Absorption: Principles, Applications, Techniques of EXAFS, SEXAFS, and XANES.** Edited by D. C. Konnigsberger

Vol. 93. **Quantitative Structure–Chromatographic Retention Relationships.** By Roman Kaliszan

Vol. 94. **Laser Remote Chemical Analysis.** Edited by Raymond M. Measures